Chemical Marine Monitoring

Water Quality Measurement Series

Series Editor
Philippe Quevauviller
Vrije Universiteit Brussel (VUB), Belgium

Published Titles in the Water Quality Measurements Series

Hydrological and Limnological Aspects of Lake Monitoring
Edited by Pertti Heinonen, Giuliano Ziglio and André Van der Beken

Quality Assurance for Water Analysis
Philippe Quevauviller

Detection Methods for Algae, Protozoa and Helminths in Fresh and Drinking Water
Edited by Franca Palumbo, André Van der Beken and Giuliano Ziglio

Analytical Methods for Drinking Water: Advances in Sampling and Analysis
Edited by Philippe Quevauviller and K. Clive Thompson

Biological Monitoring of Rivers
Edited by Giuliano Ziglio, Maurizio Siligardi and Giovanna Flaim

Wastewater Quality Monitoring and Treatment
Edited by Philippe Quevauviller, Olivier Thomas and André van der Beken

Rapid Chemical and Biological Techniques for Water Monitoring
Edited by Catherine Gonzalez, Richard Greenwood and Philippe Quevauviller

The Water Framework Directive – Ecological and Chemical Status Monitoring
Edited by Philippe Quevauviller, Ulrich Borchers, K. Clive Thompson and Tristan Simonart

Groundwater Monitoring
Edited by Philippe Quevauviller, Anne-Marie Fouillac, Johannes Grath and Rob Ward

Chemical Marine Monitoring: Policy Framework and Analytical Trends

Edited by

PHILIPPE QUEVAUVILLER
Vrije Universiteit Brussel (VUB), Brussels, Belgium

PATRICK ROOSE
Royal Belgian Institute of Natural Sciences, Oostende, Belgium

GERT VERREET
Flemish Government – Environment Nature and Energy Department (LNE), Brussels, Belgium

WILEY
A John Wiley & Sons, Ltd., Publication

This edition first published 2011

Registered office
John Wiley & Sons Ltd., The Atrium, Southern Gate, Chichester, West Sussex, PO19 8SQ, United Kingdom

For details of our global editorial offices, for customer services and for information about how to apply for permission to reuse the copyright material in this book please see our website at www.wiley.com.

Library of Congress Cataloging-in-Publication Data

Chemical marine monitoring : policy framework and analytical trends / [edited by] Philippe Quevauviller, Patrick Roose, Gert Verreet.
p. cm.
Includes bibliographical references and index.
ISBN 978-0-470-74765-0 (cloth)
1. Marine pollution–Measurement. 2. Marine pollution–Prevention–International cooperation. 3. Chemical oceanography. 4. Water quality bioassay. 5. Environmental monitoring. I. Quevauviller, Ph. II. Roose, Patrick. III. Verreet, Gert.
GC1085.C475 2011
628.1′68–dc22 2011011526

A catalogue record for this book is available from the British Library.

Print ISBN: 9780470747650
ePDF ISBN: 9781119990833
oBook ISBN: 9781119990826
ePub ISBN: 9781119977599
Mobi ISBN: 9781119977605

Set in 9/11pt Times by Laserwords Private Limited, Chennai, India.

Contents

Series Preface

Water is a fundamental constituent of life and is essential to a wide range of economic activities. It is also a limited resource, as we are frequently reminded by the tragic effects of drought in certain parts of the world. Even in areas with high precipitation, and in major river basins, overuse and mismanagement of water have created severe constraints on availability. Such problems are widespread and will be made more acute by the accelerating demand on freshwater arising from trends in economic development.

Despite the fact that water-resource management is essentially a local, river-basin-based activity, there are a number of areas of action that are relevant to all or significant parts of the European Union and for which it is advisable to pool efforts for the purpose of understanding relevant phenomena (e.g. pollutions, geochemical studies), developing technical solutions and/or defining management procedures. One of the keys for successful cooperation aimed at studying hydrology, water monitoring, biological activities and so on is to achieve and ensure good water quality measurements.

Quality measurements are essential to demonstrate the comparability of data obtained worldwide and they form the basis for correct decisions related to management of water resources, monitoring issues, biological quality and so on. Besides the necessary quality control tools developed for various types of physical, chemical and biological measurements, there is a strong need for education and training related to water quality measurements. This need has been recognized by the European Commission, which has funded a series of training courses on this topic, covering aspects such as monitoring and measurements of lake recipients, measurements of heavy metals and organic compounds in drinking and surface water, use of biotic indexes, and methods to analyse algae, protozoa and helminths. This was the initial input for launching the *Water Quality Measurement* series in the year 2000, which originally aimed to cover monitoring features of various aquatic environments with focus on freshwater resources. It was, however, found of high added value to also include marine considerations to complete the overall discussions about water quality measurements.

This book series, therefore, includes the topics of the above-mentioned courses and the outcome of recent scientific advances, as well as other aspects related to quality control tools (e.g. certified reference materials for the quality control of water analysis) and monitoring of various types of water (river, wastewater, groundwater, seawater).

The book *Chemical Marine Monitoring – Policy Framework and Analytical Trends* is the tenth volume and the last one of this series; it has been written by policy-makers and scientific experts in issues related to chemical marine monitoring as required by the international conventions and the EU Marine Strategy Directive. It offers the reader an overview of technical issues related to chemical marine monitoring, including policy considerations, types of parameters and up-to-date monitoring/analytical techniques.

Ph. Quevauviller
Series Editor

Preface

Wide-scale chemical monitoring programmes are required by international conventions and European Union policies such as the Water Framework Directive (2000/60/EC) and the new EU Marine Strategy Framework Directive (MSFD). In the case of the WFD, EU member states have to undertake monitoring programmes for all waters, including transitional and coastal waters, sediments and biota, in order to develop river basin management plans and programmes of measures aiming to achieve 'good status' objectives by 2015. Marine monitoring is more specifically pursued in the framework of international conventions (e.g. OSPAR, HELCOM, Barcelona Convention), each of which has contributed to the international approach to marine environmental monitoring, and whose experience can be harnessed for broader application in other sea areas. The MSFD will pave the way for a thorough ecological marine assessment and require the definition of ecological objectives expressing 'good environmental status', with management targets and associated indicators.

Monitoring programmes will need to follow up the status of the marine environment in the light of the policy objectives. This will generate a wide array of collaborations among R&D and policy communities, training activities, educational materials and so on. Monitoring requirements are often prone to various interpretations. The policy-making and scientific communities, along with industrial stakeholders and NGOs, have recognized this and have worked altogether to develop guidance documents reflecting common understanding.

This book is all about these ongoing features. It is very timely, in that it should be published at the time of implementation of WFD river basin management plans (which operationally started in early 2010) and anticipates identified needs for reviews of monitoring and assessment methods planned in 2012. The timeframe for the MSFD foresees that the first assessment and target formulation should be done by 2012, whereas the marine monitoring programmes will need to be recast by 2014, followed by programmes of measures from 2015 onward.

The three editors have been striving to collect state-of-the-art information on chemical marine monitoring from the international community, providing further stimulation to the work of all parties involved in the huge challenges on the way to a ensure a sound quality assessment of the marine environment.

Philippe Quevauviller *Patrick Roose* *Gert Verreet*

The Series Editor – Philippe Quevauviller

Philippe Quevauviller began his research activities in 1983 at the University of Bordeaux I, France, studying lake geochemistry. Between 1984 and 1987 he was Associate Researcher at the Portuguese Environment State Secretary, where he performed a multidisciplinary study (sedimentology, geomorphology and geochemistry) of the coastal environment of the Galé coastline and of the Sado Estuary, which was the topic of his PhD degree in oceanography gained in 1987 (at the University of Bordeaux I). In 1988 he became Associate Researcher in the framework of a contract between the University of Bordeaux I and the Dutch Ministry for Public Works (Rijskwaterstaat), in which he investigated organotin contamination levels of Dutch coastal environments and waterways. From this research work he gained another PhD in chemistry at the University of Bordeaux I in 1990. From 1989 to 2002 he worked at the European Commission (DG Research) in Brussels, where he managed various Research and Technological Development (RTD) projects in the field of quality assurance and analytical method development for environmental analyses in the framework of the Standards, Measurements and Testing Programme. In 1999 he obtained an HDR (Diplôme d'Habilitation à Diriger des Recherches) in chemistry at the University of Pau, France, from a study of the quality assurance of chemical species' determination in the environment.

In 2002 he left the research world to move to the policy sector at the EC Environment Directorate-General, where he developed a new EU Directive on groundwater protection against pollution and chaired European science-policy expert groups on groundwater and chemical monitoring in support of the implementation of the EU Water Framework Directive. Since 2008 he has been at the EC DG Research, where he is managing research projects on climate change impacts on the aquatic environment, while ensuring strong links with policy networks.

Philippe Quevauviller has published (as author and co-author) more than 220 scientific and policy publications, 80 reports and 6 books for the European Commission and has acted as an editor and a co-editor for 22 special issues of scientific journals and 10 books. Finally, he is Associate Professor at the Free University of Brussels and promoter of Master theses in an international Master on water engineering (IUPWARE programme), and he also teaches integrated water management issues and their links to EU water science and policies to Master students of the EuroAquae Erasmus Mundus master programme at Polytech' Nice (France).

List of Contributors

AMOUROUX David	Laboratoire de Chimie Analytique Bio-Inorganique et Environnement, CNRS et Université de Pau et des Pays de l'Adour, Pau, France
ASMUND Gert	National Environmental Research Institute, Aarhus University, Roskilde, Denmark
BELZUNCE-SEGARRA J. Maria	Marine Research Division, AZTI-Tecnalia, Pasajes (Gipuzkoa), Spain
BENOIT-BONNEMASON Christelle	Ultra Traces Analyses Aquitaine, Hélioparc Pau-Pyrénées, Pau, France
BOOIJ Kees	NIOZ Royal Netherlands Institute for Sea Research, Texel, The Netherlands
DE BOER Jacob	Vrije Universiteit Amsterdam, Institute for Environmental Studies Environmental Chemistry and Toxicology, Amsterdam, The Netherlands
DONARD F.X. Olivier	Laboratoire de Chimie Analytique Bio-Inorganique et Environnement, CNRS et Université de Pau et des Pays de l'Adour, Pau, France
FONES R. Gary	University of Portsmouth, School of Earth and Environmental Sciences, Portsmouth, UK
FRYER Rob	Marine Scotland, Marine Laboratory, Aberdeen, UK
GREENWOOD Richard	University of Portsmouth, School of Biological Sciences, Portsmouth, UK
HAARICH Michael	Johann Heinrich von Thünen-Institut, Federal Research Institute for Rural Areas, Forestry and Fisheries, Institute of Fishery Ecology, Analytical Department, Hamburg, Germany
HEININGER Peter	Federal Institute of Hydrology, Koblenz, Germany
HEISE Susanne	Hamburg University of Applied Sciences, Hamburg, Germany
KOWALEWSKA Grazyna	Institute of Oceanology, Marine Pollution Laboratory, Polish Academy of Sciences, Sopot, Poland
KRAMER J.M. Kees	Mermayde, Bergen, The Netherlands

LARSEN M. Martin	National Environmental Research Institute, Aarhus University, Roskilde, Denmark
LAW J. Robin	The Centre for Environment, Fisheries and Aquaculture Science, Cefas Lowestoft Laboratory, Suffolk, UK
McGOVERN Evin	Marine Institute, Oranmore, Co.Galway, Ireland
MENDIGUCHIA Carolina	Department of Analytical Chemistry, Faculty of Marine and Environmental Science, University of Cádiz, Cádiz, Spain
MILLS A. Graham	University of Portsmouth, School of Pharmacy and Biomedical Sciences, Portsmouth, UK
MOFFAT F. Colin	Marine Scotland, Marine Laboratory, Aberdeen, UK
MONPERRUS Mathilde	Laboratoire de Chimie Analytique Bio-Inorganique et Environnement, CNRS et Université de Pau et des Pays de l'Adour, Pau, France
PANNIER Florence	Laboratoire de Chimie Analytique Bio-Inorganique et Environnement, CNRS et Université de Pau et des Pays de l'Adour, Pau, France
PARMENTIER Koen	ILVO-Visserij, Oostende, Belgium
PELLIZZATO Francesca	Marine Scotland, Marine Laboratory, Aberdeen, UK
QUEVAUVILLER Philippe	Vrije Universiteit Brussel (VUB), IUPWARE, Department of Hydrology and Hydrological Engineering, Brussels, Belgium
ROCHA Carlos	Geography Department, School of Natural Sciences, Trinity College Dublin, Dublin, Ireland
ROOSE Patrick	Royal Belgian Institute of Natural Sciences, Management Unit of the North Sea Mathematical Models and the Scheldt(MUMM), Oostende, Belgium
RUMNEY S. Heather	The Centre for Environment, Fisheries and Aquaculture Science, Cefas Lowestoft Laboratory, Suffolk, UK
SCHUBERT Birgit	Federal Institute of Hydrology, Koblenz, Germany
SEBY Fabienne	Ultra Traces Analyses Aquitaine, Hélioparc Pau-Pyrénées, Pau, France
SØNDERGAARD Jens	National Environmental Research Institute, Aarhus University, Roskilde, Denmark
THEOBALD Norbert	Federal Maritime and Hydrographic Agency, Hamburg, Germany
VERMAERCKE Peter	SCK.CEN-Belgian Nuclear Research Centre, Boeretang, Belgium
VERREET Gert	Flemish Government – Environment, Nature and Energy Department (LNE), International Policy Division, Brussels, Belgium
WEBSTER Lynda	Marine Scotland, Marine Laboratory, Aberdeen, UK
WOODWARD Malcolm	Plymouth Marine Laboratory, Plymouth, UK

Part 1

Setting the Scene

1

Monitoring of Pollutants: A Historical Perspective for the North-East Atlantic Region

Kees J.M. Kramer

Mermayde, Bergen, The Netherlands

Chemical Marine Monitoring: Policy Framework and Analytical Trends, First Edition.
Edited by Philippe Quevauviller, Patrick Roose, and Gert Verreet.

1.1 INTRODUCTION

The general public had, until the mid 1950s, little or no idea about the 'environment'; it was just there, as it had always been. It was there to be used and exploited. That chemical products entering the sea could be hazardous to man became shockingly apparent in 1956 when in Minamata Bay (Japan) people suffered from a neurological syndrome as a result of eating methyl mercury-contaminated tuna and swordfish. Rachel Carson published *Silent Spring* in 1962 (Carson, 1962). In this book she focused attention on the environment itself by discussing the biological effects of pesticides. It is considered the start of the environmental movement.

In March 1967 the *Torrey Canyon*, a 120 000 tons oil tanker, was wrecked on Seven Stones off the Isles of Scilly (UK). Two years later, in 1969, there was a blow-out on Union Oil's Platform A in the Santa Barbara Channel 9 km off the Californian coast (USA). Similar to the April 2010 disaster with BP's Deepwater Horizon in the Gulf of Mexico, the environmental effects were huge in all these accidents. In contrast to many pollutants that are dissolved in water or adsorbed to particles, and thus unseen by the naked eye, crude oil and its biological effects are very visible, also to the public. Newspapers reported on these and other pollution events with evident biological effects, like fish kills in the River Rhine and massive bird mortality in Scotland, also in 1969. Public opinion did help to create a climate in which effective legislation was possible and scientific activities in research and monitoring were encouraged. Environmental laws, like the implementation of 'Statement of environmental effects', or banning of chlorinated pesticides were imminent. The US Environmental Protection Agency (EPA) was founded within 2 years after the Santa Barbara blow-out (1970).

Scientific research had started to get better information on the input, transport, fate (sinks) of pollutants and of their biological effects (Pearce, 1998). In response to this and to 'help coordination of research by rapid dissemination of information relating to pollution of the sea', the first issue of *Marine Pollution Bulletin* was published in January 1970. It intended to be an information bulletin ('spreading news of pollution') rather than a scientific journal, and clearly aimed also to inform policy (Anon, 1970).

1.1.1 Definition of Monitoring

In 1977 the International Council of Scientific Union's (ICSU) Scientific Committee on Problems of the Environment (SCOPE) defined 'monitoring' as (Holdgate and White, 1977):

> the collection, for a predetermined purpose, of systematic, inter-comparable measurements or observations in a space–time series, of any environmental variables or attributes which provide a synoptic view or a representative sample of the environment (global, regional, national, or local). Such a sample may be used to assess existing and past states, and to predict likely future trends in environmental features.

This definition still seems valid today. Monitoring is thus a systematic method of collecting data needed for environmental problem solving; it is linked to environmental policy. (National) monitoring in this sense started in most areas not before the early 1970s.

Very often, monitoring programmes are understood as linked to the concentration of harmful substances only. However, for a proper assessment, different variables shall not only include measuring of levels of pollutants (in various compartments, including biota), but also include physical attributes (e.g. salinity, turbidity, fluxes) and biological effects (possibly at several biological organization levels). Species distribution and density have been measured for over a century for fundamental biological research; although in the past the objective was never 'monitoring' (at most surveys), today these biological variables are the corner stones of describing the biodiversity of a region.

Basic reasoning for monitoring changed over the years. In the mid 1970s it was argued that the first concern was to avoid hazards to human health. It was considered of no importance to monitor, for instance, Hg in coastal waters unless it presented an unacceptable hazard to human health or produced an ecological change. Reduction in the monitoring effort to the barest essentials was thus advocated (Anon, 1975). Today, reasoning stems from a more holistic approach. The concern is reaching – in Water Framework Directive (WFD) and Marine Strategy Framework Directive (MSFD) terminology (see Chapter 3) – Good Environmental Status (GES) and the focus is on the environment. Evaluation and assessment of biological and chemical status and trends have become key environmental management tools.

In this chapter, the centre of attention will be on the historic development of marine monitoring, with a focus on the situation in the North-East Atlantic Ocean, including the North Sea and Baltic seas. This will relate to the definition and purpose of monitoring, relevant international treaties, the context and developments in international institutions, and general concepts like monitoring of different environmental compartments and the need for quality-assured data.

Because of the limited space available, this contribution had to be restricted to the monitoring of hazardous substances in water, sediment and biota. Other fields, such as biological effect monitoring or monitoring of species richness (biodiversity), are not included. As an example for many other groups of pollutants, long-term trends for selected trace metals will be briefly discussed, also in view of the improvement of analytical methods.

1.1.2 Stockholm Conference 1972

The United Nations Conference on the Human Environment (Stockholm, 5–16 June 1972) is considered a landmark in (marine) monitoring. A Declaration and an Action Plan were adopted. According to the Declaration (Principle 7):

> States shall take all possible steps to prevent pollution of the seas by substances that are liable to create hazards to human health, to harm living resources and marine life, to damage amenities or to interfere with other legitimate uses of the sea.

The Action Plan (Recommendation 73) 'recommended that governments actively support, and contribute to international programmes to acquire knowledge for the assessment of pollutant sources, pathways, exposures and risks' (UNEP, 1972).

The 1972 Stockholm Conference strengthened the partially existing efforts of environmental (marine) monitoring in national and international programmes. Many international organizations, all (becoming) active in the marine monitoring field, each with their own objectives and plans and possibly afraid of competition, used the outcome to better harmonize and structure future monitoring plans and activities.

1.2 INTERNATIONAL CONVENTIONS

Supra-national policies were required to allow combating pollution of the world's oceans and seas to become successful. Hence, a number of international treaties were drafted, subsequently signed and ratified. Focusing mainly on the European situation, the following lists a number of these conventions, that would become instrumental in environmental protection, and that would install monitoring programmes to monitor status and trends of the marine environment.

1.2.1 London Dumping Convention

In 1972 the Convention on the Prevention of Marine Pollution by Dumping of Wastes and Other Matter (1972, in force 1975), in short the London Dumping Convention (LDC), now called London Convention (LC 72), was signed. This treaty intended to cover the world's oceans and seas. It followed a 'black list/grey list' approach to regulate ocean dumping: Annex I (black) chemicals were banned (unless in trace amounts), Annex II (grey) listed chemicals for which dumping was restricted. A permanent secretariat is hosted by the International Maritime Organization (IMO).

1.2.2 Oslo and Paris Conventions, OSPAR Convention

The same year the Convention for the Prevention of Marine Pollution by Dumping from Ships and Aircraft, the Oslo Convention (1972, in force 1974), was signed. The area covered consists of the North-East Atlantic and part of the Arctic Ocean, but excluding the Baltic seas. There was a distinction between 'black' and 'grey' list chemicals. The limitations of the Oslo Convention by not including land-based sources were taken away 2 years later, when the Convention for the Prevention of Marine Pollution from Land-based Sources, the Paris Convention (1974, in force 1978), was signed. It covered the same marine area as the Oslo Convention. The Oslo Commission (OSCOM) and the Paris Commission (PARCOM) shared a joint secretariat in London (OSPARCOM) (OSPAR, 1984).

The Convention for the Protection of the Marine Environment of the North-East Atlantic, or OSPAR Convention (1992, in force 1998), is the current legislative instrument regulating international cooperation on environmental protection in the North-East Atlantic. It combines and updates the 1972 Oslo Convention and the 1974 Paris Convention, but decisions and other agreements adopted under those conventions remained applicable unless they are terminated by new measures adopted under the OSPAR Convention. Work carried out under the convention is managed by the OSPAR Commission. The developments of the Oslo and Paris Commissions over the first 25 years (and their relationship with the North Sea Ministerial Conferences) were detailed by Tromp and Wieriks (1994). The OSPAR Convention now regulates (for its geographic region) European standards on marine biodiversity, eutrophication, the release of hazardous and radioactive substances into the seas, the offshore oil and gas industry and baseline monitoring of environmental conditions.

1.2.3 Helsinki Convention

In 1974 the Convention on the Protection of the Baltic Sea Area, the Helsinki Convention (1974, in force 1980), was adopted. It employed the 'black' and 'grey' lists of hazardous substances. The Helsinki Commission (or HELCOM: Baltic Marine Environment Protection Commission), located in Helsinki, is acting as its governing body. Also, in the light of political changes in the late 1980s, a new Helsinki Convention was signed in 1992 (in force 2000). Its first 20 years were reviewed by Helsinki Commission (1994). The convention now covers the whole of the Baltic Sea area, including inland waters and the water of the sea itself and the sea-bed. Measures are also taken in the whole catchment area of the Baltic Sea to reduce land-based pollution.

All these conventions aimed and aim at the regulation of inputs, to carry out baseline studies (present status), to monitor for trends and to carry out inter-calibrations between contracting parties to warrant quality data.

1.2.4 Bonn Agreement

Focused on discharges of oil and other substances into the North Sea region in 1969 is the Agreement for Co-operation in Dealing with Pollution of the North Sea by Oil, the Bonn Agreement (1969, in force 1969). This treaty was superseded by the Agreement for Cooperation in Dealing with Pollution of the North Sea by Oil and other Harmful Substances (1983, Bonn Agreement, in force 1989). Now parties were required to jointly develop and establish guidelines for joint action and to provide information on pollution incidents. Developments were discussed on the occasion of its 40th anniversary (Bonn Agreement, 2009). One of the implementation instruments of the Bonn Agreement is the ongoing aerial surveillance programme, which started in 1986 to monitor and assess trends in levels of oil inputs into the marine environment (Carpenter, 2007).

1.2.5 MARPOL

Recognizing the threat of pollution of the seas by oil from shipping, in 1954 the UK organized a conference on oil pollution which resulted in the adoption of the International Convention for the Prevention of Pollution of the Sea by Oil, the OILPOL Convention (1954, in force in 1958). It primarily addressed pollution resulting from routine tanker operations. The Intergovernmental Maritime Consultative Organization (IMCO, since 1982 the IMO) organized in 1973 the International Conference on Marine Pollution in London, which led to the International Convention for the Prevention of Pollution from Ships, MARPOL (1973, which did, however, not get into force). In 1978, it was revised by the MARPOL Protocol, and the combination of the convention and protocol led to the MARPOL 73/78 treaty (1978, in force 1983). Its worldwide objective was to preserve the marine environment through the complete elimination of pollution by oil and other harmful substances and the minimization of accidental discharge of such substances. The initial focus on oil was expanded in later years with the inclusion of other substances: noxious liquid substances carried in bulk, harmful substances carried in packaged form, sewage, garbage and air pollution.

1.2.6 Other Conventions

The concept of the Oslo and Paris Conventions was used as a basis for developing a framework for the protection of other (European) sea areas. In 1976, the Convention for the Protection of the Mediterranean Sea against Pollution, the Barcelona Convention (1976; in force 1978) was adopted. In 1992, the Convention on the Protection of the Black Sea against Pollution, the Bucharest Convention (1992, in force 1994) was signed.

1.3 INTERNATIONAL INSTITUTIONS

In the years before the 1972 Stockholm Conference, quite a number of international institutions started to take initiatives related to environmental protection and to 'monitoring of the environment'.

A certain competition was apparent, as several sought to take the lead in international monitoring. The Organization for Economic Co-operation and Development (OECD) already in 1965 developed a plan to coordinate and programme pollution research. From 1966 onwards, the International Council for the Exploration of the Sea (ICES) formalized work on marine pollution (see below). In 1967, the Biologische Anstalt Helgoland (BAH) organized the symposium 'Biological and hydrographical problems of water pollution in the North Sea and adjacent waters'. The UN Food and Agriculture Organization (FAO) convened a 'Technical conference on marine pollution and its effects on living resources and fishing' in 1969. In the same year, a joint Group of Experts on the Scientific Aspects of Marine Pollution (GESAMP), co-sponsored by IMCO, FAO, the United Nations Educational, Scientific and Cultural Organization (UNESCO) and the World Meteorological Organization (WMO) met for the first time. The purpose of GESAMP was to advise the various agencies and subsidiary bodies within the UN that were concerned with marine pollution. Later, the World Health Organization (WHO), the International Atomic Energy Agency (IAEA), the UN and the UN Environmental Programme (UNEP) were added, thus totalling eight organizations (Windom, 1991). Much later, in 2005, GESAMP would produce a strategic vision on the scientific aspects of marine environmental protection (GESAMP, 2005).

The Scientific Committee on Problems of the Environment (SCOPE) was established by the ICSU in 1969. The Commission on Monitoring of SCOPE published a report 'Global environmental monitoring' (SCOPE, 1971) as input for the Stockholm Conference. In this report it was argued that, for the marine area, a strong coordination should be built between a marine pollution monitoring system and the Integrated Global Ocean Station System (IGOSS) then under development by the Intergovernmental Oceanographic Commission (IOC) for monitoring the physical conditions of the oceans. Studies, limited to monitoring of water, top sediments and biota for the levels of a few critical substances, were proposed to be conducted in selected pilot areas, such as the North Sea, Baltic, Mediterranean and Puget Sound (USA). The organisations considered best to handle the development and implementation were considered in the report to be the IOC (policy aspects) and the Scientific Committee on Oceanic Research (SCOR), representing international science. For global monitoring, suitable (national) data should be collected by a monitoring office, and for coordination purposes a central monitoring co-ordinating unit should be set up. The ideas were further worked out in the Action Plan for Phase I of the Global Environmental Monitoring System (GEMS) (Munn, 1973), and priority pollutants were defined (Andersen *et al*., 1988).

1.3.1 ICES

ICES, founded in 1902, has been an internationally recognized player in the many scientific aspects dealing with the northern Atlantic Ocean (Griffith, 2003). Initially, ICES mainly focused on fish and fisheries, but a task also included the production of 'standard seawater' for calibration of salinity measurements (Culkin and Smed, 1978). Went (1972), reviewing the first 70 years of ICES, hardly touched the subject of monitoring, an indication that the subject was nearly ignored in the period 1902–1972. This changed considerably in the third quarter of ICES' existence, as detailed by Rozwadowski (2002) and Griffith (2003). The great experience and science network of ICES was considered beneficial at the time for setting up monitoring work (ICES, 1974c).

In 1965 the Organization for Economic Cooperation and Development (OECD) mentioned to ICES leaders its intention to coordinate an ambitious plan to programme pollution research. ICES, fearing dominance by an institution with limited practical oceanographic experience, sought a broader scope, which led in the end to a common ICES–IOC–OECD cooperation. The ICES Fisheries Improvement Committee (FIC) was established in 1966 (from which in 1978 the Marine Environmental Quality Committee (MEQC) was formed). Despite its name, their tasks included marine pollution (biological effects) issues. Soon, there was a need for more specialized working groups (WGs), and in 1968 the ICES WG on Pollution of the North Sea convened for the first time; it was followed in 1971 by the ICES/SCOR joint WG on the Study of the Pollution of the Baltic.

Practical surveys were initiated by ICES in carrying out baseline studies, a kind of precursor to regular monitoring programmes. The first was carried out in the North Sea in 1972 for trace contaminants in fish and shellfish (as potential hazard to human health) (ICES, 1974a). A few years later, in 1974, a similar study was coordinated in the Baltic (ICES, 1977a) and, upon a request from the Oslo Commission, in 1975 extended towards the northern Atlantic Ocean (ICES, 1977b). After the North Sea baseline study, annual North Sea monitoring was initiated in 1974 (ICES Coordinated Monitoring Programme; ICES, 1977c); the Baltic monitoring programme started in 1979.

In 1973 ICES established the Advisory Committee on Marine Pollution (ACMP). Amongst its early activities the ACMP, together with the ICES Working Group on Marine Pollution Baseline and Monitoring Studies in the North Atlantic (WGMPNA), developed guidelines on the sampling, sample preparation, analytical procedures and data reporting required to obtain good-quality data. It paid particular attention to the various designs of sampling programmes to meet different monitoring objectives – public health assurance, environmental protection or trend monitoring. ACMP reported annually until 1992, when it was transformed into the Advisory Committee on the Marine Environment (ACME), again producing annual reports until 2003.

In support of the exchange of scientific information, method development and intercalibration exercises, ICES installed WGs, several of which have influenced monitoring studies; for example, the Marine Chemistry WG (1978), the Marine Sediments WG (1980) and the WG on the Biological Effects of Contaminants (1986).

From 1984 onwards, the data from the (OSPAR) Joint Monitoring Programme (JMP; see below) are compiled and quality checked by ICES; HELCOM data have been submitted to the ICES database since 1998.

Thus, ICES has been (and is) instrumental in the development of monitoring programmes, initially in the North Sea, later extending towards the North-East Atlantic and the Baltic. Although ICES coordinated the early baseline surveys, it recognized that (being essentially a science oriented organization) the actual monitoring would be better in the hands of the users of monitoring data: policy. Today, it is the responsibility of the respective national governments (contracting parties to OSPAR and HELCOM) to carry out the monitoring work, including the sampling and analyses. They report their annually collected data to ICES, who, after a quality check, makes them available to OSPAR and HELCOM for reporting purposes.

1.3.2 UNEP

The United Nations General Assembly decided in 1972, as a follow up of the Stockholm Conference, to establish UNEP to serve as a focal point for environmental action and coordination within the United Nations system. The Governing Council of UNEP chose 'oceans' as one of the priority areas. At the onset, UNEP's tasks involved the assessment and control of marine pollution. As resources initially were limited, the initial plans were considered rather ambitious (Birnie, 1974). Recognizing that pollution of the marine environment is generally most severe in coastal areas and semi-enclosed seas, efforts were concentrated on promoting regional marine pollution monitoring, leading to the initiation of its Regional Seas Programme in 1974. Most of these regions are located in the less-developed areas of the world, but partner programmes, like the Baltic (HELCOM) and the North-East Atlantic (OSPAR), are members of the Regional Seas family. In 1975, 16 Mediterranean countries and the European Community adopted the Mediterranean Action Plan (MAP). The MAP was the first-ever plan adopted as a Regional Seas Programme under UNEP; currently,

MAP Phase IV (2006–2013) is operable (www.unepmap.org). The Regional Seas Programme's approach and strategy includes a regional environmental assessment, and involves the development and implementation of regional seas monitoring programmes (Gerges, 1994). In 1984, UNEP published its 'Prospects for global ocean pollution monitoring' report. A first observation was that there was no demonstrated need for open ocean monitoring. In addition, since levels of contaminants were very low, biological effects were considered impossible to identify (UNEP, 1984). Considering the logistical problems related to true open ocean monitoring (e.g. by using WMO ocean weather ships as a monitoring platform) and the associated costs of such a programme, it is not surprising that it never materialized.

1.3.3 IOC

Established in 1960, the IOC of UNESCO promotes international cooperation and coordinates programmes in marine research, services, observation systems, hazard mitigation and capacity development in order to learn more and better manage the nature and resources of the ocean and coastal areas. In response to the recommendations of the Stockholm Conference (1972) the IOC in 1974 developed the Programme of Global Investigation of Pollution in the Marine Environment (GIPME), co-sponsored by UNEP and IMO. Its objectives were: to provide authoritative evaluations of the state of the marine environment at both regional and global levels, to identify the requirements for measures to prevent, or correct, marine pollution and to develop/implement procedures for assessing and improving compliance and surveillance monitoring of conditions and effects in the marine environment. Its operational regionally based Marine Pollution Monitoring System (MARPOLMON, since 1979) was to provide information on contamination in the marine environment which ideally would lead to an assessment of pollution. The developments and implementation of the Comprehensive Plan for GIPME (MARPOLMON) have been detailed by Kullenberg (1986). MARPOLMON constitutes a marine chemical component of GEMS (Andersen *et al*., 1988), and incorporated the Marine Pollution (Petroleum) Monitoring Pilot Project (MAPMOPP, 1974–1979). Important GIPME priorities were the baseline studies (status) and the standardization of methods and techniques. A great deal of emphasis has been placed on developing, testing and calibrating methodologies to ensure the quality of data for the major classes of contaminants measured in a variety of marine compartments and to attempt to determine fluxes in the marine environment (Dawson *et al*., 1988).

The day-to-day work within GIPME is conducted by three expert scientific groups: the Group of Experts on Methods, Standards and Intercalibration (GEMSI) deals with the assessment and methodology for measuring levels and flux of contaminants; the Group of Experts on the Effects of Pollutants (GEEP) deals with the biological effects of contaminants, pollution assessment and indicators of biological and ecosystem condition in the marine environment; the Group of Experts on Standards and Reference Materials (GESREM) deals with the assurance of data quality and comparability of measurements (Ibe and Kullenberg, 1995).

1.3.4 European Union

In 1972 the European Economic Community (EEC) formulated its first formal environment policy, with the associated Programme of Action of the European Communities on the Environment being adopted the following year. The current Sixth Community Environment Action Programme, 'Environment 2010: Our Future, our Choice', provides a strategic framework for the EEC's environmental policy for 2002–2012 and is the central environmental component of the EEC's sustainable development strategy.

The European Union (EU) is not involved in operational monitoring activities, neither in the planning nor in the execution. It is the task of the member states to implement the legal frameworks, such as directives, in their national laws and to take care of their implementation. Already from the 1970s there were legal instruments to limit pollution (such as the Council Directive on Pollution Caused by Certain Dangerous Substances Discharged into the Aquatic Environment of the Community, Directive 76/464-EEC). The EU WFD is a combination of (and replaces) a number of such older directives. It was adopted in 2000 (European Commission, 2000). Although its main purpose is to establish a framework for the protection of inland surface waters and ground waters, the WFD explicitly also covers surface 'transitional waters' (like estuaries) and coastal waters, generally extending to one nautical mile from the coastline. Monitoring of these waters has been addressed by Ferreira *et al.* (2007). Member states shall ensure the establishment of programmes for the monitoring of water status in order to establish a coherent and comprehensive overview of water status, including its ecological and chemical status. Amending Annex X of this directive, the Decision 2455/2001/EC established a list of 33 priority substances in the field of water policy. This list was further elaborated in the 'Directive on Priority Substances' (Environmental Quality Standards, EQS Directive 2008/105/EC; European Communities, 2008a) which sets in its Annex I the limits on concentrations of these 33 priority substances and an additional nine substances in surface waters. Its Annex III lists substances subject to review for possible identification as priority substances or priority hazardous substances. Interestingly, this directive allows member states to apply EQS for sediment and/or biota instead of water.

The EU Directive 2008/56/EC, the MSFD, was signed in 2008 (European Communities, 2008b). Under this directive each member state should develop a marine strategy for its marine waters. Marine strategies should culminate in the execution of programmes of measures designed to achieve or maintain GES by the year 2020 at the latest. For coastal waters there is some overlap with the WFD. Under the MSFD, provision should be made for the preparation of an appropriate framework, including marine research and monitoring operations. Member states should undertake an analysis of the features or characteristics of, and pressures and impacts on, their marine waters. On the basis of such analyses, member states should determine for their marine waters a set of characteristics for GES. Achieving GES should be underpinned by the establishment of environmental targets and monitoring programmes for ongoing assessment. The MSFD recognizes a regional approach and defines regions that are covered by existing organisms: the Baltic Sea (HELCOM), the North-East Atlantic Ocean (OSPAR), the Mediterranean Sea (Barcelona Convention) and the Black Sea (Bucharest Convention). The interplay between OSPAR, North Sea conferences and MSFD policies was discussed by Skjærseth (2006) and Borja (2006).

1.4 MONITORING IMPLEMENTATION

In the 1960s, and even earlier, surveys were conducted to investigate the distribution of nutrients, trace metals and selected organic pollutants. Regular monitoring started in the early 1970s, primarily as a national activity. National governments installed national water acts in support of pollution prevention and national monitoring programmes, usually in the form of surveys or surveillance programmes. For example, in the Netherlands, surveys of the water quality of the Wadden Sea started in 1971 and of the North Sea in 1975. The earliest surveys of the Western Scheldt estuary had already begun in 1964 (Beukema *et al.*, 1986). National monitoring strategies were developed and grids of monitoring stations and sampling frequencies were defined. The strategies were usually ambitious, and over the years the number of locations and the frequencies were reduced, mainly because governments realized the high costs involved, but (in the 1990s) also because of the tendency to apply modelling 'we have measured enough' was an issue already raised in the mid 1970s (Anon, 1975). Such changes in (the number of) monitoring locations and frequencies continue to pose problems for the statistical (trend) assessment of the data (Nicholson and Fryer, 1992).

It became evident that the results of national monitoring programmes for areas subject to different national policies – as is the case for the North Sea – were very difficult to interpret without sharing information and without harmonization of strategies, methods and data reporting. Thus, initiatives were taken to combine efforts; for example, in the Oslo, Paris and Helsinki conventions areas.

1.4.1 OSPAR

The Oslo and Paris Commissions recognized their duty to examine the condition of the sea covered by the conventions. Hence, the Commissions established a permanent Joint Monitoring Group (JMG). After asking ICES for advice and guidance for monitoring aim status and trends, the two Commissions set up a JMP in 1978 (OSPAR, 1984). OSPARCOM adopted many of the principles of the ICES programme in defining the JMP. The monitoring programme of JMG had four main objectives (Portman, 1986); i.e. the assessment of:

1. possible hazards to human health;
2. harm to living resources and marine life;
3. the existing levels of marine pollution (spatial distribution); and
4. the effectiveness of measures taken for the reduction of marine pollution within the framework of the conventions (temporal trend assessment).

The JMP was to be based on the national programmes of the contracting parties, with their national laboratories responsible for the sampling and analyses. To ensure comparability of data, calibration of methods should be supported by participation in (e.g. ICES) interlaboratory comparison studies. The actual monitoring programme started in 1979, and was initially limited to mercury and cadmium in seawater and in organisms, and polychlorinated biphenyls (PCBs) in organisms. In the early 1980s a strategy was developed to set sampling frequencies according to monitoring objectives. For purpose (1), in organisms the frequency was set at every second year (implemented in 1984) and for sediments every 5th year ($\geq$1985). Purpose (3) had frequencies for organisms and seawater (per 5 years, $\geq$1985), while for objective (4) the frequencies were defined annually for organisms and seawater ($\geq$1983) and sediments ($\geq$1986) (Portman, 1986). Since 1984 the JMP data are compiled by ICES in their database (and checked for validity).

In 2003 the Ministerial Meeting of the OSPAR Commission adopted a strategy for the Joint Assessment and Monitoring Programme (JAMP), a combination of the national monitoring programmes of the contracting parties. This provided a framework for work to prepare and produce a series of thematic assessments (quality status reports, QSRs). Thus, OSPAR is coordinating repeated measurement and assessment of the marine environment over a decadal time frame. The organizational structure of OSPAR changed in 1995. Monitoring and assessment became a task of the Assessment and Monitoring Committee of OSPAR (ASMO). Monitoring has since been split into several domains in several WGs under ASMO, such as the Working Group on Concentrations, Trends and Effects of Substances in the Marine Environment (SIME), WG INPUT, dealing with atmospheric inputs (via the Comprehensive Atmospheric Monitoring Programme (CAMP) and riverine inputs and discharges (via the Comprehensive Study of Riverine Inputs and Direct Discharges (RID)), and WG MON, e.g. via CEMP, the Co-ordinated Environmental Monitoring Programme (de Jong, 2006).

1.4.2 North Sea Conferences

The baseline studies, other surveys and surveillance programmes and serious concern at the political level led to a series of North Sea conferences starting with the First North Sea Conference in 1984 (Bremen). Assessments were made, not only on the status and trends of pollutants, but also on the use of the seas as a resource for human activities. In later years several international North Sea conferences were organized in London (1987), The Hague (1990), Esbjerg (1995) and Bergen (2002) (Skjærseth, 2006).

In support of these North Sea conferences and OSPAR ministerial meetings, QSRs were prepared. Implementing the ecosystem approach, these reports provide an integrated assessment of the cumulative and relative impact of all human pressures on the marine environment, identifying where action needs to be taken. Marine pollution is one of the pressures for which monitoring provides the factual information. QSRs are compiled contributions by experts, usually from the respective governmental institutions of the different states bordering the region concerned (initially the North Sea, later the North-East Atlantic). The 1st QSR served the Bremen (1984) conference (Carlson, 1986).

One of the outcomes of the Second North Sea Conference in 1987 and QSR 1987 was the recognition that, despite the large number of contaminants measured, the spatial coverage of the North Sea was rather limited. As a result, the North Sea Task Force (NSTF) was set up in 1987, co-sponsored by OSPARCOM and ICES. The objectives of NSTF included to advise on research, to plan, develop and implement a quality-assured monitoring programme and to develop the next QSR for the North Sea (Hoogweg *et al.*, 1991). The aims of the NSTF Monitoring Master Plan were to enhance scientific knowledge and understanding of the North Sea environment, and to overcome shortcomings in data on the distribution of contaminants (Reid, 1990). An assessment of the North Sea area, the QSR 1993 was subsequently published as input to the Third North Sea Conference (NSTF, 1993). NSTF existed from 1988 to 1994 and its approach was then incorporated into ASMO.

The QSR 2000 (OSPAR Commission, 2000) was based on the combined efforts of JMP and JAMP. The geographic coverage was expanded to a larger area, the North-East Atlantic (the OSPAR convention area) which was subdivided into five regions (Arctic waters, Greater North Sea, Celtic Seas; Bay of Biscay and Iberian Coast, and the Wider Atlantic). For each region a separate report was prepared, as well as a holistic synthesis report for the entire area. QSR 2010, the most recent report, was published on the occasion of the Ministerial Meeting of the OSPAR Commission in Bergen (2010). Again, a regional approach was followed in a printed and electronically available version (http://qsr2010.ospar.org), albeit in one volume (OSPAR, 2010).

1.4.3 HELCOM

For the period 1974–1980, when the Helsinki Convention entered into force, a WG referred to as the Interim Commission (IC) was installed. It established a Scientific Technological Working Group (STWG), which had as one of its tasks 'monitoring and assessment of the state of the marine environment' (BSEP56; HELCOM, 1994). Since 1979, the Baltic Sea Environment Proceedings (BSEP) report series has been published (list and pdf files at www.helcom.fi/publications/bsep/en_GB/bseplist/).

Under the Helsinki Convention, monitoring of physical, chemical and biological variables of the open sea started in 1979 (of radioactive substances in 1984), but was considered a national obligation. It was called the Baltic Monitoring Programme (BMP) and was revised several times. The first pilot period covered 1979–1983; the second phase (1984–1988) had a larger coverage.

The third stage started in 1989. For political reasons, the coastal areas of the sea were only poorly covered by the BMP, and the programme focused on the open sea. The aim of the BMP was to monitor the long-term changes of selected indicators in the Baltic ecosystem. In support of these programmes, several guidelines were published, such as for the BMP second stage (BSEP12, 1984) and in 1988 for the BMP third stage on, for example, physical and chemical determinands in sea water (BSEP27B) and harmful substances in biota and sediments (BSEP27C).

Under the revised 1992 Helsinki Convention, the Cooperative Monitoring in the Baltic Marine Environment (COMBINE) was instituted in 1992. The HELCOM monitoring system consists of several complementary programmes: PLC-Air and PLC-Water to quantify inputs to the sea via water and air; COMBINE, to quantify the state, impacts and changes in the various compartments (water, biota, sediment); and MORS, to quantify the sources, inputs and state of artificial radionuclides in these compartments.

A first ministerial meeting was organized in 1984, at the occasion of the 10th anniversary of HELCOM; a second meeting was held in 1988. At the 1990 Prime Ministers Meeting (Ronneby) it was suggested to install a HELCOM Programme Implementation Task Force (PITF), active since 1992. A first task for PITF was the preparation of the Baltic Sea Joint Comprehensive Environmental Action Programme (JCP), projected to last for 20 years (1993–2012) (BSEP48).

In 1981 an ad hoc Group of Experts on the Assessment of the Marine Environment of the Baltic Sea (GEA, later to become GESPA) was installed. Their task was to compile the first periodic assessment report. A series of such 'Periodic Assessment of the State of the Marine Environment of the Baltic Sea Area' reports was published. The first, covering 1980–1985 (BSEP17A-B), was issued in 1986. A second periodic assessment covered 1984–1988 (BSEP35A-B, 1990), followed by a third assessment for 1989–1993 (BSEP64A-B, 1996) and the fourth for 1994–1998 (BSEP82AB, 2002). The 'ecosystem approach' adopted by the Joint HELCOM/OSPAR Ministerial Meeting in 2003 led to a different type of assessment. The objective now was to assess the pressures of human activities and the resulting impacts on, and state of, the marine environment. For the Stakeholder Conference held in Helsinki in 2006, thematic assessments were published on six issues. Next to thematic assessments for eutrophication, maritime transport, climate change, ecological objectives, and biodiversity and nature protection, the hazardous substances report gives evidence on status, trends and strategies for monitoring of the Baltic (www.helcom.fi). The HELCOM Baltic Sea Action Plan (BSAP) was adopted in November 2007 (Krakow, Poland) by the HELCOM Extraordinary Ministerial Meeting.

1.5 MONITORING OF WHAT?

1.5.1 Selection of Compartments

When monitoring started, fish and shellfish (biological tissue) were the matrix of interest. This clearly stemmed from the focus on potential hazards to human health from the consumption of this seafood. The initial North Sea, Baltic and North Atlantic baseline studies, in the early 1970s, exclusively used these compartments for their assessment. The concept of 'Mussel Watch' was presented by Goldberg (1975) as a method to assess the health of the ocean. Bivalves are sedentary organisms; they concentrate many pollutants in their tissue to a level where they are in equilibrium with the surrounding water. He argued that by utilizing *Mytilus edulis* and similar species from coastal and open ocean sites they could be annually analysed for their concentrations of halogenated carbons, transuranics, heavy metals and petroleum. Since then the Mussel Watch approach has been implemented in the USA under the NOAA National Status and Trends programme and in many other countries around the world (O' Connor *et al*., 1994). Concentrations of trace metals and other pollutants, such as polycyclic aromatic hydrocarbons (PAHs) and PCBs, are routinely monitored in the OSPAR and HELCOM areas.

In the compartment water, notably trace metals (since the late 1970s) and nutrients (from 1990 onwards) were routinely monitored under the ICES and HELCOM programmes. The (third-stage) BMP in 1988 routinely monitored heavy metals, petroleum hydrocarbons and chlorinated hydrocarbons in seawater.

The OSPARCOM JMP started routine sediment monitoring in 1982. Although many organic pollutants have as a measure of hydrophobicity an octanol–water partition coefficient higher than 10^5 (log $K_{ow} > 5$, e.g. hexachlorobenzene, HCB), indicating that they are predominantly present in sediment and suspended particulate matter, those with a log $K_{ow} < 3$ (e.g. atrazine) are preferably analysed in water. This has been recognized by, for example, the WFD amendment EQS Directive, which states that hydrophobic compounds may be monitored in water and/or biota (instead of water).

1.5.2 Selection of Compounds

At the beginning of environmental protection the now common ecosystem approach was yet to be developed. At the first GESAMP meeting in 1969, a 'list of substances carried as cargo' might be selected, as if disasters at sea were the only input route for pollution of the marine environment. Chemicals were rated on the basis of human toxicity, aquatic toxicity and aesthetic effect (GESAMP, 1969). This was refined at a later stage when GESAMP reviewed the harmful substances which may have deleterious effects on human health and on economic and cultural activities in the marine environment and coastal areas, and with regard to (harm to) living resources in the sea as a major source of protein-rich food (GESAMP, 1976). Important pollutants, for which restrictive or preventive measures were recommended, were domestic sewage, pulp and paper mill wastes, organochlorine pesticides, PCBs, mercury and its compounds, organo-lead compounds, elemental phosphorus, silver and several organic chemicals, often solvents.

ICES, with its background in fisheries research, in its initial monitoring days limited their monitoring activities to toxicants in fish and shellfish (at first primarily being foodstuff for humans, but later also related to the health of the marine environment; i.e. higher trophic levels). Initially, the compounds selected were the heavy metals Cd and Hg and selected organochlorine pesticides and PCBs accumulated in a few indicator species, such as mussels, herring, cod and plaice. For the North Atlantic baseline study, as well as the research programme for investigations of the Baltic Sea, the number of substances had increased, although subject to analytical capabilities: 'whenever it seems feasible with the present sensitivity of methods' (ICES, 1974b). Nearly the same lists of compounds were used: organohalogen compounds (HCB, PCBs, pesticides such as the DDTs, dieldrin, chlordane and hexachlorocyclohexane (HCH) isomers), the metals Hg, Pb, Cu, Cd, Cr (replaced by As for the Baltic) and Zn, petroleum hydrocarbons and nutrients (nitrogen compounds and total phosphorus), a list that formed the basis for HELCOM's pilot-phase BMP. The second-phase BMP listed, next to basic oceanographic variables, nutrients, heavy metals and petroleum and chlorinated hydrocarbons in seawater, and heavy metals and chlorinated hydrocarbons in fish and shellfish (HELCOM, 1983). At about the same time, ICES' ACMP reviewed the occurrence and effects of (new) contaminants, like Hg, Pb, Cd, Zn, PCBs, toxaphene, hexachloro-1,3-butadiene, polychlorinated terphenyls (PCTs), alkyl-tin, furans and dioxins in the marine environment (McIntyre, 1985).

Gradually, as more chemicals were identified as being harmful to the (marine) environment, more and more chemical compounds became part of routine monitoring programmes. Examples include many more pesticides, organo-tin compounds (like tributyltin (TBT), affecting oysters and marine gastropods) and brominated compounds (flame retardants).

The EU Directive on Priority Substances (Environmental Quality Standards Directive) supports the WFD and applies to the member states' coastal waters. It defines the limits on concentrations of

33 priority substances and nine other pollutants in EC surface waters, including the coastal waters (European Communities, 2008a). The list includes trace metals (Cd, Hg, Ni, Pb, TBT), different pesticide groups, seven PAHs, (chlorinated) solvents and other compounds such as nonylphenol and brominated diphenylether.

Focusing solely on the marine environment, OSPAR priority groups of chemicals currently include trace metals (Cd, Pb, Hg), organometals (of Pb, Hg and Sn), organohalogens (including short-chained chlorinated paraffins (SCCPs), perfluorooctane sulfonates (PFOSs), polychlorinated dibenzodioxins and dibenzofurans (PCDDs, PCDFs), PCBs, brominated flame retardants and other polybromodiphenyl ethers (PBDEs), hexabromocyclododecane (HBCD) and tetrabromobisphenol-A (TBBP-A)). Furthermore, seven pesticides are included, such as endosulfan, HCH isomers (including lindane), pentachlorophenol (PCP) and trifluralin. Other groups include phenols, phthalates (such as DBP and DEHP), PAHs, pharmaceuticals and personal care and other substances (OSPAR, 2010).

Usually in support of biological research, nutrients have been analysed since the beginning of the 20th century (Brandt, 1927), albeit initially not always of high analytical quality. Nutrients, mainly the nitrogen compounds nitrate, nitrite and ammonia, and phosphorus, are not considered as hazardous substances, but their enrichment may lead to adverse effects upon the marine environment. Monitoring of nutrients in the OSPAR area started rather late. As a result of decisions to reduce nutrient inputs to the OSPAR area by 50%, OSPARCOM's JMP, NSTF and NUT (Nutrient Working Group, under the Oslo/Paris Commissions) agreed to start a mandatory monitoring programme in 1990. Results would provide input to the QSR (de Jong, 2006).

1.5.3 Background Concentrations

Heavy metals are natural and, subject to the hydro-geological situation, are present in lower or higher concentration; thus, this background concentration may vary from place to place. Pollution may add to this background concentration, creating harmful concentrations in the environment. Background concentration is subject to many debates; for example, for the correct implementation of environmental quality standards. For man-made compounds, like PCBs and TBT, the background concentration is zero. For a few groups of compounds, such as PAHs, some occur naturally; others are anthropogenic.

Owing to the regional differences, global averages will not work, and attempts have been made to define regional (North Sea) background concentrations for metals in water sediment and biota (Kersten *et al*., 1994). OSPAR, in 2005, agreed on background concentrations for contaminants in seawater, sediments and biota (OSPAR, 2005). In view of the new demands by the WFD, Tueros *et al*. (2008) discussed ways to define background concentrations for trace metals in coastal seawater.

1.6 QUALITY OF DATA

During the early 1970s there was increased concern about the quality of the analytical results from monitoring programmes. Within the laboratories one was confident about one's own data, but data sets from different laboratories often did not match. Interlaboratory comparisons were organized to investigate this. They showed that, for example, the results for the analysis of seawater for trace metals differed by two or even three orders of magnitude. Quality data will lead to quality information. Hence, Goldberg and Taylor (1985) argued that only validated data should be included in databases. In order to avoid expensive environmental management actions, monitoring data should not be published without demonstrable evidence of quality practices in all aspects of the monitoring exercise (Batley, 1999).

Several instruments were developed to improve the quality of monitoring data. They include the building up of the laboratory's quality assurance (QA) measures and a continuous analytical quality control (QC) programme. This included the development and implementation of guidelines on sampling, analysis and QA/QC, the implementation of quality control by the use of certified reference materials (CRMs) and QC materials (QCMs) in the analytical process, and the participation in laboratory performance studies (proficiency tests, PTs). Within-laboratory QC and between-laboratory QC shall be demonstrated throughout the monitoring programme (Nicholson, 1989).

Recently, European policies recognized the need for quality data from monitoring programmes. The European Commission Directive 2009/90/EC (European Community, 2009) underpins the WFD in laying down technical specifications for chemical analysis and monitoring of water. It establishes minimum performance criteria for methods of analysis to be applied by member states when monitoring water status, sediment and biota, as well as rules for demonstrating the quality of analytical results for other directives dealing with monitoring. It explicitly addresses QA and QC, and makes it mandatory that laboratories demonstrate their competences by participation in proficiency testing programmes and analysis of representative reference materials.

1.6.1 Interlaboratory Comparisons

The initial interlaboratory comparison studies aimed at demonstrating that sufficient agreement was present to enable intercomparison of monitoring results by different laboratories from different countries. The studies showed that, for nearly all analytes and matrices, insufficient agreement was present and considerable work was to be done to arrive at these objectives. Interlaboratory comparisons thus started to become a training tool for improvement of laboratory skills.

Following the development of the compounds and matrices that were monitored in the period 1970–1990s, the first ICES intercomparison was for trace metals (Hg only) and chlorinated hydrocarbons in biological tissue in 1972. Many would follow. For biological tissues, 9 exercises would be conducted for trace metals (1972–1989), 3 for hydrocarbons (1984–1990) and no less that 15 for organochlorine compounds (1972–1993). The second matrix tested was seawater, which spanned two decades for 11 interlaboratory comparisons on trace metal analysis (1976–1996), and 5 studies for nutrients (1989, 1993) (ICES, 1992). Sediment intercomparison studies started in 1980. There was one for hydrocarbons (1980) and two for chlorinated hydrocarbons (1980, 1993). For heavy metals in sediments there were nine studies (1983–1993).

In particular, the exchange of information at meetings where the outcomes of the tests were discussed proved beneficial, as experts could support each other on how to further improve their methods. Usually, the lower the concentration, the higher the variance in the analytical results, as postulated by Horwitz (1982): the 'Horwitz trumpet'. Still today, the variance (coefficient of variation, CV) between laboratories for groups of analytes follows CV(nutrients)<CV(trace elements)<CV(chlorinated hydrocarbons).

Several times (as, for example, in the case of chlorinated organic compounds in biological tissue) it initially proved too difficult to start with the matrix: analytical results varied too much. A holistic approach was followed by starting on a less complex analysis level, using calibration solutions and/or extracts that were centrally prepared. Only when harmonization could be shown for these samples were natural matrices reconsidered in the interlaboratory tests.

Although the ICES/SCOR Baltic WG joined the ICES intercomparison studies, HELCOM also organized interlaboratory comparisons, such as the biological intercalibration workshops which included a component on nutrients (first in 1979; second in 1982 (BSEP09) and third in 1990 (BSEP38)), and in 1981 on the analysis of hydrocarbons in seawater (BSEP06). In 1992, HELCOM reported on an interlaboratory comparison for airborne pollutants (BSEP41). Other organizations coordinating interlaboratory comparisons were the IAEA, initially often for the analysis of radionuclides, but later including many other organic and inorganic compounds. The WHO/FAO/UNEP Joint Project MED POL, a concerted effort to monitor contaminants and to assess pollution in the Mediterranean Sea (since 1975, part of the Barcelona Convention) has been the cornerstone of the MAP and has been instrumental in developing the capabilities of the (Mediterranean) countries to measure and assess marine pollution (Carvalho and Civili, 2001). The IAEA Marine Environment Laboratory in Monaco acted as the analytical and training centre for MED POL.

In order to define the baseline concentrations for selected trace metals in open Atlantic seawater, IOC–GIPME conducted an intercalibration in 1990. One of the prerequisites was the mandatory use of CRMs (Landing *et al*., 1995).

The role of (ICES) interlaboratory comparisons has been gradually taken over by the participation in PTs, such as the QUASIMEME programme (see below), which was a combined ICES/QUASIMEME exercise.

Since the implementation of QA/QC procedures in analytical laboratory activities, comparability of data has improved considerably. Although intercomparisons were not only devoted to the analysis itself, but also included, for example, trace metals sampling (Bewers and Windom, 1982) and filtration procedures (Bewers *et al*., 1985), Kramer (1994) argued that most QA/QC procedures were limited to the activities within the laboratory, and that they should be extended to the activities before samples reached the laboratory, such as sampling, sample treatment and storage. This can be achieved by using written standard operating procedures, agreement on a specifically developed sampling plan and safeguarding full documentation at every stage of the operation, including unexpected events (Wagner, 1995). Davies and Wells (1997) concluded that QA and the interpretation of international marine monitoring programmes had improved considerably, in part as a result of the series of interlaboratory comparison studies and laboratory performance studies. There still was (and is) an urgent need for (certified) reference materials (CRMs, QCMs) as instruments for further improvement of the analytical quality.

1.6.2 Guidelines

Initially, national guidelines and standard operating procedures were implemented. When the JMP resulted in poor overlap in the respective results, there was a need for international guidance for sampling and analyses. Without trying to be complete, a few organizations that produced guidelines are discussed below.

Since its foundation, ICES has had a tradition in method development by its Central Laboratory. Already in the early 1970s, guidelines for sampling and analysis of physical, chemical and biological parameters had become available (e.g. ICES, 1972). In support of sampling and analysis for biological and chemical monitoring programmes, ICES started the series of publications 'Techniques in Marine Environmental Sciences' (TIMES) in 1987. Each issue treated a particular aspect related to a specific analytical procedure; for example, on sampling and storage methods for trace metals in seawater (TIMES 2; Yeats, 1987), good laboratory practice and QA (TIMES 6; Vijverberg and Cofino, 1987) or the review of methods for analysis of hydrocarbons in seawater, biota and sediments (TIMES 12; Ehrhardt *et al*., 1991).

When OSPAR started to coordinate monitoring under JAMP it also started to produce monitoring guidelines. Monitoring guidance is regularly reviewed in collaboration with ICES. Where necessary, they are updated to take into account new developments, including the inclusion of new monitoring parameters in the CEMP. Examples of JAMP guidelines are the 'JAMP Guidelines for Monitoring Contaminants in Biota (agreement 1999-2)' or 'JAMP Guidelines for Monitoring Contaminants in Sediments (agreement 2002-16)' (latest updated versions are electronically available at www.OSPAR.org).

Similarly, HELCOM has published guidelines in support of its monitoring programme; for example, from the Baltic Sea Environment Proceedings series (BSEP27A-D): Harmful Substances in Biota and Sediments (HELCOM, 1988). Following new developments since the initiation of the COMBINE programme, the 'Manual for Marine Monitoring in the COMBINE Programme of HELCOM' is maintained up to date on the HELCOM website (www.helcom.fi), which includes, for example, the 'General guidelines on quality assurance for monitoring in the Baltic Sea'.

All three GIPME Expert Scientific Groups, already indicated above (GEMSI, GEEP and GESREM), supported the UNEP Regional Seas Programme by the publication of guidelines, in development and testing of reference methods, in preparation of reference materials, in QC of monitoring data, in the organization of intercalibration exercises and in QA/QC training (Ibe and Kullenberg, 1995).

1.6.3 (Certified) Reference Materials

'Standard seawater', used to calibrate chlorinity/salinity measurements, has been produced since the beginning of the twentieth century as a primary standard and is probably the first environmental reference material (Culkin and Smed, 1978). In support of pollution monitoring programmes, new materials were urgently needed. The National Research Council Canada (NRCC) was the first institution that recognized the need for a series of environmental CRMs to answer to the increasing needs of monitoring programmes. In support of trace metal analyses, they started their production with MESS-1 and BCSS (marine sediments, in 1981), NASS-1 (seawater, in 1982) and TORT-1 (lobster hepatopancreas, in 1984). For PCBs, poor agreement between laboratories for organochlorine compounds in matrix reference materials in those days forced laboratories to develop their methods; hence, standard iso-octane solutions were produced (CLB-1-A and -D, 1985) (Waldichuck *et al.*, 1987).

In Europe, the EC Community Bureau of Reference (BCR) was installed in 1973 (van der Eijk, 1977). It existed until 2003, when its activities were transferred to the JRC–IRMM (Quevauviller, 2003) and BCR remained only a brand name. Also, the BCR recognized the monitoring needs and started CRM production; for example, for trace metals BCR-279 (sea lettuce, in 1987), BCR-277 (estuarine sediment), BCR-278 (mussel tissue), both in 1988, and BCR-403 (seawater, in 1991). In 1988, PCBs in cod liver oil (BCR-349) served to further improve analytical performance of the analytical laboratories. Only in 2001, when analyses were generally under control, did PCBs in a matrix CRM (mussel tissue, BCR-682) come on the market. Producers followed the trends in policy and monitoring practice, where more (and more complex) hazardous substances had to be analysed. As a result, CRMs for the analysis of methylmercury in tuna (BCR-463, in 1994) and of the antifouling agent TBT and other organotin compounds in mussel tissue (BCR-477) and in coastal sediment (BCR-462) were introduced in 1997 and 1998 respectively; CRMs for the analysis of brominated flame retardants (PBDEs) were introduced only recently.

Other early CRM producers were the Japanese National Institute for Environmental Studies (NIES), such as mussel tissue (NIES No. 6, 1984) and Sargasso seaweed (NIES No. 9, 1988), both for trace element analyses. Early IAEA marine reference materials produced by their Marine Environment Laboratory (Monaco) include IAEA-SD-N-2 (marine sediment, 1986) and IAEA-MA-B-3/RN (fish flesh, 1987), each for the analysis of radionuclides.

The US National Institute of Standards and Technology (NIST) started the production of environmental matrix CRMs in the 1990s. In recent years they set a new trend by producing matrix materials certified for a wide range of analytes, thus serving the needs of laboratories that have to respond to the monitoring requirements, such as the US Environmental Protection Agency (EPA) and the EC WFD. Examples are the SRM-1944, a marine sediment certified for PAHs, PCBs, chlorinated pesticides and trace elements, or SRM-2974, a freeze-dried mussel tissue, certified for about the same set plus selected PBDE congeners. With these wide-range CRMs laboratories may reduce costs, and the more frequent use of the CRMs for different analyses (thus earlier replacement) is considered beneficial.

Nutrient concentrations in natural seawater are not stable owing to the activity of microorganisms. This inherent stability problem was solved; and soon after monitoring of nutrients under the JMP became mandatory, a reference material for nutrients was developed (Aminot and Kérouel, 1991).

Several overviews of available marine CRMs were produced in the past (e.g. Cantillo, 1992; Kramer, 1998). Today, web-based databases are readily accessible, such as the Virtual Institute for Reference Materials (www.VIRM.net). Unfortunately, not all analytes in all matrices requested by monitoring programmes are fully covered. Notably, reference materials for organics in water are still lacking due to problems in production (stability of the analytes) (Wells, 1993).

1.6.4 Laboratory Performance Studies

Laboratory performance studies, also known as PTs, aim to verify the proficiency of a laboratory, namely its capability to analyse a given analyte in a given matrix with a sufficiently high accuracy and precision. This is usually done by coordinated sending out of samples with unknown composition to a large group of laboratories. Their analytical results are inter-compared and compared with a reference value and (accepted) deviation unit, and a measure of their proficiency (usually as a z-score) is confidentially reported back to the laboratory (Thompson *et al*., 2006). A framework for PTs is guided by, for example, ISO/IEC 17 043 standard (ISO/IEC, 2010), which replaces ISO/IEC Guide 43-1:1997.

In the early 1990s European laboratories were participating in marine pollution monitoring programmes, organized by the OSPARCOM JMG and the Intergovernmental NSTF. In order, for example, to prepare the QSR 1993 (NSTF, 1993) it was essential that data, collected by different countries, would match in one overview of the region concerned, the North Sea. Both NSTF and JMG, but also other international programmes conducting similar monitoring programmes in other regional seas, had expressed concern about data quality (Kullenberg, 1986; Topping, 1992). In view of developing a collaborative approach to serve these monitoring programmes in the (further) development of their QA and internal and external QC, the European Community funded the project Quality Assurance in Marine Monitoring in Europe (QUASIMEME), which started in 1992 (Wells *et al*., 1993).

QUASIMEME was instrumental in the development of laboratory performance studies in support of monitoring of the (initially European) marine environment. A wide range of PTs was carried out, including nutrients in seawater, trace metals, PCBs and organochlorine pesticides in sediment and biological tissue, and PAHs in sediment (extract), thus covering nearly all analytes and matrices of monitoring interest (Wells *et al*., 1997; Topping, 1998). Since the project has ended, the PT programme has continued as an independent sustainable activity. Today, other, sometimes commercial, parties also develop and implement PTs for specific analyte–matrix combinations in support of monitoring laboratories.

Other international organizations active in laboratory performance studies include the IAEA, for organic contaminants (PCBs, sterols) and trace metals, often including methylmercury (de Mora *et al*., 2006).

Today, monitoring laboratories need to be accredited. They demonstrate their competence by internal and external QA and QC measures, complying with the ISO 17 025 standard (ISO, 2005). All institutes/laboratories submitting data to OSPAR or HELCOM databases should participate in regular intercomparison exercises and proficiency testing schemes arranged, for example, under QUASIMEME, ICES, HELCOM and IAEA. They shall use CRMs and QCMs and new such reference materials as they become available; these should be used by all participating institutes or laboratories.

1.7 EXAMPLE: MONITORING OF TRACE METALS IN SEAWATER

Trace metals concentrations in seawater have been reported since the beginning of the twentieth century (As, Au, Cu, Fe, Hg) or even before (Ag, Zn), as reviewed by Johnston (1965). For elements at the nanogram/litre level, like cadmium and lead, earliest reported concentrations arrive in the late 1950s. Considering the low concentrations in seawater, sophisticated analytical methods were not available in the early periods. As a result of lack of attention to contamination control and QA/QC, concentrations tended to show a large variation. As a result of analytical improvements, this spread became smaller with time and concentrations dropped. Temporal trends of reported

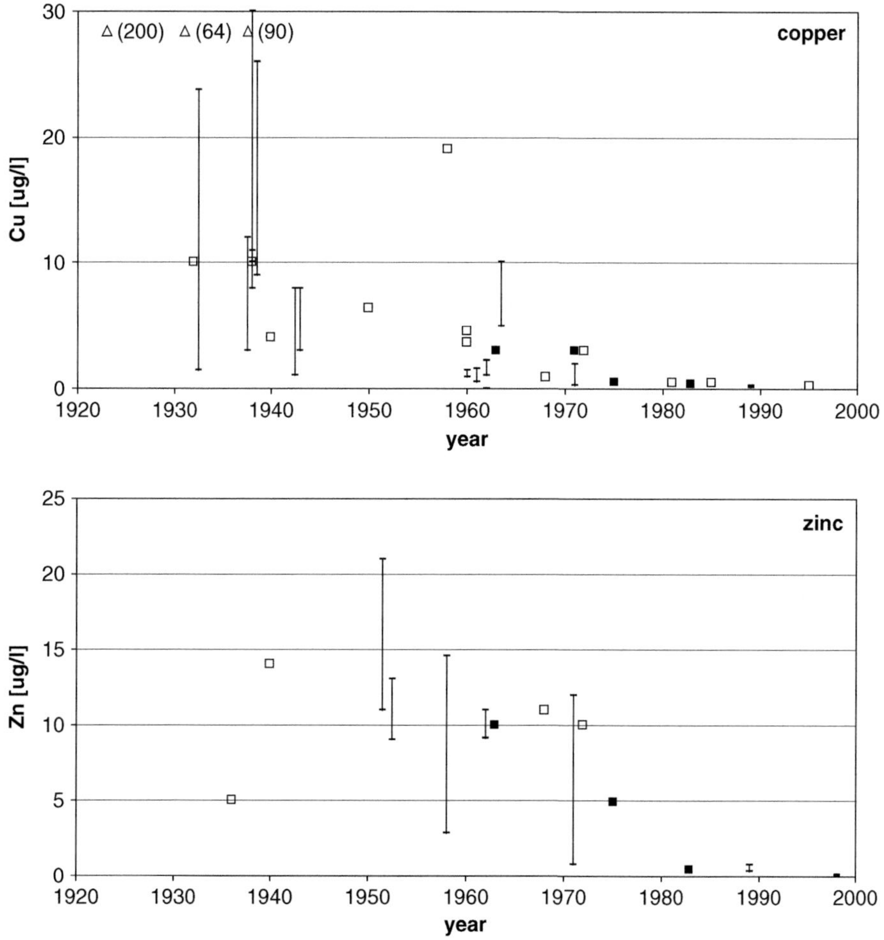

Figure 1.1 Concentration of dissolved copper (Cu) and zinc (Zn) reported for open sea waters since the 1930s. Closed symbols refer to the 'accepted reference values' (Goldberg, 1965; Riley and Chester, 1971; Brewer, 1975; Bruland, 1983). Data prior to 1963 are from Johnston (1965); later data are from Fabricand *et al.* (1962), Alexander and Corcoran (1967), Turekian (1968), Dutton *et al.* (1973), Ivanoff (1972), Brügmann (1981), Bruland and Franks (1983), Kremling (1983), Balls (1985), Mart and Nürnberg (1986), Nolting and de Baar (1994) and Nolting *et al.* (2000).

concentrations (and concentration ranges) for open sea waters, e.g. copper and zinc, both since the 1930s, give (very) high observations with a large spread in results within data series until the mid 1970s; see Figure 1.1. Only after proper attention was given to the combat of contamination, also during sampling, and the implementation of QC procedures (see Section 1.6) did the results come to a closer agreement and the spread in results becomes much smaller. The 'accepted' average concentrations for seawater (Goldberg, 1965; Riley and Chester, 1971; Brewer, 1975; Bruland, 1983), identified by filled symbols in Figure 1.1, clearly demonstrate the advancement of analytical chemistry in monitoring programmes (such large decreases in concentrations for open sea waters are not expected to have resulted from a decrease in pollution). The 'typical' concentration values defined around 1980–1985 for the various trace elements in open sea water have hardly changed since.

Another illustration of the temporal trend for the dissolved fraction of selected trace metals is given in Figure 1.2. The data shown are results of the Dutch national monitoring programme (MWTL) for one sampling location at the mouth of the Scheldt estuary (Vlissingen boei SSVH), taken from the DONAR database (Rijkswaterstaat, 2010). This location was selected because of the long time series available. Despite there being a few gaps in the copper and lead data, the data cover the period 1971–2009. At first sight there is an apparent decline in the concentration of all four metals shown (Cu, Zn, Pb, Cd), seemingly in line with the reduction of riverine inputs of the River Scheldt in the past decades. Again, there is an additional argument: improvement of the analyses. The period 1971–1982 is characterised by high concentrations for all metals with a large

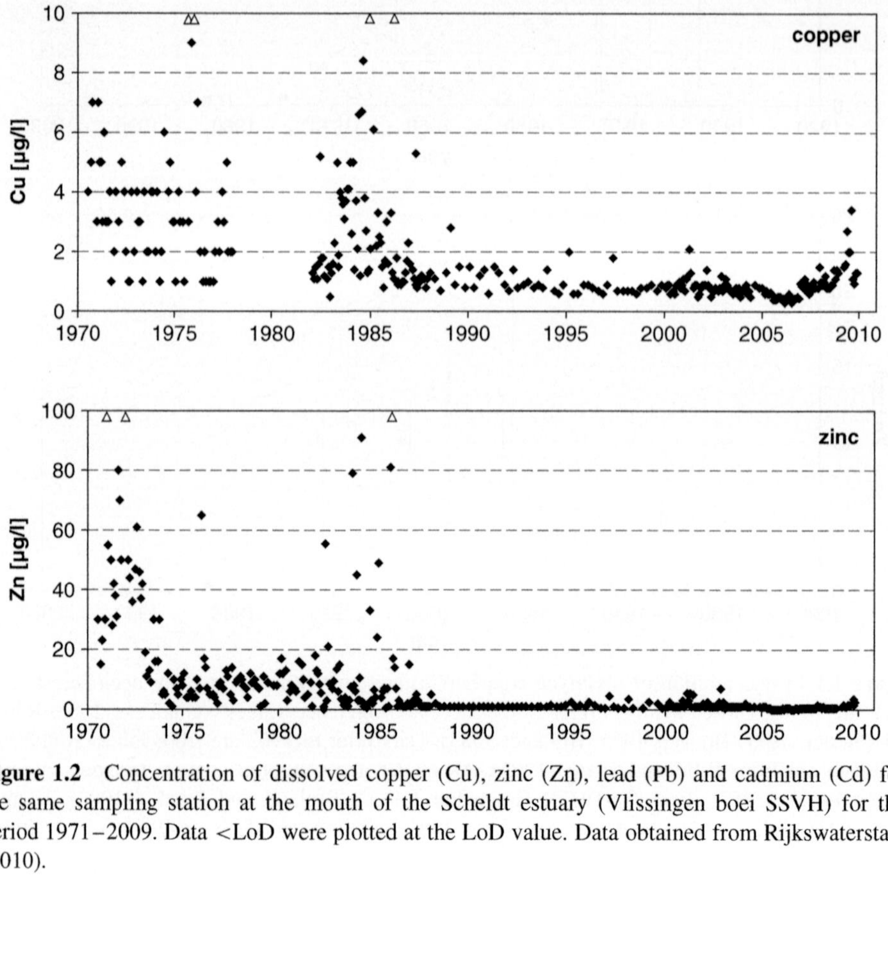

Figure 1.2 Concentration of dissolved copper (Cu), zinc (Zn), lead (Pb) and cadmium (Cd) for the same sampling station at the mouth of the Scheldt estuary (Vlissingen boei SSVH) for the period 1971–2009. Data <LoD were plotted at the LoD value. Data obtained from Rijkswaterstaat (2010).

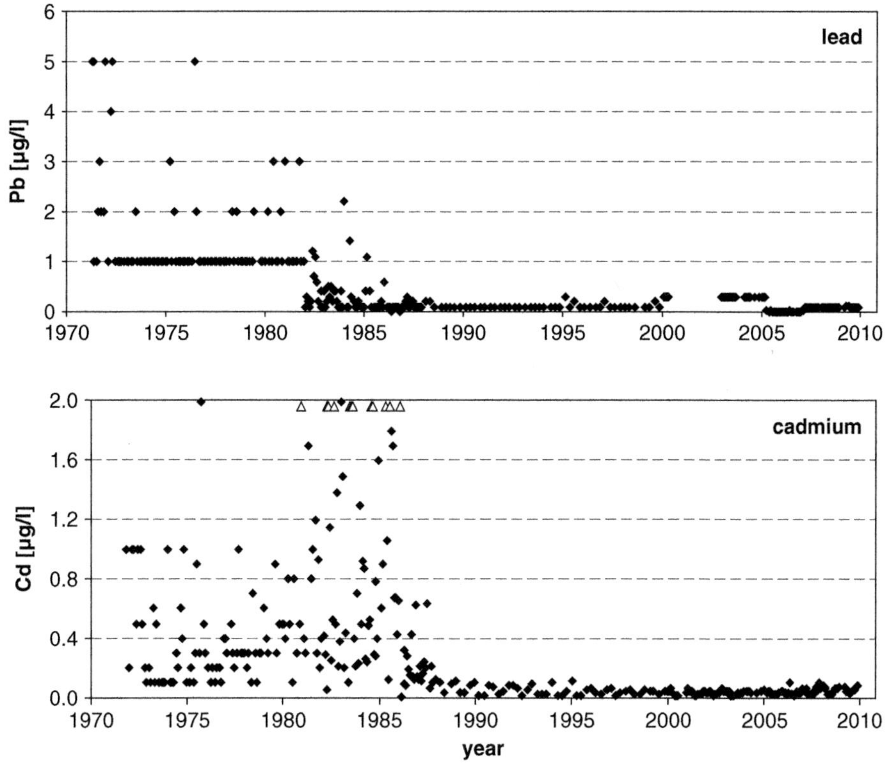

Figure 1.2 *(continued)*

(sometimes extreme) range in the results. This was the period before rigorous QA/QC methodologies were introduced. The analytical method (graphite furnace atomic absorption spectrometry (GF-AAS), direct in water samples) was not sensitive, as can be seen from the reported values: in this period results for dissolved Cu and Pb were presented in full digits only.

By 1982 the methods had improved and the results for these elements were now reported to one decimal place. In 1986 the laboratory introduced the metal complexation/extraction technique with subsequent analysis by GF-AAS. To ensure inter-comparability of the two methods, both techniques were applied for alternating samples for 2 years; from 1988 onwards, only extracts were analysed by GF-AAS, to be replaced by inductively coupled plasma mass spectrometry of water samples in 2005. Starting in 1990, values below the limit of detection (LoD) were identified in the database (in Figure 1.2, data <LoD are presented at the LoD value). Notably, the analyses for Pb often are subject to values reported <LoD, a value that in fact changes with time: 1990–1999: 0.1 µg/l; 2000–2005: 0.3 µg/l; 2005–2007: 0.01 µg/l; and from 2007: 0.1 µg/l.

Using all available data for different areas of the North Sea from the period 1980–1989, Scholten *et al.* (1998) found considerable differences in concentration between the first and second halfs of this period and concluded this in line with the reduction in river inputs. Clearly, the improvements in analytical procedures have had an impact on the reported results, a factor complicating statistical trend analyses on long data sets. Because of the uncertainties in the 'older' analytical results, OSPAR, for example, in evaluating trends in riverine inputs, has used data since 1990 only. Even for the period 1990–2006, trend analyses proved cumbersome due to changes in analytical techniques and in LoDs (OSPAR, 2009).

1.8 CONCLUSION

This chapter focused mainly on the situation for the North-East Atlantic. As such, it is an example for developments in the rest of the world; albeit, the developments may not have taken place at the same time and/or at the same speed.

In response to public as well as policy awareness of the intrinsic quality of the (marine) environment in the late 1960s, international institutions started to focus on its protection. (Inter)national legal instruments were developed and implemented. The assessment of the status and temporal trends of, for example, harmful substances required the implementation and harmonization of international monitoring programmes.

Since routine monitoring started in the early 1970s, many developments have taken place. Early problems in the quality of the monitoring data have been solved by interlaboratory comparisons and rigorous implementation of QA and QC programmes. The initial use of fish and shellfish as a matrix for monitoring studies was soon expanded with monitoring of seawater and sediments. From monitoring only a few trace metals and PCBs, the list of hazardous substances has grown considerably, answering to the need to reduce the inputs of (often new and man-made) substances and to minimize their threat to ocean life.

Over the past 40 years the objectives of monitoring harmful substances in fish and shellfish as a potential hazard to human health have changed into an ecosystem approach where, in Europe under the WFD and MSFD, GES shall be reached. Monitoring is one of the management tools used for assessing the quality status and the temporal trends.

REFERENCES

Alexander J. E. and Corcoran E. F. (1967). The distribution of copper in tropical seawater. *Limnol. Ocean.*, **12**, 236–242.

Aminot A. and Kérouel R. (1991). Autoclaved seawater as a reference material for the determination of nitrate and phosphate in seawater. *Anal. Chim. Acta*, **248**, 277–283.

Andersen N. R., Dawson R., Duinker J. C. *et al.* (1988). Objectives, components and experiences in the development of the IOC Marine Pollution Monitoring System (MARPOLMON). *Environ. Mon. Assess.*, **11**, 299–314.

Anon. (1970). Spreading news of pollution. *Mar. Pollut. Bull.*, **1**, 1.

Anon. (1975). Do we monitor too much? *Mar. Pollut. Bull.*, **6**, 17–18.

Balls P. W. (1985). Trace metals in the northern North Sea. *Mar. Pollut. Bull.*, **16**, 203–207.

Batley G. E. (1999). Quality assurance in environmental monitoring. *Mar. Pollut. Bull.*, **39**, 23–31.

Beukema A. A., Hekstra G. P. and Venema C. (1986). The Netherlands' environmental policy for North Sea and Wadden Sea. *Environ. Mon. Assess.*, **7**, 117–155.

Bewers J. M. and Windom H. L. (1982). Comparison of sampling devices for trace metal determinations in seawater. *Mar. Chem.*, **11**, 71–86.

Bewers J. M., Yeats P. A., Westerlund S. *et al.* (1985). An intercomparison of seawater filtration procedures. *Mar. Pollut. Bull.*, **16**: 277–281.

Birnie P. W. (1974). United Nations Environment Programme: the marine environment. *Mar. Pollut. Bull.*, **5**, 90–93.

Bonn Agreement (2009). *40 Years of the Bonn Agreement*. Brochure, Bonn Agreement Secretariat, London, pp. 54.

Borja A. (2006). The new European Marine Strategy Directive: difficulties, opportunities, and challenges. *Mar. Pollut. Bull.*, **52**, 239–242.

Brandt K. (1927). Stickstoffverbindungen der Meere. *Wiss. Meeresunters. Kiel*, **20**, 203–292.

Brewer P. G. (1975). Minor elements in sea water, in *Chemical Oceanography*, vol. 1, 2nd edn (eds J. P. Riley and G. Skirrow), Academic Press, London, pp. 415–498.

Brügmann L. (1981). Heavy metals in the Baltic Sea. *Mar. Pollut. Bull.*, **12**, 214–218.

Bruland K. W. (1983). Trace elements in seawater, in *Chemical Oceanography*, vol. 8 (eds J. P. Riley and R. Chester), Academic Press, London, pp. 157–220.

Bruland K. W. and Franks R. P. (1983). Mn, Ni, Cu, Zn and Cd in the western North Atlantic, in *Trace Metals in Seawater* (eds C. S. Wong, E. Boyle, K. W. Bruland *et al.*), Plenum, New York, pp. 395–414.

Cantillo A. Y. (1992). *Standard and Reference Materials for Marine Science*, 3rd edn, NOAA National Status and Trends program, US Dept of Commerce, Rockville, MD, USA, 577 pp.

Carlson H. (ed.) (1986). *Quality Status of the North Sea*, Deutsche Hydrographische Zeitschrift, Erganzungsheft, B, No. 16, Deutsches Hydrographisches Institut, Hamburg, pp. 1–424.

Carpenter A. (2007). The Bonn Agreement aerial surveillance programme: trends in North Sea oil pollution 1986–2004. *Mar. Pollut. Bull.*, **54**, 149–163.

Carson R. (1962). *Silent Spring*, Houghton Mifflin, Boston, MA.

Carvalho F. P. and Civili F. S. (2001). Monitoring of the Mediterranean Sea pollution (MED POL) and data quality assurance. *Int. J. Environ. Studies*, **58**, 139–158.

Culkin F. and Smed J. (1978). The history of standard seawater. *Oceanol. Acta*, **2**, 355–364.

Davies I. M. and Wells D. E. (1997). Quality assurance and the interpretation of international marine monitoring data: the potential contribution of international laboratory performance studies. *Mar. Pollut. Bull.*, **35**, 133–139.

Dawson R., Bewers J. M., Andersen N. R. and Kullenberg G. (1988). Global strategies for the assessment of pollution in marine environments and progress achieved in the IOC–GIPME programme. *Aquat. Toxicol.*, **11**, 345–356.

De Jong F. (2006). *Marine Eutrophication in Perspective: On the Relevance of Ecology for Environmental Policy*, Springer, Berlin, 336 pp.

De Mora S., Villeneuve J.-P. and Wyse E. (2007). An overview of laboratory performance studies conducted by the IAEA for marine pollution studies of metals and organic contaminants. *Accred. Qual. Assur.*, **12**, 587–592.

Dutton J. W. R., Jefferies D. F., Folkard A. R. and Jones P. G. W. (1973). Trace metals in the North Sea. *Mar. Pollut. Bull.*, **4**, 135–138.

Ehrhardt M., Klungsoyr J. and Law R. J. (1991). Hydrocarbons: review of methods for analysis in seawater, biota and sediments, in *ICES Techniques in Marine Environmental Sciences*, No. TIMES-12, Copenhagen, 47 pp.

European Commission (2000). Directive 2000/60/EC of the European Parliament and of the Council of 23 October 2000 establishing a framework for Community action in the field of water policy. *Off. J. Eur. Comm.*, **L327**, 1–72.

European Communities (2008a). Directive 2008/105/EC of the European Parliament and of the Council of 16 December 2008 on environmental quality standards in the field of water policy, amending and subsequently repealing Council Directives 82/176/EEC, 83/513/EEC, 84/156/EEC, 84/491/EEC, 86/280/EEC and amending Directive 2000/60/EC of the European Parliament and of the Council. *Off. J. Eur. Comm.*, L 348/84, 24.12.2008, 14 pp.

European Communities (2008b). Directive 2008/56/EC of the European Parliament and of the Council of 17 June 2008 establishing a framework for community action in the field of marine environmental policy. (Marine Strategy Framework Directive). *Off. J. Eur. Comm.*, L 164/19, 25.6.2008, 22 pp.

European Community (2009). Commission Directive 2009/90/EC of 31 July 2009 laying down, pursuant to Directive 2000/60/EC of the European Parliament and of the Council, technical specifications for chemical analysis and monitoring of water status. *Off. J. Eur. Comm.*, L 201/36 EN 1.8.2009, pp. 36–38.

Fabricand B. P., Sawyer R. R., Ungar S. G. and Adler S. (1962). Trace metal concentrations in the ocean by atomic absorption spectroscopy. *Geochim. Cosmochim. Acta*, **26**, 1023–1027.

Ferreira J. G., Vale C., Soares C. V. *et al.* (2007). Monitoring of coastal and transitional waters under the E.U. Water Framework Directive. *Environ. Mon. Assess.*, **135**, 195–216.

Gerges M. A. (1994). Marine pollution monitoring, assessment and control – UNEP approach and strategy. *Mar. Pollut. Bull.*, **28**, 199–210.

GESAMP (1969). *Report of the first session*. Joint IMCO/FAO/UNESCO/WMO Group of Experts on the Scientific Aspects of Marine Pollution, GESAMP, I/11, UNESCO, Paris, 57 pp.

GESAMP (1976). *Review of Harmful Substances*, Reports and Studies 2, UNESCO, Paris, 88 pp.

GESAMP (2005). *The New GESAMP: Science for Sustainable Oceans*, IMO, London, 21 pp.

Goldberg E. D. (1965). Minor elements in sea water, in *Chemical Oceanography*, vol. 1 (eds J. P. Riley and G. Skirrow), Academic Press, London, pp. 163–196.

Goldberg E. D. (1975). The Mussel Watch – a first step in global marine monitoring. *Mar. Pollut. Bull.*, **6**, 111.

Goldberg E. D. and Taylor J. K. (1985). The VD conspiracy. *Mar. Pollut. Bull.*, **16**, 1.

Griffith D. de G. (2003). The evolution of ICES, in *Stockholm 1999 Centenary Lectures*, ICES Cooperative Research Report, No. CRR-260, Copenhagen, pp. 3–19.

HELCOM (1983). Guidelines for the Baltic monitoring programme for the second stage. In *Baltic Sea Environment Proceedings, BSEP 12*, HELCOM, Helsinki, 247 pp.

HELCOM (1988). Guidelines for the Baltic monitoring programme for the third stage; Part D. Biological determinands. In *Baltic Sea Environment Proceedings, BSEP 27D*. HELCOM, Helsinki.

HELCOM (1994). Intergovernmental activities in the framework of the Helsinki Convention 1974–94. In *Baltic Sea Environment Proceedings, BSEP 56*, Helsinki Commission, Helsinki, 235 pp.

Helsinki Commission (1994). *HELCOM: 20 Years of International Cooperation for the Baltic Marine Environment 1974–94*, Helsinki Commission, Helsinki.

Holdgate M. W. and White G. F. (eds) (1977). *Environmental Issues*, SCOPE Report 10, John Wiley & Sons, Inc., New York, 224 pp.

Hoogweg P., Ducrotoy J. P. and van de Wetering B. (1991). The North Sea Task Force – the first two years. *Mar. Pollut. Bull.*, **22**, 328–330.

Horwitz W. (1982). Evaluation of analytical methods used for regulation of foods and drugs. *Anal. Chem.*, **54**, 67A–76A.

Ibe A. C. and Kullenberg G. (1995). Quality assurance/quality control (QA/QC) regime in marine pollution monitoring programmes: the GIPME perspective. *Mar. Pollut. Bull.*, **31**, 209–213.

ICES (1972). *New Baltic manual, with methods for sampling and analysis of physical, chemical and biological parameters*, ICES Cooperative Research Report CRR-029, ICES, Copenhagen.

ICES (1974a). *Report of the Working Group for the international study of the pollution of the North Sea and its effects on living resources and their exploitation*, Cooperative Research Report CRR-039, ICES, Copenhagen.

ICES (1974b). *Research programmes for investigations of the Baltic as a natural resource with special reference to pollution problems*, ICES Cooperative Research Report CRR-042, ICES, Copenhagen, 51 pp.

ICES (1974c). ICES in the marine pollution field. *Mar. Pollut. Bull.*, **5**, 161–163.

ICES (1977a). *Studies of pollution of the Baltic Sea by the ICES/SCOR Working Group on the Study of Pollution of the Baltic*, Cooperative Research Report CRR-063, ICES, Copenhagen.

ICES (1977b). *A baseline study of the level of contaminating substances in living resources of the North Atlantic*, Cooperative Research Report CRR-069, ICES, Copenhagen, 82 pp.

ICES (1977c). *ICES Coordinated Monitoring Programme in the North Sea* 1974, Cooperative Research Report CRR-058, ICES, Copenhagen.

ICES (1992). *ICES nutrient I/Cs – the first 25 years*, ICES leaflet, Kopenhagen, 2 pp.

ISO (2005). ISO/IEC 17025. *General requirements for the competence of testing and calibration laboratories*, International Organization for Standardization, Geneva, Switzerland.

ISO (2010). ISO/IEC 17043. *Conformity assessment – general requirements for proficiency testing*, International Organization for Standardization, Geneva, Switzerland.

Ivanoff A. (1972). *Introduction a l'Océanographie Propriétés Physiques et Chimiques des Eaux de Mer*. tome 1, Vuibert, Paris, 208 pp.

Johnston R. (1965). *Serial atlas of the marine environment. Folio 8. The trace elements*. American Geographical Society, New York, 5 pp.

Kersten M., Balls P. W., van Enk R. J. *et al*. (1994). Background concentrations for metals in the North Sea: sediment, water, mussels and atmosphere, in *Circulation and Contaminant Fluxes in the North Sea* (ed. J. Sündermann), Springer, Berlin, pp. 290–316.

Kramer K. J. M. (1994). What about quality assurance before the laboratory analysis? *Mar. Pollut. Bull.*, **29**, 222–227.

Kramer K. J. M. (1998). Inorganic certified reference materials in 'water': what do we have, what do we need? *Analyst*, **123**, 991–995.

Kremling K. (1983). Trace metal fronts in European shelf waters. *Nature*, **303**, 225–227.

Kullenberg G. (ed.) (1986). The IOC programme on marine pollution. *Mar. Pollut. Bull.*, **17**, 341–352.

Landing W. M., Cutter G. A., Dalziel J. A. *et al*. (1995). Analytical intercomparison results from the 1990 intergovernmental oceanographic commission open-ocean baseline survey for trace metals: Atlantic Ocean. *Mar. Chem.*, **49**, 253–265.

Mart, L. and Nürnberg, H. W. (1986). Cd, Pb, Cu, Ni and Co distribution in the German Bight. *Mar. Chem.*, **18**, 197–213.

McIntyre, A. D. (1985). ACMP activities. *Mar. Poll. Bull.*, **16**, 123–124.

Munn R. E. (1973). *Global Environmental Monitoring System (GEMS)*. Action Plan for Phase I. Consultant appointed by SCOPE to the UN Inter-Agency Working Group on Monitoring, Commission on Environmental Monitoring and Assessment, SCOPE Report 3, Toronto, Canada, 67 pp.

Nicholson M. D. (1989). Analytical results: how accurate are they? How accurate should they be? *Mar. Pollut. Bull.*, **20**, 33–40.

Nicholson M. D. and Fryer R. J. (1992). The statistical power of monitoring programmes. *Mar. Pollut. Bull.*, **24**, 146–149.

Nolting R. F. and de Baar H. J. W. (1994). Behaviour of nickel, copper, zinc and cadmium in the upper 300m of a transect in the Southern Ocean (57 degrees–62 degrees S, 49 degrees W). *Mar. Chem.*, **45**, 225–242.

Nolting R. F., Heijne M., de Jong J. T. M. *et al*. (2000). The determination and distribution of Zn in surface water samples collected in the northeast Atlantic Ocean. *J. Environ. Monit.*, **2**, 534–538.

NSTF (1993). *North Sea Quality Status Report 1993*, NSTF, OSPARCOM and ICES, Olsen and Olsen, Fredensborg, 132 pp.

O'Connor T. P., Cantillo A. Y. and Lauenstein G. G. (1994). Monitoring of temporal trends in chemical contamination by NOAA National Status and Trends Mussel Watch Project, in *Biomonitoring of Coastal Waters and Estuaries* (ed. K. J. M. Kramer), CRC Press, Boca Raton, FL, pp. 29–50.

OSPAR (1984). *The First Decade: International Cooperation in Protecting our Marine Environment*, Oslo and Paris Commissions, London.

OSPAR (2005). *Agreement on background concentrations for contaminants in seawater, biota and sediment*, OSPAR Agreement 2005–6, OSPAR Commission, London, 4 pp.

OSPAR (2009). *Trends in Waterborne Inputs. Assessment of Riverine Inputs and Direct Discharges of Nutrients and Selected Hazardous Substances to OSPAR Maritime Area in 1990–2006*, OSPAR Commission, London, 114 pp.

OSPAR (2010). *Quality Status Report 2010*, OSPAR Commission, London, 176 pp.

OSPAR Commission (2000). *Quality Status Report 2000*, OSPAR Commission, London, 108 pp.

Pearce J. B. (1998). A short history of marine environmental monitoring. *Mar. Pollut. Bull.*, **37**, 1–2.

Portman J. (1986). The development of the Joint Monitoring Programme of the Oslo and Paris Commissions, in *Baltic Sea Monitoring Symposium*, Baltic Sea Environment Proceedings BSEP19, HELCOM, Helsinki, pp. 317–340.

QSR (1987). *Quality Status Report of the North Sea* 1987. *Summary*, Department of the Environment, 10/87 (Dd. 8032340), HMSO, London, 25 pp.

Quevauviller Ph. (2003). BCR, turning over a new leaf. *Trends Anal. Chem.*, **22**, 187–190.

Reid Ph. C. (1990). The work of the North Sea Task Force, in *The North Sea: Perspectives on Regional Environmental Co-operation* (eds D. Freestone and T. IJlstra), Graham and Trotman, London, pp. 80–88.

Rijkswaterstaat (2010). *Waterbase, on-line DONAR database* (http://www.live.waterbase.nl/waterbase wns.cfm?taal=en).

Riley J. P. and Chester R. (1971). *Introduction to Marine Chemistry*, Academic Press, London, 465 pp.

Rozwadowski, H. M. (2002). *The Sea Knows No Boundaries. A Century of Marine Science under ICES*, University of Washington Press, Seattle, WA, 410 pp.

Scholten M. C. T., Kramer K. J. M. and Laane R. W. P. M. (1998). Trends and variation in concentration of dissolved metals (Cd, Cu, Pb, and Zn) in the North Sea (1980–1989). *ICES J. Mar. Sci.*, **55**, 825–834.

SCOPE (1971). *Global environmental monitoring*. A report submitted to the United Nations Conference on the Human Environment, Stockholm 1972 by the Commission on Monitoring of the Scientific Committee on Problems of the Environment (SCOPE) of the International Council of Scientific Unions (ICSU) International Council. SCOPE 1, Stockholm, 67 pp.

Skjærseth J. B. (2006). Protecting the North-East Atlantic: enhancing synergies by institutional interplay. *Mar. Policy*, **30**, 157–166.

Thompson M., Ellison S. L. R. and Wood R. (2006). The International Harmonized Protocol for the proficiency testing of analytical chemistry laboratories (IUPAC Technical Report). *Pure Appl. Chem.*, **78**, 145–196.

Topping G. (1992). The role and application of quality assurance in marine environmental protection. *Mar. Pollut. Bull.*, **25**, 61–66.

Topping G. (1998). Introduction to the special issue of Marine Pollution. *Mar. Pollut. Bull.*, **35**, 1–2.

Tromp D. and Wieriks K. (1994). The OSPAR convention: 25 years of North Sea protection. *Mar. Pollut. Bull.*, **29**, 622–626.

Tueros I., Rodríguez J. G., Borja A. *et al*. (2008). Dissolved metal background levels in marine waters, for the assessment of the physico-chemical status, within the European Water Framework Directive. *Sci. Total Environ.*, **407**, 40–52.

Turekian K. K. (1968). *Oceans*, Prentice-Hall, Englewood Cliffs, NJ, 120 pp.

UNEP (1972). *Declaration of the United Nations Conference on the Human Environment*, Stockholm. Conference, 5–16 June 1972.

UNEP (1984). *Prospects for global ocean pollution monitoring*, UNEP Regional Seas Reports and Studies No. 47, 55 pp.

Van der Eijk W. (1977). Standards. The Community Bureau of Reference (BCR). *Proc. Anal. Div. Chem. Soc.*, **3**, 53–54.

Vijverberg F. A. J. M. and Cofino W. P. (1987). *Control procedures: good laboratory practice and quality assurance*, ICES Techniques in Marine Environmental Sciences, TIMES-06, ICES, Copenhagen, 33 pp.

Wagner G. (1995). Basic approaches and methods for quality assurance and quality control in sample collection and storage for environmental monitoring. *Sci. Total Environ.*, **176**, 63–71.

Waldichuk M., Jamieson W. D. and Berman S. S. (1987). Reference materials as assurance of quality in analysis of seawater, marine sediments, and biological tissues. *Mar. Pollut. Bull.*, **18**, 477–481.

Wells D. E. (1998). Development of reference and test materials for organic contaminants in water. *Analyst*, **123**, 983–989.

Wells D. E., Cofino W. P., Quevauviller Ph. and Griepink B. (1993). Quality assurance of information in marine monitoring – a holistic approach. *Mar. Pollut. Bull.*, **26**, 368–375.

Wells D. E., Aminot A., de Boer J. *et al*. (1997). A review of the achievements of the EU project 'QUASIMEME' 1993–1996. *Mar. Pollut. Bull.*, **35**, 3–17.

Went A. E. J. (1972). *Seventy years agrowing. A history of the International Council for the Exploration of the Seas 1902–1972*, Rapports et Proces-Verbaux des Reunions du Conseil International pour l'Exploration de la Mer, 165 pp.

Windom H. L. (1991). *GESAMP: Two Decades of Accomplishments*, IMO, London, 52 pp.

Yeats P. A. (1987). *Trace metals in sea water; sampling and storage methods*, ICES Techniques in Marine Environmental Sciences, TIMES-02, ICES, Copenhagen, 8 pp.

Part 2

Policy Settings

2

International Conventions

Patrick Roose

Royal Belgian Institute of Natural Sciences, Management Unit of the North Sea Mathematical Models and the Scheldt estuary (MUMM), Oostende, Belgium

2.1 INTRODUCTION

Although international collaboration on marine chemistry followed in the wake of the foundation of the International Council of the Exploration of the Seas (see Chapter 1) in 1902, it was the advent of an entirely new problem that resulted in today's international marine conventions. The oceans, covering 71% of the Earth's surface, have long seemed immense, inexhaustible and impervious

Chemical Marine Monitoring: Policy Framework and Analytical Trends, First Edition.
Edited by Philippe Quevauviller, Patrick Roose, and Gert Verreet.

to human influence. However, during the 1960, scientists became aware that the technological progress that boomed in the Western Hemisphere after the Second World War had contaminated our environment at a scale and level previously unseen. The growing economy, novel agricultural practices and global population boom caused increasing and widespread pollutant emissions. DDT proved to be responsible for the decline of bird of prey populations, such as the bald eagle in North America (Broley, 1958) and the white-tailed eagle in the Baltic (Bernes, 1998). Scientists demonstrated the ubiquitous presence of this and other contaminants, particularly in the higher trophic levels, in areas as far away as the Arctic. Other studies showed a greatly increased input of nutrients to coastal seas, causing eutrophication and leading to highly undesirable changes in ecosystem structure and function (Smith *et al*., 1999).

This general awareness, heightened by a number of incidents (Minimata incident, Japan 1961; Yusho incident, Japan 1968), urged for measures to reduce or eliminate the release of contaminants to the environment in general and the marine environment in particular. As a result, treaties were signed at the regional and global levels, committing countries to prevent and eliminate the pollution of the marine environment by human activities. The marine conventions were born.

The implementation of these policies had to be followed by observation and assessment of the evolving situation in the field. National and international workshops and study groups were convened to discuss the monitoring and assessment of micro-contaminants and to outline strategies for terrestrial and aquatic programmes (Pearce, 1998). Clearly, any policy regarding the marine environment should be based on a sound scientific basis. An assessment assembles this knowledge in a form useful for decision-making. It can tease out the relative significance of different ocean's problems and their causes, and it can analyse response measures showing what has worked and the likely consequences of various options for future action (UNEP and IOC–UNESCO, 2009). Clearly, evaluating the policy against its achievements and shortcomings through an assessment process allows adjusting it as necessary. Ideally, such a cycle must be continuously repeated. As such, monitoring and assessment are fundamental to a sound policy; they must be correctly designed from the start and rigorously implemented. Yet, at the same time there should be a feedback mechanism that allows modifying an existing monitoring programme if new scientific information becomes available.

Today, we know that marine ecosystems are experiencing unprecedented environmental change, driven by human activities. Pressures such as pollution from land-based and sea-based sources, but also from fishing, marine debris, the loss and degradation of valuable habitat and invasions by non-native species are recognized worldwide. However, the initial conventions, and hence research and monitoring, were heavily focused on measurements related to the two major problems being, at that time, eutrophication and contamination. Furthermore, frequent sampling at sea also became a source of hydrological data in a broader sense, as contaminant data is generally supported by metadata such as salinity, temperature, dissolved oxygen and others. This has been going on since the late 1960s and early 1970s. Not surprisingly, marine chemical monitoring is presently one of the technically most advanced branches of marine environmental monitoring. Gradually, the focus shifted from pollution to a more holistic approach, often referred to as the ecosystem approach. Chemical monitoring is nowadays only one aspect, albeit an important one, of marine environmental monitoring. Yet, new threats such as climate change and ocean acidification will certainly give a new emphasis on chemical measurements.

Marine monitoring and research have been ongoing for several decades and have become the basic tools for understanding what is happening in the oceans, why and how effective response measures have been. Regular assessments have become an integral part of adaptive management that can respond to changing conditions. This chapter aims to give an overview of the existing internal organisations and conventions that are at the heart of this process.

2.2 THE GLOBAL LEVEL

2.2.1 UN Convention on the Law of the Sea

At the global level, the UN Convention on the Law of the Sea (UNCLOS) provides a legal framework and basic principles for the management of the oceans (UNEP and IOC–UNESCO, 2009). Ocean issues are considered in a comprehensive manner in the United Nations General Assembly (UNGA) and its processes. The international rules and standards that implement UNCLOS provisions are further developed at specialized global organizations like the Food and Agriculture Organization (FAO) and International Maritime Organization (IMO). The instruments can be both conventions (e.g. MARPOL 73/78, see Chapter 1) and normative instruments (e.g. Code of Conduct for Responsible Fisheries). A large number of multilateral environmental agreements also apply to the oceans, covering themes like climate change, hazardous substances, biodiversity and protection of species and habitats.

UNCLOS also provides the framework for regional seas collaboration. The regional level is appropriate for responding to the many problems that occur at larger than national scales. Regional organizations can bring together coastal states adjacent to the same oceans and seas, and sometimes also other states that use the areas. The most important regional seas conventions and organizations will be discussed later, but one should realize that in some oceans and seas there are no strong instruments or collaboration. Alternatively, in areas such as North America, bilateral cooperation can be appropriate.

2.2.2 United Nations Environment Programme

Monitoring activities on a global scale are, as was demonstrated above, inevitably linked to the United Nations Environment Programme (UNEP). UNEP was established as a follow-up to the 1972 Stockholm Conference on the Human Environment, as the environmental conscience of the UN system (see http://www.unep.org). With UNEP, a basis was created for comprehensive consideration and coordinated action within the UN on the problems of the human environment. UNEP particularly attempts to nurture partnerships with both other UN bodies and, for example, the scientific community and non-governmental organizations such as OSPAR.

UNEP has several water-related programmes. For the marine environment, the *Regional Seas Programme* is of particular importance. Initiated in 1974, it provides a legal, administrative, substantive and financial framework for the implementation of Agenda 21 (UN, 1992), the Plan of Implementation of the World Summit on Sustainable Development (UN, 2002) and for the Bali Strategic Plan (UNEP, 2004). It is an action-oriented programme and focuses not only on the mitigation or elimination of the consequences, but also on the causes of environmental degradation (http://www.unep.ch/regionalseas). The action plans are usually adopted by high-level intergovernmental meetings and implemented, in most cases, within the framework of a legally binding Regional Seas Convention and its specific protocols, under the authority of the respective contracting parties (e.g. MEDPOL; see further). The focus has gradually shifted from protecting the marine environment from pollution to striving at sustainable development of the coastal and marine environment through integrated management. Currently, 17 members of the regional seas family are reflected in the Assessment of Assessments (AoA) regions (see below). Altogether, more than 140 countries participate in at least one Regional Seas Action Plan or convention. In 12 of the regions, states have also adopted a legally binding convention.

The new Regional Seas strategy encourages the Regional Seas programmes to increase monitoring and assessment activities, and to facilitate a science-based decision-making system, including participation in such processes of the UN General Assembly known as the Global Assessment of the State of the Marine Environment (GMA). This concept (i.e. establishing regular marine environment assessments) was initiated in 1999 at the seventh session of the Commission on Sustainable Development. It has led to the AoA, which is described in greater detail in the next section. UNEP is one of the lead agencies responsible for taking forward the AoA through the implementation of UN General Assembly Resolution 60/30. It is worthwhile mentioning that the AoA builds heavily on earlier UNEP-supported actions in this field, namely the UNEP (2003, 2007) 'Global Marine Assessments'.

UNEP is also responsible for the secretariat set up to implement the 1995 *Global Programme of Action* for the Protection of the Marine Environment from Land-based Activities, and, one of the implementing agencies for the *Global Environment Facility* (GEF). This is an independent international financing entity with the long-term goal to ensure progress toward global environmental security. The UNEP portfolio of GEF-funded activities in international waters includes global assessments, transboundary diagnostic analyses of shared water bodies, support for the implementation of strategic action programmes for marine and freshwater areas, and support for integrated management of shared freshwater bodies. Because water issues play an important and increasing role in international development cooperation, GEF has designated international waters as one of its four focal areas.

The *Global International Waters Assessment* (GIWA), led by UNEP and funded for 50% by GEF, will provide the information needed for GEF's work in this particular area. The aim of GIWA is to produce a comprehensive and integrated global assessment of international waters, the ecological status of (and the causes of environmental problems in) 66 water areas around the world and to focus on the key issues and problems facing the aquatic environment in transboundary waters. The assessment is designed not merely to analyse the current problems, but also to develop scenarios of the future conditions of the world's water resources and analyse policy options with a view to providing sound scientific advice to decision-makers and managers concerned with water resources. In the near future, GIWA activities will be linked and coordinated with the monitoring programmes described elsewhere in this section, such as, for example, OSPAR and HELCOM.

UNEP should essentially be seen as a facilitator and does not play the same role as organizations such as OSPAR and HELCOM (see further). UNEP's role is creating the conditions that make marine monitoring feasible through capacity building projects, technical and scientific advice (e.g. in the form of technical guidelines) and by bringing organizations together that have common goals, facilitating exchange of practices and information.

2.2.3 The Assessment of Assessments

During 2001–2002, the feasibility of establishing a regular global process for assessing the marine environment was explored leading to the 2002 World Summit on Sustainable Development in Johannesburg. There, actions at all levels were supported to establish by 2004 a Regular Process under the United Nations for global reporting and assessment of the state of the marine environment, including socio-economic aspects, both current and foreseeable, building on existing regional assessments. This was endorsed at the UNGA later in 2002 (Resolution 57/141).

The General Assembly Resolution launched further preparatory work by UN bodies, member states and international organizations in 2003–2005. In November 2005, UNGA launched the AoA as a preparatory stage towards the establishment of the Regular Process (Resolution 60/30) and an Ad Hoc Steering Group (AHSG) was established to oversee implementation of the AoA (UNEP, 2007). The actual work was undertaken by a Group of Experts and supported from a secretariat in the two lead agencies, UNEP and the Intergovernmental Oceanographic Commission (IOC) of UNESCO (see further).

The mandate of the AoA was elaborated by the AHSG at its first meeting in 2006 (UNEP and IOC–UNESCO, 2009). In essence, the mandate was to prepare a report for the General Assembly on its work to:

1. Assemble information about marine assessments relevant to the Regular Process, carried out under the purview of UN bodies, global treaty organizations, regional organizations, national governments and other relevant organizations where appropriate.
2. Undertake a critical appraisal of the assessments in order to evaluate their scientific credibility, policy relevance, legitimacy and usefulness. The appraisal should in particular identify: best practices and approaches; thematic and geographic assessment gaps and needs; uncertainties in scientific knowledge, data gaps and research needs; and networking and capacity-building needs in developing countries and countries with economies in transition.
3. Identify a framework and options to build the Regular Process, including potential costs, based upon current relevant assessment processes and practices.

It is important to mention that the AoA was not to produce any new assessments and that the appraisal should essentially be science based and subject to review by experts and governments.

The AoA was published in 2009 and represents the most comprehensive initiative undertaken to date by the UN system to better coordinate ocean governance. Its central recommendation calls for a mechanism that builds on existing global, regional and national institutions and processes while integrating all available information, including socio-economic data, on how our seas and oceans are actually being used. It should be seen as a start-up phase towards a Regular Process for global reporting and assessment of the state of the marine environment.

The AoA instigated the development of the global and regional assessments of the marine environment database (GRAMED). GRAMED is a dynamic online, fully searchable tool which provides access to the largest existing collection of information on assessments, scientific research studies and data holdings of relevance to the marine and coastal environment at the national, regional and supra-regional scale. To date, it is the only database of its kind (UNEP and IOC–UNESCO, 2009). Within the AoA process, the database has been used as an online mechanism to consolidate information on assessments and other activities and to facilitate access to this information by the Group of Experts. GRAMED is expected to play a useful role in the future Regular Process.

2.2.4 UNESCO–IOC

Established in 1960, the IOC of UNESCO (UNESCO–IOC) promotes international cooperation and coordinates programmes in marine research, services, observation systems, hazard mitigation and capacity development in order to learn more and better manage the nature and resources of the ocean and coastal areas. Through the application of this knowledge the Commission aims to improve management practices and the decision-making process of its member states, foster sustainable development and protect the marine environment. In addition, the Commission strives to further develop ocean governance, which necessitates strengthening the institutional capacity of member states in marine scientific research and of ocean management. (www.ioc-unesco.org; also see Chapter 1). In short, it is the designated UN entity for coordinating global ocean sciences.

The IOC will collaborate with international organizations concerned with the work of the Commission, and especially with those organizations of the United Nations (UN) system that are willing and prepared to contribute to the purpose and functions of the Commission and/or to seek advice and cooperation in the field of ocean and coastal area scientific research, related services and capacity-building. The collaboration with UNEP, mentioned above, in establishing a process for global reporting and assessment of the state of the marine environment is a good example of this.

Furthermore, the IOC manages the Global Ocean Observing System (GOOS), the ocean component of the Global Climate Observing System (GCOS). As such, it helps to improve operational oceanography, weather and climate forecasts and monitoring and supports the sustained observing needs of the UN Framework Convention on Climate Change (UNFCCC). The vision guiding the development of a GOOS is one of a world served by a unified global network providing the information needed by governments, industry, science and the public to deal with marine-related issues, including environmental issues and the influence of the ocean upon climate.

The International Oceanographic Data and Information Exchange (IODE) programme enhances the IOC marine research and management programmes by facilitating the exploitation, development and exchange of oceanographic data and information between participating member states and by meeting the needs of users for data and information products.

IOC capacity development seeks to strengthen scientific, legal and institutional structures through leadership, bid-writing and team building in marine and coastal science institutes and harnesses the drive of directors of marine and coastal science institutes to break the spiral of nonsustainable dependence on external donors. IOC partners with member states to develop the capacities of their national institutes to conduct the science and observations needed to underpin informed decision-making for sustainable use of the ocean and coastal seas.

The IOC aids and advises policy makers and managers in the reduction of risks to coastal communities, their infrastructure and service-providing ecosystems from tsunamis, storm surges and other coastal hazards by focusing on implementing adaptation measures to strengthen the resilience of vulnerable.

2.2.5 GESAMP

The *Joint Group of Experts on the Scientific Aspects of Marine Environmental Protection* (GESAMP) is an advisory body, established in 1969, that advises the UN system on the scientific aspects of marine environmental protection (www.gesamp.org; see Chapter 1).

Sponsored by eight UN organizations with responsibilities relating to the marine environment, GESAMP is utilized as a mechanism for coordination and collaboration. GESAMP conducts and supports marine environmental assessments, undertakes in-depth studies, analyses and reviews of specific topics, and attempts to identify emerging issues regarding the state of the marine environment. Today, GESAMP consists of 16 experts, drawn from a wide range of relevant disciplines, who act in an independent and individual capacity. Studies and assessments are usually carried out by dedicated working groups, most of whose members are not sitting members of GESAMP but part of the broader GESAMP network. Examples of GESAMP studies related to the marine environment are: 'A sea of troubles' (GESAMP, 2001a); 'Protecting the oceans from land-based activities – land-based sources and activities affecting the quality and uses of the marine, coastal and associated freshwater environment' (GESAMP, 2001b) and 'Global Marine Assessments: a survey of global and regional assessments and related activities of the marine environment' (GESAMP, 2009), a study written in support of the AoA above.

Recently, the group underwent a revitalization process following an independent, in-depth review of its structure and functioning (GESAMP, 2001c). The goals set forth in this new structure are:

- increasing the number of experts from developing countries participating in GESAMP activities;
- extending and consolidating GESAMP's networks at the regional and global level;
- supporting GESAMP's participation in the UNGA Regular Process.

As such, the role and goals of GESAMP are similar to that of ICES (see further); that is, providing sound scientific advice, be it that its focus is on the global scale.

2.2.6 ICES

ICES, established in 1902, claims to be the oldest intergovernmental organization in the world concerned with marine and fisheries science (see Chapter 1). ICES is a leading, multidisciplinary, scientific forum for the exchange of information on the sea and its living resources, and for the promotion and coordination of marine research by scientists in its member countries. It is supra-regional in the sense that its activities pertain to the North Atlantic (East and West), the Baltic Sea and the North Sea.

Since the 1970s, a major area of ICES work has been to provide information and nonbiased, nonpolitical scientific advice to member country governments (BE, CA, DE, DK, EE, ES, FI, FR, IE, IS, LV, NL, NO, PL, PT, RU, SE, UK and the US)[1] and international regulatory commissions on the protection of the marine environment and for fisheries conservation. In support of these activities, the ICES Secretariat in Denmark maintains three databanks: the oceanographic databank, the fisheries databank and the environmental (marine contaminants) databank. ICES operates as a major data repository and includes a substantial amount of the information collected through the OSPAR and HELCOM system.

The Council's scientific work is accomplished by various committees and working groups. Recently, the structure and organization was altered to become more in line with the two pillars of ICES work. Nowadays, an Advisory Committee (ACOM) provides advice to clients on fisheries and marine ecosystem issues, and a Science Committee (SCICOM) oversees all aspects of the scientific work. Both can rely on the more than 100 expert groups covering most aspects of the marine ecosystem that work under them.

ICES advice and studies have been very influential in the development of monitoring programmes such as those of OSPAR and HELCOM (see further) and those of the USA and Canada. More recently, ICES also became involved in European initiatives such as the Marine Strategy Framework Directive (see Chapter 3).

2.3 THE REGIONAL SCALE

2.3.1 Introduction

Pollution of our seas and oceans is never a matter that can be adequately dealt with at the level of an individual country. Environmental protection is best handled at a regional level and has proven to be a profoundly unifying issue (UNEP and IOC–UNESCO, 2009). Hence, all over the world regional organizations were founded to implement joint strategies protecting the marine environment. For the purpose of this chapter, marine regions are those indentified by the UNEP Regional Seas and in the AoA (UNEP and IOC–UNESCO, 2009). An overview of the major marine regions is graphically illustrated in Figure 2.1. A more extensive description can be found in the AoA (UNEP and IOC–UNESCO, 2009). For most of these regions, agreements were made between the parties involved and environmental monitoring was initiated in support of these policies. An overview of prominent regional organizations is given Table 2.1.

Not surprisingly, the extent and coverage of this monitoring has reached the highest level in the industrialized world. Table 2.2 gives an idea about the extent of that information, which is

[1] ISO 3166 code for countries.

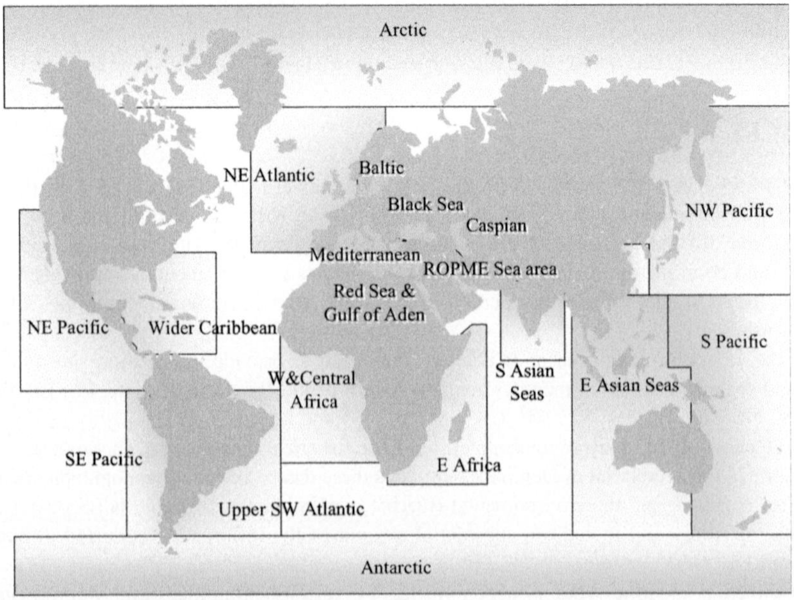

Figure 2.1 UNEP regional seas (from Roose (2010)).

Table 2.1 Overview of the major regional commissions or organizations and their web site.

Regional commission or organization	Web site
Permanent Commission for the South Pacific (CPPS)	http://www.cpps-int.org
South Asia Co-operative Environment Programme (SACEP)	http://www.sacep.org
South Pacific Regional Environment Programme (SPREP)	http://www.sprep.org
Convention on the Conservation of Antarctic Marine Living Resources (CCAMLR)	http://www.ccamlr.org
Caribbean Environment Programme (CEP)	http://www.cep.unep.org
Protection of the Arctic Marine Environment (PAME)	http://www.pame.is
Baltic Marine Environment Protection Commission (Helsinki Commission or HELCOM)	http://www.helcom.fi
Commission for the Protection of the Black Sea Against Pollution	http://www.blacksea-commission.org
Caspian Environment Programme (CEP)	http://www.caspianenvironment.org
East African Coastal Database	http://www.unep.org/eafatlas
Mediterranean Action Plan (MAP)	http://www.unepmap.org
OSPAR Commission (OSPAR)	http://www.ospar.org
The Regional Organization for the Conservation of the Environment of the Red Sea and Gulf of Aden (PERSGA)	http://www.persga.org
Regional Organization for the Protection of Marine Environment (ROPME)	http://www.ropme.net

almost entirely the result of chemical measurements: water quality. Water quality attributes include: physical conditions (sediments, temperature, suspended solids, etc.); pH and related constituents such as CO_2; pathogens; oxygen-consuming substances; toxic substances and substances that accumulate in food chains; nutrients, etc. Clearly, for only about a third of the regions, there are time series and the data are regularly analysed in periodic assessments. For 20% of the regions there are time series on several parameters and some assessments are available. However, for half of the regions the time series of water quality data are only available for a portion of the region and covering only some important attributes. An extensive discussion of all the regions in Table 2.2 would be beyond the scope of this chapter. For this, the reader is referred on the excellent overview in the Assessment of Assessments (UNEP and IOC-UNESCO, 2009). Nevertheless, for the purpose of this book, the regions with the best coverage of chemical parameters – the top 30% in Table 2.2 – are briefly discussed below, in order to give an idea about how and what data is being collected and the way it is assessed.

Table 2.2 Extent of the assessment information available on ecosystem status and trends for water quality attributes[a].

Region	Available information[b]
Southern Ocean	+
Arctic Ocean	+++
Baltic Sea	+++
Black Sea	+
East Asian Seas	++
Eastern African Seas	+
Mediterranean Sea	+++
North Central Pacific Ocean	+
North East Atlantic Ocean	+++
North East Pacific Ocean	+++
North West Atlantic Ocean	+++
North West Pacific Ocean	++
Red Sea and Gulf of Aden	+
ROPME/RECOFI Area	++
South Asian Seas	+
South East Pacific Ocean	+
Southern Indian Ocean	+
Southern Pacific Ocean	+
South West Atlantic Ocean	+
Western African Seas	+
Wider Caribbean Region	++

[a]Water quality attributes include physical conditions (sediments, temperature, suspended solids etc); pH and related constituents such as CO_2; pathogens; oxygen-consuming substances; toxic substances and substances that accumulate in food chains; nutrients, etc.

[b]+++: multiple attributes (see footnote *a*) are regularly analyzed in periodic assessments; ++: time series on multiple attributes (see footnote *a*) and some analyses are available; +: some time series of water quality data are available for a substantial portion of the region covering some important attributes.

2.3.2 Arctic Ocean: AMAP

For the Arctic Ocean, data is gathered and assessed through the Arctic Monitoring and Assessment Programme (AMAP). AMAP was established in 1991 to implement certain parts of the Arctic Environmental Protection Strategy, primarily 'providing reliable and sufficient information on the status of, and threats to, the Arctic environment, and providing scientific advice on actions to be taken in order to support Arctic governments in their efforts to take remedial and preventive actions relating to contaminants' (see http://www.amap.no). The Arctic Council, established in 1996 by the eight Arctic countries (CA, DK, FI, IS, NO, RU, SE and the US), coordinates AMAP activities. AMAP was conceived as a programme which integrates both monitoring and assessment activities in relation to pollution issues and provides information and reports on the state of the Arctic environment. The AMAP Trends and Effects Monitoring Programme (AMAP, 2000) is designed to monitor the levels of pollutants and their effects in all compartments of the Arctic environment.

There are five sub-programmes. These deal with atmospheric, terrestrial, freshwater and marine environments and with human populations with respect to human health. The sub-programmes are defined in terms of essential and recommended parameters (see above) and media (matrices) to be monitored on a circumpolar or sub-regional level. The programme includes both monitoring and research components, and special studies that yield information, which is vital for the valid interpretation of monitoring data.

AMAP has produced comprehensive and scientifically sound assessments of contaminants in the Arctic environment, discussing their levels, trends and effects (both based on comparison with current literature and the observation of biological effects as such) (AMAP, 2004). An AMAP assessment, therefore, should not be considered as a formal environmental risk assessment. Rather, it constitutes a compilation of current knowledge. The emphasis is very much on science as the basis for relevant policy and the programme has no regulatory purpose as such. The set-up, therefore, is much like a scientific programme, and there seem to be no legally binding obligations or a rather strict set of guidelines, such as, for instance, for the OSPAR CEMP (see further). Specific for this programme is the assessment of human health through the study of dietary intake and body burdens (AMAP, 2009).

2.3.3 Baltic Sea: HELCOM

International cooperation in the study of the Baltic Sea environment can be traced back to the establishment of ICES in 1902, mentioned earlier. The deteriorating marine environment mandated the development of a specific convention for the Baltic Sea. Currently, all sources of pollution, as well as monitoring and assessment activities around the Baltic, are subject to a single convention, the Convention on the Protection of the Marine environment of the Baltic Sea Area, signed in 1992 (see www.helcom.fi and Chapter 1).

The *Baltic Marine Environment Protection Commission or the Helsinki Commission* (HELCOM) is its governing body. HELCOM's main goal is to protect the marine environment of the Baltic Sea from all sources of pollution and to restore and safeguard its ecological balance. The present contracting parties to HELCOM are DE, DK, EE, EC, FI, LV, LT, PL, RU and SE. The set-up is very similar to that of OSPAR (see below), and many of the principles (such as the 'best environmental practices', 'best available technologies' and 'the polluter pays') are adopted and applied by HELCOM. Monitoring and assessment are an integral part of the convention and, according to the convention, 'Emissions from both point sources and diffuse sources into water and the air should be measured and calculated in a scientifically appropriate manner by the Contracting Parties'. Every 5 years, the Commission publishes a 'Periodic Assessment of the State of the Environment of the Baltic Marine Area' based on monitoring activities going on in the area.

HELCOM has a well-developed and coordinated monitoring programme that shows large similarities to that of OSPAR. The monitoring data include hydrographical and hydrochemical, contamination of water, sediments, fish and benthic animals, as well as biological data. As for OSPAR, the data are stored in the ICES database.

HELCOM's Monitoring and Assessment Group (HELCOM MONAS) looks after one of its key tasks by assessing trends in threats to the marine environment, their impacts, the resulting state of the marine environment and the effectiveness of measures adopted. This forms the basis for the work of HELCOM's other main groups and helps to define the need for additional measures. Both HELCOM and OSPAR have relied on, often the same, scientific advice from ICES (see above). The result is a set of instructions, guidelines and common procedures (HELCOM Combine manual) that deals with sampling, sample handling, analysis, quality assurance, data handling, data reporting and interpretation. This has resulted in high-quality assessments focused, for contaminants, on their impact and trends. Specific for HELCOM is the development of Indicator Fact Sheets that provide information on the recent state of and trends in the Baltic marine environment.

HELCOM has also developed Ecological Quality Objectives (EcoQOs), associated indicators and target levels of these indicators as central tools in identifying, striving towards and achieving a healthy ecosystem; that is, the implementation of the ecosystem approach. Finally, although contracting parties are required to meet their HELCOM obligations, there are no legal sanctions if this is not the case. This often results in gaps in the datasets, weakening the assessments.

2.3.4 Mediterranean Sea: Barcelona Convention

In 1976, the Mediterranean countries (AL, DZ, BA, HR, CY, EG, ES, FR, GR, IL, IT, LB, LY, MT, MC, MA, SI, SY, TN, TR) and the EU adopted the Barcelona Convention for the Protection of the Mediterranean Sea Against Pollution, overarching the Mediterranean Action Plan (MAP), approved the year before (see http://www.unepmap.org/). Being a framework convention, it envisages the preparation of technical protocols, like the ones for the protection against pollution from land-based sources, namely the LBS Protocol (Barcelona Convention, 1980), and from hazardous wastes disposal (Barcelona Convention, 1996). As the environmental assessment component of MAP and the associated protocols, the Programme for the Assessment and Control of Pollution in the Mediterranean region (MEDPOL) was established. When started in 1975 its main aim was the establishment of a network of institutions undertaking marine pollution work and the collection of information regarding the level of pollution in the Mediterranean Sea. The monitoring activities covered heavy metals in marine biota (mainly mercury and cadmium), halogenated hydrocarbons in marine biota (mainly PCBs and DDTs), and petroleum hydrocarbons in seawater. The development and maintenance of these national monitoring programmes was the aim of the second phase (1981), whereas more recently (1996) the emphasis shifted from pollution assessment to pollution control. In parallel, MEDPOL provides assistance in the formulation and implementation of regional and national action plans addressing pollution from land-based sources and activities. It also formulates and carries out capacity-building programmes related to the analysis of contaminants and treatment of data and to technical and management training.

Although monitoring of the Mediterranean environment is firmly in place for several countries, a coordinated and well-developed monitoring programme has not been realized, as is the case for HELCOM and OSPAR. Emphasis has been mainly on capacity building and setting up the conditions that will eventually result in a sustainable monitoring programme for the Mediterranean. For instance, guidelines have been developed covering various aspects of monitoring and efforts were made to enhance the quality of the processes. Nevertheless, according to the AoA (UNEP and IOC–UNESCO, 2009), the current network of participating scientific and professional institutions and individuals is very competent to carry out assessments. The available data and information were satisfactory for the preparation of a certain number of competent assessment documents. To continue and improve such a process, however, it would be necessary to allocate more funds and apply greater effort to systematically collect reliable data and information.

2.3.5 North East Atlantic Ocean: OSPAR

As for the Baltic, international cooperation in the study of the marine environment of the North East Atlantic region goes back to the establishment of ICES in 1902 (see Chapter 1). For many decades, the emphasis on this work was on oceanography and fisheries. In 1974, the 1972 Oslo Convention (also called the Convention for the Prevention of Marine Pollution by Dumping from Ships and Airplanes) entered into force. The convention regulated dumping operations involving industrial waste, dredged material and sewage sludge (North Sea Task Force, 1993). The Paris Convention (or Convention for the Prevention of Marine Pollution from Land-Based Sources) was established in 1974 and came into force in 1978. Its principal aim was to prevent, reduce and, if necessary, eliminate pollution of the convention area from land-based sources, which are discharges from rivers, pipelines and the coast, but also offshore installations and the atmosphere.

The tasks set forth in both conventions were originally handled by two individual commissions: the Oslo Commission and Paris Commission. In 1978, both commissions established a Joint Monitoring Programme (JMP), obliging contracting parties to initiate monitoring activities for a number of parameters in their water bodies.

In the 1980s, the policy of the Oslo and Paris Commissions evolved with the general evolution of environmental policy in western Europe, voiced at the Ministerial Conferences for the Protection of the North Sea (Roose and Brinkman, 2005). It was soon recognized that the existing Oslo and Paris Conventions did not adequately control some of the many sources of pollution and that a revision was warranted. This resulted not in a revision of the initial conventions, but, more importantly, in the merger of both commissions into a new convention, the Convention for the Protection of the Marine Environment of the North-East Atlantic or OSPAR (see http://www.ospar.org). The new convention was opened for signature at the Ministerial Meeting of the Oslo and Paris Commissions, in September 1992. The key objective of the strategy is the cessation of discharges, emissions and losses of hazardous substances by 2020 with the aim of achieving concentrations in the marine environment close to background values for naturally occurring substances and close to zero for man-made synthetic substances. To monitor environmental quality throughout the North-East Atlantic, a Joint Assessment and Monitoring Programme (JAMP) has been established, which has recently been revised (OSPAR, 2006).

As with HELCOM, OSPAR has, with the JAMP, a well-developed monitoring and assessment programme that shows large similarities to that of the former following ICES advice. The level of coordination is high, particularly for the Coordinated Environmental Monitoring Programme (CEMP), which can be seen as exemplary. The CEMP can be described as that part of monitoring within JAMP where the national contributions overlap and are coordinated. The CEMP has guidelines, quality assurance tools and assessment tools for all its mandatory parameters (which include biological effects techniques). Important here is that the three elements above have to be in place before monitoring is undertaken, and that sufficient indications have to be present that the chemical in question needs to monitored on a routine basis. ICES operates as the data repository for the majority of OSPAR data. OSPAR holds most of the remaining data it collects, with air pollution data being managed by a specialist centre that also services the European air-pollution convention. Data reporting is also coordinated and quality assured. All data are publicly available when its quality has been assured.

OSPAR has taken, with its assessment tools, the next step in data assessment by using novel statistical approaches to test against set reference values such as background assessment concentrations and environmental assessment criteria. OSPAR also has well-defined statistical tools for temporal trend analysis that take the quality of the data used into account. This has resulted in innovative and high-quality assessments that appear at regular intervals describing, for example, the spatial distribution and trends of contaminants in the OSPAR area. Like HELCOM, OSPAR has developed EcoQOs for the implementation of the ecosystem approach, and there are also no legal sanctions if contracted parties do not meet their obligations, resulting in gaps in the datasets.

2.3.6 North East Pacific Ocean

There are extensive chemical data series that cover the North East Pacific region's biophysical environment during much of the past 50 years, particularly off the coasts of the USA and Canada (Rosenberg, 2009). These data are, amongst others, the result of directed research and monitoring efforts of governments and academia that cover both ocean physics and chemistry. There is also a well-developed science programme for the region. However, there is no formal convention and accompanying monitoring programme covering the entire region, as is the case, for example, for the North East Atlantic.

For the northern part of the North East Pacific region, periodic monitoring and assessments are conducted by each of the accountable government institutions, including Fisheries and Oceans and Environment Canada (DFO), the USA National Oceanic and Atmospheric Administration (NOAA), the US Environmental Protection Agency (US-EPA) and their counterparts. A good example is NOAA's *National Status and Trends* (NST) monitoring programme that has been going on since 1986 and covers the Atlantic, Pacific and Gulf coasts of the USA. Canada, Mexico and the USA also have formal science advisory processes that conduct assessments and provide advice to the respective governments on policy and management (Rosenberg, 2009). For the remainder of the region, a UNEP Regional Seas Programme has been based on the Convention for Cooperation in the Protection and Sustainable Development of the Marine and Coastal Environment of the Northeast Pacific (Antigua/Guatemala Convention). This convention was signed in 2002 by CR, SV, GT, HN, NI and PA (Rosenberg, 2009).

Finally, although it does not play a formal advisory role to governments on policy or management questions, the North Pacific Marine Science Organization, or PICES, also conducts scientific assessments of ecological issues that are relevant to policy and management (www.pices.int). PICES is an intergovernmental scientific organization that was established in 1992 to promote and coordinate marine research in the northern North Pacific and adjacent seas. Its present members are CA, JP, CN, KR, RU and the US. In 2004, it synthesized the major assessments that cover several issues in the northern part of the North East Pacific region, many of which have a direct link to policy making by national and international decision-making bodies.

2.3.7 North West Atlantic Ocean

As the North West Atlantic Ocean region is only bordered by the shores of three countries (CA, DK (Greenland) and the US), no regional convention and accompanying monitoring and assessment programme have been set up. Nevertheless, there are extensive data series covering most of the past 50 years on the biophysical environment and there is a well-developed science programme for the region, including cooperative Canada–USA studies on various topics (Rice and Rosenberg, 2009). Very extensive data collection systems are operating in both countries, including research surveys operating year round and newly developed ocean observing systems which are coming on stream. As for the previous region, a substantial body of assessment work is carried by the responsible national institutions (DFO, NOAA, US-EPA) or in collaboration with regional organizations. This is particularly the case for the northern part of the region that is also covered by AMAP. There are also strong links with European efforts through ICES (see before). Still, despite its size, the region is mostly approached on a national level. It is not surprising, therefore, that there has been no overall synthesis of the information on the North West Atlantic Ocean region ecosystems (AoA). This does not mean that there are no major assessments for the ecosystems along the coast under the auspices of government agencies. For instance, there are Canadian series on the State of the Oceans or the United States Coastal Condition report. Synthesis studies such as the latter give a broader overview of the entire coast, but at much lower resolution. Integrated ecosystem assessments are in development in both countries.

2.4 CONCLUSIONS

Today, there is a large (and, to the outsider, sometimes bewildering) number of international organizations and national, international, regional and sometimes global conventions which all deal with the study (primarily the monitoring) of chemicals in the marine environment (Roose, 2010). Some of these studies or monitoring programmes have been running for many years or even decades, dealing, not surprisingly, mostly with the industrialized Northern Hemisphere. The output of these programmes has been used to identify areas or regions of concern, estimate the hazards caused by chemicals to man and the marine environment and assess the effectiveness of the measures taken. For instance, it has allowed obtaining an idea about the geographical distribution of chemicals in regions such as the Arctic (AMAP, 2004; Muir and de Wit, 2010), Baltic (HELCOM, 2002–2010), North Sea (North Sea Task Force, 1993; OSPAR, 2000a), North-East Atlantic (OSPAR, 2000b, 2010) and the Atlantic and Pacific coasts of North America (Sericano *et al*., 1995; Lauenstein and Daskalakis, 1998; EPA, 2005, 2008).

As a policy tool, these long-term monitoring programmes are expected and able to demonstrate changes in the levels of chemicals in the marine environment resulting from policy actions. For instance, monitoring in the Baltic clearly showed the decline of several organic pollutants following actions to prevent their release (Bignert *et al*., 1993, 1998). Similarly, assessment of the data collected through NST and the International Mussel Watch (IMW) programme, which covers Central and South America, including Mexico and the Caribbean (Sericano *et al*., 1995), showed a general decrease in the total DDT concentrations in molluscs for the northern part of the Gulf of Mexico, presumably as a result of the ban on DDT in the USA in 1972. More recently, in preparation of the Quality Status Report of the North-East Atlantic (OSPAR, 2010), a major temporal trend assessment was undertaken, covering a 10-year period (OSPAR, 2009). The study shows still a significant number of downward trends (e.g. PCBs in biota and sediment, PAHs in biota, Pb in biota and Hg in sediment), but also a number of upward trends (Cd in sediment of the southern UK coast, Hg in biota of the southern Norwegian coast) and, more importantly, a large number of trend series that are not consistently going downward or upward (e.g. Pb in sediment, PAHs in sediment). This is illustrated in Figure 2.2. Furthermore, these trend series were used to assess the status of the environment by statistically comparing the fitted values of the last year of monitoring against assessment criteria, allowing a more solid evaluation (Webster *et al*., 2009).

Despite this, the situation is far from ideal. Collecting and reporting of the data are often incomplete, which hampers the assessment process, and this particular aspect deserves special attention at all the levels involved. Clearly, it is one thing to conceive a monitoring programme, it is quite another to implement it. This requires time and effort of all parties involved and one should not forget that monitoring of the marine environment is an expensive business. This is true already because of the large cost of sampling at sea, but much more so because of the time required in coming to meaningful conclusions. Calculations from the OSPAR CEMP dataset indicate that, in order to detect an annual trend of 10% (a rather high figure in real life), a 10-year sampling

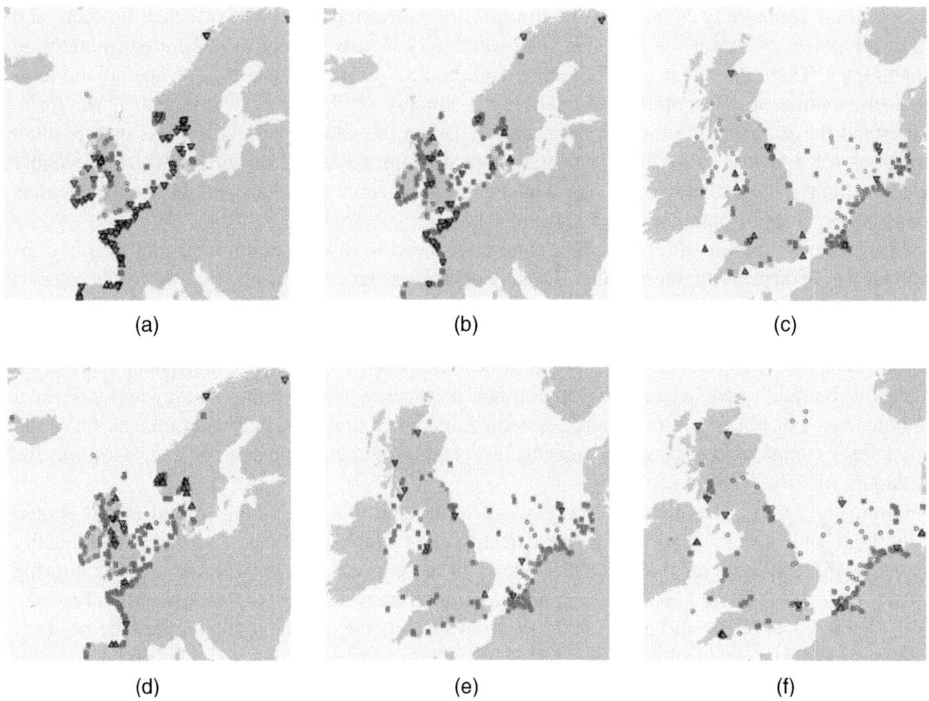

Figure 2.2 OSPAR temporal trend assessment for the period 1998–2007. Significant trends are indicated by the triangles ∇ (downward) and Δ (upward). Colour indicates status in last year of monitoring. Open circles indicate where insufficient data are available to assess trends (OSPAR, 2009). (a) PCBs in biota; (b) Pb in biota; (c) Cd in sediment; (d) Hg in biota; (e) Pb in sediment; (f) PAHs in sediment. See plate section for a colour version of this image.

programme is required (Nicholson *et al.*, 2001). The considerable number of series for which no trend was observed during the CEMP assessment above hint that this may not suffice. It could point towards a stabilizing situation in the field (i.e. inputs of contaminants are balanced by their removal) or, more importantly, indicate that the power of the trend series is insufficient to predict trends. To hamper matters further, there is often a considerable time lag between the inception of a programme and its execution. The CEMP was conceived in the early 1990s as the result of a discussion that took several years even though it was based on an earlier programme that ran for more than a decade. An inception phase of a decade is, unfortunately, not unusual. The HELCOM convention was signed in 1974, but came into force only in 1980. Monitoring was then started immediately, but it required another 7 years before, in 1987, the first assessment of the 5-year period 1980–1985 could be made – 13 years after the signing of the convention. If regular assessments based on monitoring are to be an integral part of an adaptive management that can respond to changing conditions, this type of time lag is far from ideal. Furthermore, modifying an existing monitoring programme if new scientific information becomes available is often hampered in a similar fashion, as it can take many years to add or remove parameters from the programme.

A further fundamental difficulty for an efficient interface between research and policy arises from the fact that research and policy have different and varying agendas (Roose *et al.*, 2011). Whereas policy tends to focus on the short-term perspective, science envisages a long-term perspective. Moreover, while policy tries to involve the development of acceptable compromises, the scientific community aims to work towards the collection of objective scientific facts and the development of reasonable theories. This ambiguity is also present in the current monitoring programmes. There is a clear need for more data, both in terms of quantity (e.g. spatial and temporal distribution of data points) and quality (e.g. number of chemicals determined) if we are to understand the status of the marine environment. Given the natural variability, less data is often disastrous for a proper assessment, whereas more data implies more resources, which inevitably meets restraints. Nevertheless, one can wonder if the potential consequences of limited datasets outweigh the costs of obtaining the right information.

On the more practical side, there is a continuous need to discuss common technical questions applying to all aspects of monitoring. These discussions have been ongoing within the various commissions and organizations, but there is a definite need to involve all parties in the discussions. For assessments, these can include methodologies such as trend analysis techniques and other statistical evaluation methods. For monitoring, these can include the temporal and spatial distribution of sites to be monitored, selection of appropriate matrices, sampling methodology and analytical methodology. For both monitoring and assessment, data handling will be important, and there can be interests common to different sectors in the creation and management of databases and the techniques of data exchange.

Fortunately, there is a general awareness that the data that are produced and stored should be of high and, specifically, well-defined quality. The emphasis on and awareness of quality assurance and quality control (QA/QC) is one of the major achievements of chemical monitoring to date. Successful intercomparison exercises and programmes such as QUASIMEME (Topping, 1997; Wells, 2000; Wells and de Boer, 2006; www.quasimeme.org) have highlighted the need for rigorous QA/QC in chemical monitoring. QA/QC is by now omnipresent and an essential element of all well-defined monitoring programmes. Amongst other things, this implies the obligation of laboratories to participate in international intercomparison exercises or proficiency testing schemes and to have documented QA/QC procedures, if not outright accreditation.

To conclude, a solid basis has been laid for the use of chemical monitoring as a tool to assess the quality of the marine environment and the changes thereof. Future developments should focus on a continued discussion of common technical questions and approaches, even more international collaboration and coordination and a move towards more integrated, more holistic monitoring by combining chemical monitoring with other relevant observations such as, for example, biological effects monitoring.

REFERENCES

AMAP (2000). *AMAP Trends and Effects Programme: 1998–2003*, Arctic Monitoring and Assessment Programme Report 99:7, Oslo, Norway, 9 pp.

AMAP (2004). *AMAP Assessment 2002: Persistent Organic Pollutants (POPs) in the Arctic*, Arctic Monitoring and Assessment Programme (AMAP), Oslo, Norway, 310 + xvi pp.

AMAP (2009). *AMAP Assessment 2009: Human Health in the Arctic*. Arctic Monitoring and Assessment Programme Report 99:7, Oslo, Norway, 256 pp.

Barcelona Convention (1980). Protocol on the Protection of the Mediterranean Sea against Pollution from Land-Based Sources (LBS Protocol), 17 May 1980, Athens, Greece.

Barcelona Convention (1996). Protocol on the Prevention of Pollution of the Mediterranean Sea by Transboundary Movements of Hazardous Wastes and their Disposal (Hazardous Wastes Protocol). 1 October 1996, Izmir, Turkey.

Bernes C. (1998). *Persistent Organic Pollutants. A Swedish View of an International Problem*. Swedish Environmental Protection Agency, Monitor 16.

Bignert A., Göthberg A., Jensen S. *et al*. (1993). The need for adequate biological sampling in ecotoxicological investigations: a retrospective study of twenty years pollution monitoring. *Sci. Total. Environ*., **128**, 121–139.

Bignert A., Olsson M., Persson W. *et al.* (1998). Temporal trends of organochlorines in northern Europe, 1967–1995. Relation to global fractionation, leakage from sediments and international measures. *Environ. Pollut.*, **99**, 177–198.

Broley C. E. (1958). The plight of the American bald eagle. *Audubon Mag.*, **60**, 162–163.

EPA (2005). *United States National Coastal Condition Report 2005*, EPA Report 620/R-03/002, Washington, DC, 273 pp.

EPA (2008). *National Coastal Condition Report III*, EPA/842-R-08-002, Washington, DC, 273 pp.

GESAMP (2001a). *A Sea of Troubles*. Reports and Studies GESAMP No. 70, UNESCO, Paris, 35 pp.

GESAMP (2001b). *Protecting the Oceans from Land-based Activities – Land-based Sources and Activities Affecting the Quality and Uses of the Marine, Coastal and Associated Freshwater Environment*. Reports and Studies GESAMP No. 71, UNESCO, Paris, 162 pp.

GESAMP (2001c). *Independent and In-Depth Evaluation of GESAMP*. IMO/FAO/UNESCO-IOC/UNIDO/WMO/IEAE/UN/UNEP Joint Group of Experts on the Scientific Aspects of Marine Environmental Protection, IMO, London, UK, 64 pp.

GESAMP (2009). *Pollution in the Open Ocean: A Review of Assessments and Related Studies*. Reports and Studies GESAMP No. 79, UNESCO, Paris, 64 pp.

HELCOM (2002). Environment of the Baltic Sea area 1994–1998. *Baltic Sea Environment Proceedings* 82B, Helsinki, Finland, 216 pp.

HELCOM (2003). Environment of the Baltic Sea area 1999–2002, *Baltic Sea Environment Proceedings* No. 87, Helsinki, Finland, 48 pp.

HELCOM (2010). Hazardous substances in the Baltic Sea – an integrated thematic assessment of hazardous substances in the Baltic Sea. *Baltic Sea Environment Proceedings* 120B, Helsinki, Finland, 20 pp.

Lauenstein G. G. and Daskalakis K. D. (1998). U.S. long-term coastal contaminant temporal trends determined from mollusk monitoring programs, 1965–1993. *Mar. Poll. Bull.*, **37**, 6–13.

Muir D. and de Wit C. (2010). Levels, trends and effects of legacy and new persistent organic pollutants in the Arctic: an AMAP Assessment. *Sci. Total. Environ.* **408**, 2851–3052.

Nicholson M., Fryer R., Law R. and Davies I. (2001). Criteria for screening data for analytical quality in contaminant trend studies, in *Report of the Advisory Committee on the Marine Environment*, Annex 8, ICES Cooperative Research Report, Copenhagen, Denmark, pp. 211–217.

North Sea Task Force (1993). *North Sea Quality Status Report 1993*, Oslo and Paris Commissions, London, UK, 132 pp.

OSPAR (2000a). *Quality Status Report 2000, Region II – Greater North Sea*, OSPAR Commission, London, UK, 136 + xiii pp.

OSPAR (2000b). *Quality Status Report 2000*, OSPAR Commission, London, UK, 108 + vii pp.

OSPAR (2006). *Strategy for a Joint Assessment and Monitoring Programme (JAMP) (2006 Revision)*, OSPAR Commission Reference number 2003-22, London, UK, 28 pp.

OSPAR (2009). *CEMP assessment report: 2008/2009. Assessment of trends and concentrations of selected hazardous substances in sediments and biota*, OSPAR Commission publication number: 309/2009, London, UK, 80 pp.

OSPAR (2010). *Quality Status Report 2010*, OSPAR Commission, London, UK, 176 pp.

Pearce J. B. (1998). A short history of marine environmental monitoring. *Mar. Pollut. Bull.*, **37**, 1–2.

Rice J. and Rosenberg A. A. (2009). AoA region: North West Atlantic Ocean, in *An Assessment of Assessments, Findings of the Group of Experts. Start-up Phase of a Regular Process for Global Reporting and Assessment of the State of the Marine Environment including Socio-economic Aspects*, Annex IV regional summaries, United Nations Environment Programme, Nairobi, Kenya, and the Intergovernmental Oceanographic Commission of UNESCO, Paris, France, pp. 141–145.

Roose P. (2010). *Volatile Organic Compounds and Related Microcontaminants*, Lambrecht Academic Publishing, Saarbrücken, Germany, 226 pp.

Roose P. and Brinkman U. A. Th. (2005). Monitoring organic microcontaminants in the marine environment: principles, programmes and progress. *TrAC Trends Anal. Chem.*, **24**, 897–926.

Roose P., Albaigés J., Bebianno M. J. *et al.* (2011). Monitoring of existing and emerging chemicals in the European marine and coastal environment, in *Marine Board Position Paper* 16 (eds J. B. Calewaert and N. McDonough) Marine Board-ESF, Ostend, Belgium, 163 pp.

Rosenberg A. A. (2009). AoA region: North West Pacific Ocean, in *An Assessment of Assessments, Findings of the Group of Experts. Start-up Phase of a Regular Process for Global Reporting and Assessment of the*

State of the Marine Environment including Socio-economic Aspects, Annex IV regional summaries, United Nations Environment Programme, Nairobi, Kenya, and the Intergovernmental Oceanographic Commission of UNESCO, Paris, France, pp. 141–145.

Sericano J. L., Wade T. L., Jackson T. J. *et al*. (1995). Trace organic contamination in the Americas: an overview of the US National Status & Trends and the International 'Mussel Watch' programmes. *Mar. Pollut. Bull*., **31**, 214–225.

Smith V. H., Tilman G. D and Nekola J. C. (1999). Eutrophication: impacts of excess nutrient inputs on freshwater, marine, and terrestrial ecosystems – principles, problems, restoration. *Environ. Polluuuuut*., **100**, 179–196.

Topping G. (1997). QUASIMEME: quality measurements for marine monitoring. Review of the EU project 1993–1996. *Mar. Pollut. Bull*., **35**, 1–201.

UN (1992). *The Agenda 21 Plan of Action*, UN Conference on Environment and Development (Earth Summit), Rio de Janeiro, Brazil, United Nations, New York.

UN (2002). *Report of the World Summit on Sustainable Development*, Johannesburg, South Africa, 26 August–4 September 2002, A/CONF.199/20, United Nations, New York, 173 pp.

UNEP (2003). *Global Marine Assessments: a survey of global and regional marine environmental assessments and related scientific activities*, UNEP-WCMC/UNEP/UNESCO-IOC, 132 pp.

UNEP (2004). *Bali Strategic Plan for Technology Support and Capacity-building*, UNEP/GC.23/6/Add.1, Governing Council of the United Nations Environment Programme, Nairobi, Kenya, 9 pp.

UNEP (2007). *Global Marine Assessments: a survey of global and regional assessments and related activities of the marine environment*, UNEP/UNESCO-IOC/UNEP-WCMC, 56 pp.

UNEP and IOC-UNESCO (2009). *An Assessment of Assessments, Findings of the Group of Experts. Start-up Phase of a Regular Process for Global Reporting and Assessment of the State of the Marine Environment including Socio-economic Aspects*, United Nations Environment Programme, Nairobi, Kenya, and the Intergovernmental Oceanographic Commission of UNESCO, Paris, France.

Webster L., Fryer R., Davies I. *et al*. (2009). *Background Document on CEMP Assessment Criteria for QSR 2010*, OSPAR Commission Publication Number: 461/2009, London, UK, 25 pp.

Wells D. E. (2000). QUASIMEME-QUASH. Quality Assurance of Information for Marine Environmental Monitoring in Europe–Quality Assurance of Sampling and Sample Handling. *J. Environ. Monit*., **2**, 69N–70N.

Wells D. E. and de Boer J. (2006). Evaluation of the quality of measurement of organochlorine contaminants in the marine environment: the QUASIMEME experience. *Trends Anal. Chem*., **25**, 350–363.

3

EU Marine Strategy

Gert Verreet

Flemish Government – Environment, Nature and Energy Department (LNE), International Policy Division, Brussels, Belgium

3.1 INTRODUCTION: CONTEXT

Concerns about marine pollution have made marine chemical monitoring one of the technically most advanced branches of marine environmental monitoring. International action in this field became intense in western Europe from the 1970s, as described in more detail in Chapters 1 and 2 by Roose and Kramer respectively. The needs for common approaches to analytical and quality issues, monitoring programme design and assessment criteria were addressed in this framework of

Chemical Marine Monitoring: Policy Framework and Analytical Trends, First Edition.
Edited by Philippe Quevauviller, Patrick Roose, and Gert Verreet.

regional conventions from the mid 1970s: OSCOM and PARCOM (now OSPAR), HELCOM or the MED POL programme. Systematic scientific advice from international scientific cooperation organizations such as the International Council for the Exploration of the Sea (ICES) supported this process. Thus, a body of internationally validated tools (monitoring guidelines, quality assurance techniques, proficiency schemes, etc.) was established in the context of the different policy regimes of the marine regions around Europe, principally to ensure that monitoring data had comparable reliability so that they could be assessed collectively at that international level.

The policy framework for marine monitoring developed in response to the gradually better assessment of the impacts of contaminants/pollutants. At international level or in the European Community, a range of legal instruments was created to deal with the diversity of pollution problems and sources. Oil pollution from ships was being addressed through the International Maritime Organization since the 1950s, and this intensified in Europe after the catastrophe with the *Torrey Canyon* in 1967. All efforts combined led, over time, to a reduction of acute forms of marine pollution (OSPAR, 2000; HELCOM 2004). There was often also a direct connection between the regulatory regime and the associated monitoring requirements.

The acute forms of marine pollution were never more than the tip of the iceberg. The hydrological cycle 'conveyor belt' carries substances discharged to freshwater away from their point of release towards the seas. The atmosphere also acts as a transport mechanism. International cooperation was established to deal with the chronic and land-based sources of marine pollution (Hassan, 2006). It was often assumed to be convenient that the aquatic and atmospheric pathways resulted mostly in the substance removal from the biosphere, either through dilution, burial or degradation; it was soon realized that there are also pathways whose properties are such that damage can result far from the original source(s) of the substance (e.g. bioaccumulation in biota, biomagnification in the food chain, or other pathways selectively enhancing substance concentrations in specific media). Causal chains connecting the impact with all the sources become longer and more complicated. This clearly requires a more ambitious and comprehensive approach to monitoring and assessment, and also to policy making.

Policy makers have at times recognized the limitations of the existing regulatory situation that was a patchwork of different, largely disconnected, legal instruments.

For instance, in the early 1980s, countries around the North Sea area started to organize a series of Ministerial Conferences for the Protection of the North Sea[1] which set out ambitious objectives and actions for dealing with North Sea pollution, mobilizing a range of competent bodies and fostering more coherent action in a range of frameworks and instruments.

Whilst a number of broader policy frameworks are now in place to handle widespread and diffuse pollution of toxic substances (e.g. at global level the Stockholm Convention on Persistent Organic Pollutants, 2001), whether the regulatory regime has now become successful to assure such a comprehensive approach encompassing all sources of pollution and processes over larger geographical scales remains to be determined.

Several international collaborative programmes have contributed to the dissemination of information to foster marine chemical monitoring and assessment. What can be reasonably well (and repeatedly) observed are levels of concentrations of substances in inputs and in marine media. Marine environmental monitoring, therefore, has focused mostly on what happens with substances in the marine-receiving media. When inputs and pathways to the marine environment are not monitored equally well, a full understanding of the relative strength of pathways contributing to the existing levels of marine contamination remains elusive.

To strengthen the knowledge base for effective policy measures, marine chemical monitoring and assessment should be broadened to allow for a stronger causal contextualization. This involves:

- 'going upstream' in the causal chains to determine with more certainty the main pressures;

[1] Bremen in 1984, London in 1987, The Hague in 1990, Esbjerg in 1995, Bergen in 2002 and Gothenburg in 2006.

- understanding better the overall ecosystem dynamics – of which the contaminant dynamics are a part – to understand the impacts and degree of change that can be effectuated by measures taken at source.

This causal contextualization of marine monitoring is not only driven by the scientific needs of the ecosystem-based approach underpinning current policy. There is a political and economic rationale as well. Dealing with widespread and diffuse pollution involves more countries, sectors, and so on, and the reality is that devising effective measures will, despite all the merits of the precautionary approach, often require a strong backing by scientific evidence of risks and harm in the marine environment.

3.2 EU MARINE ENVIRONMENTAL POLICY

Protection of water is a long-standing part of the European Communities environmental policy.[2]

Like the international legal framework generally, the European Communities' policy regarding the protection of the marine environment has grown in a relatively piecemeal fashion over the years. Quality standards for different types of water use (e.g. shellfish water, bathing water) or seafood were set in a first wave of EC directives on water quality. It was also attempted at EC level to elaborate common discharge standards for point sources of listed substances on land (Directive 76/464/EC). Over the period 1975–1993, the European Community had become, as a regional economic integration organization, a party to the regional sea conventions in three of the four main sea basins around Europe,[3] with discussions regarding membership of the convention dealing with the pollution of the Black Sea (Bucharest Convention) progressing only slowly since the EU membership of Romania and Bulgaria in 2007.

Aquatic environmental policy and biological resource policy are becoming more connected under an ecosystem-based approach. Environmental policy is a shared competence between the EU and its member states. In contrast, the 'conservation of marine biological resources under the common fisheries policy' (Treaty on the Functioning of the European Union, Art. 3 (1)(d)) has been an exclusive Community policy since the early 1980s, following international developments on access to fisheries resources during the 1970s when coastal states started to declare their exclusive economic zones (EEZs) under the UN Convention on the Law of the Sea. EU member states agreed that the management of the fisheries stocks in their EEZ could be better managed in a common European regime.

An important milestone and precedent for a comprehensive approach to the protection of the marine environment was the adoption of the Water Framework Directive 2000/60/EC. Its objective is to establish a framework for the protection of inland surface waters, transitional waters, coastal waters and groundwater, recognizing that this contributes to the achievement of international objectives of international agreements which aim to prevent and eliminate pollution of the marine environment. Its geographical limit of application (as defined by the term 'surface waters') at the seaward side extends fully to the coastal waters (with a limit of one nautical mile from the baseline) and to the territorial waters only as regards 'chemical status' (most often these cover waters up to 12 nautical miles from the coast).

---

[2] With the entry into force of the Lisbon Treaty, the European Union is the successor of the European Communities. An overview of the current body of EU legislation in the field of water protection is at http://europa.eu/legislation_summaries/environment/water_protection_management/index_en.htm.

[3] In the Baltic: the EC became a party to the 1992 Helsinki Convention in 1993. In the North-East Atlantic Ocean: the EC became a party to the 1974 Paris Convention in 1975; this convention preceded the 1992 OSPAR Convention currently in force. In the Mediterranean Sea, the EC became a party to the 1976 Barcelona Convention in 1977.

Even after the adoption of the Water Framework Directive, EU environmental policy makers considered there was a lack of strategy underpinning the policies to protect the marine environment. A commitment to formulate such a strategy was included by the European Parliament and the Council in the sixth Environmental Action Programme adopted in July 2002 and applicable for the period 2002–2012 (EC, 2002a). The European Marine Strategy developed over the period 2002–2005 (EC, 2006a) and resulted in the adoption of the Marine Strategy Framework Directive (Directive 2008/56/EC).

Important steps during this process were:

- Commission Communication 'Towards a strategy to protect and conserve the marine environment' (EC, 2002b).
- A period of regular informal discussions with member states and stakeholder organizations in working groups and stakeholder conferences (2002–2007), including the setting up of an informal working group on monitoring and assessment aspects of the strategy, the so-called European Marine Monitoring and Assessment group (EMMA).[4]
- August 2004: then newly appointed Commission President Barroso announces the elaboration of an integrated EU maritime policy, which would also encompass marine environmental issues.
- October 2005, Commission final proposals: Communication 'Thematic Strategy on the Protection and Conservation of the Marine Environment' (EC, 2005) with a proposal for a directive and an impact assessment.
- Negotiations on the proposal for the directive in the Council and the European Parliament in 2006 and 2007 (see below).
- Formal adoption of the Marine Strategy Framework Directive (MSFD) by its signature on 17 June 2008. It became Directive 2008/56/EC and entered into force on 15 July 2008, an important date that started the clock for the elaboration of member states' marine strategy elements.
- Progressive integration of the marine strategy as the 'environmental pillar' of the developing EU maritime policy:[5]
  - June 2006, Green Paper Maritime Policy starting a broad consultation;
  - October 2007, 'Blue Book' with European Commission policy proposals and an action plan;
  - December 2007, European Council recognizes forthcoming Marine Strategy Framework Directive as environmental pillar;[6]
  - October 2009, the first EC progress report on the integrated maritime policy gives clear recognition that environmental sustainability and cumulative effects of different pressures on the marine environment will be addressed under the MSFD (EC, 2009a):

    > The definition of the boundaries of sustainability of human activities that have an impact on the marine environment in the years ahead, in the framework of the Marine Strategy Framework Directive, will provide clarity and design a platform

---

[4] Documentation accessible via http://circa.europa.eu/Public/irc/env/marine/library?l=/&vm=detailed&sb=Title [last accessed 14 January 2010].

[5] http://ec.europa.eu/maritimeaffairs/policy_documents_en.html [last accessed 14 January 2010].

[6] See http://www.consilium.europa.eu/ueDocs/cms_Data/docs/pressData/en/ec/97 669.pdf paragraph 58 [last accessed 15 October 2009].

> for the successful development of all maritime activities, paying due attention to their cumulative impacts. Hence, the implementation of this Directive will remain a key objective of the IMP, which should also develop the necessary cooperation between all relevant sectors and services to this end, including inter-alia between marine science and the marine environment policy.

This account of recent history indicates that the implementation of the MSFD will be an important instrument in the EU to move towards a more sustainable use of marine goods and services within a policy context of intensified attention for the economic opportunities provided by maritime activities.

Marine environmental and maritime policy are, however, not the only relevant policy frameworks for chemical monitoring of the seas. Prevention of pollution by chemicals is also addressed by extensive legislation on production, marketing and use of substances, such as the REACH Regulation (EC, 2006b) for example. An analysis within the framework of the OSPAR Commission concluded in 2008 that most of the EU chemicals legislation seemed to address adequately the possible risk to the marine environment, although this judgement was not fully reached for substances used as veterinary medicines and pesticides (OSPAR, 2008).

3.3 MARINE STRATEGY FRAMEWORK DIRECTIVE 2008/56/EC

3.3.1 Key Issues during the Negotiations in the Council and European Parliament

The MSFD was adopted by the European co-legislators under the co-decision procedure which gives the Council and the European Parliament an almost equal role (under the Lisbon Treaty, this procedure is now called the 'ordinary legislative procedure'). Negotiations of the draft directive in the Council and the European Parliament were especially intense in the second half of 2006 (first reading) and 2007 (second reading), periods during which the EU Council was presided over by Finland and Portugal respectively.[7]

Key issues that were addressed during these negotiations, and which remain important considerations during implementation, include:

- Stronger environmental ambitions (many amendments in the European Parliament).
- Clear recognition of regional natural differences of the marine environment. In view of the overall philosophy that the application of an ecosystem approach (ICES, 2005) should allow managers to target their efforts primarily to those questions which require specific attention and that, on the scale of the European continent, natural characteristics of marine ecosystems vary significantly from north to south and from east to west, these provisions were felt to be more than simply justified by the subsidiarity principle.
- More subsidiarity and flexibility for implementation by the member states (many amendments in the Council), including the use of 'exceptions' (MSFD Article 14) when conditions do not allow the achievement of targets, including that

  > Member States shall not be required, except in respect of the initial assessment described in Article 8, to take specific steps where there is no significant risk to the marine

[7] The evolution of the decision-making process can be followed on the Legislative Observatory facility of the European Parliament; for this particular file, see: http://www.europarl.europa.eu/oeil/FindByProcnum.do?lang=&procnum=COD/2005/0211 [last accessed 14 January 2010].

environment, or where the costs would be disproportionate taking account of the risks to the marine environment, and provided that there is no further deterioration.

This provision, which may have an important bearing on the future extent of monitoring and assessment, will require further interpretation as the meaning of 'disproportionate' is not further explained in the directive;

- Clear recognition for the role of the existing regional sea conventions around Europe. Most member states indicated that they intended to utilize the work undertaken in the context of the OSPAR, HELCOM, Barcelona and Bucharest Conventions also for the purpose of their MSFD implementation. This echoed a development that was already ongoing earlier concerning the maximization of synergies between the work of the regional sea conventions and the EC; see, for example, in OSPAR (2005).

That the legislative process was concluded with the adoption of the directive at the end of the second reading, despite significant different orientations taken by the European Parliament and the Council, attests to the willingness of the co-legislators to find pragmatic compromises. As a framework directive, the practical meaning given to its provisions by the member states and the European Commission will be all-determining; these meanings will only become clear during implementation. Any conflicts of interpretation will be settled before the European Court of Justice.

3.3.2 MSFD Objectives: The Notion 'Good Environmental Status'

The objectives of the MSFD are clearly to achieve a certain desired degree of environmental quality: 'This Directive establishes a framework within which Member States shall take the necessary measures to achieve or maintain good environmental status in the marine environment by the year 2020 at the latest' (MSFD Article 1(1)).

However, the rest of Article 1 associates important, other linked objectives with this central environmental status objective:

- the protection, preservation of marine environment, prevention of deterioration, restoration where practicable;
- the prevention and reduction of inputs to the marine environment;
- the application of an ecosystem-based approach, ensuring that collective pressure of activities is kept within levels compatible with 'good environmental status', maintaining resilience, enabling sustainable use of marine goods and services;
- that the implementation of the Directive should contribute to coherence between, and aim to ensure the integration of environmental concerns into, different policies which have an impact on the marine environment.

The concept of 'good environmental status' is the central innovation of the MSFD; see Box 3.1. This marks an important shift towards a more proactive and forward-looking marine environmental protection regime. It embodies an important aspect of the ecosystem-based approach. The environmental management will be geared to achieving explicit objectives rather than be driven only by the principles and concepts of the past, such as 'prevention of pollution'. It can be argued that these past concepts were in themselves insufficient to secure the maintenance of healthy, clean and productive seas. This shift is not without its own challenges; it hinges on the ability to ensure that the objectives are well formulated and that they provide the correct drivers for management action. Being able to measure and monitor the right variables is a key consideration underlying environmental objective formulation.

BOX 3.1

'Environmental status' is defined in the MSFD as:

> the overall state of the environment in marine waters, taking into account the structure, function and processes of the constituent marine ecosystems together with natural physiographic, geographic, biological, geological and climatic factors, as well as physical, acoustic and chemical conditions, including those resulting from human activities inside or outside the area concerned

'Good environmental status' is defined in the MSFD as:

> the environmental status of marine waters where these provide ecologically diverse and dynamic oceans and seas which are clean, healthy and productive within their intrinsic conditions, and the use of the marine environment is at a level that is sustainable, thus safeguarding the potential for uses and activities by current and future generations, i.e.:
>
> (a) the structure, functions and processes of the constituent marine ecosystems, together with the associated physiographic, geographic, geological and climatic factors, allow those ecosystems to function fully and to maintain their resilience to human-induced environmental change. Marine species and habitats are protected, human-induced decline of biodiversity is prevented and diverse biological components function in balance;
> (b) hydro-morphological, physical and chemical properties of the ecosystems, including those properties which result from human activities in the area concerned, support the ecosystems as described above. Anthropogenic inputs of substances and energy, including noise, into the marine environment do not cause pollution effects;
>
> Good environmental status shall be determined at the level of the marine region or subregion as referred to in Article 4, on the basis of the qualitative descriptors in Annex I. Adaptive management on the basis of the ecosystem approach shall be applied with the aim of attaining good environmental status;

MSFD Annex I sets out qualitative descriptors of 'good environmental status', addressing physical, biological and chemical features; see Box 3.2.

BOX 3.2

1. Biological diversity is maintained. The quality and occurrence of habitats and the distribution and abundance of species are in line with prevailing physiographic, geographic and climatic conditions.
2. Non-indigenous species introduced by human activities are at levels that do not adversely alter the ecosystems.

3. Populations of all commercially exploited fish and shellfish are within safe biological limits, exhibiting a population age and size distribution that is indicative of a healthy stock.
4. All elements of the marine food webs, to the extent that they are known, occur at normal abundance and diversity and levels capable of ensuring the long-term abundance of the species and the retention of their full reproductive capacity.
5. Human-induced eutrophication is minimised, especially adverse effects thereof, such as losses in biodiversity, ecosystem degradation, harmful algae blooms and oxygen deficiency in bottom waters.
6. Sea-floor integrity is at a level that ensures that the structure and functions of the ecosystems are safeguarded and benthic ecosystems, in particular, are not adversely affected.
7. Permanent alteration of hydrographical conditions does not adversely affect marine ecosystems.
8. Concentrations of contaminants are at levels not giving rise to pollution effects.
9. Contaminants in fish and other seafood for human consumption do not exceed levels established by Community legislation or other relevant standards.
10. Properties and quantities of marine litter do not cause harm to the coastal and marine environment.
11. Introduction of energy, including underwater noise, is at levels that do not adversely affect the marine environment.

To determine the characteristics of good environmental status in a marine region or subregion as provided for in Article 9(1), Member States shall consider each of the qualitative descriptors listed in this Annex in order to identify those descriptors which are to be used to determine good environmental status for that marine region or subregion. When a Member State considers that it is not appropriate to use one or more of those descriptors, it shall provide the Commission with a justification in the framework of the notification made pursuant to Article 9(2).

The definitions and descriptors are in themselves insufficiently 'self-executing' and require further methodological underpinning, as recognized in MSFD Article 9:

> Criteria and methodological standards to be used by the Member States, which are designed to amend non-essential elements of this Directive by supplementing it, shall be laid down, on the basis of Annexes I and III, in accordance with the regulatory procedure with scrutiny referred to in Article 25(3) by 15 July 2010 in such a way as to ensure consistency and to allow for comparison between marine regions or sub-regions of the extent to which good environmental status is being achieved. Before proposing such criteria and standards the Commission shall consult all interested parties, including Regional Sea Conventions.

Really operational 'criteria and methodological standards', by which 'good' status will be determined, are under elaboration within the context of a 2010 Commission Decision (EC, 2010) that sets out a framework. They should provide the further guidance on adequate methods how member states can ensure that the normative aspects of 'good' are correctly applied in their concrete, operational, expression of what 'good environmental status' means for their marine waters. The incorporation, under EU law, of these components of the ecosystem-based approach creates a significant driver for the further development and application of quality assessment techniques which were hitherto only applied in a 'soft law' context, such as the ecological quality objectives promoted earlier in the frameworks of the North Sea Conferences, OSPAR and their equivalents under HELCOM.

The 'determination' of good environmental status (MSFD Article 9) by the member states will be accompanied by the setting of targets and associated indicators (MSFD Article 10). These will need to be designed so as to guide the selection of the necessary management measures. It is good practice to formulate policy objectives and indicators together, so that there is a clear understanding of how success or failure will be evaluated.

3.3.3 Marine Strategies of the Member States

The initial assessment, the determination of 'good environmental status' and the setting of associated targets and indicators are an extensive prelude to the real substance of the member states' marine strategies, which should be their programme of measures that are necessary to achieve the environmental objectives; see Table 3.1.

'Adaptive management', reflecting the progress in available knowledge and changing challenges and circumstances, is factored into the MSFD mechanisms through a 6-yearly review cycle of each of the marine strategy elements (Article 17 on updating).

In view of the large possible synergies to member states' activities under the regional sea conventions and for the purpose of the MSFD implementation, actions to achieve these deliverables will be programmed at EU, (sub-)regional and national levels. Vertical and lateral exchange of data, information and best practices can significantly facilitate the MSFD implementation. MSFD Article 19 links requirements on public consultation (Article 19(1) and 19(2)) to the access of the general public to data and with sharing of data between the member states, the European Commission and the European Environment Agency (Article 19(3)). An efficient infrastructure underpinning exchange of data and information between all the relevant actors will be an important contribution to the MSFD implementation.

3.3.4 EU-Level Work to Foster the MSFD Implementation

At the time of writing, work is under way to underpin the MSFD implementation with the necessary more-detailed rules and guidance necessary for a sufficient level of comparability across the member states, using where necessary the official further decision-making procedure foreseen in

Table 3.1 Calendar for progressive elaboration of member states' marine strategy elements.

Marine strategy element	First version ready
A description and assessment of current environmental status, including the environmental impact of human activities	By 15 July 2012
The determination of good environmental status	By 15 July 2012
The establishment of environmental targets and associated indicators	By 15 July 2012
The establishment and implementation of a monitoring programme for ongoing assessment and regular updating of targets	By 15 July 2014, except where specified otherwise in the relevant Community legislation
The programme of measures towards good environmental status	Established by 2015, entry into operation by 2016

the directive (Articles 24 and 25 on technical adaptations and on the Regulatory Committee). In order to ensure that, apart from the EU member states who have members on the Regulatory Committee, relevant international governmental and nongovernmental organizations can also contribute to the preparatory work at EU level, an informal cooperative working structure has been set up by the European Commission and by the member states. There will be regular meeting of so-called 'Marine Directors' as a main body giving the strategic directions, assisted by a Marine Strategy Coordination Group and several working groups. Important areas of collective work currently (2010) undertaken in this cooperation include:

- the development of further practically and locally applicable criteria and methodological standards which member states can take into account in their determination of 'good environmental status', following the adoption, by the European Commission, of the Commission Decision 2010/477/ EU of 1 September 2010 on criteria and methodological standards on good environmental status of marine waters (EC, 2010);
- the development of the necessary data and information sharing and reporting infrastructure, in line with the principles set out for the Shared Environmental Information System (SEIS);
- the development of common approaches to operate the socio-economic considerations required under the various stages of marine strategy development.

The EU-level methodologies will already be informed by the existing regional tools. The MSFD, however, has additional requirements, such as for the aspects of marine environmental quality which have hitherto been barely covered by monitoring and assessment (e.g. seabed integrity, biodiversity, food web, marine litter, noise). Beyond reusing pre-existing regional methodologies for the MSFD, member states have an interest in ensuring that the ongoing and future work of the regional sea conventions covering their marine waters can be used to a maximum extent also for the MSFD implementation. The two conventions where EU member states are forming the large majority of the parties are, so far, most advanced in the internal process designed to achieve this.

3.4 MONITORING IN THE MSFD POLICY CONTEXT

3.4.1 Use of Data and Information, Indicators

During the development of the EU Marine Strategy many stakeholders and the European Commission emphasized the need for 'evidence-based policy' and a 'knowledge-based approach' (EC, 2005). This requires *inter alia* a strong basis in science for all elements of marine strategies. An adequate science–policy interface is needed so that science continuously informs policy. Only then can the information value derived from observation and monitoring be maximized.

Monitoring is not explicitly defined in the MSFD, but its Article 11, as seen in Box 3.3, clearly specifies that the purpose of monitoring is the 'ongoing assessment of the environmental status of [...] marine waters'. In order to do so, an understanding is needed of the types of signal to be monitored that are necessary to underpin such an assessment.

BOX 3.3

Article 11-Monitoring programmes

1. On the basis of the initial assessment made pursuant to Article 8(1), Member States shall establish and implement coordinated monitoring programmes for the ongoing assessment of

the environmental status of their marine waters on the basis of the indicative lists of elements set out in Annex III and the list set out in Annex V, and by reference to the environmental targets established pursuant to Article 10.
Monitoring programmes shall be compatible within marine regions or subregions and shall build upon, and be compatible with, relevant provisions for assessment and monitoring laid down by Community legislation, including the Habitats and Birds Directives, or under international agreements.

2. Member States sharing a marine region or subregion shall draw up monitoring programmes in accordance with paragraph 1 and shall, in the interest of coherence and coordination, endeavour to ensure that:

   (a) monitoring methods are consistent across the marine region or subregion so as to facilitate comparability of monitoring results;
   (b) relevant transboundary impacts and transboundary features are taken into account.

3. Member States shall notify the Commission of the monitoring programmes within three months of their establishment.

4. Specifications and standardized methods for monitoring and assessment which take into account existing commitments and ensure comparability between monitoring and assessment results, and which are designed to amend non-essential elements of this Directive by supplementing it, shall be adopted in accordance with the regulatory procedure with scrutiny referred to in Article 25(3).

These are based, to an important degree, on indicators updated on a regular basis. The environmental status assessments' methodologies evolve continuously in relation to advances in their constituent disciplines (chiefly the abilities to measure, express, treat statistically and evaluate significance of the assessed signals). Although an environmental assessment should not be reduced to 'indicators' only, for the sake of this discussion the notion of 'indicator' is most useful to clarify a number of points.

An environmental indicator is defined by the OECD (n.d.(a)) as:

> A parameter, or a value derived from parameters, that points to, provides information about and/or describes the state of the environment, and has a significance extending beyond that directly associated with any given parametric value. The term may encompass indicators of environmental pressures, conditions and responses.

Indicators can have several functions under the MSFD marine strategies and can occur in different components of the strategies playing a different role:

- Indicators purely expressing 'how it is'; that is, factual indicators – marine environmental assessments (pursuant MSFD Article 8) will include a number of descriptive indicators of environmental quality and related factors.
- Indicators expressing 'how it should be'; that is, indicators with a reference basis that represents a desirable status and which, under law, can be normative – 'Good environmental status' (determined pursuant to MSFD Article 9) will be made operational with concrete and measurable expressions that can be represented in the form of an indicator.

- Indicators answering the question 'Are we going in the right direction?' That is, indicators of the (direction of) change. MSFD Article 10 spells out that EU Member States shall:

  > On the basis of the initial assessment made pursuant to Article 8(1), [...] in respect of each marine region or subregion, establish a comprehensive set of environmental targets and associated indicators for their marine waters *so as to guide progress* [emphasis added] towards achieving good environmental status in the marine environment, taking into account the indicative lists of pressures and impacts set out in Table 2 of Annex III, and of characteristics set out in Annex IV.

  Indicators of direction and speed of change will be especially important for those aspects of environmental quality for which no satisfactory determination of 'good' can be given (yet).

The DPSIR model (drivers, pressures, state, impact, response) is a convenient schematic structure for the categorization of indicators, which is extensively used in Europe, for instance by the European Environment Agency (OECD, n.d.(b)). A matrix showing the combination of MSFD Articles 8, 9 and 10 and the functional relation to elements of the DPSIR framework is in Table 3.2. 'State' is the central component on which MSFD effectiveness should ultimately be assessed.

Given the (often slow) speed of change of some phenomena in the marine environment, a case can be made for the selection, by the authorities in charge of elaborating the marine strategies, of a judicious mix of indicators that provides sufficient coverage of those elements that matter particularly strongly for the purposes of the MSFD, including attention for the following.

- Sufficient coverage of the drivers of change; that is, D-type indicators (covering, for example, intensity of drivers, i.e. human activities that exert significant pressures on the environment) and P-type indicators (covering a synoptic view of the change in pressures from various human activities). Rogers and Greenaway (2005) found that, in the UK (but this conclusion would also hold true in many other European countries) there are many indicators of state for ecosystem components, but relatively few for pressure of human activities or of the socio-economic response to those pressures. Additional considerations for their selection include the folloing.

  - Such indicators can be especially useful for policy makers as they express features of managed reality for which they have a higher degree of accountability than the marine environmental reality, which is also influenced by many other factors besides. For many obstinate marine environmental issues, the likelihood of success of a marine strategy will be noticeable first in these types of indicator.
  - D- and P-type indicators are, however, somewhat more difficult to construct owing to the disparity of information needed for a comprehensive and robust indicator. They often require aggregation of economic statistics and other human activity data over specific spatial and temporal scales that may not be the scale of the original data collection. However, they are not all extremely difficult: a relatively easy type indicator that is commonly used from this category is the regular quantification of land-based inputs of contaminants by rivers to the sea.

- A critical choice regarding S- and I-type indicators, which are based on environmental monitoring, reflecting priority information requirements. This requires a thorough analysis of information needs and the means necessary to meet them. Countries will need to match their monitoring and assessment resources to an adequate (representative, etc.) set of parameters and to achieve an optimum. Arriving at such a choice will often involve difficult trade-offs; for example, resolving the friction between today's most pressing questions, any emergent question and the maintenance of long-term stewardship of existing monitoring data time series.

Table 3.2 MSFD Articles 8, 9 and 10 and additional requirements as a basis for indicators across the DPSIR spectrum.

MSFD Article	Additional requirement	Driving forces	Pressures	State	Impact	Response
8. (Initial) assessment	On basis mainly of Annex III	Descriptive	Descriptive	Descriptive[a]	Descriptive	Descriptive
9. Good environmental status	On basis of EU common criteria and methodological standards	(Unlikely)	"'Take into account'	Normative	'Take into account'	(Unlikely)
10. Environmental targets and associated indicators	On basis of Articles 8 and 9 output[b]	Potentially all, they are 'established so as to (...) guide progress towards achieving good environmental status'				

[a]Likely to be progressively adapted to reflect normative approach following implementation of Article 9.
[b]Taking account of needs of the programme of measure.

Composing an aggregate assessment on the basis of a series of discrete indicators is also an important question that will need to be addressed. Unlike the Water Framework Directive, the MSFD regulatory framework does not include a specific aggregation algorithm for a compound judgement of overall environmental quality. So far, few methodologies exist to integrate all the elements into a single evaluation of a water body (Borja *et al.*, 2008). A model representation of ecosystem structure, functioning and processes seems to be necessary as a background canvas to allow a systemic approach to such integration.

3.4.2 Operational Link between Policy and Science for the Purpose of the MSFD Implementation

There is a growing body of interpretation of the 'ecosystem approach', a term that shot to prominence in the marine policy context owing to it being mentioned in the conclusions of the 2002 Johannesburg Summit on Sustainable Development in the context of fisheries management (UN, 2002):

> Encourage the application by 2010 of the ecosystem approach, noting the Reykjavik Declaration on Responsible Fisheries in the Marine Ecosystem and decision 5/6 of the Conference of Parties to the Convention on Biological Diversity.

In the marine environmental policy contexts of OSPAR and HELCOM (OSPAR, 2003), the notion of an 'ecosystem approach' was interpreted as being synonymous with 'ecosystem approach to the management of human activities', stressing that it is the latter that are being managed, not the environment as such. During the elaboration of the EU Marine Strategy, ICES (2005) advised the European Commission to interpret the ecosystem approach as:

> a comprehensive integrated management of human activities based on the best available scientific knowledge about the ecosystem and its dynamics, in order to identify and take

action on influences which are critical to the health of the marine ecosystems, thereby achieving sustainable use of ecosystem goods and services and maintenance of ecosystem integrity.

Despite the policy shift caused by the adoption of the ecosystem approach a decade ago, there is still much to do. The UN Food and Agriculture Organization (FAO) indicates on its web page on the ecosystem approach that (FAO, n.d.)

> Despite these good intentions and some progress in implementation in many parts of the world, in most if not all countries much progress in ecosystem research and institutional development is still needed before the implications of the approach are fully understood and credible management strategies are adopted and effectively implemented.

The ecosystem-based approach has a price. It rests on the assumption that scientific understanding and observation of the marine environment in all its components should be an important, if not the main, basis for environmental management measures. This choice is not uncontested, as evidenced by the intense debate with the environmental stakeholder organizations during the MSFD negotiations, who pleaded for an approach that would be guided more strongly by pressure minimization (Greenpeace *et al*, n.d.) rather than by environmental status (and impact) descriptors that require more sophisticated knowledge for their operation.

If the role of scientific knowledge has become even more important under the MSFD than it was without it, great care should be taken to organize (or optimize), within the marine environmental management cycle related to the elaboration and updating of marine strategies, a strong two-way and interactive relation between the policy and scientific worlds.

Institutionally, such arrangements will be required at EU, (sub-)regional, national and possibly even lower levels. The question of the science–policy interface underpinning the EU maritime policy is addressed in the EU marine and maritime research strategy (EC, 2008), where the work of the European Commission is intended to contribute to facilitating the implementation of the MSFD in this regard. Each regional sea convention around Europe has arrangements in place for scientific advice and review of its work; whether these arrangements will match the enhanced science–policy dialogue requirements described here will be a matter for consideration by the parties to those conventions.

A structured dialogue between policy makers, science programme managers and those responsible for commissioning and programming environmental monitoring should:

- set out strategic directions that reflect identified policy needs;
- take account of the state of play as regards scientific understanding of the existing marine environmental issues requiring attention and identify means to address any weaknesses of the knowledge base;
- bridge the divides, where necessary, between people involved in policy, institutionalized environmental monitoring and assessment, and scientific research.

3.4.3 Monitoring

Against the above-described background of MSFD information needs, monitoring and assessment play a vital role.

Despite available methodologies and decades of experience, its coverage still tends to be focused on coastal areas and can still, even in advanced economies, and seen from a pan-European perspective, lack sufficient coherence: 'data on concentrations and impacts of hazardous substances on the marine environment are scarce and fragmented' (EEA, 2007).

Monitoring programmes will be embedded in member states' marine strategies – see Box 3.3 – linking in with the cycle of monitoring and assessment activities (which are often managed regionally) and that of policy making and effectiveness evaluation. With a strong focus on environmental status (MSFD Annexes I and III), the monitoring programmes are likely to include a strong marine environmental monitoring component. As explained above, they will go beyond environmental monitoring to the extent that indicators require other types of data, such as socio-economic data.

Marine environmental monitoring is relatively costly. Cost is a strong driver for maximizing the information benefit that can be derived from the data produced. This consideration brought the European Commission to include a strong element on data and information sharing in the development of the EU integrated maritime policy (development of a European Marine Observation and Data Network, EMODNET; EC, 2007).

Embedding the management of monitoring and assessment in economic reasoning may not come naturally to environmental scientists. Given the broad range of monitoring needs, it is extremely important, however, that data creation and management receives a sound financial and economic backing, including demonstration of the benefits derived from monitoring. As a framework directive, the MSFD does not automatically 'force' the member states to monitor any given substance or compound. There is no straightforward 'European obligation' for any of the monitoring particulars. Member states will have to fill in responsibly the subsidiarity that they negotiated for.

The broader ongoing development of economic analysis of the use and value of environmental goods and services can assist in strengthening the basis for deciding the rationale and justification of assessment and monitoring programmes.

MSFD Annex V, see Box 3.4, gives further specifications that member states should take into account in designing their monitoring programmes. During the negotiation process, relatively little change was deemed necessary by the co-legislators to this particular annex. It reflects almost exactly what the European Commission had proposed in 2005 (one element, (8), was added).

BOX 3.4 ANNEX V

Monitoring programmes (referred to in Articles 11(1) and 24)

1. Need to provide information for an assessment of the environmental status and for an estimate of the distance from, and progress towards, good environmental status in accordance with Annex III and with the criteria and methodological standards to be defined pursuant to Article 9(3).
2. Need to ensure the generation of information enabling the identification of suitable indicators for the environmental targets provided for in Article 10.
3. Need to ensure the generation of information allowing the assessment of the impact of the measures referred to in Article 13.
4. Need to include activities to identify the cause of the change and hence the possible corrective measures that would need to be taken to restore the good environmental status, when deviations from the desired status range have been identified.

5. Need to provide information on chemical contaminants in species for human consumption from commercial fishing areas.
6. Need to include activities to confirm that the corrective measures deliver the desired changes and not any unwanted side effects.
7. Need to aggregate the information on the basis of marine regions or subregions in accordance with Article 4.
8. Need to ensure comparability of assessment approaches and methods within and between marine regions and/or subregions.
9. Need to develop technical specifications and standardized methods for monitoring at Community level, so as to allow comparability of information.
10. Need to ensure, as far as possible, compatibility with existing programmes developed at regional and international level with a view to fostering consistency between these programmes and avoiding duplication of effort, making use of those monitoring guidelines that are the most relevant for the marine region or subregion concerned.
11. Need to include, as part of the initial assessment provided for in Article 8, an assessment of major changes in the environmental conditions as well as, where necessary, new and emerging issues.(12) Need to address, as part of the initial assessment provided for in Article 8, the relevant elements listed in Annex III

including their natural variability and to evaluate the trends towards the achievement of the environmental targets laid down pursuant to Article 10(1), using, as appropriate, the indicators established and their limit or target reference points.

The annex indicates that monitoring programmes are likely to become very sophisticated to meet all the needs mentioned.

3.4.4 Environmental Quality Descriptors Addressing Substances

The MSFD coverage of chemical water quality cannot be completely dissociated from that of the Water Framework Directive (like the MSFD cannot, in the sphere of nature and biodiversity, be dissociated from the EU legislation in those areas). In proposing the MSFD, the European Commission had stressed the possible synergies of surface water management under the two regimes.

The Water Framework Directive approach to chemical quality is based on the attainment of water quality objectives, underpinned by mechanisms identifying which substances should be addressed (with a common list of priority substances and of priority hazardous substances, additional substances identifiable at the level of individual river basins or districts). Separate documentation is available on monitoring for the purpose of the Water Framework Directive (EC, 2003, 2009b). Documentation on monitoring requirements for the MSFD can be developed under MSFD Article 11(4), but there are no concrete plans yet in this area at the time of writing.

The MSFD further complements it by an approach based on the identification, as an element of environmental status, of a risk of possible 'pollution effects' by substances in member states' wider marine waters. Descriptor (8) links 'good environmental status' with absence of pollution effects. Pollution effects are those deleterious effects which result or could result from the introduction of substances or energy (see definition of pollution in MSFD Article 3). Descriptor (9) puts a particular emphasis on ensuring that all biota that we harvest as seafood are safe to eat.

As the Water Framework Directive implementation is more advanced (certainly in time) than that of the MSFD, it seems possible that member states would cover the management of most of the risks of chemical substances – most of which originate from land-based sources – under the Water Framework Directive river basin management plans (which include coastal waters, if not the territorial waters of the sea), and complement this with a review of any additional needs based on elements of information derived from their initial assessment of the environmental status (including risks and threats) of their marine waters.

Novel elements regarding chemical quality are the inclusion, in MSFD Annex III, of parameters reflecting ocean acidification (pH, pCO_2), which is likely to be a basin-wide phenomenon best addressed at that level.

At the time of writing, elements on which to base the common EU 'criteria and methodological standards' (MSFD Article 9(3)) had been assembled by groups of experts. It is expected that this will result in further guidance on the application of these descriptors during 2010–2011.

3.5 CONCLUSIONS

The policy framework for marine monitoring has become progressively more complex and comprehensive, with as main actor platforms the regional sea conventions around Europe which fostered and catalysed this development.

The European Communities, now the European Union, have gradually expanded their legislative instruments dealing with water quality, including for the marine environment. The MSFD reflects this development with its ambition to achieve good environmental status in the EU member states' marine waters by 2020 and to bring human use of marine goods and services to a sustainable level so that the human pressures on the marine environment are consistent with the maintenance of a good environmental status.

Monitoring is driven by information needs that will emerge from the first implementation steps of the MSFD. The function of the information will define the monitoring attributes. MSFD monitoring strategies are likely to become more complex as a result of the many needs that will have to be addressed.

The challenges of the ecosystem approach are moving the goalposts also for traditional marine chemical monitoring. Evidence-based policy not only requires the establishment of a well-functioning science–policy interface, but also that there is a dynamic interface with the management of monitoring.

REFERENCES

Borja, A., Bricker, S. B., Dauer, D. M. *et al.* (2008). Overview of integrative tools and methods in assessing ecological integrity in estuarine and coastal systems worldwide. *Mar. Pollut. Bull.*, **56**, 1519–1537.

EC (2002a). 6th DECISION No 1600/2002/EC of 22 July 2002 laying down the Sixth Community Environment Action Programme (2002–2012), http://eur-lex.europa.eu/LexUriServ/.do?uri=CELEX:32002D1600:EN:NOT [last access 17 February 2010].

EC (2002b). Communication from the Commission. Towards a strategy to protect and conserve the marine environment, (COM)2002/539, http://eur-lex.europa.eu/LexUriServ/LexUriServ.do?uri=:2002:0539:FIN:EN:PDF [last access 17 February 2010].

EC (2003). Common Implementation Strategy for the Water Framework Directive (2000/60/EC), Guidance document no. 7, Monitoring under the Water Framework Directive, Brussels

EC (2005). Communication from the Commission. Thematic strategy on the protection and conservation of the marine environment COM(2005) 504, http://eur-lex.europa.eu/LexUriServ/.do?uri=COM:2005:0504:FIN:EN:PDF [last access on 17 February 2010]

EC (2006a). EU Marine Strategy. The story behind the strategy, European Commission, http://.europa.eu/environment/water/marine/pdf/eumarinestrategy_storybook.pdf [last access 17 February 2010].

EC (2006b). Regulation 1907/2006 of the European Parliament and of the Council of 18 December 2006 concerning the Registration, Evaluation, Authorisation and Restriction of Chemicals (REACH).

EC (2007). Communication from the Commission. An integrated maritime policy for the European Union, COM(2007) 575, http://eur-lex.europa.eu/LexUriServ/LexUriServ.do?uri=:2007:0575:FIN:EN:PDF [last access 17 February 2010].

EC (2008). Communication from the Commission. A European strategy for marine and maritime research: a coherent European Research Area framework in support of a sustainable use of oceans and seas COM(2008) 534, http://eur-lex.europa.eu/LexUriServ/LexUriServ.do?uri=COM::0534::EN:PDF [last access 17 February 2010].

EC (2009a). Communication from the Commission. Progress report on the EU's integrated maritime policy, 15.10.2009 COM(2009)540, http://eur-lex.europa.eu/LexUriServ/LexUriServ.do?uri=:2009:0540:FIN:EN:PDF [last access 17 February 2010].

EC (2009b). Common Implementation Strategy for the Water Framework Directive (2000/60/EC), Guidance document no. 19, Guidance on Surface Water Chemical Monitoring under the Water Framework Directive. Technical Report 2009-025, Brussels.

EC (2010). Commission Decision 2010/477/ EU of 1 September 2010 on criteria and methodological standards on good environmental status of marine waters, O.J. L 232/14 of 2 September 2010, http://lex.europa.eu/LexUriServ/LexUriServ.do?uri=OJ:L:2010:232:0014:0024:EN:PDF [last access 22 December 2010].

EEA (2007). State of the Environment Report 2007, Chapter 5: Marine and Coastal environment, http://www.eea.europa.eu/publications/state_of_environment_report_2007_1/chapter5.pdf [last access 17 February 2010].

FAO (n.d.). The ecosystem approach to fisheries management, http://www.fao.org/fishery/topic/13261/en [last access 17 February 2010].

Greenpeace, WWF, IFAW *et al*. (n.d.). Defining good environmental status in the context of the European Marine Strategy Directive – What constitutes a healthy marine environment? Memorandum, http://www.seas-at-risk.org/1mages/NGO%20paper%on%20Good%20%Status%20(FINAL)_2.pdf [last access 17 February 2010].

Hassan, D. (2006). *Protecting the Marine Environment from Land-based Sources of Pollution: Towards Effective International Cooperation*, (especially Chapter 6: Regional frameworks for land-based sources of marine pollution control), Ashgate Publishing Ltd, Aldershot, UK.

HELCOM (2004). 30 years of protecting the Baltic Sea, HELCOM 1974-2004, Helsinki Commission, http://www.helcom.fi/stc/files/Publications/OtherPublications/30_years_jubilee.pdf [last access 17 February 2010].

ICES (2005). *Guidance on the Application of the Ecosystem Approach to Management of Human Activities in the European Marine Environment*, International Council for the Exploration of the Sea (ICES) Cooperative Research Report, No. 273, www.ices.dk.

OECD (n.d. (a)). Glossary of Statistical Terms, http://stats.oecd.org/glossary/detail.asp?ID=830 [last access 17 February 2010].

OECD (n.d. (b)). Using the Pressure-State-Response Model to develop Indicators of Sustainability, OECD Framework for Environmental Indicators, http://destinet.ew.eea.europa.eu/media_server/files/destinet/tools/measurement_instruments/oecd_p-s-r_indicator_model.pdf/OECD_P-S-R_model.pdf [last access 1 February 2010].

OSPAR (2000). Quality Status Report 2000, OSPAR Commission, London, www.ospar.org [last access 17 February 2010].

OSPAR (2003). Statement on the Ecosystem Approach to the Management of Human Activities 'Towards an Ecosystem Approach to the Management of Human Activities'", Joint Ministerial Meeting of the OSPAR Commission and the Helsinki Commission, Bremen, www.ospar.org [last access 17 February 2010].

OSPAR (2005). Synergies in Assessment and Monitoring between OSPAR and the European Union, Analysis of synergies in assessment and monitoring of hazardous substances, eutrophication, radioactive substances and offshore industry in the North-East Atlantic, http://www.ospar.org/documents/dbase/publications/p00230_SIAM%20consolidated.pdf [last access 17 February 2010].

OSPAR (2008). OSPAR Commission meeting June 2008 Summary Record, § 9.5 and Hazardous Substances Committee meeting February 2008, Annex 4, www.ospar.org [last access 17 February 2010].

Rogers S. I. and Greenaway B. (2005). A UK perspective on the development of marine ecosystem indicators. *Mar. Pollut. Bull.*, **50**, 9–19.

UN (2002). Johannesburg Plan of Implementation, chapter 4, http://www.un.org/esa/sustdev/documents/WSSD_POI_PD/English/POIToc.htm [last access 17 February 2010].

Part 3

Marine Monitored Parameters, Matrices and Related Techniques

4

Trace Elements

Martin M. Larsen[a], Jens Søndergaard[a], Gert Asmund[a], Koen Parmentier[b], and Peter Vermaercke[c]

[a]*National Environmental Research Institute, Aarhus University, Roskilde, Denmark*
[b]*ILVO-Visserij, Oostende, Belgium*
[c]*SCK.CEN-Belgian Nuclear Research Centre, Boeretang, Belgium*

Chemical Marine Monitoring: Policy Framework and Analytical Trends, First Edition.
Edited by Philippe Quevauviller, Patrick Roose, and Gert Verreet.

4.1 INTRODUCTION

In this chapter we summarize the current status for the analytical methods for the minor trace elements causing deleterious effects in the environment, often called heavy metals. Both historically used methods that are still in use, like electrochemical and atomic absorption methods, and modern multi-element methods based on plasma sources are included, and their main advantages and disadvantages are described. Special consideration is given to the digestion of samples before analysis.

The trace elements are naturally occurring substances and, therefore, a 'background' concentration is expected. This is different from most anthropogenic organic pollutants, like hexachlorobenzene and tributyltin, and is influencing the way data assessment should be treated. The environmental impact of trace elements varies with different abiotic factors (e.g. pH, alkalinity and salinity), and in the Water Framework Directive (WFD) the trace elements, therefore, have environmental quality standards depending on alkalinity in freshwater (European Commission, 2008); see Table 4.1. Also, many trace elements are essential elements, used in biomolecules, and as such are actively taken up or kept out of the organism by exudates that form complexes; for example, algae and copper (Whitfield, 2001; Andrade *et al*., 2010).

The anthropogenic component of metals in the atmosphere can be considered mainly from the burning of fossil fuels, either directly from coal burning or as an additive to fuels (e.g. leaded fuels), and from mining and metal smelters. Run-off from fields and transport through the waterways are another major source, and many point sources with treated or untreated effluents directly to rivers or the ocean can have significant local impact (HELCOM, 2010; OSPAR, 2010b).

Table 4.1 Trace elements and their inclusion in EU directives and the conventions.

Element	Directives		Conventions	
	Water Framework	Shell fish	OSPAR (North Sea)	HELCOM (Baltic Sea)
Ag		X		
As		X		
Cd	X	X	X	X
Cr		X		
Cu		X		
Hg	X	X	X	X
Ni	X	X		
Pb	X	X	X	X
Zn		X		
Li			(X)	
Al			(X)	
Tributyltin (Sn)	X		X	X

Marine Strategy Directive: not decided yet.

Mercury, lead and cadmium have no known biological function in plants and animals. Mercury is biomagnified through the food chain, mainly as the naturally occurring organometal methylmercury. Where mercury is often concentrated in the muscles, cadmium and lead are normally concentrated in the liver or kidney (or organs with similar function), where they can cause organ failures. In mammals, mercury and cadmium also affect the nervous system. In humans, even low environmental concentrations of lead and mercury have been shown to impair learning ability in children. Zinc and copper, on the other hand, are essential trace elements and are included in many enzyme systems (Metzler, 2003).

The trace elements included in the WFD priority substances list are cadmium, lead, mercury, nickel and tributyltin (an organometallic compound), all with established environmental quality status (EQS) values for water (European Commission, 2008). For mercury, an EQS for biota to protect top predators has also been set at $20\,\mu g\ kg^{-1}$. Most monitoring programmes tend to include zinc and copper, being trace elements of biological relevance. Both are also micronutrients that become toxic in higher doses. In the directive for shellfish waters SFD (European Commission, 2006) the list of metals includes those mentioned above, but also silver, arsenic and chromium; see Table 4.1. The target is such that the individual concentration and synergic effects of the metals must not exceed a level that gives rise to harmful effects on the shellfish or their larvae. For sediments, natural background contents of sediments depend on the mineral composition, with the main content of trace elements being in the fine fractions (Loring, 1990), and in most cases sand particles ($>63\,\mu m$) will act as a diluter of the metal concentrations. Hence, when analysing sediments, lithium and aluminium are usually measured to assist in normalizing the trace element concentrations (OSPAR, 2010a: Annex 5) to a common sediment matrix of, for example, 5% Al or $50\,mg\ kg^{-1}$ Li.

In the marine conventions, trace elements have traditionally been monitored in biota and sediments (OSPAR JAMP/CEMP), while monitoring in the water phase has been discontinued in most areas since the early 1990s. This is because results were often found to be not representative for a given area due to changing winds and currents. Biota and sediments are considered more robust, integrating the contamination on a site for longer time spans of weeks to years. The background for using biota and sediment in relation to the WFD is given in European Commission (2010).

4.2 ANALYTICAL METHODS

Methods for trace element analysis can be performed in a number of independent ways, based on the electrochemical, absorption, emission, fluorescence or X-ray properties of the element in question. Examples of the current methods are given below, with a short description of their benefits and shortcomings.

Historically, electrochemical methods have been used successfully for measuring amalgamating trace metals (Cd, Pb, Cu, Zn) in seawater using a mercury-drop electrode, whereas atomic absorption spectroscopy in different forms has been the method of choice for sediments and biota. In the 1990s, inductively coupled plasma (ICP) mass spectrometry (MS) was introduced and today it is the most widely used tool for general trace elements analysis, owing to its applicability to almost the whole periodic table of elements.

For special applications, especially in sediments and for screening, different methods employing X-rays can be useful (West *et al*., 2008); and for very precise and SI-unit traceable results, only neutron activation analysis (NAA) and isotope-dilution ICP-MS are currently available.

For most of the methods, European Norm or US-EPA standard methods exists, together with guidelines from the HELCOM and OSPAR conventions and the International Council for the Exploration of the Sea (ICES) (e.g. Loring and Rantala, 1992).

4.3 ATOMIC ABSORPTION AND EMISSION SPECTROMETRY (AAS/AES) TECHNIQUES

Both atomic absorption spectrometry (AAS) and atomic emission spectrometry (AES) techniques are usually based on dissolution of samples into a clear liquid that can be aspirated or pipetted into the analytical detector system. This calls for a digestion system to be used, with possibility of losses, contamination or improper dissolution of the matrix before the analysis. While some development is still going on in the mature AAS field, the main scientific development seems to be preconcentration and speciation techniques for routine applications (Evans *et al*., 2005). The first AAS and AES techniques used flames or (graphite) furnaces to ensure the elements were atomized, but in the last decade, ICP is becoming the preferred way of ensuring atomization, either measuring atomic emission from the plasma or extraction of ions to a mass spectrometric detector.

4.3.1 Digestion Methods for Biota

Most analytical techniques require the sample to be liquid, and usually this involves heating a sample in acid, namely digestion of the sample.

Heating the sample and acid can be done in a number of ways, but the main distinction is between open and closed digestion. In open-digestion systems, temperature is limited to the boiling point of the acid mixture or acid and peroxide mixtures used, and combinations of perchloric and nitric acids were common in the 1970s. In the closed-digestion systems, the sample and acid are heated in a closed vessel, often called a bomb, and the gases released increase the pressure inside the vessel, so temperatures above the normal boiling point of the mixture are achievable. The term 'bomb' is coined from what happens if the pressure exceeds the strength of the container. The bomb system used for ordinary ovens consists of an inner Teflon cylinder, with an airtight lid, enclosed in a stainless steel container.

The main concern in using an open-digestion system is the possible loss of volatile elements, mainly Hg, which is often minimized by a reflux system. The gain is the possibility of larger sample intake (up to 15 g compared with typically 1 g for closed vessels).

Heating in open vessels can be achieved by hot plates, or specially designed blocks that fit the digestion vessels, and these heat the acid from all around. For closed vessels of stainless steel, an ordinary oven can be used, typically at 250–300 °C over 4 h and cooling overnight. Since the first introduction of microwave ovens (Abusamra *et al*., 1975), microwaves have become the standard heating source for digestions. Microwave ovens for both closed and open digestions have been developed; the closed systems bear resemblance to the ordinary kitchen microwave, but energy output is often higher, and the distribution of microwaves inside the oven is much more controlled, with a specially designed holder for the digestion vessels, both for safety and equal distribution of the microwave energy. In modern closed microwaves, such as shown in Figure 4.1a, up to 64 samples can be digested in parallel in closed microwave systems, but the norm is 6–16 samples per digestion. For open microwave systems, such as shown in Figure 4.1b, usually only one sample at a time is digested. The digestion vessel can in this case be open, with reflux or a closed, bomb-like system. The vessel is lowered into a microwave cavity where it is heated by a focused microwave, resulting in fast digestion (around 10 min per sample).

Crucial to the success of microwave ovens for digestion is the design of the magnetron. In both open and closed systems, modern magnetrons are computer controlled and feed back measurements of temperature (and pressure for closed systems) adjust the added microwave energy, to give a reproducible heating and sample treatment. In modern microwaves, a number of vessels (if not all) are monitored by infrared temperature sensors. Pressure is usually monitored in only one sample or as a maximum of a number of samples. Setting up the system, the sample(s) for pressure or

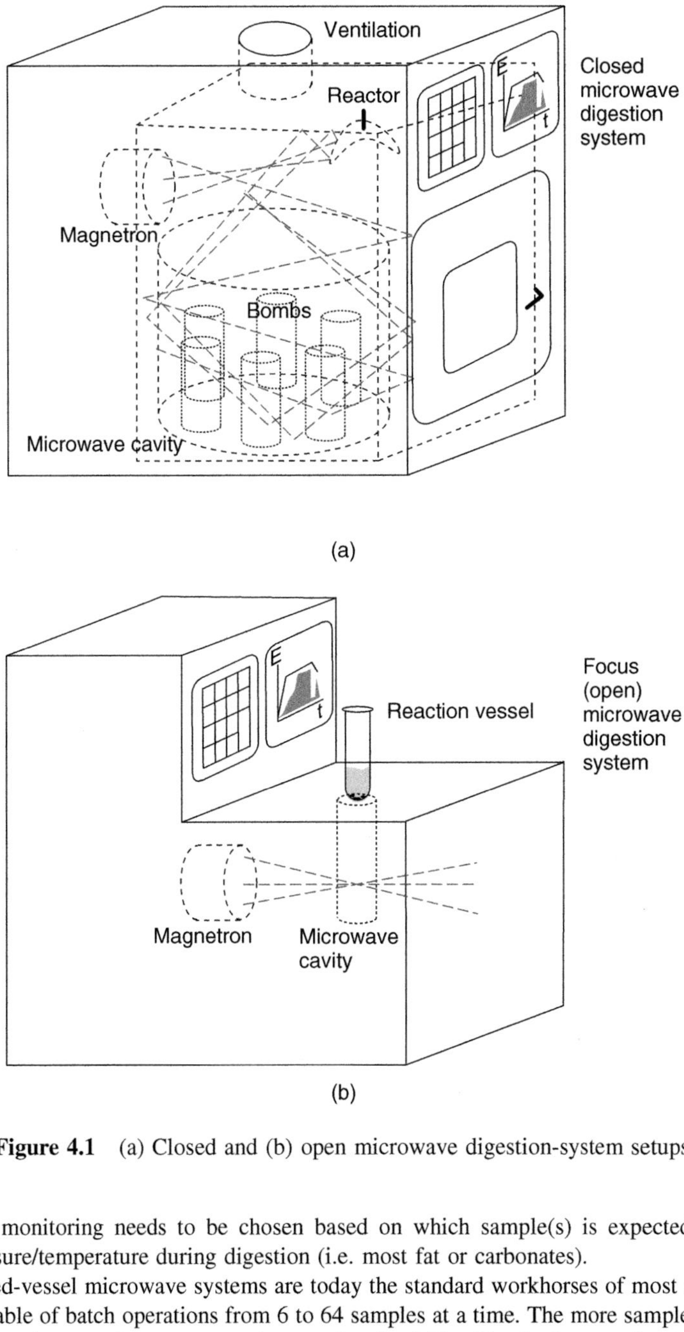

Figure 4.1 (a) Closed and (b) open microwave digestion-system setups.

temperature monitoring needs to be chosen based on which sample(s) is expected to give the highest pressure/temperature during digestion (i.e. most fat or carbonates).

The closed-vessel microwave systems are today the standard workhorses of most modern laboratories, capable of batch operations from 6 to 64 samples at a time. The more samples to digest at the same time, the more homogeneous the samples should be. For the best result, similar samples should be digested together, to avoid sample blow outs e.g. if pressure builds above safety levels in the fattiest samples while monitoring pressure on a non-fatty sample. That said, focused microwave systems can be used in on-line digestion systems or in single-sample preparations when there are large differences between individual samples.

The simplest digestion is just nitric acid, which introduces a minimum of interferences in the subsequent analysis. For metals like Hg, Cu and Zn in lean matrices, nitric acid is usually sufficient, but for samples with high fat content and for botanical samples, oxidation is boosted by addition of hydrogen peroxide and perchloric or hydrochloric acid. In cases with a high content of siliceous minerals, even hydrofluoric acid (HF) in combination with aqua regia can be necessary (Lamble and Hill, 1998).

For metals that are predominantly present as organometals in tissues, such as As, Hg and Se, special care should be taken to break the organometal bond, especially if the analytical method is hydride generation (HG)- or cold vapour (CV)-AAS, which requires a specific oxidation state of the analytes. For As, sulfuric acid has been employed to break down arsenobetaine and other organo-arsenic substances. For Hg, bromide/bromate oxidation, and for Se, nitric acid and hydrogen peroxide are often sufficient, but in the case of HG-AAS, perchloric acid and sulfuric acid is needed to digest organo-selenium compounds.

Especially problematic elements are Al, Fe and Cr, for which it is often a problem to find the correct value of plant reference materials with high silicate content. In this case, HF can be the only solution. Use of HF makes sample handling more time consuming, and is a potential health hazard for the personnel. HF is either evaporated to dryness in an effective fume hood and redissolved in nitric acid after digestion or neutralized with boric acid. Neutralization result in a very complex sample matrix that can result in high blanks and detection limits, and standard addition or matrix-matched calibration should be used in the final analysis.

In most cases, levels of Zn in biota are well above the inevitable blank levels in reagents and air, whereas Al concentrations can be very low and at the same level as the digestion blanks. In this case, clean-room facilities and ultrapure chemicals are required. It is strongly recommended to test the digestion method on a certified reference material that matches the actual samples.

The use of platinum group elements (PGEs) in automotive exhaust catalysts has opened a new interest in the fate and analysis of especially Pt, Pd and Rh (Ravindra *et al*., 2004). Mainly road dust and plants along roads have been targeted for analysis. Digestion of plants should include both nitric and hydrochloric acid (aqua regia) after grinding to a fine powder (Djingova *et al*., 2003). Also, for other biota, aqua regia is usually employed for digestion, and it can also be necessary for other precious metals, especially Ag (Daskalakis *et al*., 1997) in molluscs – although care should be taken, as too high a chloride concentration can also result in precipitation of AgCl.

4.3.2 Digestion Methods for Sediments

For sediments, the discussion of open- and closed-vessel digestions for biota are also valid. While the content of fatty acids is usually small, other substances, such as plant residues or peat soils, can give faster temperature or pressure rises in closed vessels, and high carbonate content also gives a higher pressure due to CO_2 release. The use of high-pressure bombs or microwave oven digestion vessels with glass or perfluorinated (perfluoralkoxy or polytetrafluorethylene) liners is recommended.

To measure the total content of metals in sediments in acid digestions, it is usually necessary to use HF to dissolve the silica in sand particles. HF is very harmful to humans, so, to avoid using it, many standards instead use strong mineral acids (nitric acid or aqua regia). When using HF, glass containers cannot be used, as HF etches the glass, making it more prone for carry-over effects or, in the worst case, ruptures. For some sediment types which are very fine fractions (clay/silt, $<63\,\mu m$ grain size) or very organic (humic acid), the difference between total digest and strong acid digests are small (Smedes, 1999), but for more sandy samples the differences can be very large (factor of two), especially for Cr and V. Total methods are usually prescribed in geological surveys, or if comparison with non-invasive elemental analysis (X-ray, instrumental NAA (INAA)) is needed.

The strong acid methods are more relevant for environmental studies, as metals in the sand fraction are released on a geological timescale, whereas clay–silt-bound metals can be ingested and released in the digestion tract of sediment-living organisms. It has also been suggested that the strong acid digest is more relevant to the anthropogenic input to the ocean, which in general is adsorbed to the clay–silt particles more than on sand owing to the higher surface area/weight ratio and electrostatic properties. The strong acids are expected to extract all metals bound loosely and up to iron–manganese nodules, but not in silicates (Rao *et al.*, 2008)

To verify the usefulness of strong acid digest in a given situation, it is possible to compare Li and Al concentrations in the digests of total and strong acids. If concentrations of Li is the same in the two type of digests, all of the fine material can be expected to be in solution, and any remaining fraction mainly of geological interest. Li, Sc and Al have been used as tracers for the fine fraction ($<63\,\mu m$) as they are mainly found in this fraction of different mineralogical sediment types (Loring, 1990). In some sediments from very old mountain ranges (igneous rocks) that have been glacially eroded, the grain size and normalizers for grain size distribution are unrelated to the metal content in general, as, for example, around Greenland (Loring and Asmund, 1996).

Platinium group elements analysis in sediments and geological material can be performed with AAS, ICP-MS and INAA, with detection limits in the range of $0.1\,ng\ g^{-1}$ to $5\,ng\ g^{-1}$ for the elements Au, Pt, Pd, Ru, Ir and Rh in precious metal ores (Perry *et al.*, 1995). Recovery of reference materials indicates that after aqua regia and HF dissolutions only 20–30% recovery was achieved, unless the small residue was fused with Na_2O_2 and added to the solution. This underlines the care that should be taken when analysing PGEs.

4.3.3 Flame AAS

A simple search on Web of Science returns more than 20 000 records on AAS, in the period 1963 to 2010, with a steady increasing number per year: rounding to 100 per year in 1975 and 500 per year in 1991, and since 2007 more than 1000 publications per year.

The original paper on the use of AAS in analytical chemistry is usually attributed to Walsh (1955), who laid the theoretical ground and identified the hollow cathode lamp (HCL) as the most promising owing to its line shape with a width smaller than the absorption peaks in the flame. From the beginning of the 1980s to well up in this century, AAS has been the cornerstone of elemental analysis. It is a well-matured method, and instrumentation and software nowadays can live up to both automated quality control (QC) and Code of US Federal Regulation (CFR) part 11 compliance for electronic traceability demanded for accreditation and audit trails of the calculations behind the analysis.

A history of the AAS technique was given by L'vov (2005) in a 50 years celebration and current status of the achievement of the technique. Although now surpassed in many ways by ICP-MS (see below), the detection limits and precision of AAS methods has been a very good match for the background concentrations in the environment, and for single-element studies it can still offer good precision and reproducibility of better than 1–2% over several hours of operation.

The sample is typically self-aspirated into the flame, where the sample is first made into small droplets in the nebulizer and spray chamber; see Figure 4.2a. Going up through the flame, the droplets are first desolvated and then the atoms go into the ground state; the more easily excitable atoms can even go into an ionized state, thereby reducing the absorption of ground-state wavelengths. For easily excited alkali metals an ionization buffer (e.g. adding lanthanum to the sample) can be used to suppress ionization. Alternatively, atomic emission spectroscopy can be used instead of atomic absorption, thus eliminating the use of an HCL. The atomic peak width depends on the temperature of the flame (Doppler broadening), but as the HCL is at a lower temperature, the

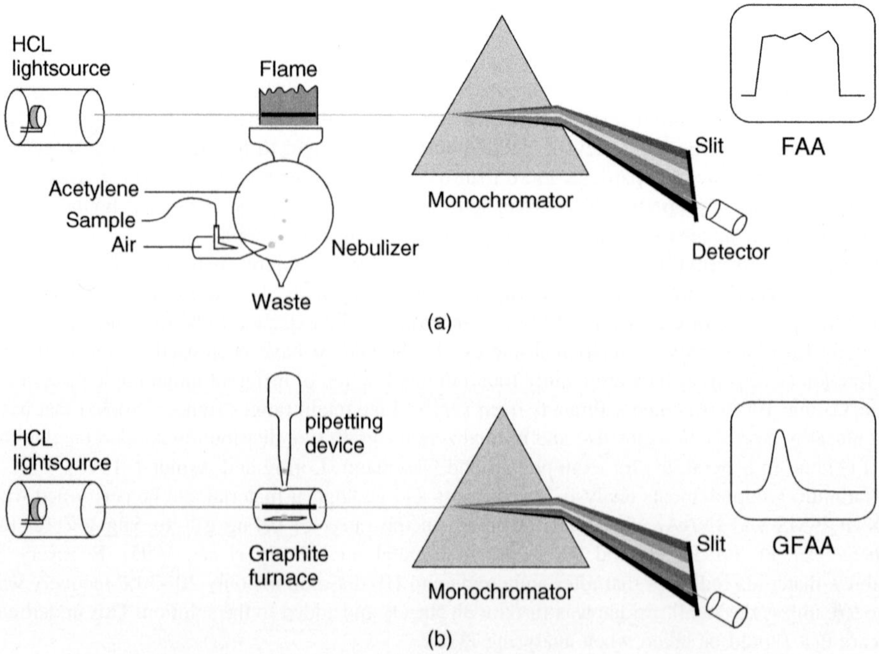

Figure 4.2 (a) Flame and (b) graphite furnace AAS setup.

absorbing width is wider than that from the light source. To account for background absorption in the flame (the flame itself, molecular species with very wide peaks), a chopper is usually employed to measure the background absorbance and subtract it (Figure 4.3a).

The viscosity of the sample determines the flow rate, so matrix-matched standard curves or standard addition calibration are the usual way to counter this physical matrix effect. Some reactions can still occur in the flame, and ionization of the analyte atoms can be controlled by an ionization buffer (i.e. easily ionizable atoms like alkali metals) to minimize the ionization of the analyte atoms.

A flame temperature of 1700–3150 °C is enough to atomize most metals, but the sensitivity is not always sufficient for measurements of environmentally relevant concentrations, except for minor constituents and the lower melting-point metals like Zn, Cu, Al and the alkali metals. The temperature of the flame is determined by its oxidation agent and fuel. Air–acetylene flames (flame temperature 2300 °C) are typically used for Zn, Cu and Li, and a wide burner head can be used to give a good sensitivity. For higher vaporization temperatures, a 2700 °C acetylene–nitrous oxide flame has to be used (e.g. for Al). The burner head is typically smaller, reducing the absorption light path compared with air–acetylene burner heads, and a carbon 'chimney' from acetylene can build up if gas flow is not adjusted properly.

The single- or multi-element lamps using an HCL design are still used; for some elements, more powerful electronic discharge lamps (EDLs) are available. The main recent development, though, has been on the detector side, where the original photomultiplier tube (PMT) has usually been exchanged to a charge-coupled devices (CCD) that offers a superior signal-to-noise ratio. A number of special CCDs have been developed for emission spectroscopy, mainly for ICP-AES (see below), but have also been implemented in modern AAS instruments. This has increased the detection limits by a factor of 10, because of the improved signal-to-noise ratio due to lower dark currents of the CCD compared with a PMT.

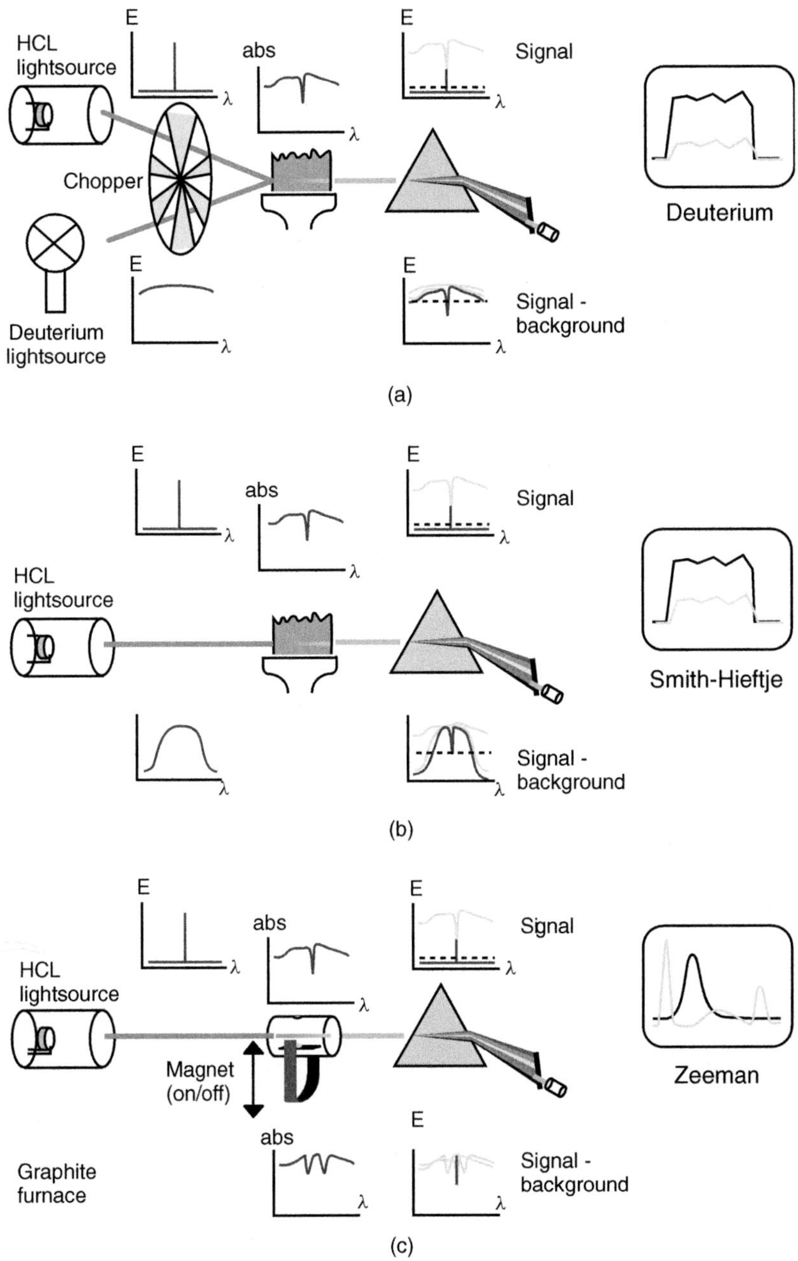

Figure 4.3 Background correction methods: (a) deuterium lamp; (b) Smith–Hieftje; (c) Zeeman effect.

Even with HCL/EDL lamps and CCD detectors, some spectral interference persists, but usually it is possible to select an atomic absorption wavelength without, or with very little, interference; most modern software has wavelength and help functions built in to avoid these. See Section 4.3.4 for more information.

For some elements, an HCL/EDL lamp is not even needed, as emission lines are as strong as or stronger than absorption, especially for the alkaline earth elements like Na and K.

4.3.4 Graphite Furnace AAS

To reduce the sample amount and to increase sensitivity, flameless atomization systems were developed. The graphite furnace is small (a cylinder a couple of centimetres long and 0.5 cm in diameter). The tube can contain a small platform (L'vov, 1990) to collect the sample (5–100 μl droplets); this helps to prevent condensation of the sample (see below). The stepwise heating allows drying (80–150 °C) to remove water, ashing (400–700 °C) to avoid organic interference in the atomization step, pre-atomization (1000–1500 °C) for removing potential interfering metals of lower volatility, and finally atomization and measurement (2000–3000 °C). The furnace is baked out at even higher temperature and cooled down before the next sample injection. Owing to a strong electrical field applied to the oven, heating rates of in excess of 1000 °C s^{-1} are possible. The limited residence time causes an absorbance peak. Both height and area of the peak allow quantification; the latter is usually preferable. The technique is especially suitable for Ag, Cd, Cr, Cu, Ni and Pb, whereas the sensitivity for Zn is too high, rendering blanks prohibitively high unless clean-room conditions are used. Recent commercially available AAS models allow simultaneous measurement of several elements by using multi-element HCLs or EDLs, at the price of higher detection limits and less optimized temperature programmes.

The theory of GF-AAS is well developed, and through the characteristic mass (concentration needed to obtain 1% absorbance in a flame or 0.0044 A s in integrated graphite furnace atomic absorption) semi-quantitative determination is possible even without standards. The direct measurement without standardization was already envisioned in the mid 1980s (Slavin and Carnrick, 1984), utilizing the stabilized temperature platform furnace (STPF) technique (L'vov, 1990) described above, together with a matrix modifier. For precise measurements, the standard method prescribes calibration though, but the characteristic mass is a very good indicator of the system performance.

The main interferences involve inter-element effects:

- intermetallic bonds occurring in the gas phase;
- metal–nonmetal bonds increase atomization temperature (e.g. Pb and Cl in seawater).

Condensation in the gas phase is usually prevented by the L'vov platform: the heating of the walls of the furnace precedes that of the inner gas; the platform in the centre reaches the atomization temperature when the gas is already hot.

Spectral interferences can often be overcome by choosing another, less-sensitive wavelength.

Background correction methods (Zeeman effect, deuterium lamp) are used to cope with nonspecific absorption, caused either by nondissociated molecules in the light path or scattering caused by particles like salt. This nonspecific absorption takes place over very broad wavelength bands, in contrast to the very narrow atomic lines. As absorbance is an additive function ($A = \sum(A_i)$), quantification of the background absorbance allows proper correction. All methods are based upon measurement of the total signal and measurement of the background-only absorbance.

Correction with a deuterium lamp is achieved by measuring alternating light coming from either the specific lamp or the deuterium lamp, pulsed through the atomization cell by means of a chopper; see Figure 4.3a. The specific absorbance (bandwidth ~0.005 nm) with the deuterium lamp is negligible compared with the total bandwidth passing the monochromator (~1 nm). The deuterium lamp is used because of the very intense radiation in the UV spectrum (190–350 nm). At higher wavelengths, tungsten halide lamps are necessary for background correction.

The Zeeman effect is based on the fact that a strong magnetic field splits the atomic spectral line in two or more. In the normal Zeeman effect, this spectral line splits in three, a π component at the original wavelength and two σ components at slightly higher and lower wavelengths. The shift depends on the magnetic field force, the sum of the three lines being equivalent to the original signal. Moreover, the components are polarized: the π component parallel and the σ component perpendicular to the magnetic field. This is true for, for example, Cd and Zn (normal Zeeman effect); for other elements, the spilt patterns can be readily more complex, with multiple π and σ lines, but following the same principle. Molecular and particle components do not undergo this phenomenon; as such, an accurate background correction can be made at the very wavelength the determination is made.

The magnetic field is usually applied to the atomic cell, because special lamps are required if the Zeeman effect were to work on the lamps. One could either use a static magnetic field and a rotating polarization crystal between lamp and cell, or an alternating magnetic field and a static polarization crystal between cell and detector. The parallel component is absorbed by the π component sticking at the same wavelength; the σ components are displaced outside the wavelength; see Figure 4.3c.

Background correction according to Smith–Hieftje is less accurate. It uses the self-absorption and broadening of atomic lines in the HCL when a high current is applied. There is, however, always a residual signal, so the background is over corrected for and the lamp life is compromised.

An automated recovery spike, or standard addition calibration, is often used, since it allows a simple way to cope with matrix effects. The sample is pipetted together with a blank or standard solution (when spiking) in one to three injections into the furnace (multiple addition). A plot of absorbance versus added concentration gives a straight line, and from the intercept the concentration of the unknown is calculated. Ideally, the spiked amount is comparable to the unknown. The method is time consuming, but usually very accurate. With unknown samples close to the detection limit, the error is often underestimated. Any background signal independent of atomic concentration is not corrected for using standard additions, so a separate determination of background in reagents and sample matrix must be subtracted after the standard addition.

The development of the STPF principle has gone from end-heated graphite tubes to transversally heated and miniaturization of the tubes, as shown in Figure 4.4. The main development is faster heating times and the removal of cold spots. The tube is protected from burning up through an

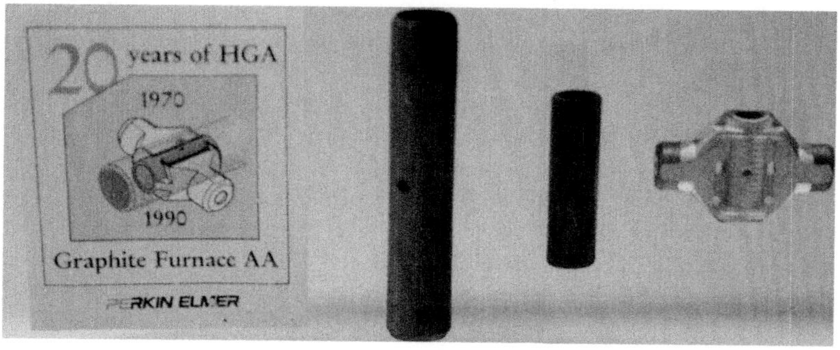

Figure 4.4 Development of graphite tubes over time – PE.

argon sheet gas, and the efficiency of this gas sheet is imperative to the number of firings achievable for each graphite tube. To increase residence time of the atomic vapour in the tube, end-caps have been employed.

GF-AAS is mainly applied to solutions. Good results are obtained, however, with solids using stirred slurry sampling for direct solid sampling (Cal-Prieto *et al.*, 2002). The fine particles analysed (from sludge, sediment or suspended matter) are mixed with a certain amount of liquid, creating a fluid mud, kept homogeneous through vigorous stirring by, for example, ultrasound. This slurry is pipetted directly into the furnace. Care must be taken that standards are representative for the sample under analysis, and reproducibility of the results needs to be checked, but can be excellent if the slurry is uniform at the applied sampling volume. More information can be found in Miller-Ihli and Baker (2006).

4.3.5 Cold-Vapour and Hydride-Generation AAS

The classical furnace is less sensitive for volatile elements, like the hydride-forming elements As, Bi, Ge, Sb, Se and Te. The hydride-generation system offers excellent sensitivity, because all these elements (As(III), As(V), Bi(III), Ge(IV), Sb(III), Sb(V), Se(IV) and Te(IV)) form volatile hydrides when reduced by atomic hydrogen (generated by addition of $NaBH_4$ to acidified solutions). Subsequently, an inert gas stream, usually argon, separates the analytes from the matrix and leads them towards an electrically heated quartz tube, where atomization is performed at $\sim$1000 °C. Detection limits for these elements are typically reduced with a factor 5–25 compared with direct AAS (typically $0.1 - 0.2\,\mu g\,L^{-1}$); additionally, matrix effects are greatly eliminated, as only matrix carried to the quartz cell by the argon stream is seen by the detector. For use in the WFD context, the total is to be determined. This indicates that different oxidation states have to be converted; for example, Se by treatment with strong HCl at elevated temperature to ensure oxidation of Se(II) species to the hydride-generating Se(IV), As after treatment with ascorbic acid and HCl for reduction of As(V) to the more sensitive As(III) species.

The determination of mercury via cold-vapour AAS (CV-AAS) involves an analogous mechanism, making use of the easy reduction of both Hg^{2+} and Hg_2^{2+} to metallic Hg in both acid and alkaline media. Using Sn^{2+} as a reducing agent, this is a fast and quantitative reaction. The Hg(0) formed is removed via a stream of argon bubbles and lead towards a quarts cell. Since the vapour contains Hg in atomic form, heating is redundant (hence the name cold vapour), except to avoid water vapour in the cell, and the absorbance can be readily determined. Sensitivity goes well beyond the microgram per litre level, a must for environmental analysis of this very toxic element. For biota, $NaBH_4$ is used often for reduction, while sediments and water samples normally are reduced with Sn^{2+}. To increase sensitivity, gold traps can be used, forming Hg–Ag amalgam, which can be thermally desorbed above 450 °C. Detection limits are typically in the 1–10 ng L^{-1} range with amalgamation, depending on the time/volume of sample preconcentrated onto the gold trap.

Solids can be determined either after digestion or by direct analysis of dried samples by thermally decomposing the sample in an oxygen atmosphere at 700–800 °C (EPA method 7473), followed by removal of halogens and pre-concentration on a gold trap and release by thermal desorption (Salvato and Pirola, 1996). Two AAS cells of different length are used to extend the linear calibration range. The detection limit for the direct thermal decomposition is lower than digestion-based methods, as digestion usually includes a 50-fold dilution. The absolute detection limit of around 5 pg Hg is usually achieved on the long cell (25 cm). In the case of silicate-bound Hg, the release could be incomplete using thermal decomposition.

4.3.6 Atomic Fluorescence Techniques

The atomic fluorescence techniques are usually based on flow injection or a similar system for online preparation, typically hydride or cold vapour generation and gas–liquid separation. The main advantage of fluorescence compared with atomic absorption techniques is the sensitivity, typically a factor of 10–100 better than AAS, and the even higher specificity, as excitation and fluorescence signal must both match the measured element.

Atomic fluorescence spectrometry (AFS) systems for Hg are commercially available for analysis of gases, liquids or solids. Pre-concentration using amalgamation and thermal desorption reduces detection limits to sub ng L^{-1} levels for total Hg.

For seawater, the background level of Hg is sub ng L^{-1} low ng L^{-1} levels, and standard methods for analysis at this level are, for example, EPA methods 1631 and 245.7 and European Standards EN-13 506 and EN-12 338 (detection limit <0.2 ng L^{-1} using 25–50 mL sample), using oxidation to Hg^{2+} by bromide/bromate in hydrochloric acid, neutralization of surplus oxidant with hydroxylamine hydrochloride, followed by reduction of Hg^{2+} to Hg(0) with $SnCl_2$. Hg(g) is collected on gold traps and, after collection, thermally desorbed. Low concentrations can be masked by high organic contents or high iodide concentrations, and water vapour in the fluorescence cell can quench the fluorescence signal. The low natural concentration level requires operation under very clean conditions.

Solid sampling systems for Hg-AFS similar to the ones described for CV-AAS above are commercially available with detection limits even lower than achieved with AAS (EPA method 7474).

Hydride-forming elements can be analysed by atomic fluorescence methods as described under hydride-generation AAS, except the quartz cell can be exchanged for a hydrogen flame. Detection limits are typically better than AAS methods, at 2–30 ng L^{-1} for As, Se, Bi, Sb, Te and Sn. Using a flame system, fluorescence techniques can also be used for Zn, Pb and Cd.

4.3.7 ICP-AES

The ICP as a source of atomizer for AES (also called optical emission spectrometry (OES)) or ions into a mass spectrometer system shares the same sample introduction system and excitation/ionization source; see Figure 4.2.

The liquid sample is nebulized using a mechanism comparable to flame spectroscopy. The sample is introduced in the plasma at the end of a torch, consisting of three concentric tubes, via the inner tube (carrier gas). The outer tube is the plasma gas, supplying gas to the plasma and protecting the outer quartz tube from the heat. The gas flow is high (tens of litres per minute). The gas in the middle tube prevents deposition on the torch, changing its radio-frequency characteristics, and adjusts the height of the plasma above the torch. A toroidal plasma occurs when the carrier gas breaks through the plasma. The observation zone is ideally located between the induction zone and the plasma tail. The torch can be placed either horizontally (especially suited for AES) or vertically. The temperature is generally 6000–10 000 °C, but can be even higher.

4.3.7.1 Plasma Physics/Chemistry: What Happens in the Plasma

The plasma is a very complex mixture of molecules, ions, atoms and electrons; the presence of charged particles is inevitable to keep the plasma in its state, because the electrical conductivity permits energy delivery to the system by inductive coupling of an alternating magnetic field (ICP). Excitation and ionization occur through collision of particles. The major processes are:

- electron collision excitation – a fast electron collides with an atom, yielding a (more) excited atom and a slow electron;
- electron collision ionization – a fast electron collides with an atom, yielding an ion, free electron and a slow electron;
- excited atoms collide with atoms in the ground state and exchange excitation;
- recombination of (excited) ions forming (excited) atoms and molecules;
- meta-stable compounds formed by high-energetic precursors.

The basis is the application of an induction coil at the end of a quartz tube, fed with argon gas. A radio-frequency alternating current (at very stable frequency) imposes circular orbits to both electrons and Ar^+ ions. Owing to collisions, new charged particles are formed constantly, the temperature increases and the plasma is kept alive. The originally neutral argon gas needs to be started with a high-voltage spark. The choice of argon is obvious, since its first ionization energy is higher than that of the analytes, while lower than their second ionization energy, assisting the ionization of matrix elements whilst blocking the formation of higher charged particles.

The stability of an ICP is outstanding; the noise is of the order 0.1%. The main uncertainty contribution is from the nebulizer and spray chamber, and is in the order of 1–3%.

4.3.7.2 *Detector System*

Originally, ICP-OES systems consisted of either a sequential wavelength scanning monochromator and one photomultiplier tube, or a polychromator and a number of photomultiplier tubes to be set at the selected first-order wavelength. Nowadays, the signal from the plasma is split up and recombined in a raster-shaped form, including higher order lines, and read out by a chip, either a CCD or a charge injection device (CID). This allows analysis of a virtually unlimited amount of elements in the same run. As the difference in concentration levels of different elements can be huge, the CCD/CID chips for ICP-OES are specifically developed to resist blooming or over lightning of areas close together (Sweedler *et al*., 1989).

4.3.7.3 *Interferences and Spectral Overlap Correction*

Spectral overlap can usually be remediated by measuring at another, less-sensitive wavelength. Modern software allows easy and fast programs to choose the ideal wavelength for a certain analysis type (with its specific interferences) automatically. Another possibility is to use several wavelengths to make more robust calibrations (e.g. Schiavo *et al*., 2009) or modelling of interferences and automated subtraction.

The original ICP-OES always observed the plasma from the side (radial view), and optimized the view height to minimize interference or optimize signal. When viewing from the end of the plasma (axial view), a longer path length and, hence, higher signals were achieved, although with more self-absorption and interferences in the cold end of the plasma (Brenner and Zander, 2000). In modern instruments this is minimized by using a shear gas or end-on gas flow to cut off the cold end of the plasma in front of the detector (Silva *et al*., 2004). The linear range of ICP-OES is usually quite large, but combining radial and axial views in the same instrument (dual view) can extend this even further.

4.3.8 ICP-MS

The ICP provides an excellent ion source for introduction into a mass spectrometer. At typical plasma temperatures, over 90% of the metals are present as monovalent ions M^+. The mass spectrum consists of:

- Monovalent metal ions, leading to several peaks at masses according to the relative abundance of the isotopes (mostly no more than three at a reasonable abundance).
- Poly- (usually di-)valent metal ions with peaks situated at the m/z value of the nuclide. The ratio of divalent/monovalent is quite low, except for alkali metals.
- Molecular ions of the metal hydrides (MH^+), oxides (MO^+) or hydroxides (MOH^+), with peaks situated at accordingly higher masses. The presence of H and O in the sample is virtually inevitable.
- Molecular ions of plasma gas and solvent constituents (Ar^+, Ar_2^+, ArO^+, OH^+, N_2^+, ArO^+, $ArCl^+$) creating problems for the detection of several masses, but no higher than m/z of 80. $ArCl^+$ interferes with the detection of the only isotope $^{75}As^+$, rendering As analysis very challenging using ICP-MS.
- Reactions between plasma gas and matrix-ions (e.g. $CuAr^+$) interfering with the isotope ^{103}Rh.

For most trace metals, mass interferences can be overcome by appropriate choice of the isotope and acid mixture, but for some it is necessary to use mass correction; for example, US-EPA 6020, as shown in Table 4.2. Some examples of other interferences encountered in low-resolution quadrupole MS (QMS) are shown in Table 4.3, and a comprehensive list of polyatomic interferences can be found in May and Wiedmeyer (1998).

The quadrupole mass spectrometer consists of four rods placed at the corners of a square, and a radio-frequency voltage is applied diagonally over each pair of rods. Superimposing a direct current on the radio-frequency voltage, a mass scan can be induced by increasing the voltage over both electrode pairs simultaneously, since, at a certain voltage, only ions of a distinct m/z range have a stable trajectory towards the detector, collide with the negatively charged pole, set free electrons accelerating towards the positive pole, inducing a multiplication effect. Originally, MS devices had a very low resolution in the order of one mass unit. For metal analysis, there is little need for better resolution, since mass differences of isotopes from different elements with the same

Table 4.2 USEPA 6 020 standard correction formulas.

Isotope	Formula	In words
^{51}V	$(51) - (53) \times 3.127 + (52) \times 0.353$	$^{35}Cl^{16}O^+$ correction of $Cl^{37}O^{16+}$ minus ^{53}Cr by the ratio of ^{35}Cl to ^{37}Cl and ^{53}Cr to ^{52}Cr
^{75}As	$(75) - (77) \times 3.127 + (82) \times 2.736 - (83) \times 2.76$	$^{40}Ar^{35}Cl^{35}$ correction of $^{40}Ar^{37}Cl$ by the ^{77} Se mass inclusive Kr correction
^{82} Se	$(82) - (83) \times 1.009$	Se corrected for ^{82}Kr
^{98}Mo	$(98) - (99) \times 0.146$	Mo corrected for the ^{98}Ru
^{111}Cd	$(111) - (108) \times 1.073 + (106) \times 0.764$	$^{95}Mo^{16}O^+$ corrected for the $^{92}Mo^{16}O^+$ minus Pd^{108} from ratio of $Pd^{108/106}$
^{115}In	$(115) - (118) \times 0.016$	^{115}In corrected for Sn^{115}

Table 4.3 Common interferences in ICP-MS (de Boer, 2000; Thermo, 2007).

Element	Interferences observed	Alternative isotopes
^{27}Al	$^{54}Fe^{2+}$, $^{11}B^{16}O^+$, $^{12}C^{14}N^1H^+$	
^{43}Ca	$^{86}Sr^{2+}$	^{44}Ca
^{44}Ca	$^{88}Sr^{2+}$	^{43}Ca
^{51}V	$^{35}Cl^{16}O^+$, $^{11}B^{40}Ar^+$, $^{33}S^{18}O^+$, $^{34}S^{17}O^+$	
^{52}Cr	$^{40}Ar^{12}C^+$, $^{40}Ar^{12}C^+$, $^{38}Ar^{14}N^+$, $^{36}Ar^{16}O^+$, $^{35}Cl^{16}OH^+$, $^{37}Cl^{15}N^+$, $^{36}S^{16}O^+$	^{53}Cr
^{53}Cr	$^{37}Cl^{16}O^+$	^{52}Cr
^{55}Mn	$^{39}K^{16}O^+$, $^{15}N^{40}Ar^+$	
^{56}Fe	$^{40}Ar^{16}O^+$, $^{40}Ca^{16}O^+$	
^{57}Fe	^{27} $Mg^{16}O_2^{2+}$, $^{40}Ca^{16}O^1H^+$	
^{58}Ni	$^{42}Ca^{16}O^+$, $^{58}Fe^+$	^{60}Ni, ^{62}Ni
^{59}Co	$^{43}Ca^{16}O^+$, $^{42}Ca^{16}OH^+$, $^{118}Sn^{2+}$	
^{60}Ni	$^{23}Na^{37}Cl^+$, $^{44}Ca^{16}O^+$, ^{24} $Mg^{40}Ar^+$, $^{40}Ca^{16}O^+$	^{58}Ni, ^{62}Ni
^{63}Cu	$^{23}Na^{40}Ar^+$, $^{31}P^{16}O_2^+$, $^{32}S^{16}O_2^1H^+$	^{65}Cu
^{64}Zn	$^{40}Ar^{24}$ Mg^+, $^{32}S^{16}O_2^+$	^{66}Zn
^{65}Cu	$^{40}Ar^{25}$ Mg^+, $^{130}Ba^{2+}$	^{63}Cu
^{66}Zn	$^{40}Ar^{26}$ Mg^+, $^{34}S^{16}O_2^+$, $^{132}Ba^{2+}$	^{64}Zn
^{68}Zn	$^{40}Ar^{28}Si^+$, $^{136}Ba^{2+}$	^{66}Zn
^{75}As	$^{35}Cl^{40}Ar^+$, $^{39}K^{35}Cl^1H^+$	None
^{76} Se	$^{60}Ni^{16}O^+$, $^{38}Ar^{40}Ar^+$	^{82}Se
^{77}Se	$^{37}Cl^{40}Ar^+$	^{82}Se
^{78}Se	$^{38}Ar^{40}Ar^+$	^{82}Se
^{82}Se	$^{40}Ar^{42}Ca^+$, $^1H^{81}Br^+$	^{77}Se
^{95}Mo	$^{55}Mn^{40}Ar^+$, $^{40}Ca^{16}O^+$	^{98}Mo
^{98}Mo	^{98}Ru	^{95}Mo
^{111}Cd	$^{95}Mo^{16}O^+$, $^{95}O^{16+}$	^{114}Cd
^{114}Cd	$^{98}Mo^{16}O^+$, $^{114}Sn^+$	^{111}Cd
^{121}Sb	$^{81}Br^{40}Ar^+$, $^{85}Rb^{36}Ar^+$	
^{204}Pb	^{204}Hg	$^{206}Pb + ^{207}Pb + ^{208}Pb$
^{205}Tl	$(^{31}P_3^{16}O_7)^+$	

mass differ very slightly. Standard methods exists for ICP-MS determinations (e.g. USEPA, 1994; DIN, 2010).

To minimize the interferences, tuning for ICP-MS is done to minimize oxide and double-charge formation, typically using a Ce standard, as Ce is prone to oxide formation and double-charge formation. Both should be tuned to below 2%, as measured as the $^{140}Ce^+$ ratio to $^{140}Ce^{16}O^+$ (mass 140/156) and $^{140}Ce^+$ ratio to $^{140}Ce^{2+}$ (mass 140/70). To reduce the multi-ion interferences, a gas-filled cell can be put inside the vacuum chamber between the inlet and the mass spectrometer part. Depending on the gas, this is either referred to as a collision cell (H_2, He_2) or a dynamic reaction cell (NH_4, CH_4). In both cases, the polytatomic interference of, for example $^{75}As^+$ and $^{78}As^+$, $^{35}Cl^{40}Ar^+$ or $^{38}Ar^{40}Ar^+$ can be reduced by factors of 1000.

A number of different MS detectors have been used with different mass resolution R, defined as the mass divided by the mass difference of 10% peak overlap for equal-height peaks. QMS has $R = 10$, time-of-flight detectors have $R = 500$ and high-resolution (HR) sector field MS has $R = 4000–10\,000$ mass unit resolutions, capable of isolating even ^{75}As from ArCl interference. The complexity of adjusting the magnetic sector field mass spectrometer has often caused HR-MS to be used only for research, although automated optimization is beginning to narrow the gap. At the top of the line, multicollector systems are capable of simultaneous measurement of 10–13 isotopes at resolution of the sector field instrument, thus removing variation from the nebulizer and sample introduction.

4.3.8.1 Isotope Dilution ICP-MS, Reference Method

A very precise and accurate, albeit expensive, analysis method involves the addition of a small amount of a certain isotope of the analyte to the sample. The isotope behaves in a chemically identical way. By measuring this isotope together with at least one other natural isotopes, and knowing the normal isotopic distribution of the analyte, the original concentration in the sample (the sample 'dilutes' the spike) can be calculated by the ratio of the added isotope to the natural isotope(s). When using isotope dilution, mass interferences need to be carefully controlled, since they can seriously affect the measurement. The combination with HR-MS leads to extraordinarily good results.

HR-ICP-MS can be used for increased sensitivity (down to $pg\,L^{-1}$) and overlapping peaks (Zheng *et al.*, 2007). In ground water, determination of Co can be determined in the presence of Ca ($^{43}Ca^{16}O$) and Na ($^{23}Na^{36}Ar$). For many radionuclides, the sensitivity of HR-ICP-MS is needed to be able to measure at natural levels. The main challenge with HR-ICP-MS is keeping from contaminating the sample, so extreme caution and ultra clean environment and reagents are needed. The complex matrix of seawater can be analysed by dilution, with the purest of reagents and low-flow desolvating nebulizers. Even then, pre-concentration and removal of interferences, as described in Section 4.3.9, can be helpful. Using HR-ICP-MS for sediments, the usual problem of dissolving the matrix before introduction into the plasma apply (see Section 4.3.3), but dilution and the superior peak resolution can solve most of the remaining matrix problems. HR-ICP-MS instruments are still used mainly for research, but improvements in the software for tuning and stability are increasing the applicability in the environmental laboratory, although costs remain high.

For isotope studies, HR-ICP-MS is superior to ICP-QMS, as the peak shape can be traced, and most elements can be resolved from overlapping peaks. For very precise isotope studies, multicollector-HR-ICP-MS should be used, to account for the fluctuation in the nebulizer-spray chamber-plasma in real time.

4.3.9 Pre-concentration Methods for Seawater Analysis

Analysis of trace metals in seawater is challenging due to the extremely low concentrations of metals in seawater and the considerable influence of matrix elements, such as Na, Mg, Ca, K and Cl. ICP-AES permits direct detection of metals at the $\mu g\,L^{-1}$ level, whereas ICP-MS detection is in the $ng\,L^{-1}$ level. For ICP-AES, pre-concentration is necessary, whereas for ICP-MS the mass-spectral and matrix interferences caused by the seawater limit applicability. Apart from the mass overlap in Table 4.3, seawater causes signal suppression due to the influence of easily ionized matrix elements on the plasma (Na and K in particular), as well as signal drift caused by accumulation of salts on the cones and lenses of the ICP mass spectrometer. Simply diluting the seawater with $18\,M\Omega$ water is a way of reducing the matrix effects that can be applied in polluted coastal waters, but most often it results in inadequate instrument sensitivity or high blank levels due to the very low metal concentrations in seawater, especially in open-ocean samples.

A common way of solving the analytical problem of seawater analysis is by pre-concentrating the metals. A column filled with a chelating resin followed by a rinse of matrix elements from the resin and elution of the metals allows one-step concentration and simplification of the matrix (Beck *et al.*, 2002; Evans *et al.*, 2005). Pre-concentration and elution can either be made 'off-line' with subsequent analysis of the collected eluate (Ferrarello *et al.*, 2002) or 'on-line' as part of the (ICP-MS) workflow (Beck *et al.*, 2002). Using the on-line method gives an elution profile (concentrations versus time) of the metals and the average peak or integrated intensity or the maximum peak intensity is used as being proportional to the sample metal concentration. Performing the pre-concentration step on-line greatly reduces the contamination problems associated with the off-line method.

A number of chelating resins have been successfully applied for pre-concentration of trace metals in seawater, including Chelex 100, Metpac CC-1, Toyopearl AF-Chelate 650 M and MAF-8HQ (Sohrin *et al.*, 1998; Beck *et al.*, 2002; Jiménez *et al.*, 2002). These resins/columns work for most trace metals of interest, including Ni, Cu, Zn, Cd and Pb. As the adsorption of metals on the chelating resins is pH dependent, a buffer, most often ammonium acetate, is added to the seawater sample to obtain a constant pH before it is sent through the column. The optimum pH for most trace metals in seawater is usually between pH 5.5 and 7.5, depending on the resin used (Beck *et al.*, 2002; Jiménez *et al.*, 2002).

Special precautions may have to be applied when working with trace metals in seawater, as contamination problems are notoriously difficult to overcome, particularly with more ubiquitous metals such as Zn and Ni. That means that all equipment in contact with the water (bottles, filters, vials, pipette tips, etc.) has to be thoroughly acid cleaned prior to use and the handling of samples and so on should take place only in clean-room or clean-bench facilities.

4.3.9.1 Liquid–liquid Extraction

Pre-concentration of divalent metals after complexation with ammonium pyrrolidine dithiocarbamate (APDC) and extraction to methyl isobutyl ketone (MIBK) followed by back-extraction to nitric acid was all ready in use in the 1960s (e.g. Brooks *et al.*, 1967), making it possible to measure background levels of selected metals in seawater. MIBK often contained metals and redistillation was often necessary, so other solvents and chelates were investigated. Another classical method used is buffering to optimum pH (4.5), complexation with APDC and diethylammonium diethyldithiocarbamate (DDDC), followed by extraction to 20 mL Freon or chloroform. The organic phase is then separated, and either evaporated and the metals redissolved, or the complexation broken by nitric acid and back-extracted to water phase (Grasshoff *et al.*, 1999; Komjarova and Blust, 2006). Pre-concentration by a factor of 40–400 is achievable with 400 mL sample to 10 mL final volume for flame/ICP or 1–2 mL final volume for graphite-furnace analysis. All solvents and glassware need to be scrupulously cleaned before use to avoid contamination. The method is applicable to Cd, Co, Cu, Fe, Ni, Pb and Zn. This being said, a lot of variations on the original method exist, and other complexation agents can be used for extending the range of analytes.

Recent comparisons of different solvent extraction versus solid-phase extraction and co-precipitation can be found in Komjarova and Blust (2006). Usually, good agreement is found between the different methods. Possibilities for automation and the achievable detection limits for the available methods are, therefore, the main selection criteria, together with environmental safety (e.g. restrictions on the use of Freon).

4.3.9.2 Micro-Liquid–Liquid Extractions

To avoid the large amounts of solvents usually employed in liquid–liquid extractions, micro-liquid extractions have been employed for a number of species. A recent review of the techniques by Dadfarnia and Shabani (2010) concluded that a lot of potential is there, with enrichment factors of 20 to 560 for metals, but up to 1000 to 5000 for organotins and organomercury. The method most similar to standard liquid–liquid extraction is dispersive liquid–liquid extraction, where, for example, Cd is extracted from a sample by injection of APDC in ethanol and carbon tetrachloride (CCl_4), after centrifugation, 20 μL CCl_4 can be injected into a graphite furnace, and Cd determined. Similar methods have also been employed for As, Sb, V, Au, Pb, Cr, Cu Pb, Co, Pd, Ni, Se and Hg, using APDC, 1-(2-pyrldylazo)-2-naphthol and other chelating agents together with methanol, ethanol or acetone as disperser solvent and primarily CCl_4 and chloroform as extraction phase.

4.4 (INSTRUMENTAL) NEUTRON ACTIVATION ANALYSIS ((I)NAA)

NAA is a sensitive nondestructive analytical technique for multi-elemental analysis of major, minor and trace elements in samples from almost every field of interest or for any matrix (from water samples, to any organic matrix, to sludge or sediment or solid samples). NAA allows discrete sampling of elements, as it disregards the chemical form of a sample and focuses solely on its nucleus. Several applications are well known in the marine environment: major inorganic constituents and rare earth elements in marine sediments, trace element analysis in marine organisms such as fish and shellfish, elemental composition in several marine bio-indicators (e.g. for pollution) or as a tool in the determination of extractable organohalogens (EOX: EOCl, EOBr and EOI) in organisms from the marine ecosystem.

Hevesy and Levi, who found that samples containing certain rare earth elements became highly radioactive after exposure to a source of neutrons, discovered NAA in 1936. Although the technique is quite old (also in marine environment studies; Fukai and Meinke, 1959), and previously mainly used as a high-precision research tool for academia, the focus has shifted nowadays to a reference technique recognized by the International Bureau of Weights and Measures (www.BIPM.org) as a primary method for validation of some chemical techniques. This includes certification of reference materials for elemental analysis by NIST, IAEA and IRMM (see Chapter 11), but also as a well-established commercial analysis for multi-elemental analysis of mainly trace elements with high precision.

NAA is very sensitive for the determination of halogens, which are difficult to determine by other techniques: fluorine (using a fast pneumatic system down to 2 mg kg^{-1}), chlorine (milligram per kilogram level), bromine (μkg^{-1} level), iodine (50 μg kg^{-1} level using epithermal NAA) and of volatile elements (e.g. arsenic, selenium, mercury), or of elements that are hard to dissolve (e.g. chromium and tin). When used in combination with appropriate separation techniques, NAA can also provide valuable information about trace element speciation for, for example, As, Se, Cr and Sb (Salbu and Steinnes, 1992).

In NAA, the sample is bombarded with neutrons, causing the elements to form radioactive isotopes (Glascock, 2010), as shown in Figure 4.5. The analysis is then performed using γ-ray spectrometry as follows:

- Samples are homogenized, dried or freeze-dried (depending on the matrix and different, for example, for sediments or organic matrices (sea lettuce, fish or mussels) and weighed (a few milligrams to several hundred milligram samples). The samples are packed in irradiation vials (typically poly-ethylene) together with flux monitors and QC material – no additional sample preparation (extraction, dissolving or dilution) is needed.

- Once packed, the samples are placed near the core of a nuclear research reactor and irradiated for a predetermined time span, so that the neutrons emitted by the reactor interact with the nuclei rendering them radioactive; for example: ^{48}Ca (stable) $+^{1}n \rightarrow ^{49}Ca$ (unstable, $t_{1/2} = 8.716$ min) $\rightarrow ^{49}Sc + \gamma$ (at 3084.54 keV).

- The compound nuclei relax to more stable configurations through the emission of nuclide-specific γ-rays, unique in energy and half-life time $t_{1/2}$ (varying between seconds and several

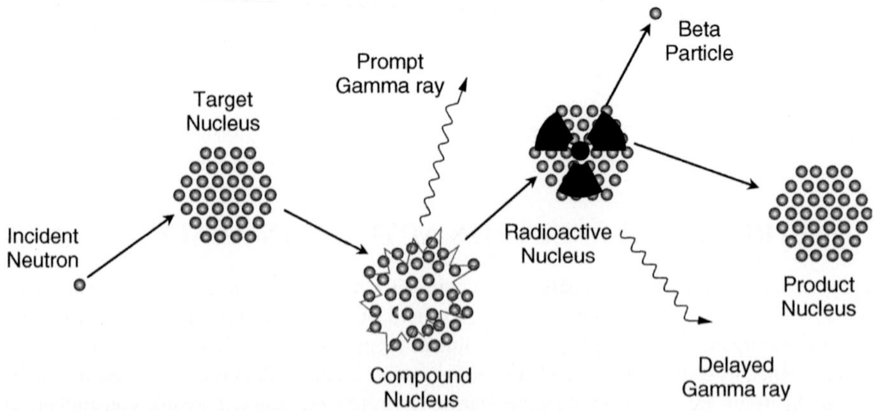

Figure 4.5 Principle of NAA.

years). This emitted radiation is a fingerprint and allows for any identification of the targeted element(s) in the sample, and the radiation intensity allows quantification. Since neutrons have no charge and will pass through any material, they will interact with all nuclei of the sample. In this respect, NAA is capable of providing a comprehensive determination of the sample's total elemental content regardless of oxidation state, chemical form or physical state of the desired element(s), so all sources of systematic or random uncertainty are well controlled.

- The sample is pulled out of the reactor and allowed to decay for a predetermined length of time. Counting is performed using liquid-nitrogen-cooled high purity Ge spectrometers (scintillation counters). The sample may be counted immediately, for elements with short $t_{1/2}$, or delayed for samples with longer $t_{1/2}$. The reason for the delay for isotopes with longer $t_{1/2}$ is to allow the isotopes with a short $t_{1/2}$ to decay, preventing interference during quantification of isotopes with longer $t_{1/2}$. The method is nondestructive and the sample can be reanalysed, if necessary. In most cases, the radioactivity level after 1 year's storage is sufficiently low to return the sample to the customer.
- The last step is the calculation of the concentrations. Nowadays, most laboratories in Europe use the k_0-standardization of NAA (k_0-NAA). This approach does not require a simultaneous irradiation of a standard for every element to be determined, but relies on independently determined k_0-factors that are traceable to an Au standard, which make the method quasi-absolute and allows us to determine about 50 to 70 elements simultaneously. The full range of elements (a full panoramic analysis) requires about two irradiations.

The method depends on the availability of a nuclear reactor. Today in Europe, NAA laboratories are still active in the Netherlands, Belgium, Germany, Austria, Italy, Slovenia, Portugal, Romania, Poland and Hungary. The cost of irradiations in the nuclear reactor and the cost of the computation make this technique relatively expensive for one element compared with most chemical techniques. Furthermore, as fast, cheap and accurate rarely match, analysis times for long-lived radionuclides may require several weeks. Figure 4.6 shows that, next to low-Z elements, it is impossible to determine some critical elements present in the marine environment, such as Pb, Bi and Tl, and that the detection limit for Cd is rather high.

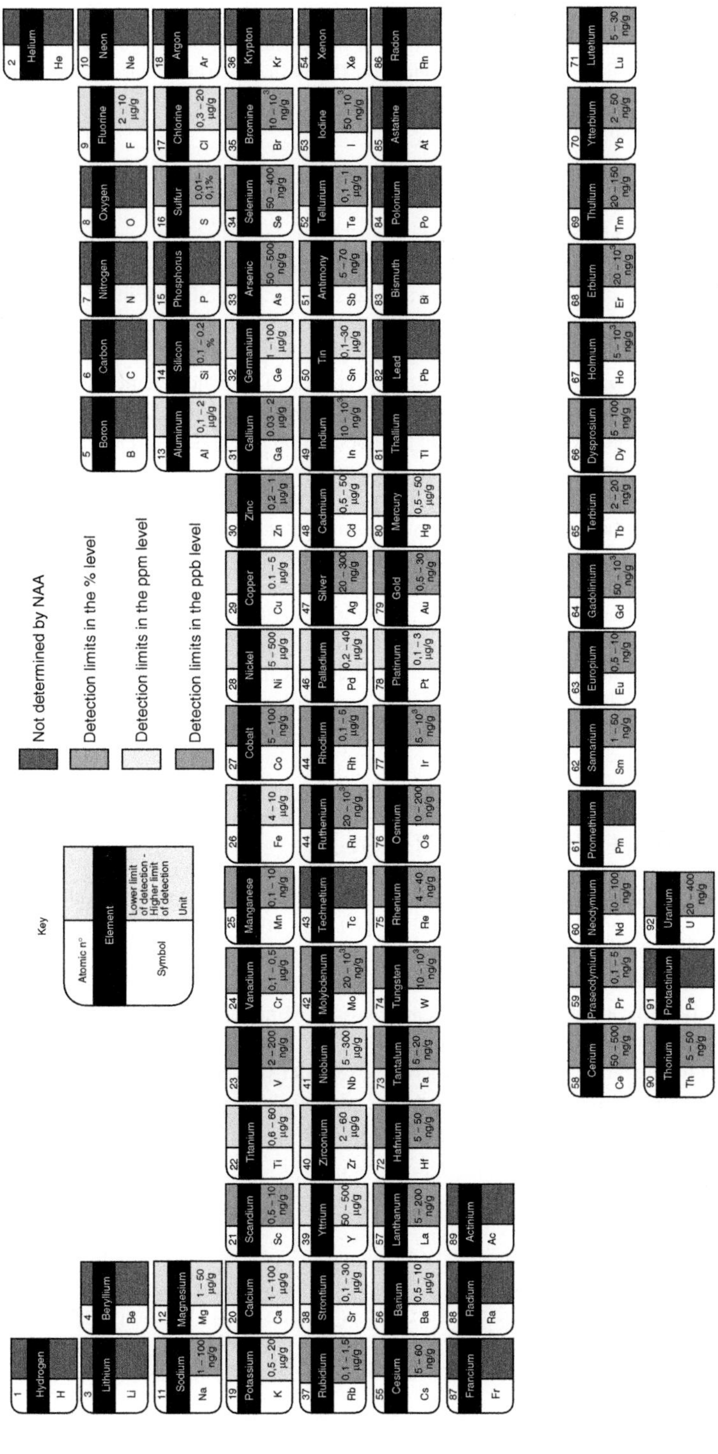

Figure 4.6 Detection limits in NAA (the lower value can be obtained in typical organic matrices such as sea lettuce – the higher value is representative for marine sediments). See plate section for a colour version of this image.

On the other hand, the very high accuracy of the method (in most cases only 4–5% (95% confidence level)) and the limited need for sample pretreatment, minimizing recovery problems, loss of volatile components or sample contamination, makes the method as good as bias free.

Figure 4.6 gives an indication of typical detection limits in a marine environment – the lower values correspond to an organic matrix (both sea lettuce and salmon); the higher values are for a solid matrix (estuarine sediment).

4.5 X-RAY TECHNIQUES

A number of techniques based on X-rays are available, and the possibility of in-situ measurements is used routinely in determining the extent of soil pollution. One problem for use in connection with sediments is the water content, and although the methods are noninvasive, they often require some pretreatment, like drying and tablet pressing before use. The major elements (Ca, Cu, Fe, K, Rb, Sr and Zn) can often be determined with X-ray methods (West *et al.*, 2008).

Detection limits and precision for use on site are not as good as for standard analysis for portable instruments (Melquiades and Appoloni, 2004), but field-portable X-ray fluorescence can be used to characterize and perform semi-quantitative analysis for determining the extent of pollution at contaminated sites. For processing large amounts of samples with minimum sample pretreatment, energy dispersive X-ray fluorescence can achieve more precise information and has also been used for surface sediments in rivers.

Other methods based on proton-induced X-ray emission (PIXE) have been used for sediment analysis, and water analysis after freeze-drying of the water or dried filter samples. PIXE can be applied to very small sample sizes with minimum sample preparation. The combination of electron microscopy and X-ray methods makes it possible to scan sediment cores or look at single particles in inhomogeneous samples.

4.6 ELECTROCHEMICAL TECHNIQUES

Electrochemical methods can be used for a variety of elements and samples. However, the real strength lies with seawater analysis. The method is cheap (the equipment cost is an order of magnitude lower than for AAS and ICP), but needs skilled operators and is slower. It is a multi-element method, as up to five elements can be determined in the same run. The equipment for it is light and can easily be brought onboard a research vessel. Miniaturization allows nano-electrodes to be deployed in otherwise inaccessible places.

Nearly all electrochemical methods use mercury, so strict procedures for collection and reuse of the mercury are imperative. Also, environmental mercury analysis requires a room separated from the one used for electrochemical analysis.

4.6.1 Polarography

Polarography is the forerunner for anodic stripping voltammetry (ASV) and cathodic stripping voltammetry (CSV). A dropping-mercury electrode is immersed in the sample. The voltage of the electrode is changed linearly and the current through the mercury is measured. There is a sudden current change at a potential characteristic for each element; this step is proportional to the concentration of that element.

The method was invented by Jaroslav Heyrovsky (Heyrovsky and Kuta, 1966), who received the 1959 Nobel Prize in Chemistry for his discovery of polarography.

4.6.2 Anodic Stripping Voltammetry

Three electrodes are immersed in the sample that has been adjusted to a specific pH. The working electrode concentrates the elements in the sample, an auxiliary electrode supplies current through the sample and a reference electrode measures the potential of the working electrode.

The reference electrode can be silver/silver chloride; the auxiliary electrode can be anything, but is usually either carbon or platinum to prevent any interferences.

The working electrode is the critical part of the method. In the first use of ASV, a hanging mercury-drop electrode was used. Later, the mercury-film electrodes came and, as the most sensitive electrode, the rotating mercury-film electrode. The mercury film is electrochemically plated on either graphite or glassy carbon; the latter seems the preferential choice.

The analyses start with a concentration step. The working electrode is kept at a sufficient low voltage (e.g. -1.4 V) during stirring of the sample and oxygen removal by a stream of nitrogen gas.

During the concentration step, dissolved elements are electroplated on the working electrode and amalgamated with the mercury film. The longer the plating time, the lower the detection limit of the method.

During the stripping step, the potential of the working electrode is gradually increased and the electric current associated with the dissolution of the individual elements is recorded. Refinements such as differential pulse or square wave measuring techniques can improve the accuracy and detection limit of the method.

The method is perfect for measuring cadmium and lead in seawater, where detection limits of a few nanograms per litre can easily be achieved. It is also very good for zinc, copper, and bismuth. ASV can be used to study the degree of complexation of an element in solution. If the element is complexed in the solution, a lower voltage reduces the element on the mercury surface. Thus, by measuring the overvoltage needed to determine the element, one can calculate the strength of the complexation in the sample (e.g. Bordin *et al*., 1988). The theory of ASV is thoroughly described, for example, in Bong (2009) and Protti (2001).

4.6.3 Cathodic Stripping Voltammetry

In CSV, the equipment is the same as in ASV, but the concentration step is performed at a more positive potential (e.g. at 0 V). At this potential, several organic metal compounds are adsorbed on the mercury film. In the stripping step, the voltage is lowered gradually and the current associated with the reduction of the metal is recorded. The method needs an organic complexation agent for the metal to be analysed (dimethylglyoxime for Ni and Co). CSV can be used for many elements, notably Fe, U, V, Cu, Zn, Mo, Ni and Co (Van Den Berg, 1986).

4.7 TECHNIQUES FOR ORGANOMETAL DETERMINATIONS

The main environmental organometallic substances covered in the WFD are methylmercury, naturally produced in anoxic sediments and bio-magnified in the food chain, and the antifouling agent tributyltin. A historic organometallic pollutant found in sediments is tetraethyl-lead, used as anti-knocking agent in leaded fuels, but not in current use in Europe. This is the subject of Chapter 5.

4.8 QUALITY ASSURANCE

The quality of an elemental analysis is determined both by the quality of the digestion/pretreatment of the sample and the quantification on the instrument. The AAS and ICP-MS methods are usually capable of performing analysis within 2–5% of reproducibility, and the main source of error is the digestion and environmental sampling steps. All analytical techniques are required to be fully validated, and routine internal QC programmes should be applied to all analytical methods. The QC procedures should be well defined and include details of maintenance, calibration of equipment and precautions to account for drift of the sensitivity during analysis. Procedural blanks, appropriate reference materials and some sample replicates should also be included in each batch of analyses. QC procedures should also cover conditions for storage, data handling and reporting.

A wide range of QC certified reference materials (CRMs) for heavy metal analysis is available (see Chapter 11), mainly for total analysis of trace elements, but also for stepwise extraction techniques and strong acid digestions, employing nitric acid, hydrofluoric acid or aqua regia. Laboratory reference materials (LRMs) can be used to assure the day-to-day quality of the analyses, especially where the elemental range or matrix matching is difficult to obtain from a single CRM. CRMs should at least be used when validating and re-evaluating standard methods to check method bias and robustness, ideally for the matrix types that the laboratory are analysing (for further details, see Chapter 11).

Several interlaboratory studies are also currently in action, and for accredited analysis regular participation in such studies is required. An international proficiency-testing provider is QUASIMEME (Wageningen UR, Alterra CWK, PO Box 47, 6700 AA Wageningen, The Netherlands), which covers trace elements in sediment, biota and water with bi-yearly test samples of each matrix, but other national proficiency testing schemes are available or recommended by the national accreditation authority. The use of proficiency testing schemes ensures that the methods are not tuned to the bias of the LRMs or CRMs in use, thus encouraging robust methods to be developed.

4.9 SUMMARY

The preferred method of choice for modern laboratories today is ICP-MS, owing to its versatility and low detection limits. The development of reaction/collision cells has made it possible to circumvent most of the interferences in the QMS detectors, but still some corrections are needed after the analysis. When only a few elements or especially Se and As (the elements most prone to mass interference in ICP-MS), other techniques like atomic fluorescence or atomic absorption can be a more cost-effective solution. For seawater samples, pre-concentration ICP-MS or electrochemical methods are the methods of choice, whereas for fast screening of sediments X-ray techniques can have advantages. The only methods currently available with direct traceability to SI units are isotope-dilution MS and neutron activation, with the possibility of high-precision analysis. For most applications, though, use of LRMs or CRMs is sufficient to ensure the quality of the analysis.

The instrumental limits of detection for the atomic absorption, emission and fluorescence techniques given by instrument vendors are shown in Table 4.4. The electrochemical techniques vary in detection limit, depending on the time of plating, but are generally on the level of GFAAS or ICP-MS, and X-ray detection limits depend on the soil matrix and specific method, but detection limits are usually comparable to flame AAS.

Table 4.4 Instrumental detection limits ($\mu g\,L^{-1}$) for atomic absorption, fluorescence and ICP methods.

Element	FAA	Hg/hydride	AFS[a]	GFAA	ICP-OES	ICP-MS
Ag	1.5		2 (F)	0.005	0.6	0.002
Al	45			0.1	1	0.005[b]
As	150	0.03	0.01	0.05	2	0.0006[c]
Au	9		2 (F)	0.15	1	0.000 9
B	1 000			20	1	0.003[d]
Ba	15			0.35	0.03	0.000 02[e]
Be	1.5			0.008	0.09	0.003
Bi	30	0.03	0.01	0.05	1	0.000 6
Br						0.2
C						0.8 (^{13}C)
Ca	1.5			0.01	0.05	0.000 2[e]
Cd	0.8		0.001	0.002	0.1	0.000 09[e]
Ce				1.5	0.2	0.000 2
Cl					0.2	12
Co	9		5 (F)	0.15		0.000 9
Cr	3			0.004		0.000 2[e]
Cs	15					0.000 3
Cu	1.5		2 (F)	0.014	0.4	0.000 2[d]
Dy	50				0.5	0.000 1 (^{163}Dy)
Er	60				0.5	0.000 1
Eu	30			0.2		0.00009
F						372
Fe	5			0.06	0.1	0.000 3[e]
Ga	75				1.5	0.000 2
Gd	1 800				0.9	0.000 8 (^{157}Gd)
Ge	300		0.05		1	0.001 (^{74}Ge)
Hf	300				0.5	0.000 8
Hg	300	0.009	0.0002		1	0.016 (^{202}Hg)
Ho	60				0.4	0.000 06
I						0.002
In	30				1	0.000 7
Ir	900			3	1	0.001
K	3			0.005	1	0.000 2[e]
La	3 000				0.4	0.000 9
Li	0.8			0.06	0.3	0.001[d]
Lu	1 000				0.1	0.000 05
Mg	0.15			0.004	0.04	0.0003[d]
Mn	1.5			0.005	0.1	0.000 07[e]
Mo	45			0.03	0.5	0.001
Na	0.3			0.005	0.5	0.000 3[d]

(*continued overleaf*)

Table 4.4 *(continued)*

Element	FAA	Hg/hydride	AFSa	GFAA	ICP-OES	ICP-MS
Nb	1 500				1	0.000 6
Nd	1 500				2	0.000 4
Ni	6			0.07	130	0.5
Os					6	
P	75 000				4	0.1[b]
Pb	15		0.01	0.05	1	0.000 04[e]
Pd	30			0.09	2	0.000 5
Pr	7 500				2	0.000 09
Pt	60			2	1	0.002
Rb	3			0.03	5	0.000 4
Re	750				0.5	0.000 3
Rh	6				5	0.000 2
Ru	100			1	1	0.000 2
S					10	28 (^{34}S)
Sb	45	0.15	0.01	0.05	2	0.000 9
Sc	30				0.1	0.004
Se	100	0.03	0.002	0.05	3	0.000 7[c]
Si	90			1	10	0.03[b]
Sm	3 000				2	0.000 2
Sn	150		0.01	0.1	2	0.000 5[b]
Sr	3			0.025	0.05	0.000 02[e]
Ta	1 500				1	0.000 5
Tb	900				2	0.000 04
Te	30	0.03	0.01	0.1	2	0.000 8 (^{125}Te)
Th					2	0.000 4
Ti	75			0.35	0.4	0.003 (^{49}Ti)
Tl	15			0.1	2	0.000 2
Tm	15				0.6	0.000 06
U	15 000				10	0.000 1
V	60			0.1	0.5	0.000 5
W	1 500	0.02				0.005
Y	75					0.000 2
Yb	8					0.000 2 (^{173}Yb)
Zn	1.5		1			0.000 3[e]
Zr	450				0.5	0.000 3

Elements from Table 4.1 are in italic. Footnote citations following an ICP-MS detection limit value refer to the use of specialized conditions (or the m/z of monitored ion).
[a]F: flame-AFS.
[b]Standard mode using Pt cones and quartz sample-introduction system.
[c]ICP-MS in DRC mode using Pt cones and quartz sample-introduction system.
[d]Standard mode in Class-100 clean room using Pt cones and quartz sample-introduction system.
[e]DRC mode in Class-100 clean room using Pt cones and quartz sample-introduction system.

4.10 ABBREVIATIONS

Conventions and legislative units

OSPAR	Oslo–Paris Convention for the North Sea
HELCOM	Helsinki Convention for the Baltic Sea
EU	European Union
WFD	Water Framework Directory
MSD	Marine Strategy Directory
SFD	Shellfish Directory
DIN	Deutsche (German) Industrial Norm
CFR	Code of (US) federal regulation

Atomic absorption techniques

AAS	Atomic absorption spectrometry
CVAAS	Cold-vapour atomic absorption spectrometry
HGAAS	Hydride-generation atomic absorption spectrometry
GFAAS	Graphite furnace atomic absorption spectrometry
AFS	Atomic fluorescence spectrometry
STPF	Stabilized temperature platform furnace
HCL	Hollow cathode lamp
EDL	Electronic discharge lamp

Plasma techniques

ICP-MS	Inductively coupled plasma mass spectrometry
ICP-HR-MS	Inductively coupled plasma-high resolution MS
ICP-QMS	Inductively coupled plasma quadrupole MS
ICP-AES/OES	Inductively coupled plasma atomic/optical emission spectrometry (synonyms)

X-ray techniques

ED-SEM	Energy-dispersive scanning electron microscopy
EDXRF	Energy dispersive X-ray
XAS	X-ray absorption spectroscopy
XPS	X-ray photon spectroscopy
XRF	X-ray fluorescence
XRD	X-ray diffraction
PXRF	Portable X-ray fluorescence
PIXE	Particle-induced X-ray emission

Electrochemical techniques

ASV	Anodic stripping voltammetry
CSV	Cathodic stripping voltammetry
PSA	Potentiometric stripping analysis

Quality control

LRM	Laboratory reference material (same as IRM)
IRM	Internal reference material
CRM	Certified reference material

REFERENCES

Abusamra A., Morris J. S. and Koirtyohann S. R. (1975). *Anal. Chem.*, **47**(8), 1475–1477.
Andrade S., Pulido M. J. and Correa J.A., (2010). *Chemosphere*, **78**, 397–401.
Beck N. G., Franks R. P. and Bruland K. W. (2002). *Anal. Chim. Acta*, **455**(1), 11–22.
Bong H., 2009. Polarography and voltammetry, http://www.drhuang.com/science/chemistry/electrochemistry [last access 1 January 2011].
Bordin G., Perttila M. and Scheinen H. (1988). *Mar. Pollut. Bull.*, **19**(7), 325–327.
Brenner I. B. and Zander A. T. (2000). *Spectrochim. Acta Part B: At. Spectrosc.* **55**(8), 1195–1240.
Brooks R. R., Presley B. J. and Kaplan I. R. (1967). *Anal. Chim. Acta*, **38**, 321–326.
Cal-Prieto M. J., Felipe-Sotelo M., Carlosena A. *et al.* (2002). *Talanta* **56**(1), 1–51.
Dadfarnia S. and Shabani A. M. H. (2010). *Anal. Chim. Acta*, **658**(2), 107–119.
Daskalakis K. D., O'Connor T. P. and Crecelius E. A. (1997). *Environ. Sci. Technol.*, **31**(8), 2303–2306.
De Boer J. L. M. (2000). *J. Anal. At. Spectrom.*, **15**(9), 1157–1160.
DIN (2010). *Foodstuffs – determination of trace elements – determination of arsenic, cadmium, mercury and lead in foodstuffs by inductively coupled plasma mass spectrometry (ICP-MS) after pressure digestion*, EN15763.
Djingova R., Kovacheva P., Wagner G. and Markert B. (2003). *Sci. Total Environ.*, **308**(1–3), 235–246.
European Commission (2006). Directive 2006/113/ec of the European Parliament and of the Council of 12 December 2006 on the quality required of shellfish waters, EU, Brussels, 7 pp.
European Commission (2008). Directive 2008/105/ec of the European Parliament and of the Council of 16 December 2008 on environmental quality standards in the field of water policy, EU, Brussels, 14 pp.
European Commision (2010). Guidance document no. 25 on chemical monitoring of sediment and biota under the Water Framework Directive, European Union, Luxembourg, 82 pp.
Evans E. H., Day J. A., Palmer C. D. *et al.* (2005). *J. Anal. At. Spectrom.*, **20**(6), 562–590.
Ferrarello C. N., de la Campa M. R. F. and Sanz-Medel A, (2002). *Anal. Bioanal. Chem.*, **373**(6), 412–421.
Fukai R. and Meinke W. W. (1959). *Limnol. Oceanogr.*, **4**(4), 398–408.
Glascock M. D. (2010) Overview of neutron activation analysis, http://archaeometry.missouri.edu/naa_overview.html [last access 8 February 2011].
Grasshoff K., Kremling K. and Ehrhardt M. (1999). *Methods of Sea-Water Analysis*, Wiley–VCH Verlag GmbH, Weinheim.
HELCOM (2010). Ecosystem health of the Baltic Sea, Helsinki Commision, Helsinki, 63 pp.
Heyrovsky J. and Kuta J. (1966). *Principles of Polarography*, Academic Press, 581 pp.
Jiménez M. S., Velarte R. and Castillo J. R. (2002). *Spectrochim. Acta Part B: At. Spectrosc.*, **57**(3), 391–402.
Komjarova I. and Blust R. (2006). *Anal. Chim. Acta*, **576**(2), 221–228.
Lamble K. J. and Hill S. J. (1998). *Analyst*, **123**(7), 103R–133R.
Loring D. H. (1990). *Mar. Chem.*, **29**, 155–168.
Loring D. H. and Asmund G. (1996). *Environ. Geol.*, **28**(1), 2–1.
Loring D. H. and Rantala R. T. T. (1992). *Earth-Sci. Rev.*, **32**(4), 235–283.
L'Vov B. V. (1990). *Spectrochim. Acta Part B: At. Spectrosc.*, **45**(7), 633–655.
L'Vov B. V., (2005). *J. Anal. Chem.*, **60**(4), 382–392.
May T. and Wiedmeyer R. (1998). *At. Spectrosc.*, **19**(5), 150–155.
Melquiades F. L. and Appoloni C. R. (2004). *J. Radioanal. Nucl. Chem.*, **262**(2), 533–541.
Metzler D. E. (2003). *Biochemistry. The Chemical Reactions of Living Cells*, Academic Press, San Diego, CA, 1900 pp.
Miller-Ihli N. J. and Baker S. A. (2006). *Slurry Sampling Graphite Furnace Atomic Absorption Spectrometry in Environmental Analyses*, John Wiley & Sons, Ltd., Chichester.
OSPAR (2010a). JAMP guidelines for monitoring contaminants in sediments, OSPAR Commission, London, 98 pp.
OSPAR (2010b). *Quality status report 2010*. London,Oslo–Paris Commission, 176 pp.
Perry B. J., Barefoot R. R. and Van Loon J. C. (1995). *Trends Anal. Chem.*, **14**(8), 388–397.
Protti P. (2001). Introduction to modern voltammetric and polarographic analysis techniques. IV, http://www.amelchem.com/download/items/voltammetry/manuals/eng/manual_eng.pdf.

Rao C. R. M., Sahuquillo A. and Sanchez J. F. L. (2008). *Water Air Soil Pollut.*, **189**(1–4), 291–333.
Ravindra K., Bencs L. and Van Grieken R. (2004). *Sci. Total Environ.*, **318**(1–3), 1–43.
Salbu B. and Steinnes E. (1992). *Analyst*, **117**(3), 243–249.
Salvato N. and Pirola C. (1996). *Mikrochim. Acta*, **123**(1–4), 63–71.
Schiavo D., Trevizan L. C., Pereira-Filho E. R. and Nûbrega J. A. (2009). *Spectrochim. Acta Part B: At. Spectrosc.*, **64**(6), 544–548.
Silva J. C. J., Santos D. M., Cadore S., *et al*. (2004). *Microchem. J.*, **77**(2), 185–190.
Slavin W. and Carnrick G. R. (1984). *Spectrochim. Acta Part B: At. Spectrosc.*, **39**(2–3), 271–282.
Smedes F. (1999). *Analyst* **124**(11), 1711–1718.
Sohrin Y., Iwamoto S.-i., Akiyama S., *et al*. (1998). *Anal. Chim. Acta*, **363**(1), 11–19.
Sweedler J. V., Jalkian R. D., Pomeroy R. S. and Denton M. B. (1989). *Spectrochim. Acta Part B: At. Spectrosc.*, **44**(7), 683–692.
Thermo (2007). *Determination of trace elements in clinical samples by high resolution ICP-MS*. *Application Note: 30003*, T. Scientific.
USEPA (1994). Method 6020. Environmental Protection Agency, Washington, DC.
Van Den Berg C. M. G. (1986). *Sci. Total Environ.*, **49**, 89–99.
Walsh A. (1955). *Spectrochim. Acta*, **7**(2), 108–117.
West M., Ellis A. T., P., Kregsamer P. *et al*. (2008). *J. Anal. At. Spectrom.*, **23**(10), 1409--1437.
Whitfield M. (2001). Interaction between phytoplankton and trace metals in the ocean, in *Advances in Marine Biology*, vol. 41 (eds A. J. Southward, P. A. Tyler, C. M. Young and L. A. Fuiman), Academic Press, New York, pp. 1–128.
Zheng J. C., Shotyk W., Krachler M. and Fisher D. A. (2007). *Glob. Biogeochem. Cycle* **21**(2), 11

5
Chemical Species

David Amouroux[a], Fabienne Seby[b], Mathilde Monperrus[a], Florence Pannier[a], Carolina Mendiguchia[c], Christelle Benoit-Bonnemason[b], and Olivier F.X. Donard[a]

[a] *Laboratoire de Chimie Analytique Bio-Inorganique et Environnement, CNRS et, Université de Pau et des Pays de l'Adour, Pau, France*

[b] *Ultra Traces Analyses Aquitaine, Hélioparc Pau-Pyrénées, Pau, France*

[c] *Department of Analytical Chemistry, Faculty of Marine and Environmental Science, University of Cádiz, Cádiz, Spain*

Chemical Marine Monitoring: Policy Framework and Analytical Trends, First Edition.
Edited by Philippe Quevauviller, Patrick Roose, and Gert Verreet.

5.1 INTRODUCTION

The fate of metals and metalloids in the environment is of primary importance owing to their potential role in the development of biological organisms and toxicity towards living cells. It is now accepted that the presence of metals and metalloids in different ecosystems is a key factor in their ecological balance, either by their stimulating effect on biological processes or by their inhibitory action when their concentration exceeds the toxicity threshold of living organisms.

On the one hand, the metallic species are present in a variety of chemical formulations, such as free ions, inorganic complexes, organometallic species or large organic complexes. The fate between different compartments of the environment and their impact on biological organisms are regulated mainly by their chemical forms.

On the other hand, biological organisms are able to accumulate within cells (bioaccumulation), transform chemically (biotransformation) or neutralized by specific ligands (biocomplexes) forms of more or less toxic metals and metalloids. Mobility, reactivity and toxicity of metals and metalloids in the environment are explicitly dependent on their molecular forms (inorganic, organometallic and so on). The understanding of biogeochemical cycles and processes of contamination of metals and metalloids is mainly related to the knowledge of physicochemical forms in which the element is engaged and its interaction with the environment. In natural environments, the thermodynamically stable element species mostly exhibit low biological activity and low toxicity. The transfer and the impact of metals and metalloids in ecosystems and biological organisms are often related to the formation of metastable species that cannot be predicted by thermodynamic laws.

In addition, methodological limitations are often encountered owing to the presence of these compounds at trace levels in marine environmental compartments (nanograms per kilogram to micrograms per kilogram). The methods developed to study the fate of these trace chemical forms of metals and metalloids are thus usually based on specific and sensitive 'atomic' detection techniques, which severely limits the molecular information available. This limitation can nowadays be solved by developing new hyphenated methods for coupling separation techniques and specific techniques for sensitive and selective detection by atomic spectrometry (optical, mass); see Figure 5.1. The application of these methods in environmental studies can now afford to get much information on the fate of metals and metalloids and better understand issues related to the phenomena of pollution, poisoning and remediation in the marine environment.

Nowadays, the speciation of inorganic and organometallic compounds in the environment is usually performed by a combination of sensitive analytical techniques, in particular chromatographic separation (gas chromatography (GC) or high-performance liquid chromatography (HPLC)) coupled on-line with a multielement detection system; for example, inductively coupled plasma mass spectrometry (ICP-MS), electron impact ionization mass spectrometry (EI-MS) and electrospray ionization tandem mass spectrometry (ESI-MS–MS). While ICP-MS enables element identification and quantification, structural identification is obtained via fragmentation using EI-MS and ESI-MS–MS techniques (Lobinski *et al.*, 2006). This instrumental equipment has to be supplemented by gel electrophoretic techniques, laser ablation (LA-ICP-MS) and matrix-assisted laser desorption and ionization mass spectrometry (MALDI-MS) techniques to be able to analytically trace the interaction of organometallic compounds with biochemical components (Szpunar, 2005;

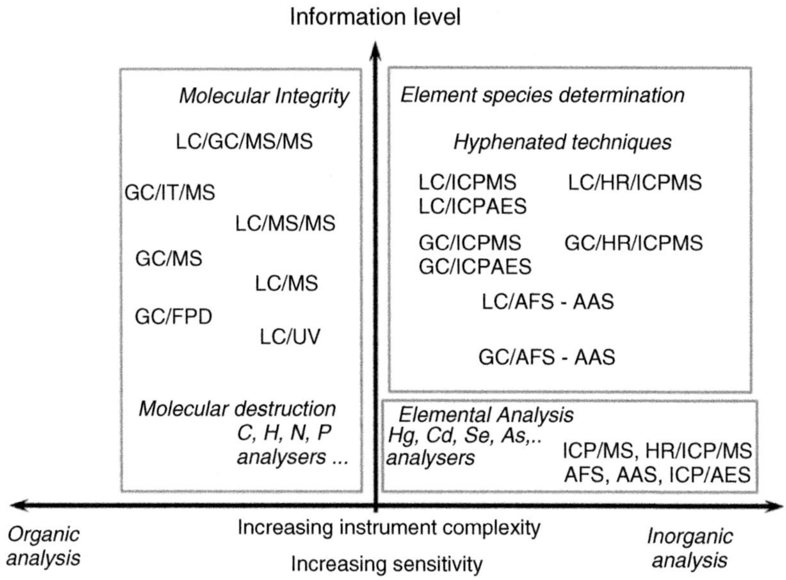

Figure 5.1 Complementarities and available techniques for element species determination by hyphenated methods: a gap between organic and inorganic analysis.

Sanz-Medel *et al*., 2003). The ideal solution would be to obtain both elemental and structural information on identical sample fractions; that is, to realize parallel detection by both systems.

5.2 LABILE/COMPLEXED FRACTIONATION OF METAL SPECIES

5.2.1 Seawater Samples

It is well known that the toxicity of metals can be related to their lability and mobility in the marine environment, which determines their availability to the biota. In this sense, several studies have demonstrated that the uptake is only related to the free-ion activity of the metal or to the labile metal forms rather than to the total concentration (Pandey *et al*., 2000; Lorenzo *et al*., 2002; Meylan *et al*., 2004). Although some attempts have been made to determine free-metal concentrations, the available analytical methodology is usually capable of determining a group of species better than a unique form. The most general division includes free-metal and inorganic complexes as a labile fraction and organic complexes, which are usually more stable, as non-labile fraction. Nevertheless, this classification is operationally defined and strongly dependent on the analytical technique used. This is a serious inconvenience, owing to the variety of techniques developed for speciation studies of dissolved metals, and it is necessary to study each one in depth to obtain a correct assessment of the data. An exhaustive revision of these techniques and its applications to metal speciation in environmental waters can be found in the literature (Buffle and Tercier-Waeber, 2005; Pesavento *et al*., 2009; López-López *et al*., 2010). Table 5.1 includes some significant examples and recent applications in the marine environment.

As shown in the Table 5.1, the methodology can be classified in two main groups: electroanalytical and non-electroanalytical techniques.

Table 5.1 Methods used for metal fractionation in seawater samples.

Sample preparation	Separative method	Detection[a]	Elements	Species/fraction	LOD	References
Filtration and acidification (total Pb)	–	DPASV	Pb	Inorganic/total	10 pM	Capodaglio *et al*. (1990)
Filtration and freezing	–	DPASV	Cd, Pb	Labile/total	–	Scarponi *et al*. (1995)
Filtration	–	DPASV and DPCSV	Cu	Weak and strong ligand groups	–	Gerringa *et al*. (1996)
–	–	ASV	Cu	Weak and strong ligand groups	–	Andrade *et al*. (2006)
–	–	DPASV	Cu	Labile/total	–	Jones and Bolam (2007)
Refrigeration	Ligand exchange	ACSV	Co, Cu, Ni	Labile/total	–	Braungardt *et al*. (2007)
Buffering	Ligand exchange	ACSV	Ni	Labile/total	0.3 nM	Martino *et al*. (2004)
Continuous flow analysis, addition of ligands	Ligand exchange	CSC (in situ)	Cu, Zn	Labile/total	5 nM	Van den Berg (1991)
Filtration and freezing	Ligand exchange	CSV	Co	Free Co^{2+}/complexed Co	–	Ellwood and Van den Berg (2001)
–	DGT	ICP-MS	Ni, Cu, Pb, Zn	Labile/total	$2\,ng\,L^{-1}$ (Pb)	Dunn *et al*. (2003)
–	DGT	ICP-MS	Cu, Co, Cd, Pb	Labile/total	$0.5\,ng\,L^{-1}$ (Co)	Munksgaard and Parry (2003)

–	DGT	ETAAS	Cu, Cd, Ni, Pb	Labile/total	–	Schintu *et al*. (2008)
–	DGT	FAAS	Cu, Cd	Labile/total	–	Li *et al*. (2005a)
Filtration	3-column SPE	ETAAS	Cu, Cd, Mn, Pb	Neutral organic metal complexes/free ions	–	Groeschner and Appriou (1994)
Filtration and freezing	3-column SPE	ICP-AES	Cu, Cd, Pb	Anionic, neutral and cationic species	1.3 nM (Cd)	Abollino *et al*. (2000)
Filtration	2-column SPE	ICP-AES/ETAAS	Mn, Cd, Cu, Zn, V, Ni, Pb, Fe	Organic and inorganic complexes	–	Abasse *et al*. (2003)
Filtration	SPE	ICP-MS	Mn, Cd, Cu, Zn, V, Ni, Pb, Fe	Organic/total	0.02 $\mu g\,L^{-1}$ (Mn)	Abasse *et al*. (2002)
Buffering at different pH	pH	DPASV	Zn, Cd, Pb, Cu	Labile inorganic, labile organic, moderately labile inorganic, moderately labile organic and inert fractions	–	Vicente-Martorell *et al*. (2009)
Buffering	CHELEX	DPASV	Cd, Pb, Zn	Very labile, moderately labile, slowly labile and inert fractions	–	Scoullos and Pavlidou (2003)

[a]DPASV: differential pulse anodic stripping voltammetry; ASV: anodic stripping voltammetry; DPCSV: differential pulse cathodic stripping voltammetry; ACSV: adsorptive cathodic stripping voltammetry; CSC: cathodic stripping potentiometry.

Electroanalytical techniques are extensively used owing to their early development to speciation studies in seawater in the mid 1970s (Nurnberg *et al*., 1976). In this sense, anodic stripping voltammetry (ASV) may be used to determine the electroactive fraction of metal (ASV-labile) and, with an additional measurement of total metal concentration, the metal fraction associated with organic ligands may also be determined. In addition, ligand concentrations and the related stability constants can be calculated. ASV has been successfully applied to differentiate the species of lead, copper, cadmium and zinc in different marine and estuarine areas (Capodaglio *et al*., 1990; Scarponi *et al*., 1995; Gerringa *et al*., 1996; Andrade *et al*., 2006; Braungardt *et al*., 2007; Jones and Bolam, 2007). Other electroanalytical techniques, such as cathodic stripping voltammetry or cathodic stripping chronopotentiometry, have also been used in speciation studies in seawater samples (Van den Berg, 1991; Ellwood and Van den Berg, 2001; Martino *et al*., 2004; Braungardt *et al*., 2007). In addition, selective electrodes were initially developed as a potential alternative to determine the free-metal concentration, but their operational capability remains limited. They are only available for a number of metals and their relatively high detection limits avoid the determination of the very low metal concentrations found in seawater (Pesavento *et al*., 2009). A recently developed technique, absence of gradients and nernstian equilibrium stripping (AGNES), seems to be a better alternative to measure free-metal concentration in seawater (Buffle and Tercier-Waeber, 2005; Galceran *et al*., 2007; Pesavento *et al*., 2009).

A common advantage of the electrochemical techniques is their capability for in-situ speciation, enabling contamination during the storage and samples preparation to be reduced. For an in-depth review of in-situ measurements with voltammetric techniques, the reader is invited to consult the paper of Buffle and Tercier-Waeber (2005).

Although electroanalytical techniques offer a very interesting alternative to speciation studies in seawater, a main drawback is that only a few elements can be determined with these techniques (mainly copper, cadmium, lead and zinc). For this reason, alternative non-electroanalytical devices based on ion exchange, diffusion gradients in thin film (DGT) gels or permeation liquid membranes (PLMs) may be used.

DGT has been successfully applied to in-situ metal speciation in freshwater and seawater since its development in the mid 1990s (Davidson and Zang, 1994; Zang and Davidson, 1995). Applications to the monitoring of several trace metals (Cu, Cd, Pb, Zn, Ni) in estuarine and coastal water have also been described (Dunn *et al*., 2003; Munksgaard and Parry, 2003; Schintu *et al*., 2008). The system is based on the transport of labile species (free ion, inorganic and also some labile organic species) by molecular diffusion through a gel layer located between the sample and a layer of ion-exchange resin. In-situ preconcentration takes place in the resin and then metals can be eluted by nitric acid before analysis with ICP-MS. A direct analysis can also be performed by X-ray fluorescence. The species diffusion rate depends on the pore size of the gel and the type of gel and resin selected. It is also possible to separate the labile organic species from the labile inorganic species when such diffusion parameters are optimized (Zang and Davidson, 2000, 2001; Li *et al*., 2005a).

PLMs have shown their applicability to determine free-metal concentration as well as organic and inorganic metal complexes in natural waters (López-López *et al*., 2010). Nevertheless, the methodology has been developed mainly to separate labile and nonlabile fractions of copper and, to a lesser extent, lead and cadmium (Slaveykova *et al*., 2004; Parthasarathy *et al*., 2004; Romero *et al*., 2005; Bayen *et al*., 2006; Gunkel-Grillon and Buffle, 2008). Some comparison has been made between this methodology and DGT or voltammetry in estuarine and seawater samples (Ndungu *et al*., 2005; Slaveykova *et al*., 2009). Although PLMs present some interesting advantages, such as the possibility for on-line speciation and simultaneous preconcentration, their application to real seawater samples is to date rather scarce, owing to their recent development as an environmental analytical tool.

Finally, ion exchange appears as a simple technique to perform speciation studies in the marine environment. The main advantages are related to its simplicity, its easier automatization, multielemental analysis, simultaneous separation and preconcentration and the possibility of in-situ speciation. The technique is based on the adsorption of some specific species onto a resin and their later elution with the appropriate solvent. Depending on the resin characteristics, the species retained are different and speciation can be operationally performed. The resins most extensively used are Chelex-100 and Chelamine to retain free ions and weak inorganic complexes, C_{18} to retain organic complexes and Dowex and AG MP-1 to retain anionic species. Many workers have used a method with several columns to allow the separation of several fractions (Groeschner and Appriou, 1994; Abollino *et al.*, 2000; Abasse *et al.*, 2003; Alberti *et al.*, 2008). Furthermore, simple procedures with only one type of resin could be used for speciation analysis, including a previous treatment of the samples (Abasse *et al.*, 2002). In spite of the great versatility of these systems thanks to the variety of sorbents available, speciation is strongly dependent on the type of resin used, which has to be taken into account when comparisons between different areas are performed. Nevertheless, the possibility of automatization of these systems combined to in-situ sampling, and even speciation, makes them an attractive alternative for marine monitoring of metals (Point *et al.*, 2007).

Although the methodology mentioned before is the most extensively used for the monitoring of heavy metals in seawater, there are other schemes based on metal fractionation at different pH (Fischer *et al.*, 1984; Vicente-Martorell *et al.*, 2009), and combinations of several methods can be performed to improve the separation (Scoullos and Pavlidou, 2003).

5.2.2 Marine Sediment Samples

It is well known that sediments play an important role as a sink of metals in marine environments, but they can also act as a source of these elements, depending upon the physicochemical conditions. For this reason, speciation of metals in sediments is focused on the establishment of metal contents which can be extracted under specific environmental conditions. In this sense, sediments can be divided into operationally defined fractions which can be selectively extracted using the appropriate reagent. Although simple extraction procedures for the analysis of one fraction can be performed, it is more interesting to carry out sequential extraction procedures which allow the determination of metals contents associated with each phase. In these procedures, the residual phase from each simple extraction is used in the next extraction step following a sequence until the last fraction (residual) is extracted. Owing to the vast literature available about sequential extraction, this text only includes the more relevant developments and recent applications on the marine environment. For an in-depth review, the reader is invited to consult papers by Filgueiras *et al.* (2002), Bacon and Davidson (2008) and Rao *et al.* (2008).

As previously stated, sequential extraction procedures are interesting for evaluating the mobility of metals under different environmental conditions. In this sense, the sequential extraction scheme proposed in the late 1970s has initiated this research on trace metals fractionation in solid samples (Tessier *et al.*, 1979). The method was first applied to the partitioning of trace metals (Cd, Co, Cu, Ni, Pb, Zn, Fe and Mn) into five fractions: exchangeable, bound to carbonates, bound to Fe–Mn oxides, bound to organic matter (including sulfide forms) and residual. The exchangeable fraction represents the more labile fraction adsorbed on sediments, which could be released by ion exchange or desorption processes, and the fraction bound to carbonates would be susceptible to changes of pH. The metals that could be released under more reducing or oxidative conditions correspond to the fraction bound to Fe–Mn oxides and those bound to organic matter respectively. Finally, the residual fraction represents metals from the mineral structures. From this starting point, it is usual to find some modifications in order to improve the fractionation of metals, differentiating metals bound to the organic matter or trapped in the sulfidic fraction (Campanella *et al.*, 1995) as well as between the easily reducible fraction and the moderately reducible one (Kersten and Förstner, 1986). In addition to the modifications in the reagent sequences and in the operational conditions, changes in the extraction of residual fraction are usual in order to compare the results with those obtained from a total extraction procedure. In these cases the same extraction methodology is used in both extraction of residual fraction and total extraction.

The variety of available methodologies makes it difficult to compare data and develop quality control tools such as certified materials (see Chapter 11 of this book). For this reason, the European Commission through its Community Bureau of Reference (BCR), now run by the Institute for Reference Materials and Measurements (IRMM), proposed in 1993 a three-step sequential extraction methodology, later re-evaluated and revised, to harmonize the available procedures (Ure *et al.*, 1993; Rauret *et al.*, 1999). The BCR methodology defines three fractions: acid soluble (exchangeable and bound to carbonates), reducible (bound to Fe/Mn oxides) and oxidizable (bound to organic matter and sulfides). Nevertheless, speciation studies carried out by this method usually includes a fourth step to extract residual fraction (Yuan *et al.*, 2004; Cuong and Obbard, 2006; Gao *et al.*, 2008). In addition, some certified reference materials (CRMs) have been prepared based on the BCR methodology, improving the quality control in sequential extraction (López-Sánchez *et al.*, 1998; Pueyo *et al.*, 2001). This is the main reason for the preferable use of the BCR extraction procedure in recent years. A review of the work carried out in this field has been published (Quevauviller, 2002).

Many comparisons between the different procedures have showed some unsolved problems associated with the sequential extractions, as lack of specificity of reagents, re-distribution of analytes among phases, incomplete extractions and precipitation of new mineral phases or the dependence on sample pretreatment (Usero *et al.*, 1998; Filgueiras *et al.*, 2002; Joksic *et al.*, 2005; Bacon and Davidson, 2008; Nemati *et al.*, 2009). In spite of these drawbacks, nowadays, sequential extraction procedures mainly based on the BCR methodology have been extensively applied in environmental studies to determine the bioavailability of metals in sediments and particulate matter from marine ecosystems. Some of the most recent applications can be seen in Table 5.2.

Finally, sequential extraction procedures have also been applied to suspended solid matter, although only a few examples in the marine environment can be found in the literature (Magi *et al.*, 2005), being most usually applied in estuarine systems (Waeles *et al.*, 2005a, 2005b), where interaction between particulate and dissolved fractions has important effects on the transport and availability of metals.

5.2.3 Marine Biota Samples

Speciation of metals in marine biota is mainly related to the determination of biogenic organometallic complexes and compounds which can be particularly synthesized under high metal concentrations exposure and/or uptake, such as metallothioneins (MTs) and metallothionein-like proteins (MTLPs). Nevertheless, other non-thionein cytosolic ligands can bind heavy metals, and information about their role in the storage and sequestering of metals is really scarce (Ferrarello *et al.*, 2000). This lack of information is being solved in recent years thanks to the development of hyphenated techniques combining a separation technique such as chromatography with a sensitive and element-specific detector such as mass spectrometry. The application of these techniques to bio-inorganic speciation and to characterization and quantification of MT isoforms, including some marine examples, has been extensively described in the literature (Szpunar, 2000; Prange and Schaumlöffel, 2002). Some examples of applications are given in Table 5.3.

Table 5.2 Methods used for metal fractionation in marine sediments.

Sample preparation	Separative method	Detection[a]	Elements	Fraction	LOD (μg g^{-1})	References
Air drying and grinding/wet and not grinding	3-step sequential extraction	ICP-MS	Ba, Cd, Co, Cu, Mn, Mo, Ni, Pb, Sc, Sr, U, Zn	Acid soluble, reducible, oxidizable and residual	–	Gao *et al.* (2008)
Sieving (63 μm)	BCR sequential extraction	ICP-MS	V, Cr, Mn, Co, Ni, Cu, Zn, Mo, Cd, Sn, Pb	Acid soluble, reducible, oxidizable and residual	–	Yuan *et al.* (2004)
Air drying, grinding and sieving (63 μm)	BCR sequential extraction	ETAAS	Cd, Cr, Cu, Ni, Pb, Zn	Acid soluble, reducible, oxidizable and residual	–	Cuong and Obbard (2006)
Air drying and sieving (1 mm)	4-step and 3-step sequential extraction	FAAS	Zn, Fe , Cu, Mn, Pb, Cr, Cd, Ca	Exchangeable, reducible, oxidizable and residual	–	Nemati *et al.* (2009)
Filtration (0.4 μm), filters washed with HCl and air dried	3-step sequential extraction	ETAAS/ ICP-AES	Cd, Cu, Fe, Mn, Pb	Acid soluble, reducible and oxidizable	0.01	Magi *et al.* (2005)
Filtration (0.45 μm) and filters oven dried	3-step sequential extraction	PSA	Cd	Weakly adsorbed metals, iron–manganese oxides, bound metals and organic/sulfide-bound metals	–	Waeles *et al.* (2005a)
Filtration (0.45 μm) and filters oven dried	3-step sequential extraction	CCSA	Cu	Weakly adsorbed metals, iron–manganese oxides, bound metals and organic/sulfide-bound metals	–	Waeles *et al.* (2005b)

[a]PSA: potentiometric stripping analysis; CCSA: constant-current stripping analysis.

Table 5.3 Methods used for metal fractionation in marine biota.

Sample preparation	Separative method[a]	Detection	Elements	Fraction	LOD (pg)	References
Homogenization, sonication and centrifugation	SEC	ICP-MS	Cu, Zn, Ca, U, Ni, Mo, Mn, Cr, V, Cd, Al, Sb	High, medium, low and very low molecular weight	–	Ferrarello *et al*. (2000)
Homogenization and centrifugation	Centrifugation and heat denaturation	AAS	Cd, Cu, Zn	Insoluble fraction, heat-stable fraction (including MTLP) and heat-denaturated fraction	–	Hamza-Chaffai *et al*. (1997)
Homogenization and centrifugation	Centrifugation and heat denaturation	AAS	Cd, Cu, Zn	Soluble and insoluble fractions	–	Mouneyrac *et al*. (2001)
Homogenization and centrifugation	Heat denaturation	AAS	Cd, Cu, Zn	Soluble, insoluble and heat-stable fractions	–	Hamza-Chaffai *et al*. (2000)
Homogenization and centrifugation	SE-HPLC	ICP-TOF-MS	Cd, Cu, Pb, Ni, Zn	Metallothioneins, higher molecular weight ligands and lower molecular weight ligands	–	Van Campenhout *et al*. (2008)
Homogenization and centrifugation	SE-IE-HPLC	ICP-MS			–	Mason and Meraz (1998)
Homogenization and centrifugation	SE-FPLC/ AE-FPLC	ICP-MS	Cu, Zn, Cd, Hg	Metallothionein isoforms	–	Rodríguez Cea *et al*. (2006)
Homogenization and centrifugation	SE-HPLC	ICP-TOF-MS	Cd, Zn, Cu	Metallothionein isoforms	19.9	Goneaga Infante *et al*. (2002)

[a]SEC: size-exclusion chromatography; SE-IE-HPLC: size-exclusion ion exchange HPLC; SE-FPLC: size-exclusion fast protein liquid chromatography; AE-FPLC: anion exchange FPLC.

A simple speciation scheme is based on the determination of metal contents in different cytosolic fractions from different target organs. In these cases a basic separation in soluble and insoluble fractions is performed using centrifugation after addition of Tris buffer (pH 8.6) and β-mercaptoethanol (Hamza-Chaffai *et al.*, 1997; Mouneyrac *et al.*, 2001). Then, a heat denaturation of the soluble fraction allows one to obtain MTs and MTLPs, which can be measured by differential pulse polarography. Metals are determined in each fraction by conventional techniques such as atomic absorption spectrometry. Finally, the relationships between MTs and metal concentrations allow the use of MTs in marine biomonitoring as an early warning system for the metal pollution threshold towards living organisms.

The use of hyphenated techniques allows a better speciation of metal binding to different cytosolic ligands. A typical approach is based on separation depending on molecular weight, although applications to marine vertebrates are very limited. As an example, Van Campenhout *et al.* (2008) used size-exclusion HPLC (SE-HPLC) combined with inductively coupled plasma time-of-flight mass spectrometry (ICP-TOF-MS) to separate metals (Cd, Cu, Ni, Pb, Zn) associated with MTs (10 kDa), higher molecular weight ligands (>10 kDa) and lower molecular weight ligands (<10 kDa) in the cytosolic fraction of European eel (*Anguilla anguilla*) livers (Hamza-Chaffai *et al.*, 2000). The results indicated that speciation was metal specific and dependent on environmental and tissue concentration, demonstrating the relation between speciation of metal in cytosolic fractions with the environmental conditions. This kind of application is usually used in invertebrate organisms, which have demonstrated their applicability as biomarkers in several situations. Mason and Meraz (1998) used a size exclusion/ion-exchange HPLC system interfaced to ICP-MS for studying the speciation, turnover and distribution of cytosolic Cu and other metals within the visceral complex of marine mollusc *Littorina litorea*. For these species, Cu concentration in control animals was related to the high molecular weight fraction; meanwhile, the exposure of animals to Cu induced a low molecular weight pool of Cu-binding proteins. In a similar way, Ferrarello *et al.* (2000) used the combination of SEC and ICP-MS for multielemental trace metal speciation in cytosols of the mussel *Mytilus edulis*.

In addition to the characteristics of each molecular weight fraction, several MT isoforms can present a different behaviour in the presence of metals (Ferrarello *et al.*, 2000). Many research efforts have subsequently been performed in order to determine metals associated with the different isoforms in different organisms (Goneaga Infante *et al.*, 2002; Álvarez-Llamas *et al.*, 2003; Rodríguez Cea *et al.*, 2006), although applications to marine biota are really scarce (Ferrarello *et al.*, 2000). Again, hyphenated techniques appear as the better alternative to achieve this objective.

5.3 INORGANIC CHROMIUM SPECIES

Chromium can exist in several chemical forms, but only trivalent and hexavalent chromium are stable enough to occur in the environment. Cr(III) is considered as an essential micronutrient for humans and animals. In contrast, Cr(VI) is much more toxic than Cr(III) for both acute and chronic exposures.

The toxicological disparity between Cr(III) and Cr(VI) is closely related to their chemical characteristics, which also conditions the stability, mobility and bioavailability of these species in the marine environment (Pettine and Capri, 2005). It is thus essential to have analytical methods allowing the determination of each species in various types of marine sample. Analytical techniques generally used for Cr speciation in marine samples can be separated into two groups. The first group brings together methods that can determine Cr(VI) or Cr(III) after a pretreatment of the sample and the second group corresponds to on-line hyphenated techniques allowing the determination of both species simultaneously. When applied to solid samples, most of these methods require a previous extraction step. These will be discussed in detail for marine sediment samples.

Spectrophotometric and colorimetric methods, for example EPA Method 7196 (USEPA, 1995b) and Method 7199 (USEPA, 1996), have been recommended for the determination of Cr(VI). Both methods are based on the diphenylcarbazide (DPC) spectrophotometric method, in which Cr(VI) reacts with DPC in strongly acid solution giving the Cr–diphenylcarbazone complex which is detected spectrophotometrically at 540 nm. Despite its simplicity, the DPC method suffers from the presence of interfering compounds which can react with DPC, resulting in the formation of complexes that absorb at the same analytical wavelength. Ion chromatography (IC) can be used to separate Cr(VI) from these compounds. In Method 7199, Cr(VI) is chromatographically separated from these positive interferences before the derivatization with DPC. In addition, the low pH required for the DPC–Cr(VI) complex formation and the presence of Fe(II), sulfide, sulfite and some organic compounds can reduce Cr(VI), leading to the underestimation of Cr(VI) concentrations (Tirez *et al.*, 2007). Although the presence of these compounds is not very usual in water samples, it becomes a problem when the DPC method is applied to sediment extracts (Pettine and Capri, 2005).

To a lesser extent, electroanalytical methods are also employed for the direct determination of Cr(VI). The most common method is stripping voltammetry, which allows one to reach nanogram per litre levels (Grabarczyk *et al.*, 2006; Grabarczyk, 2008).

Atomic spectrometric techniques such as flame atomic absorption spectrometry (FAAS), electrothermal atomic absorption spectrometry (ETAAS) and inductively coupled plasma-atomic emission spectrometry (ICP-AES) have been used for chromium speciation after a separation or isolation technique that provides selectivity for one species relative to the other. Depending on the use of a preconcentration technique, the limits of detection (LODs) can range from micrograms per litre to milligrams per litre. For samples with low levels of chromium, ETAAS and ICP-AES are more suitable than FAAS.

On-line separation and detection methods are increasingly used in order to minimize contamination and losses of Cr species or redox conversion (Cornelis *et al.*, 2003). In recent years, owing to its high sensitivity and selectivity, ICP-MS has received most attention as a detection technique for chromium. Coupled to ICP-MS, IC is the most widely used separation method. Several approaches have been used to achieve suitable chromatographic separation of chromium species. Anion-exchange columns (Barnowski *et al.*, 1997; Tirez *et al.*, 2003; Rahman *et al.*, 2005) and columns with both anion- and cation-exchange capabilities (Vassileva, 2000; Séby *et al.*, 2003) have been explored. The use of complexing agents (Andrle *et al.*, 1997), high salt concentration and eluents with carbon or chlorine generate polyatomic interferences that may disturb measurement of the most abundant chromium isotopes ^{52}Cr and ^{53}Cr. To overcome the spectral interferences caused by carbon, higher mass resolution and cool plasma conditions can be employed (Vanhaecke *et al.*, 2000). High-resolution ICP-MS efficiently removes most interferences, but its high cost makes its use difficult in routine laboratories. Cool plasma conditions enable the removal of interferences from carbon, but those generated by chloride ions are not reduced. Now, most interferences can be removed by using an ICP-MS equipped with a collision/reaction cell pressurized with an appropriate reaction gas, NH_3 or He in the case of Cr. Nevertheless, in most cases the sensitivity is reduced by increasing the reaction gas flow, and an adequate flow should be chosen in order to get the best signal-to-noise ratio (Hagendorfer and Goessler, 2008). Another advantage of using ICP-MS is to correct analytical biases by speciated isotope dilution mass spectrometry (SIDMS) (Huo and Kingston, 2000; Tirez *et al.*, 2003; Rahman *et al.*, 2005).

5.3.1 Seawater Samples

Total chromium concentration in ocean waters falls mainly in the range 50–500 ng L^{-1} (Sirinawin *et al.*, 2000) and the dominating Cr oxidation state is found to be Cr(VI) (Auger *et al.*, 1999; Sander and Koschinsky, 2000; Sirinawin *et al.*, 2000; Connelly *et al.*, 2006). However, most studies also report the presence of the thermodynamically unexpected reduced species Cr(III), and in some areas the Cr(III) concentrations were up to, or greater than, 50% of the total Cr concentrations (Auger *et al.*, 1999; Sander and Koschinsky, 2000; Connelly *et al.*, 2006).

Before Cr speciation analysis, it is recommended simply to filter seawater samples and store them at −18 °C to avoid Cr redox changes (Sirinawin and Westerlund, 1997). Acidification has to be avoided because Cr(VI) is immediately reduced into Cr(III). Different methods have been developed for the speciation analysis of Cr in seawater; see Table 5.4. Only those allowing the reaching of concentration levels usually found in seawater are presented herein.

Table 5.4 shows that sample pretreatment techniques including separation and/or preconcentration are required in order to determine the low levels of individual Cr species, even with the most sensitive techniques. Many of these methods have the disadvantage that one of the species is determined as the difference between total Cr (often obtained after reduction or oxidation) and the other chemical form of the element. In the past, coprecipitation techniques have been used for the isolation of Cr(III) and Cr(VI). Iron(III) oxides are generally used to isolate Cr(III) (Cranston and Murray, 1980; Ahern *et al.*, 1985; Kaczynski and Kieber, 1993; Pettine *et al.*, 1997; Connelly *et al.*, 2006). Only Auger *et al.* (1999) have used gallium(II) oxides to preferentially retain trivalent Cr. The concentration of Cr(VI) is obtained after coprecipitation of both Cr(III) and Cr(VI) onto iron(II) oxides and calculation of the difference between this fraction and the Cr(III) concentration. Ahern *et al.* (1985) have used the selectivity of cobalt(III) oxides for Cr(VI) to specifically determine this species. Coprecipitation is generally followed by determination of total Cr with atomic absorption spectrometry methods. Detection limits obtained are among the lowest for Cr speciation in seawater samples because of preconcentration during the coprecipitation step and the removal of the saline matrix. Furthermore, the simplicity of this procedure allows an in-situ separation of Cr species. However, coprecipitation is a time-consuming procedure and is often associated with a bad reproducibility. Liquid–liquid extraction has also been applied to Cr speciation in seawater both off-line and on-line when a flow-injection analysis (FIA) is used. This treatment is based on the complexation of Cr(VI) with ammonium pyrrolidine dithiocarbamate (APDC) (Nielsen *et al.*, 1999; Som-Aum *et al.*, 2002) or tricaprylmethylammonium chloride (Aliquat-336) (Sirinawin and Westerlund, 1997; Sirinawin *et al.*, 2000) and the complex is extracted with an organic solvent (methyl isobutyl ketone) before analysis, which is generally performed by ETAAS. Solid-phase extraction (SPE) is also mentioned in several studies on Cr speciation in seawater in combination with FIA techniques. The objective of SPE is to separate a single analyte or group of analytes from interfering sample components or matrices, often simultaneously achieving some degree of preconcentration and, therefore, gaining sensitivity at the expense of separation power. Preconcentration methods involving solid sorbents are considered superior to liquid–liquid extraction systems in terms of simplicity, rapidity and ability to obtain a high enrichment factor. In sorption preconcentration methods, solid sorbents such as TiO_2 (Vassileva, 2000) or Muromac A-1 (Hirata *et al.*, 2000), anion-exchange column (Yang and Huang, 2001) or a complexing agent immobilized on a solid sorbent (Prasada Rao *et al.*, 1998; Filik *et al.*, 2003; Scindia *et al.*, 2004; Sumida *et al.*, 2006; Yang L. *et al.*, 2009) were used. Depending on the pH, TiO_2 can selectively adsorb Cr(III) (pH 9) or Cr(VI) (pH 2) and Muromac A-1 was used to preconcentrate only Cr(III). Determination of Cr(III) species was also performed with palmitoyloxine immobilized on an XAD copolymer (Filik *et al.*, 2003) and iminodiacetate chelating resin (Sumida *et al.*, 2006). DPC immobilized on silica has been employed for on-line preconcentration of Cr(VI) (Scindia *et al.*, 2004). In order to retain the two redox forms of Cr, dithiocarbamate fixed on a C_{18} column has been used (Prasada Rao *et al.*, 1998). At low pH (1.5), only Cr(VI) is sorbed and Cr(III) is retained at pH 7. SPE is followed by determination with spectrophotometry, AAS, ICP-AES, amperometry or ICP-MS. Owing to its efficient preconcentration, detection limits obtained when SPE is used are low enough to determine Cr species concentrations in seawater, but, owing to their different physico-chemical properties, the two Cr species are not easily determined simultaneously.

Table 5.4 Analytical methods used for chromium speciation in seawaters.[a]

Matrix	Pretreatment	Detection	System type	Species	LOD (ng L^{-1})	Preconc. Factor	References
Surface seawaters	Coprecipitation	ETAAS	Off-line	Cr(III) Cr(VI)[b]	–	–	Connelly *et al.* (2006)
Coastal seawater	Coprecipitation	XRF	Off-line	Cr(III) Cr(VI)	–	–	Ahern *et al.* (1985)
Seawater	Coprecipitation	ETAAS	Off-line	Cr(III) Cr(VI)[b]	1	–	Kaczynski and Kieber (1993)
Estuary waters	Coprecipitation	FAAS	Off-line	Cr(III) Cr(VI)[b]	12	–	Cranston and Murray (1980)
Seawater	Coprecipitation	ETAAS	Off-line	Cr(III)	2	–	Auger *et al.* (1999)
Coastal waters	Coprecipitation	ETAAS	Off-line	Cr(III) Cr(VI)[b]	52	–	Pettine *et al.* (1997)
Seawater	LLE	ETAAS	Off-line	Cr(VI) Cr(III)[b]	5.2	–	Sirinawin and Westerlund (1997)
Seawater	LLE	ETAAS	Off-line	Cr(VI) Cr(III)[b]	9	–	Sirinawin *et al.* (2000)
Synthetic seawater	LLE	ETAAS	On-line (FIA)	Cr(VI)	3.3	18.4	Nielsen *et al.* (1999)

Synthetic seawater	LLE	ETAAS	On-line (FIA)	Cr(VI)	9	30.1	Som-Aum *et al*. (2002)
Seawater	SPE	Spectro-photometry	Off-line	Cr(III)	484	60	Filik *et al*. (2003)
Seawater	SPE	AAS	On-line (FIA)	Cr(III) Cr(VI)	20 20	500 450	Prasada Rao *et al*. (1998)
Synthetic seawater	SPE	ICP-AES	On-line (FIA)	Cr(III) Cr(VI)	11 17	 100	Vassileva (2000)
Deep and surface seawaters	SPE	ICP-AES	On-line (FIA)	Cr(III) Cr(VI)^{b}	40 90	–	Sumida *et al*. (2006)
Coastal seawater	SPE	Ampero-metry	On-line (FIA)	Cr(VI)	4	40	Yang and Huang (2001)
Natural seawater	SPE	ICP-MS	On-line (FIA)	Cr(III)	20	11.5	Hirata *et al*. (2000)
Certified reference seawater	SPE	SF-ICP-MS (IDMS)	On-line (FIA)	Total Cr	5	–	Yang L. *et al*. (2009)
Seawater	Complexation (DPC)	Sensor	Off-line	Cr(VI)	1.36 ng	–	Scindia *et al*. (2004)
Coastal seawater	Complexation (DPC)	Spectro-photometry	Off-line	Cr(VI)	2 600	–	Pettine and Millero (1990), Pettine *et al*. (1991)

(*continued overleaf*)

Table 5.4 *(continued)*

Matrix	Pretreatment	Detection	System type	Species	LOD (ng L^{-1})	Preconc. Factor	References
Seawater and hydrothermal fluids	UV irradiation	CSV	Off-line	Cr(VI)	52	–	Sander and Koschinsky (2000)
				Cr(III)[b]			
Synthetic seawater		HPLC–USN-ICP-AES	On-line	Cr(III)	4600	–	Posta *et al*. (1996)
				Cr(VI)	3700		
	Preconcentration	HPLC–USN-ICP-AES		Cr(VI)	200		
	Preconcentration	HPLC–ICP-MS		Cr(VI)	120		
Certified reference seawater	Derivatization (TFA) and SPME	GC-ID SF-ICP-MS	On-line	Total Cr	9	–	Yang *et al*. (2004)
Certified reference seawater	Derivatization (HTFA) and SPME	GC–ECD	On-line	Total Cr	11	–	Abranko *et al*. (2004)
		GC–MS (IDMS)			13		
		GC–ICP-MS			15		
Synthetic seawater	Microfluidic device	Chelumi-nescence	On-line	Cr(III)	8×10^{-6}	–	Som-Aum *et al*. (2007)

[a]LLE: liquid–liquid extraction; SF: sector field; USN: ultrasonic nebulization; ID: isotope dilution; ECD: electron-capture detector.
[b]Concentration calculated by difference between the sum [Cr(III) + Cr(VI)] and the fraction determined ([Cr(III)] or [Cr(VI)]).

Modern analytical techniques, such as HPLC–ICP-MS, for Cr speciation have not been very often applied to seawater samples, mainly because Cr species are present at the nanogram per litre range in a highly complex matrix. Even with matrix separation/analyte preconcentration and the use of an ultrasonic nebulizer, detection limits are higher than those obtained by other methods (Posta *et al.*, 1996). This is particularly true with a detection with low-resolution ICP-MS. Better results could be obtained using sector field (Yang L. *et al.*, 2009). Another possible approach is the application of vapour generation, wherein initially nonvolatile/ionic analyte compounds are converted to gaseous metal species which can be further analysed by GC–ICP-MS (Abranko *et al.*, 2004; Yang *et al.*, 2004). The main advantages include efficient matrix separation and virtually 100% sample introduction efficiency. For the moment, this technique is not used for Cr speciation analysis. However, as only Cr(III) can be derivatized using trifluoroacetylacetonate (TFA) to form $Cr(TFA)_3$, which is further extracted into hexane, some kind of speciation could be considered.

Finally, determination of Cr(III) using microfluidic device (glass chip) with chemiluminescence reaction has allowed one to reach very low detection limits in seawater samples. Owing to the miniaturization of the system, in-situ speciation analysis could be obtained (Som-Aum *et al.*, 2007).

5.3.2 Marine Sediment Samples

Analytical methods found in the literature for Cr speciation in sediment samples are presented in Table 5.5. The number of studies dealing with the specific determination of Cr species is low for this type of matrix, and most of the time the chromium bioavailability and mobility are still studied using the sequential extraction procedure (Pempkowiak *et al.*, 1999; Hursthouse *et al.*, 2003). The main difficulty in determining Cr(VI) in solid matrices arises from the possible interconversion of oxidation states of the different Cr species.

It is now recognized that Cr(VI) extraction is facilitated by using alkaline conditions; unfortunately, Cr(III) is precipitated and cannot be determined simultaneously. Therefore, in most studies, only the most toxic species, hexavalent Cr, is determined. It was shown that an extraction protocol using 0.28 M Na_2CO_3 and 0.5 M NaOH (or 3% Na_2CO_3 and 2% NaOH) with continuous swirling and heating at 90–95 °C for 60 min effected maximal dissolution of soluble and insoluble Cr(VI) and reduced method-induced oxidation and reduction (James *et al.*, 1995). Based on this work, the most prevalent method for determining Cr(VI) in soils, sediments and sludges is the EPA revised method 3060A (USEPA, 1995a), which is based on this alkaline digestion procedure and adopted as a basis for the European Standard EN 15 192 (2006). In this method, 2.5 g of sample is digested using 50 mL of a solution of 0.28 M Na_2CO_3 and 0.5 M NaOH (pH about 12), 400 mg of $MgCl_2$ and 0.5 mL of 1 M phosphate buffer (0.5 M K_2HPO_4/0.5 M KH_2PO_4) and heating at 90–95 °C for 60 min. Under alkaline conditions, minimal reduction of Cr(VI) or oxidation of native Cr(III) has been observed, except in solid matrices with high concentration of Mn, high Cr(III)/Cr(VI) ratios or freshly precipitated $Cr(OH)_3$, which is not expected to be present in environmental samples. In order to suppress the oxidation of Cr(III), this method suggests the addition of Mg^{2+}. This is supposed to reduce the oxidation of Cr(III) owing to its coprecipitation with it or to its sorption on Mn oxides making them less reactive to oxidize Cr(III). Recently, EPA Method 3060A has been applied to sediment samples with some modifications (Rahman *et al.*, 2005; Graham *et al.*, 2009). The main modifications included omission of phosphate and $MgCl_2$ and use of microwave (MW)-assisted alkaline extraction.

Although EPA Method 3060A minimizes method-induced oxidation and reduction, species transformation may still occur (Pettine and Capri, 2005). EPA Method 6800 (USEPA, 1998) addresses SIDMS for the correction of such degradations or conversions. This method recommends the alkaline digestion of samples following the EPA Method 3060A.

Table 5.5 Analytical methods used for chromium speciation in sediments.

Matrix	Sample preparation	Separative method	Detection	System type	Species	LOD	References
Marine sediments	Alkaline extraction	SPE (alumina)	ETAAS	On-line	Cr(VI) Cr(III)	0.45 µg g^{-1} 2.9 µg g^{-1}	Pazos-Capeans *et al*. (2006)
Harbour sediments	NaOH 0.5 м/Na_2CO_3 0.28 м, 60 min at 90 °C, MW heating	HPLC reversed phase	ICP-MS	On-line	Cr(VI)	0.5 µg L^{-1}	Graham *et al*. (2009)
Harbour sediments	0.05 м $(NH_4)_2SO_4$/0.05 м NH_4OH buffer at pH 9.2, 90 min, ultrasonic bath	HPLC	ICP-MS	On line	Cr(VI)	2 µg kg^{-1}	Séby *et al*. (2009)
River sediments and soils	0.04 м $(NH_4)_2SO_4$/0.5 м NH_4OH at pH 8, 14 min, 300 W, MW heating	Complexation DPC	Spectrophotometry	On-line	Cr(VI)	0.12 µg g^{-1}	Morales-Munoz *et al*. (2004)
Soils, sludges, sediments and similar waste materials	2% Na_2CO_3/3% NaOH, $MgCl_2$, K_2HPO_4/KH_2PO_4 buffer, 90–95 °C 60 min, hot plate	Complexation DPC	Spectrophotometry	Off-line	Cr(VI)	–	USEPA (1995a)
		HPLC anion exchange	ICP-MS (IDMS)	On-line	Cr(VI)	–	USEPA (1998)
Soil materials, chromite ore and anoxic river sediment	2% Na_2CO_3/3% NaOH, 90–95 °C 60 min, hot plate	Complexation DPC	Spectrophotometry	Off-line	Cr(VI)	–	James *et al*. (1995)
River sediment, chromite ore	2% Na_2CO_3/3% NaOH, 93 °C, 60 min, MW heating	HPLC anion exchange	ICP-MS (IDMS)	On-line	Cr(VI)	–	Huo and Kingston (2000)
Aggregate materials, soil and river sediment	0.28 м Na_2CO_3/0.5 м NaOH, 95 °C, 60 min, MW heating	HPLC anion exchange	ICP-MS (IDMS)	On-line	Cr(VI)	–	Rahman *et al*. (2005)

SIDMS proposed by EPA Method 6800 (USEPA, 1998) has been applied as a definitive diagnostic tool for the potential species interconversion during and after extraction steps of several reference materials of sediment and soil (Rahman *et al.*, 2005). SIDMS analysis was performed by adding both ^{50}Cr(III) and ^{53}Cr(VI) (double-spiking), applied before and after the extraction step. All the results were compared with those obtained with the traditional HPLC–ICP-MS method. In this study, substantial transformation of spiked Cr(III) to Cr(VI) was observed during pre-extraction stages of sample preparation and it occurs only nominally during post-extraction activity. It was also concluded that Cr(VI) and Cr(III) transformations representing both depletion and amplification of each species were significant and were affected by sample preparation activities. They were quantified accurately and precisely by the applied SIDMS methodology and corrected to enable the precise determination of Cr(VI).

Ammonium sulfate/ammonium hydroxide buffer solution has also been used for Cr(VI) extraction in sediments (Morales-Munoz *et al.*, 2004; Séby *et al.*, 2009). Even this method has been reported to extract only soluble or sparingly soluble Cr(VI) (James *et al.*, 1995); its use allows the extraction of less organic matter, making speciation analysis with HPLC–ICP-MS more comfortable, particularly when Cr(VI) is present at very low levels. This method was compared with the EPA method 3060 and similar results were obtained for sediments (Morales-Munoz *et al.*, 2004); see Figure 5.2.

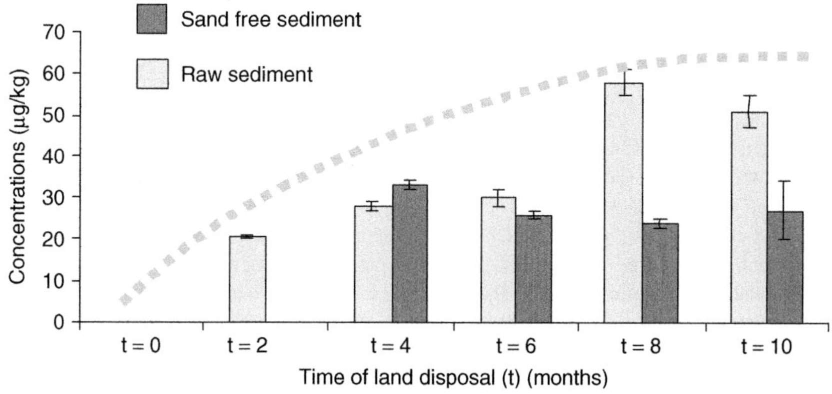

Figure 5.2 Cr(VI) concentration in dredged marine sediments with time of land disposal (Séby *et al.*, 2009). When dumping at sea is impossible, land disposal can be considered as an alternative solution. In these conditions, sediment aging can generate chemical, physical and/or biological transformations which could transform the structure of some toxic molecules into less-dangerous compounds. The Cr(VI) concentrations were thus determined in sediments stored on land for 10 months. Whereas no Cr(VI) is detected in freshly dredged sediments, trace amounts are present after 2 months in raw sediments and after 4 months in sand-free sediments. The very low concentrations determined explain the variability in the results, but the formation of Cr(VI) seems stabilized after 4 months. This behaviour is characteristic of a global oxidation of the sediments that occurs in the first 2 months after dredging. Chromium is initially present under the trivalent chromium reduced form and is slowly oxidized into Cr(VI). Reprinted with kind permission from Revue Paralia, Etude de l'évolution des formes chimiques des métaux danes des sédiments marins dragués stockés à terre by F. Seby, C. Benoit-Bonnemason, E. Tessier, C. Alzieu, J.-L. Aqua, L. Sannier and O. F. X. Donard, 10.5150/revue-paralia.2009.s03.

5.3.3 Marine Biota Samples

To the authors' knowledge, no study describing the specific determination of Cr species in marine biota is available.

5.4 INORGANIC AND ORGANIC ARSENIC SPECIES

Arsenic is widely present in the marine environment as a consequence of both natural sources (mainly by leaching from geological formations) and anthropogenic sources (smelters, arsenic pesticides, herbicides, etc.). In seawater, the average concentration of total arsenic is about 1 to $2\,\mu g\,L^{-1}$ (Nakazato *et al.*, 2000; Ng, 2005) whereas in marine animals the arsenic level ranges from a few micrograms of As per gram to a few tens micrograms of As per gram (Francesconi and Kuehnelt, 2002).

This element occurs in different forms presenting varying toxicity. It is now admitted that the toxicity of As species varies in order As(III) > As(V) > monomethyl arsonate (MA) > dimethylarsinate (DMA) > organic forms (Penrose and Woolson, 1974; Jain and Ali, 2000). Recent studies have shown that the reduced forms of MA and DMA, namely methylarsonite and dimethylarsinite, could be more toxic than As(III) (Petrick *et al.*, 2000; Styblo *et al.*, 2002; Dopp *et al.*, 2004; Hirano *et al.*, 2004). Contrarily to inorganic forms, organic forms such as arsenobetaine (AB), arsenocholine (AC), trimetylarsine oxide (TMAO), tetramethylarsonium ion (TETRA) or arsenosugars are toxic only with high doses (Kaise *et al.*, 1992; Neff, 1997). Arsenic toxicity appears strictly related to its chemical form, and plants and animals are known to accumulate this element mostly as nontoxic organoarsenic molecules.

The need for arsenic speciation arose at the beginning of the twentieth century when high arsenic concentrations were found in foods from marine origins. It was only in the 1970s that the hydride-generation (HG) technique coupled to atomic absorption spectrometry was used to separate the two inorganic forms As(III) and As(V) and the two simple methylated forms MA and DMA (Braman and Foreback, 1973). This method has since been widely used to study arsenic speciation, but is mainly limited to aqueous matrices such as waters or urine. Several techniques have successfully been used for speciation analysis of As and many reviews have been issued dealing with arsenic speciation analysis (Guérin *et al.*, 1999; Goessler and Kuehnelt, 2002; Gong *et al.*, 2002; B'Hymer and Caruso, 2004; Leermakers *et al.*, 2006; Niegel and Matysik, 2010) and it will be impossible to summarize in this paragraph the hundreds of articles published on arsenic speciation in the past 30 years (the reader is invited to refer also to these reviews complementary to this summary).

A number of specific detectors were used to conduct arsenic speciation analysis, each providing a certain number of advantages and limitations in the results. The first major limitation is linked to sensitivity. As such, the detectors used in the past (FAAS, ICP-AES, etc.) appear to be insufficient to analyse the traces present in natural samples. The second limitation is linked to the interferences which may occur during the detection. This point proves to be particularly critical to one of the detectors used most; that is, ICP-MS. Even though this detection offers the best sensitivity as well as a multielementary capability, it suffers of polyatomic interferences in the case of arsenic analysis. With regard to arsenic, the main interference arises from the association of $^{40}Ar^{35}Cl$, which is particularly disruptive for analysing marine products if the chromatographic conditions do not allow separation of chloride ions from arsenical species, as this element presents only one isotope. This problem is now solved by the use of ICP-MS equipped with a collision/reaction cell, allowing interferences removal.

In the following we will focus on the most widely used techniques for As speciation in marine samples, from seawater to marine biota, illustrated by the most recent papers.

5.4.1 Seawater Samples

In seawater, arsenate is the major arsenic form and As(III), MA and DMA are present at lower levels. However, arsenic species may be subject to seasonal changes, particularly in surface waters where primary productivity is high (Braman and Foreback, 1973; Andrae, 1978).

Seawater samples are usually simply filtrated and stabilized with HCl, EDTA or hydrazine chloride to prevent oxidation of As(III) into As(V) before storage at 4 °C. However, the stabilization step remains questionable, as several studies have demonstrated the risk of species interconversion with acidification (Francesconi and Kuehnelt, 2004). Moreover, the use of HCl should be avoided when analysis is performed by ICP-MS not equipped with a collision/reaction cell able to remove the ArCl interferences.

Several methods are available for arsenic speciation, some of which are designed specifically for inorganic species (and sometimes methylated species), while others can determine both organic and inorganic forms. Some of these techniques and recent publications are presented in Table 5.6.

The most popular technique is based on HG, because arsenic can be separated from the seawater matrix and consequently the detection limit can be lowered with the use of larger samples. The method relies on the difference between the species able to generate hydride (arsine): inorganic forms As(III) and As(V), mono- and di-methylated forms and the non-hydride-forming species, i.e. organic species. Volatile arsines are produced by reaction with sodium borohydride in acidic media and flushed to a specific detector (AAS, atomic fluorescence spectrometry (AFS)) by an inert gas. By carefully controlling the pH for the hydride formation reaction it is possible to differentiate As(III) from As(V), as for pH <1 As (III), As(V), MA and DMA can generate arsines, whereas As(III) is the only one reduced to volatile arsine for pH about 5. Another possibility to distinguish between the two redox forms is to use coprecipitation of As(V) with aluminium hydroxide ($Al(OH_3)$). As(III) is calculated from the difference between the total As content and As(V) content. Reported detection limits vary from 1 to $100\,ng\,L^{-1}$ (Cabon and Cabon, 2000; Tuzen *et al.*, 2009, Anninou and Cave, 2009). However, it has been shown that arsenosugars can also generate arsines (Schmeisser *et al.*, 2004a); therefore, the use of HG should be questioned if some arsenosugars are present in water samples as algae decomposition products.

Another group of methods dealing with electrochemical speciation for As(III) and As(V) in seawater is based on As(III) electroactive species analysis and total As analysis after reduction of As(V) into As(III) by different reagents such as L-cysteine (Vandenhecke *et al.*, 2007), a mixture of $Na_2S_2O_5$ and $Na_2S_2O_3$ (De Carvalho *et al.*, 2008) or sodium iodide (Dugo *et al*, 2005). Here again, results for As(V) are obtained by difference. The modification of a working electrode by gold plating allows one to decrease the detection limits to a few tens of nanograms per litre (Dugo *et al.*, 2005; Vandenhaecke *et al.*, 2007).

Actually, these methods have been in use for a long time for redox speciation, but they tend now to be replaced by hyphenated techniques in which a separation method is coupled to a selective elemental detector and allowing full screening of the species present in the sample. The coupling of HPLC, GC and capillary electrophoresis with ICP-MS are accepted hyphenated techniques for such speciation analysis. Campillo *et al.* (2008) describe the application of GC–atomic emission detector (AED) coupling for As speciation after derivatization with methylthioglycolate. In this case, inorganic species are analysed together, but it is possible to separate MA and DMA from inorganic As. The detection limits are in the microgram per litre range.

Based on several reviews about arsenic speciation methods (Guérin *et al.*, 1999; Gong *et al.*, 2002; Francesconi and Kuehnelt, 2004; B'Hymer and Caruso, 2004; Le *et al.*, 2004; Leermakers *et al.*, 2006) it appears clearly that most of the studies imply HPLC interfaced with element-specific detection. Liquid chromatography (LC) is preferred, as it does not require a previous derivatization step to produce volatile species, which is not always feasible for all compounds. In the case of seawater analysis, anion exchange is usually chosen, allowing the separation of As(III), As(V), MA and DMA. However, the highly saline samples may deteriorate the separation processes and

Table 5.6 Analytical methods used for arsenic speciation in seawater.

Matrix	Sample preparation	Separative method	Detection[a]	System type	Species	LOD (ng L^{-1})	References
Seawater	Addition hydrazine dichloride after sampling, extract filtered 0.45 μm Acidification to pH 2 with HCl, storage dark 4 °C	Reduction As(V) into As(III) with cysteine	SCP	Off-line As(III) measurement and sum As(III) + As(V) after reduction	As(III), As(V)	22	Vandenhecke *et al*. (2007)
Seawater	Dilution 1/4	Reduction As(V) into As(III) with metabisulfite/thiosulfate	ACSV	Off-line As(III) measurement and sum As(III) + As(V) after reduction	As(III), As(V)	4 200	De Carvalho *et al*. (2008)
Seawater	Addition hydrazine dichloride, storage dark 4 °C	Reduction As(V) into As(III) with KI in strong HCl	dASCP	Off-line As(III) measurement and sum As(III) + As(V) after reduction	As(III), As(V)	80	Dugo *et al*. (2005) La Pera *et al*. (2008)
Seawater	–	Hydride or nonhydride-forming species	FI-HG-AAS	On-line (total As measurement and species by difference)	As(III), As(V), DMA, MA	1.5	Cabon and Cabon (2000)

Seawater	–	Coprecipitation As(V)	HG-AAS	On-line (total As measurement and species by difference)	As(III), As(V)	12	Tuzen *et al*. (2009)
		Hydride- or nonhydride-forming species					
Seawater	Acidification to pH<2 with HCl, storage dark 4 °C	Hydride- or nonhydride-forming species	HG-ETAAS	Off-line (total As measurement and species by difference)	As(III), As(V), DMA, MA	100	Anninou and Cave (2009)
Seawater	Acidification to pH 2 with HCl, storage dark 4 °C	GC (derivatization with methylthioglycolate, extraction cyclohexane)	AED	On-line	DMA	1	Campillo *et al*. (2008)
					MA	10	
					inorg As	25	
Seawater	–	HPLC (ion exclusion)	HG-ICP-MS	On-line	As(III), As(V), MA	As(III) 3.4 As(V) 4.2 MA 3.5	Nakazato *et al*. (2002)
			ORS-ICP-MS			As(III) 4.3 As(V) 6.7 MA 7.3	
Seawater	–	HPLC anion exchange	HG-AFS	On-line	As(III), As(V), DMA, MA	–	Price and Pichler (2005)

[a]SCP: stripping chronopotentiometry; dASCP: derivative anodic stripping chronopotentiometry; FI-HG-AAS: flow injection HG-AAS; ORS-ICP-MS: octopole reaction system ICP-MS.

also the presence of chloride ion generates interferences by formation of ArCl when ICP-MS is used as a detector. Dilution of the samples or pretreatment for removing the matrix are necessary (Nakazato *et al.*, 2002) and, therefore, decrease the level of As species. Pretreatment may also modify the original speciation in the sample (Cabon and Cabon, 2000). An alternative to sample dilution was found with the introduction of collision/reaction cell on new generations of ICP-MS. The use of collision or reaction gases allows one to remove the interference created in the argon plasma and enable highly sensitive determination of As species at the nanogram per litre level. Another option is to insert an HG step before introduction into the ICP-MS (Nakazato *et al.*, 2002). The coupling of other elemental detectors, such as AFS, is also reported for As speciation in pore waters in hydrothermal systems with HG introduction of sample (Price *et al.*, 2007).

5.4.2 Marine Sediment Samples

Different approaches are mentioned in the literature for arsenic speciation in marine sediments. When mobility or bioavailability of As contained in sediments is studied, the sequential extraction procedure is still often used (Price and Pichler, 2005; Hung *et al.*, 2009). Arsenic is sequentially extracted from the easily extractable, carbonate, iron oxyhydroxide or hydrous ferric oxide, organic matter and residual fractions. Arsenic species can also be specifically determined, and methods cited in the literature for sediment samples are presented in Table 5.7.

Most of the time, As species are extracted with diluted phosphoric acid to be further analysed with HPLC–ICP-MS. The extraction of As methylated forms, such as MA or DMA, is quite simple, since these compounds are generally stable in several extraction media and are easily released (Demesmay and Ollé, 1997; Sanz *et al.*, 2005). However, inorganic species (As(III) and As(V)) are generally the major species in sediments and are more strongly retained by the matrix than the others. The efficiency of phosphate ions in releasing inorganic As from environmental solids is essentially based on the ion-exchange reaction between phosphates and arsenates inducing As desorption. Complementary effects of phosphoric acid could be linked to its acidic properties and complexing power, which may dissolve some mineralogic phases of the solid matrix and/or desorb species other than arsenate by modifying surface charges (Demesmay and Ollé, 1997; Vergara Gallardo *et al.*, 2001; Georgiadis *et al.*, 2006). Recoveries generally vary between 60 and 100%, depending on physico-chemical properties of sediments, and are often improved by MW heating. All studies point to the fact that the main problems encountered when using phosphoric acid as extracting reagent are a partial oxidation of As(III) to As(V) and a distortion of the ion-exchange-based chromatographic separation, particularly for As(V). This is generally avoided by diluting of the extracts before injection to obtain a phosphoric acid concentration close to 0.25 mol L^{-1} (Thomas *et al.*, 1997; Ellwood and Maher, 2003; Orero Iserte *et al.*, 2004). Dilution also allows one to preserve As(III) from oxidation (Vergara Gallardo *et al.*, 2001). To minimize this redox reaction, extraction is often performed with phosphoric acid with a concentration equal or lower than 1 mol L^{-1} (Demesmay and Ollé, 1997; Ellwood and Maher, 2003) or by adding reducing or complexing reagents in combination with phosphoric acid, such as hydroxylamine hydrochloride (Ellwood and Maher, 2003), ascorbic acid (Zheng *et al.*, 2003) or sodium diethyldithiocarbamate trihydrate (NaDDC) (Georgiadis *et al.*, 2006). In addition to As(III), As(V), MA and DMA, arsenosugars, arsenobetaine and traces of arsenocholine have been detected in marine sediments or in suspended particles. Occurrence of these compounds is directly linked to the presence of microorganisms, detritus of algae or marine organisms within the sediment (Hanaoka *et al.*, 1997; Ellwood and Maher, 2003).

With arsenic often being present with a high concentration in sediments, solid-state techniques are increasingly being used to show the presence of inorganic As species in sediments. Solid-state techniques are generally less sensitive than wet-chemistry analytical methods, but are less

Table 5.7 Analytical methods used for arsenic speciation in sediments.

Matrix	Sample preparation	Separative method	Detection	System type	Detected Species	LOD	References
River sediments	H_3PO_4 1 M, 20 min at 40 W, MW heating	HPLC anion exchange	HG-AFS	On-line	As(III) and As(V)	0.02–0.04 mg kg^{-1}	Vergara Gallardo *et al.* (2001)
Spiked lake sediments	H_3PO_4 0.3 M, 10 min at 20 W, MW heating	HPLC anion exchange	ICP-MS	On-line	As(III), As(V), MA and DMA	–	Demesmay and Ollé (1997)
	Ammonium oxalate, 10 min, magnetic stirring						
Marine and estuarine sediments	H_3PO_4 0.3 M, 60 min shaking	HPLC anion exchange	ICP-MS	On-line	As(III), As(V) and MA	0.005–0.04 mg kg^{-1}	Orero Iserte *et al.* (2004)
River and lake sediments	H_3PO_4 1 M/ascorbic acid 0.1 M, 15 min ultrasonic stirring	HPLC anion exchange	SF-ICP-MS	On-line	As(III) and an unknown 'As-complex'	–	Zheng *et al.* (2003)
Antarctic sediment	H_3PO_4 1 M, 1 min sonication probe	HPLC anion exchange	ICP-MS	On-line	As(III), As(V), MA and DMA	–	Sanz *et al.* (2005)
Marine sediments	H_3PO_4 0.5 M/hydroxylamine hydrochloride 0.1 M, 60 min shaking	HPLC anion exchange	ICP-MS	On-line	As(III), As(V), Arsenosugars 1c and 1d	–	Ellwood and Maher (2003)
River sediments	H_3PO_4 1 M, 20 min at 40 W, MW heating	HPLC anion exchange	ICP-MS	On-line	As(III) and As(V)	1–2 mg kg^{-1}	Thomas *et al.* (1997)
Micro-suspended particles (size $\leq 5\ \mu m$)	Water/chloroform-methanol (1:4), 2 min shaking	HPLC anion exchange	ICP-MS	On-line	Arsenobetaine, As(III) and traces of arsenocholine	–	Hanaoka *et al.* (1997)
Marine sediment	Phosphate 10 mM/NaDDC, 0.5%, shaking	HPLC anion exchange	HG-AFS	On-line	As(III) and As(V)	0.04 mg kg^{-1}	Georgiadis *et al.*, (2006)

prone to speciation changes, as they require almost no treatment of the sample. X-ray absorption spectroscopy techniques such as X-ray absorption near-edge spectroscopy and extended X-ray absorption fine structure are generally used for the identification and quantification of arsenic oxidation states (Kneebone *et al*., 2002; Bostick *et al*., 2004; Root *et al*., 2009). These techniques also allow one to determine the nature of the mineral phases where As is present, particularly sulfide and/or iron minerals that play an important role in the behaviour of As in marine sediments.

5.4.3 Marine Biota Samples

Most of the studies on arsenic speciation performed in the field of the marine environment are related to marine biota. Almost all the different organisms encountered in marine ecosystems are analysed: algae, shellfish, fishes, turtles, anemones, and so on. Actually, besides the two inorganic species and the two simple methylated forms, a large variety of organoarsenicals have been identified (Francesconi and Edmonds, 1996; Francesconi and Kuehnelt, 2002; Schmeisser *et al*., 2004b). However, owing to health and safety concerns, it is mainly food products that are subject to intense monitoring of As species – that is, algae and edible fishes and seafood (shrimps, seashells, etc.) – as seafood is by far the major contributor of arsenic to the European diet (European Commission, 2004). A paper by Borak and Hosgood (2007) reviews the nature and concentrations of arsenicals in dietary seafood and focuses on the possible implications of seafood arsenic on human risk assessment. Actually, the concentration of arsenic species in marine animals is much higher than the background concentrations in the surrounding water. Arsenobetaine is the most commonly reported organoarsenical in marine animals. Recently, arsenolipids have been found in cod fish oil, resulting in concerns for human health (Schmeisser *et al*., 2005). Arsenosugars are found in marine algae and in marine animals such as scallops feeding on algae, and have also been identified in fish and mussels. While arsenosugars are assumed to be relatively nontoxic to animals and humans compared with inorganic arsenic species, biotransformations of arsenosugars can result in toxic arsenicals, and As(V), As(III), MA and DMA can also be detected with low concentration (Francesconi and Kuehnelt, 2002).

Regarding arsenic speciation in marine organisms, one of the crucial points is the extraction, which should maintain the identity of the endogenous species, not only during the extraction process but also during the period of time between the extraction and the final measurement. Most of the studies use extracting solutions made of a methanol–water mixture or water only to recover As species. Other reagents have also been reported, such as tetramethylammonium hydroxide (TMAH; Ackley *et al*., 1999), alkaline alcohol (Larsen *et al*., 2005; Sloth *et al*., 2005), nitric acid (Brisbin and Caruso, 2002), acetone (Kirby and Maher, 2002), phosphoric acid (Kuehnelt *et al*., 2001) or mixtures of H_2O–MeOH–CH_3Cl for extraction of lipid-soluble As compounds (Hanaoka *et al*., 2001). These reagents allow good recovery of As compounds with efficiency between 80 and 100%, but the stability of the compounds during extraction is rarely verified. Combination of different reagents is also used to sequentially extract different classes of arsenic compounds (Foster *et al*., 2007; Cao *et al*., 2009). Enzymatic hydrolysis of the samples with trypsine has also been reported as showing good efficiency on reference materials DORM 1 (96%), TORT 1 (108%) and other seafood (82–102%) (Fitzpatrick *et al*., 2002; Rattanachongkiat *et al*., 2004). However, these studies focus on the efficiency for arsenosugars and DMA extraction, but do not give any conclusion on the efficiency for the other As species. More recently, a mixture of protease and lipase has been used for simultaneous speciation analysis of As and Se (Reyes *et al*., 2009). In some cases, workers also focus their study on the bioaccessibility of arsenic compounds to human by using enzymatic extraction simulating gastrointestinal tract conditions (Dufailly *et al*., 2008).

Assistance of extraction is usually performed by either mechanical or ultrasound agitation (Wuilloud *et al*., 2006). Accelerated solvent extraction (ASE) or pressurized liquid extraction

(PLE) has also been tested for As extraction from seafood (McKiernan *et al*., 1999; Wahlen *et al*., 2004; Mato Fernandez *et al*., 2007). The use of MW-assisted extraction is now increasing for As extraction from biological matrices (Karthikeyan *et al*., 2004; Larsen *et al*., 2005; Hirata *et al*., 2006; Reyes *et al*., 2009; Yang G.D. *et al*., 2009). A recent paper has compared the different assistance modes in term of efficiency on several seafood reference materials with recovery in the range 76–87% (Nam *et al*., 2010).

It is worth noting that the extraction efficiency of the protocols developed depends strongly on the matrix of the sample and the question of the real necessity to achieve complete extraction of As from the different samples arises (Francesconi, 2003), and more and more As speciation studies target on specific classes of compounds.

Regarding analytical methods, while few workers still use analytical approaches based on selective hydridation of inorganic As (Baeyens *et al*., 2008; Peshut *et al*. 2008; Geng *et al*., 2009), the coupling of LC with specific detectors is widely and increasingly used. The different mechanisms available in LC offer excellent possibilities for the separation of the different arsenic forms based on their chemical properties. The most commonly used mechanisms are ion-pair chromatography, reversed-phase chromatography and ion-exchange chromatography (anionic or/and cationic). Anion-exchange chromatography is suitable for the separation of inorganic and mono- and dimethylated species, which have anionic character. Other organic species, such as AB, AC, TMAO and TETRA are neutral or cationic, and may also have hydrophobic properties due to their alkyl groups. As a consequence, they should be separated by cation-exchange or reversed-phase chromatography. For the simultaneous study of organic and inorganic species, it appears that a single chromatographic mechanism is not sufficient. For this reason, combination of anion- and cation-exchange chromatography has been used, using columns in series or column-switching systems (Suner *et al*., 2001; Sakai *et al*., 2001). According to Saverwyns *et al*. (1997), a combination of anion-exchange and hydrophobic mechanisms is also suitable for the separation of anionic and organic compounds, and some workers have developed chromatographic conditions allowing the separation of up to 17 species in a single run (Kohlmeyer *et al*., 2002; Karthikeyan and Hirata, 2004; Simon *et al*., 2004; Dufailly *et al*., 2007). Separation by capillary electrophoresis has also been reported for standards, but it is rarely used for real samples analysis (Yang G.D. *et al*., 2009).

High-sensitivity detectors in use are ICP-MS, AFS and in some cases AAS, but this latter detector is used less and less as it suffers from lack of sensitivity. The latter two detectors require on-line decomposition (UV or MW) and volatilization (HG) of the separated compounds, as the organoarsenicals do not generate hydride in classical hydridation conditions. In recent decades the interest for AFS has increased, encouraged by its low cost of investment and handling and its simplicity of use. Recently, this detector has even been proved to be as efficient as ICP-MS, in terms of sensitivity and repeatability. Detection limits reached with HPLC coupled to ICP-MS or HG-AFS are below the tens of micrograms per kilogram. Recent examples of the above-mentioned techniques are presented in Table 5.8. The reader will also find useful information in the reviews already mentioned in the introduction.

Besides these 'classical' approaches, other compounds, such as arsenosugars or thio-arsenicals, are now subject to specific analytical developments (Raml *et al*., 2006; Niegel and Matysik, 2010). Owing to the number of unknown As compounds observed during analysis with elemental detectors, a new trend in As analysis has appeared in recent years with the coupling of LC to molecular mass spectrometry, able to provide structural information on the separated compounds (Nischwitz and Pergantis, 2005, 2006; Schaeffer *et al*., 2006).

Table 5.8 Analytical methods used for arsenic speciation in biota.

Matrix	Sample preparation	Separative method	Detection	System type	Species	LOD	References
Fish, shellfish	Freeze-drying, powdering, 500 mg+ 10 ml 2 M NaOH, heating 75–85 °C 12 h, extract filtered	HG	ICP-MS		Total As	2.5 ng g^{-1}	Peshut *et al*. (2008)
			QFAAS		Inorganic As ($NaBH_4$ reducible As(III) + As(V))	9.1 ng g^{-1}	
Fish, shellfish. CRMs: DORM2 dogfish muscle; TORT2 lobster hepatopancreas; DOLT2 dogfish liver	Freeze-drying, powdering, 0.1–0.2 g+ 6 ml HNO_3 65%, MW heating +1 ml H_2O_2, MW heating, dilution	HG	ICP-MS		Total As		Baeyens *et al*. (2008)
			HG-AFS		Inorganic As ($NaBH_4$ reducible As(III) + As(V)) + MA + DMA	5 ng g^{-1}	
Algae. CRMs: DORM2 dogfish muscle; TORT1 lobster hepatopancreas; DOLT2 dogfish liver; BCR422 cod muscle; NIST 1566b oyster; NMIJ 7402a cod muscle	50 mg powder + 5 ml 2 M NaOH, heating 90–95 °C 3 h	CT-HG[a]	CT-HG-AAS		As(III) + As(V), MA, DMA, AB	–	Geng *et al*. (2009)
	500 mg powder extracted with 10 ml 0.25 M H_3PO_4, mechanical shaking 12 h	HPLC anion exchange	UV-HG-AFS	On-line	As(III), As(V), MA, DMA, AB		

Oyster; shrimp CRM: DORM2 dogfish muscle	Freeze-drying, powdering, 500 mg + 10 ml MeOH, magnetical stirring 15 min, extract filtered 0.45 μm	HPLC anion exchange	UV-HG-AFS	On-line	As(III), As(V), MA, DMA, AB, AC, Arsenosugars, TMAO, TETRA	2–7.5 ng g^{-1}	Simon *et al*. (2004)
Fish. CRMs: GBW-08 751 mussel; BCR-278R mussel; DORM2 dogfish muscle	100 mg extracted with MeOH, ASE	HPLC anion exchange	UV-HG-AFS	On-line	As(III), As(V), MA, DMA, AB, AC	35–120 ng g^{-1}	Mato Fernandez *et al*. (2007)
Algae	Washing with deionized water, oven drying 60 °C, powdering 1 g extracted with 50 ml $MeOH/H_2O$ (1/1), sonication 3 h 30 °C, centrifugation, extract filtered 0.45 μm + 0.22 μm 5 μl injected	HPLC anion exchange	ICP-MS	On-line	As(III), As(V), MA, DMA, Arsenosugars	–	Wuilloud *et al*. (2006)
		cation exchange	ESI-IT-MS[b]				
Oyster, fish, shrimp	Freeze-drying, powdering, 100 mg extracted with 10 ml H_2O, MW heating 50 W 10 min, extract filtered	HPLC: ion pairing reversed phase	ICP-MS	On-line	As(III), As(V), MA, DMA, TMAO, AC, TeMAs, AB	20–100 μg L^{-1}	Hirata *et al*. (2006)
Clam, shrimp	Freeze-drying, powdering, 500 mg extracted with 5 ml $MeOH/H_2O$ (1/1), MW heating 60 °C 6 min, extract filtered 0.22 μm	CE	ICP-MS	On-line	As(III) + As(V), MA, DMA	1–2 ng g^{-1}	Yang g.D. *et al*. (2009)

(continued overleaf)

Table 5.8 *(continued)*

Matrix	Sample preparation	Separative method	Detection	System type	Species	LOD	References
CRMs: DORM2 dogfish muscle; BCR422 cod muscle; BCR 627 tuna fish	150 mg extracted with 10 ml $MeOH/H_2O$ (4/1) MW heating 75 °C 4 min, extract filtered 0.45 µm	HPLC anion exchange	ICP-MS	On-line	As(III), As(V), MA, DMA, TMAO, AC, AB	8–33 ng g^{-1}	Dufailly *et al.* (2007)
Dry seafood. CRM: TORT2 lobster hepatopancreas	200 mg for sequential extractions: • non polar As compounds • polar As compounds • inorganic As compounds	HPLC anion exchange	ICP-MS	On-line	As(III), As(V), MA, DMA, AC, AB	–	Cao *et al.* (2009)
Fish	Homogenization, freezing, defatting procedure for fat fish Extraction: MW heating 90 °C 10 min, diluted with H_2O 1 g extracted with 10 ml alkaline alcohol, MW heating 90 °C 10 min	HPLC anion exchange	ICP-MS	On-line	As(III), As(V)	3 ng g^{-1}	Larsen *et al.* (2005)
Oyster, fish, schrimp, algae	Freeze-drying, powdering 100 mg extracted with 10 ml H_2O/MeOH (1/1), MW heating 50 W 10 min, centrifugation, extract filtered 0.45 µm For algae, MW heating is replaced by 15 min sonication	HPLC anion exchange	ICP-MS	On-line	As(III), As(V), MA, DMA, TMAO, AC, AB, TeMAs	0.03–0.14 µg L^{-1}	Karthikeyan and Hirata (2004)

Candidate reference material BCR 710, oyster	100 mg extracted with 10 ml MeOH/H_2O (4/1), shaking overnight, centrifugation	HPLC reversed phase	ICP-MS	On-line	Thioarsenicals	–	Raml *et al*. (2006)
CRMs: BCR 627 tuna fish; DORM2 dogfish muscle; CRM 278R oyster; NIST 1566b Oyster	200 mg extracted with 4 ml H_2O, shaking 14 h	HPLC anion exchange and combined anion/cation exchange	ESI-MS	On-line	TETRA, TMAO, MA, DMA, AB, AC, Arsenosugars, thioarsenicals	–	Nischwitz and Pergantis (2005)
CRM: TORT2 lobster hepatopancreas	250 mg extracted with 10 ml H_2O Extraction repeated 3 more times with 5 ml H_2O, sonication 20 min, extract filtered	HPLC anion exchange and cation exchange	ESI-MS	On-line	TETRA, TMAO, DMA, TMAP, AB, AC, Arsenosugars, thioarsenicals	–	Schaeffer *et al*. (2006)

[a]CT: cold trap.
[b]IT: ion trap.

5.5 INORGANIC AND METHYLATED-MERCURY SPECIES

Owing to the important toxicological impact of mercury, the understanding of its biogeochemical cycle involving distribution, transport and transformation pathways in the natural environment has for a long time been a field of concern. Although all the chemical forms of mercury are poisonous, the ecological and human health effects of Hg are generally related to the environmental transformations of inorganic mercury (IHg) into the toxic and biomagnified methylmercury (MeHg). Therefore, mercury speciation, and particularly the assessment of IHg and MeHg in environmental matrices, is a critical and important step in the interpretation of the mercury biogeochemical cycle in aquatic environments (Stoichev *et al*., 2006).

The determination of different mercury species from the various and complex matrices present in marine ecosystems (e.g. water, sediments and biological tissues) is still considered a difficult task owing to the very low levels of the endogenous mercury species in water samples, the extremely low relative concentrations of MeHg compared with IHg in sediments and the possible methylation/reactions occurring during the analytical procedures (water, sediment and biological samples). Therefore, mercury speciation, and more precisely MeHg determination in environmental samples, is hindered not only by traditional problems related to nonquantitative recoveries, but also by questions about the occurrence of artefact formation and transformations of MeHg during the sample preparation and separation steps. The recent development of species-specific isotope dilution procedures (i.e spiking the samples with isotopically enriched species) has drastically improved the quality and accuracy of the data on mercury speciation analysis (Wilken and Falter, 1998; Demuth and Heumann, 2001; Rodriguez Martin Doimeadios *et al*., 2002, 2003; Monperrus *et al*., 2005a). In addition, the use of isotopically enriched mercury species as tracers has revealed that mercury species are subject to abiotic artificial transformations (i.e methylation and demethylation), affecting a large set of environmental matrices, including seawater, biological materials and sediments, hence leading to erroneous determinations. In the last few years, several methodologies based on the use of multiple spiking species-specific isotope dilution analysis have been developed to overcome these problems (Hintelmann, 1999; Monperrus *et al*., 2008). When several Hg species need to be simultaneously analysed, each of them can be enriched in a different isotope. In this way, the concentration of the species can be quantified even when opposite or simultaneous degradation processes occur. In the case of mercury speciation analysis, the addition of two isotopically enriched species to the sample (double spiking) provides the quantification of the extent of both methylation and demethylation processes and, therefore, the correction of the final species concentrations.

5.5.1 Seawater Samples

In seawater there are three main forms of mercury, namely elemental mercury (Hg^0), inorganic mercury (Hg(II)) and organic mercury (MeHg and dimethylmercury (DMeHg)). Hg^0 is a ubiquitous component found for all depths and is usually supersaturated, especially in surface water, where it can be nearly 50% of the total mercury in the mix layer (Mason *et al*., 1995). Hg(II) and MeHg concentrations in seawater are in the range of nanograms to picograms per litre. They mainly form chloro-complexes and the proportion of organic mercury is typically $<5\%$ of the total mercury concentration.

Owing to adsorption to the sample container and abiotic or biotic processes that could lead to a redistribution between chemical species, the sample pretreatment is a critical issue in trace mercury speciation. The water samples (filtered or nonfiltered) are stored normally at 4 °C in the dark after the addition of a small volume ultrapure concentrated acid (0.1 to 1% v/v). During the storage, Hg(II) can be lost from the sample either by sorption on the bottle walls or by chemical processes (reduction, methylation) (Krivan and Haas, 1988). The stability study of Hg(II) solutions demonstrated that HNO_3 is less effective than HCl and an H_2O_2–HCl mixture is the best to preserve Hg(II). The question of the stability of MeHg is not completely solved and depends on the matrix (Baeyens, 1992). For example, HNO_3 is effective for simple solutions of MeHg but not for seawater. Marine waters can be preserved with H_2SO_4 (Jackson, 1988), HCl (Cossa *et al.*, 1997) or frozen without acidification (Bloom, 1989). The stability of the mercury species during sample storage and the loss mechanisms were recently investigated (Yu and Yan, 2003; Parker and Bloom, 2005). Different factors, like concentration of mercury species, matrix composition, container material, pH, temperature and light, have been considered and ways of decreasing the loss and the transformation of the mercury species are proposed.

Volatile mercury forms (Hg^0, DMeHg) may be also present in water in extremely low concentrations (Coquery *et al.*, 1997; Amouroux *et al.*, 1998). For their analysis, large volumes of the samples are recommended. The samples should not be acidified because acids can enhance rapid oxidation of the volatile species (Parker and Bloom, 2005). The analytes can be collected in-situ using cryogenic traps and stored at low temperature (liquid nitrogen) until analysis in the laboratory (Amouroux *et al.*, 1998) or can be purged onto a gold traps (for Hg^0) and carbotrap (for DMeHg) (Coquery *et al.*, 1997). If samples cannot be purged and trapped in the field, they should be collected in completely full glass bottles with Teflon-lined caps, as those species are lost rapidly from Teflon and polyethylene bottles (Parker and Bloom, 2005).

Highly sensitive detection techniques are available for mercury (e.g. cold vapour AFS, ICP-MS), but speciation analyses also require effective and selective preconcentration procedures. Preconcentration of ionic mercury species from the aqueous phase is also used for separation. SPE can be applied for seawater samples (Kwokal and Branica, 2000). The MeHg is retained, while the Hg(II) is eluted with a potassium cyanide solution, then the MeHg is eluted in acid media. Mercury species can also be concentrated from waters using dithiocarbamate resin and subsequently eluted with thiourea solution (Emteborg *et al.*, 1993, 1995).

For the separation of mercury species at environmental levels, GC is most often used, since it offers quantitative transfer to the detector and better sensitivity. When analysing volatile forms (Hg^0 and DMeHg), after fast desorption from the sampling traps they are preconcentrated and separated directly without derivatization (Amouroux *et al.*, 1998). Direct analysis of MeHg in the form of MeHgX (X = Cl, Br) selectively extracted with organic solvents is carried out by GC–ECD (Puk and Weber, 1994). However, one of the drawbacks of this detector is its low selectivity for mercury and it responds to other compounds containing halogens co-extracted, causing interferences. In most of the cases, it is necessary to derivatize the ionic mercury species in order to convert them to volatile forms, which are then separated by GC and detected by specific atomic detectors. During the HG with sodium borohydride ($NaBH_4$), the Hg(II) is transformed to Hg^0 while MeHg forms MeHgH. The speciation by HG was considered impossible because the Hg–H bond is unstable. But the half-life time of MeHgH is about 2 h, which is sufficient for its determination (Filippelli *et al.*, 1992). The derivatization should be applied in inert atmosphere and should start at pH 1–2, because otherwise MeHg is reduced to Hg^0. An interferences study (De Diego *et al.*, 1998) showed that the heavy metals and metalloids hinder the mercury species determination by HG, whereas chloride anions have almost no effect. HG can, therefore, be directly applied for sea and estuarine waters (Ritsema and Donard, 1994; Tseng *et al.*, 1998, 2000; Stoichev *et al.*, 2002). During the ethylation (Eth) with sodium tetraethylborate ($NaBEt_4$) the Hg(II) is transformed to $HgEt_2$ while MeHg forms MeHgEt. This is motivated by the higher stability of MeHgEt compared with MeHgH. Interference studies (De Diego *et al.*, 1998; Demuth and Heumann, 2001) have shown that the Eth is not affected by heavy metals and metalloids, whereas the chlorides decrease the signal of the analytes strongly. Additionally, in salty water, one part of MeHg is reduced to Hg^0 during Eth. Water vapour distillation is also a possible way to rapidly recover MeHg from the sample matrix (Bowles and Apte, 1998). A new powerful approach is to use isotope dilution ICP-MS to solve the problems with the matrix and nonquantitative derivatization (Demuth and Heumann,

2001; Monperrus *et al*., 2005a). The inadequate quality of some lots of $NaBEt_4$ is involved in some analytical artefacts (Tseng *et al*., 1999; Stoichev *et al*., 2004b). Other alkylation procedures can be a remedy; for example, propylation (Pr) of mercury species with sodium tetrapropylborate. Additionally, the Pr is not sensitive to interferences from chlorides and can be directly applied to estuarine and seawaters (Demuth and Heumann, 2001; Monperrus *et al*., 2005a).

In the on-line systems the derivatization is directly connected to the main apparatus and the automation of the analysis is possible, thus reducing the risk of contamination and losses of the analytes. The principal steps are the derivatization and the purge with an inert gas of the derivatized volatile species and the trap (cryogenic or on adsorbent), followed by chromatographic separation and mercury-specific detection (De Diego *et al*., 1999). The derivatization reactions used in the on-line systems for mercury speciation analysis in water samples are the above-mentioned HG, Eth or Pr. Simultaneously with the alkylation derivatization, the volatile mercury derivatives are purged by means of a gas stream and trapped on adsorbent at ambient temperature (Bloom, 1989; Liang *et al*., 1994; Reuther *et al*., 1999; Tseng *et al*., 2004) or on cryogenic trap (Ceulemans and Adams, 1996; Tseng *et al*., 1998; Demuth and Heumann, 2001; Stoichev *et al*., 2004b). During the trapping step, an enrichment factor of 50–100 is achieved (De Diego *et al*., 1999). If HG is applied, only cryogenic traps are used (Ritsema and Donard, 1994; Reuther *et al*., 1999; Tseng *et al*., 1998, 2000; Stoichev *et al*., 2002, 2004b) because MeHgH is not very stable. A drawback of the cryogenic trapping is the water vapour interference on both chromatographic separation and the detection, especially for the HG procedure (Stoichev *et al*., 2004b). A universal method for water vapour elimination is the utilization of condensational pre-trap (approximately −15 °C) before the column (Ceulemans and Adams, 1996; Amouroux *et al*., 1998; Reuther *et al*., 1999; Tseng *et al*., 2000). Another possibility is the utilization of some specific hygroscopic substances like potassium carbonate or soda lime (Demuth and Heumann, 2001; Stoichev *et al*., 2004b). For systems with ambient temperature preconcentration, the interferences due to water vapour are less severe and could be eliminated by passing a dry gas over the trap after the preconcentration step (Bloom, 1989; Liang *et al*., 1994). The detection of the mercury species in on-line analysis of water samples has been achieved by AFS (Bloom, 1989; Ritsema and Donard, 1994; Cai *et al*., 1996; Bowles and Apte, 1998; Stoichev *et al*., 2002, 2004b; Tseng *et al*., 2004), AAS (Tseng *et al*., 1998), microwave-induced plasma (MIP)-AES (Ceulemans and Adams, 1996; Reuther *et al*., 1999; Slaets and Adams, 2000) and ICP-MS (Amouroux *et al*., 1998; Tseng *et al*., 2000; Demuth and Heumann, 2001; Bravo-Sanchez *et al*., 2004).

In the off-line system the derivatization is independent and separated from the analytical technique used. In order to have low LODs it is always necessary to preconcentrate the analytes before the derivatization. The derivatization with Grignard reagent is used for different types of water (Emteborg *et al*., 1993, 1995). The derivatized analytes are extracted with organic solvent, the separation is achieved with capillary GC (CGC) and the detection is with MIP-AES. For the determination of MeHg in waters, extraction of MeHgCl in organic solvent can be applied. The extract can be analysed for example, with capillary GC–AFS (Cai *et al*., 1996). The derivatization with iodoacetic acid is used to volatilize MeHg as MeHgI (Lansens *et al*., 1990). The MeHgI is determined by headspace GC with MIP-AES detection. With the latter method the extractions are avoided and the column is not charged with the matrix components. Solid-phase microextraction (SPME) on poly(dimethylsiloxane) fibre is used to preconcentrate the mercury species from seawater after Pr reaction (Bravo-Sanchez *et al*., 2004). In some off-line systems the volatile forms produced during Eth (Liang *et al*., 1994) and HG (Tseng *et al*., 2000) are purged and trapped and the traps are later analysed by GC. This makes it possible to preconcentrate the analytes *in situ*. Such a type of off-line system can also be connected on-line.

Some methods used for the analysis of mercury species in natural waters are summarized in Table 5.9. The LOD for MeHg is between 0.004 and 5.6 ng L^{-1} and for Hg(II) it ranges from 0.07 to 5.2 ng L^{-1}. These values are usually higher than for MeHg owing to reagents and/or ambient

Table 5.9 Analytical methods used for mercury speciation in seawater.

Matrix	Sample preparation	Separative method/detection[a]	System type	Species	LOD ($ng\,L^{-1}$)	References
Fresh- and sea-water	SPE	UV-VG-Amalgamation-AAS	Off-line	$MeHg^+$	0.007	Kwokal and Branica (2000)
Fresh- and sea-water	HG	HG-CT-GC–AAS	On-line	Hg^{2+} $MeHg^+$	0.1 0.1	Tseng *et al*. (1998)
Seawater	Eth	Eth-PTI-CT-GC–AFS (extraction/back extraction)	Off-line	$MeHg^+$	0.04	Bloom (1989)
Fresh- and sea-water	Eth	Eth-CCT-CGC–ICP-MS (isotopic dilution)	On-line	Hg^{2+} $MeHg^+$	 0.03	Demuth and Heumann (2001)
Fresh- and sea-water	Pr	Pr-CCT-CGC–ICP-MS	On-line	Hg^{2+} $MeHg^+$	 0.03	Demuth and Heumann (2001)
Fresh- and sea-water	Eth	Steam D-Eth-PTI-CT-GC–AFS	Off-line	$MeHg^+$	0.024	Bowles and Apte (1998)
Fresh- and sea-water	SPE	SPE-CGC–AFS	Off-line	$MeHg^+$	0.01	Cai *et al*. (1996)
Fresh- and sea-water	SPE	SPE-Butyl-CGC–MIP-AES	Off-line	Hg^{2+} $MeHg^+$	0.2 0.05	Emteborg *et al*. (1993)
Seawater	Eth	Eth-CT-MCGC–MIP-AES	On-line	Hg^{2+} $MeHg^+$	0.8 0.4	Slaets and Adams (2000)
Fresh- and sea-water	HG	HG-CT-GC–ICP-MS	Off-line	Hg^{2+} $MeHg^+$	0.1 0.004	Tseng *et al*. (2000)

(*continued overleaf*)

Table 5.9 (*continued*)

Matrix	Sample preparation	Separative method/detection[a]	System type	Species	LOD (ng L^{-1})	References
Estuarine water	–	Purge-CT-GC–ICP-MS	Off-line	Hg^0	0.023	Amouroux *et al*. (1998)
				Me_2Hg^+	0.0002	
Fresh- and seawater	–	Preconc-HPLC–UV-HGAAS	On-line	Hg^{2+}	0.5	Falter and Schöler (1995)
				$MeHg^+$	0.5	
Fresh and sea-water	–	FI-Preconc-HPLC–HGAAS	On-line	Hg^{2+}	0.068	Qvarnström *et al*. (2000)
				$MeHg^+$	0.034	
Seawater	Pr	Pr-SPME-CGC–ICP-MS	Off-line	Hg^{2+}	0.35	Bravo-Sanchez *et al*. (2004)
				$MeHg^+$	0.17	
Seawater	Pr	Pr-GC–ICP-MS (isotopic dilution)	Off-line	Hg^{2+}	0.063	Monperrus *et al*. (2005a)
				$MeHg^+$	0.012	

[a]VG: vapour-generation; PTI: positive thermal ionization; CCT: collision cell technology.

air contamination. Usually, GC methods have higher sensitivity than LC. Special attention should be given to methods that allow simultaneous mercury species determination.

5.5.2 Marine Sediment Samples

Although contamination risk during sample handling and natural variability for bulk sediments is less important than for waters (Mason *et al.*, 1998), major problems occur during sample preparation for mercury speciation. The sampling strategy depends on the objective of the study and could use sediment grabs for surface sediment samples (Varekamp *et al.*, 2000; Macalady *et al.*, 2000; Stoichev *et al.*, 2004a) or sediment cores. During the treatment, especially for anoxic sediments, the sample oxidation has to be minimal.

In order to accurately extract the mercury species, several types of milder or diluted agents are used, such as citrate buffer, and extraction with dithizone–chloroform (Hintelmann and Wilken, 1993; Hempel *et al.*, 1995), $KBr-CuSO_4-H_2SO_4$ (Lambertsson *et al.*, 2001), urea–$HCl-CuSO_4$ (Gilmour *et al.*, 1992), dilute $HCl-HNO_3$ mixtures (Dietz *et al.*, 2001) or HNO_3 (Tseng *et al.*, 1997a). Alkaline digestion is also used (Gagnon *et al.*, 1996; Ramalhosa *et al.*, 2001).

Different extraction techniques are used in a manual mode or automatically with supercritical fluids (Lorenzo *et al.*, 1999). Vacuum (Morrison and Weber, 1997), steam (Bowles and Apte, 2000) or gas-stream (Horvat *et al.*, 1993; Wilken and Falter, 1998) distillation of $MeHg^+$ is used to separate it from the matrix. The MW extraction of the mercury species is an extremely fast method (2–10 min). The procedure can be automated and controlled, allowing simultaneous extraction of nonvolatile mercury species. Both open (Tseng *et al.*, 1997a; Dietz *et al.*, 2001) and closed systems (Vazquez *et al.*, 1997; Ramalhosa *et al.*, 2001) are used for alkaline or acid extractions of mercury species from the sediments. The MW extraction of sediments should, however, be used with precaution, since it may also suffer from artefact formation of MeHg (Landaluze *et al.*, 2004).

Thus, questions about the possibility of artefact formation of MeHg in sediments have been raised. Different techniques commonly used to extract MeHg have been screened and tested to evaluate their potential to generate MeHg accidentally from inorganic mercury during sample preparation (Hintelmann, 1999; Rodriguez Martin Doimeadios *et al.*, 2003). The results highlighted the assumption that mercury species transformation were occurring during the sample pretreatment step and, more specifically, with distillation-based methods (Morrison and Weber, 1997; Hintelmann, 1999).

The amount of MeHg produced during extraction was related to the amount of inorganic mercury in the sample and the extraction method. Additionally, when the concentrations of Hg(II) are very high this can hinder the determination of MeHg. In these cases, a pre-separation of the organomercurials from Hg(II) by liquid–liquid extraction and back-extraction is necessary (Hintelmann and Wilken, 1993; Morrison and Weber, 1997; Ramalhosa *et al.*, 2001).

The use of SIDMS has proven to be a powerful tool to check whether any interconversion is taking place. The use of Hg(II) and MeHg labelled with multiple isotopes allows one to discover the error of the specific steps and their contribution to the overall transformations of a species known (Wilken and Falter, 1998; Hintelmann, 1999; Rodriguez Martin Doimeadios *et al.*, 2003; Rahman and Kingston, 2005). There is a recent review dealing with the artefacts during mercury speciation analysis (Leermakers *et al.*, 2005).

Most of the techniques described for the analysis of water samples can be used for sediment extracts after some modifications. HPLC allows the quantification of organomercurials at environmental levels in real samples (Hintelmann and Wilken, 1993; Wilken and Falter, 1998). A limitation is that Hg(II) is usually not determined. Detectors used with HPLC to determine organomercury species in sediments are AFS (Hintelmann and Wilken, 1993) and ICP-MS (Wilken and Falter, 1998).

Off-line and on-line derivatization methods are also used for the determination of Hg species in sediment extracts by GC. The HG is vulnerable to interferences from heavy metals present in the extracts but can be used for the determination of $MeHg^+$ after its separation from the matrix (Morrison and Weber, 1997). On the contrary, as mentioned earlier, Eth is not prone to metal ions interferences and can be directly applied for simultaneous on-line mercury species determination in sediment extracts (Tseng *et al.*, 1998, 2004; Wasik *et al.*, 1998). Eth, followed by liquid–liquid extraction, is used for off-line analysis by capillary (Landaluze *et al.*, 2004) or multicapillary GC (MCGC; Rodriguez Pereiro *et al.*, 1998). SPME is applied after Eth for the detemination of mercury species by multicapillary GC (Jitaru and Adams, 2004).

The detectors used with GC for analysis of environmental sediment samples are QFAAS (Tseng *et al.*, 1998), AFS (Morrison and Weber, 1997; Bowles and Apte, 2000; Tseng *et al.*, 2004), MIP-AES (Dietz *et al.*, 2001; Rodriguez Pereiro *et al.*, 1998; Landaluze *et al.*, 2004) and ICP-MS (Wasik *et al.*, 1998; Lambertsson *et al.*, 2001; Jitaru and Adams, 2004; Rodriguez Martin Doimeadios *et al.*, 2004). Some of the methods used for mercury species determination in sediments are summarized in Table 5.10.

5.5.3 Marine Biota Samples

To adopt a sampling strategy for the determination of mercury species in marine biota samples it is necessary to consider the trophic level of the organisms studied; see Figure 5.3. Since mercury is not uniformly distributed in the organisms it is necessary to specify the type of tissue sampled. In general, the samples are frozen immediately after the preliminary treatment. Later, in the laboratory, they are analysed directly or after freeze-drying. For further information, Cabanero *et al.* (2002) have, for example, evaluated different drying techniques: oven-dried, MW oven-dried and freeze-dried.

Analysis of biological tissues is less affected by artefacts and interferences during extraction procedures and quantification compared with sediments. The quantification requirements are easily met, since the levels of MeHg are usually high. To extract the mercury forms, similar methods as for sediments can be applied (Baeyens, 1992). Alkaline or acid digestion at ambient temperature or with conventional heating is usually a slow process. It can be accelerated by means of ultrasonic (Bloom, 1989; Rio-Segade and Bendicho, 1999; Välimäki and Perämäki, 2001; Dong *et al.*, 2004; Perna *et al.*, 2005) or MW (Tseng *et al.*, 1997b, 2004; Dietz *et al.*, 2001) treatments. The distillation is also used for MeHg extraction in biological tissues (Wilken and Falter, 1998).

The extracts from biological tissues can be analysed by methods described for waters and sediments. Since the levels of mercury species in biological tissues are usually high, less-sensitive methods such as HPLC with UV detection can also be used (Dong *et al.*, 2004). SPME can be applied for off-line preconcentration of the ethylated (Mester *et al.*, 2000) or phenylated mercury species (Rodil *et al.*, 2002; Jokai *et al.*, 2005). A preconcentration of MeHgH from the headspace into a single drop of Pd(II) solution is used for off-line determination of MeHg by HG and ETAAS (Gil *et al.*, 2005). This method is fast and requires easily available laboratory equipment. It can be a good alternative to sophisticated techniques for MeHg determination in biological tissues. Some of the methods used for mercury species determination in biological tissues are summarized in Table 5.11.

5.6 BUTYLTIN AND OTHER ORGANOTIN SPECIES

The intensive use of organotin compounds (OTCs) and their subsequent release into the environment have created a great environmental concern owing to their high toxicity (Donard *et al.*, 2001). The marine environment is the most impacted by OTCs, and particularly by tributyltin

Table 5.10 Analytical methods used for mercury speciation in sediments.

Sample preparation	Separative method/detection	System type	Species	LOD (ng g^{-1})	References
1 g sample, MW extraction, HNO_3	Eth-CT-GC–AAS	On-line	Hg^{2+}	0.5	Tseng *et al*. (1998)
			$MeHg^+$	0.5	
0.4 g sample, MW extraction, diluted aqua regia	Eth-CCT-CGC-MIP-AES	On-line	Hg^{2+}	0.08	Dietz *et al*. (2001)
			$MeHg^+$	0.06	
1 g sample, MW extraction, HNO_3	Eth-Extraction-MCGC–MIP-AES	Off-line	Hg^{2+}	0.2	Rodriguez Pereiro *et al*. (1998)
			$MeHg^+$	0.2	
1 g sample, MW extraction, HNO_3	Eth-CCT-MCGC–ICP-MS	On-line	Hg^{2+}	0.03 (2σ)	Wasik *et al*. (1998)
			$MeHg^+$	0.03 (2σ)	
0.5 g sample, $KBr/Cu^{2+}/H_2SO_4$, extraction/back extraction	Eth-PTI-CGC–ICP-MS (isotope dilution)	Off-line	$MeHg^+$	0.3	Lambertsson *et al*. (2001)
1 g wet sample, $KCl/Cu^{2+}/H_2SO_4$, steam distillation	Eth-PTI-GC–AFS	–	$MeHg^+$	0.01 (wet wt.)	Bowles and Apte (2000)
5 g sample, citrate buffer, extraction/back extraction	HPLC–VG-AFS	–	$MeHg^+$	0.1	Hintelmann and Wilken (1993)
			$EtHg^+$	0.1	
0.5 g sample, $NaCl/H_2SO_4$, N_2 distillation	Preconc-HPLC–VG-ICP-MS (isotope dilution)	–	$MeHg^+$	0.015	Wilken and Falter (1998)
0.5 g sample, $NaCl/Cu^{2+}/H_2SO_4$	HG-CT-GC–AFS	On-line	$MeHg^+$	0.2	Morrison and Weber (1997)
0.2 g sample, HNO_3	Eth-Extraction-CGC–MIP-AES	Off-line	$MeHg^+$	5	Landaluze *et al*. (2004)
Steam distillation	FI-Eth-PTI-GC–AFS	On-line	$MeHg^+$	0.01	Tseng *et al*. (2004)

(*continued overleaf*)

Table 5.10 *(continued)*

Sample preparation	Separative method/detection	System type	Species	LOD ($ng\,g^{-1}$)	References
1 g sample, MW extraction, HNO_3	Eth-CT-GC–AFS	On-line	Hg^{2+}	0.22	Stoichev *et al*. (2004b)
			$MeHg^+$	0.02	
1 g sample, $KCl/Cu^{2+}/H_2SO_4$, extraction/back extraction	Preconc-HPLC–VG-AFS	-	$MeHg^+$	0.001	Shade and Hudson (2005)
1 g sample, MW extraction, HNO_3	Eth-SPME-MCGC–ICP-MS	Off-line	Hg^{2+}	0.27	Jitaru and Adams (2004)
			$MeHg^+$	0.027	
1 g sample, MW extraction, HNO_3	Eth-CGC–ICP-MS (isotope dilution)	Off-line	Hg^{2+}	0.13	Rodriguez Martin Doimeadios *et al*. (2003)
			$MeHg^+$	0.13	
0.5 g sample, MW extraction, HNO_3	Eth-CGC–ICP-MS (isotope dilution)	Off-line	Hg^{2+}	0.2	Monperrus *et al*. (2008)
			$MeHg^+$	0.2	

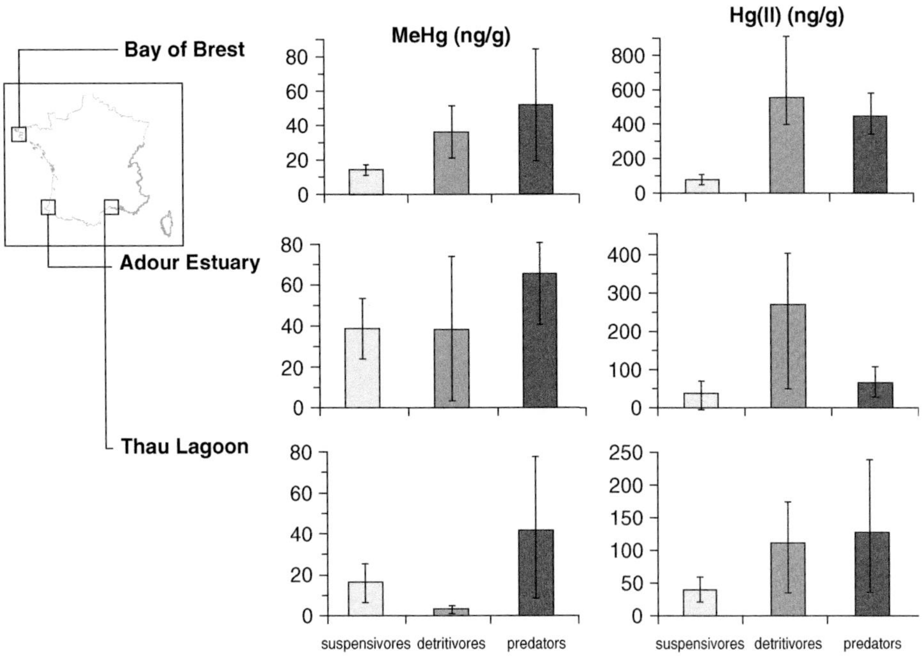

Figure 5.3 Monitoring of the bioaccumulation of mercury species in benthic fauna of French coastal environments in relation to their feeding guilds. Concentrations are the mean values for organisms classified in three trophic levels (suspensivores, detritivores, predators) collected at three sites (Bay of Brest, Adour Estuary, Thau lagoon). The biological samples were analysed by species-specific isotope dilution GC–ICPMS after a microwave extraction using TMAH as extractant. Bio-magnification along the trophic levels is highlighted for MeHg, with higher concentrations found for predators. Higher bioaccumulation of Hg(II) is observed for deposit feeders directly impacted by sediment contamination compared with suspensive feeders and predators.

(TBT), which has led to a widespread contamination of marine ecosystems and particularly coastal and harbour areas. TBT was very efficient at preventing fixation of aquatic organisms on the hulls of ships, but its use caused extensive damage to nontarget organisms. Shell deformations in oysters (Alzieu, 1991, 1998), sex changes (imposex) in whelks (Bryan *et al.*, 1986; Gibbs *et al.*, 1987) and harmful effects in marine life at very low concentrations have been reported (Evans, 1999). As a result, since the mid 1980s several countries have implemented national legislation with the specific intention of limiting the use of TBT in antifouling paint. In 2001, the International Maritime Organization (IMO) adopted the AFS Convention (Convention on the Control of Harmful Anti-fouling Systems on Ships) calling for an international ban of TBT in antifouling paints by 1 January 2008. Furthermore, the Water Framework Directive (WFD) has been supplemented by a list of 33 priority substances or group of substances, including TBT (Coquery *et al.*, 2005), for which quality standards and measurements for the reduction of emission control will be set at community levels (Directive 2008/105/EC). Thus, for each of these substances, environmental quality standards (EQSs) are established according to ecotoxicological factors. For inland and other surface water, the EQS for TBT (as cation) are expressed as an annual value ($0.0002\,\mu g\,L^{-1}$) or as maximum allowed concentration ($0.0015\,\mu g\,L^{-1}$) in the total phase. There is no limit recommended for sediments or biota.

Table 5.11 Analytical methods used for mercury speciation in biota.

Sample preparation	Separative method/detection	System type	Species	LOD (ng g^{-1})	References
0.2 g sample, MW digestion, $(CH_3)_4NOH$	HG-CT-GC–AAS	On-line	Hg^{2+} $MeHg^+$	8 8	Tseng *et al*. (1998)
0.3 g sample, MW digestion, diluted aqua regia 0.2 g tuna, $(CH_3)_4NOH$	Eth-CCT-CGC–MIP-AES	On-line	Hg^{2+} $MeHg^+$	0.11 0.08	Dietz *et al*. (2001)
0.25 g sample, ultrasound, KO/CH_3OH	Eth-PTI-CGC–AAS	On-line	Hg^{2+} $MeHg^+$	224 136	Välimäki and Perälmäki (2001)
0.1 g sample, MW digestion, $(CH_3)_4NOH$	Eth-CCT-MCGC–ICP-MS	On-line	Hg^{2+} $MeHg^+$	0.3 (2σ) 0.3 (2σ)	Wasik *et al*. (1998)
0.3 g sample, $(CH_3)_4NOH$, extraction/back extraction	HPLC–VG-AFS	–	$MeHg^+$	2.5	Hintelmann and Wilken (1993)
0.25 g sample, HCl/NaCl	Eth-SPME-ICP-MS	Off-line	$MeHg^+$	24	Mester *et al*. (2000)
0.2 g sample, MW digestion, $(CH_3)_4NOH$	Eth-Extraction-CGC–MIP-AES	Off-line	Hg^{2+} $MeHg^+$	– 0.003	Gerbersmann *et al*. (1997)
0.2 g sample, MW digestion, $(CH_3)_4NOH$	HG-CCT-CGC–MIP-AES	On-line	Hg^{2+} $MeHg^+$	– 0.13	Gerbersmann *et al*. (1997)
5 g sample wet wt., ultrasound, HCl	FI-Preconc-HPLC-UV detection	On-line	Hg^{2+} $MeHg^+$	25 (wet wt.) 19 (wet wt.)	Dong *et al*. (2004)

2 g sample wet wt., KOH/CH_3OH, extraction/back extraction	HPLC-oxidation–HG-AFS	On-line	$MeHg^+$	5 (wet wt.)	Liang *et al*. (1994)
0.25 g sample, MW digestion, $(CH_3)_4NOH$	FI-Eth-PTI-GC–AFS	On-line	Hg^{2+} $MeHg^+$	0.1 0.01	Tseng *et al*. (2004)
0.4 g sample, ultrasound, KOH/CH_3OH	Eth-PT (*i-octane*)-CGC–ICP-MS	Off-line	Hg^{2+} $MeHg^+$	18 11	Perna *et al*. (2005)
0.4 g sample, ultrasound, HCl	FI-VG-AAS	–	Hg^{2+} $MeHg^+$	5 11	Rio-Segade and Bendicho (1999)
0.05 g sample, HCl/NaCl, Eth-extraction (ultrasound)	CGC–MIP-AES	Off-line	$MeHg^+$	4.4	Tu *et al*. (2000)
0.05 g sample, HCl/NaCl, Eth-extraction (ultrasound)	CGC–ICP-MS	Off-line	$MeHg^+$	2.6	Tu *et al*. (2000)
0.2 g sample, ultrasound, $(CH_3)_4NOH$	FI-Preconc-HPLC–HGAAS	On-line	Hg^{2+} $MeHg^+$	0.017 0.009	Qvarnström *et al*. (2000)
KOH/CH_3OH, 70 °C	Eth-PTI-GC–AFS	On-line	$MeHg^+$	1.4	Saouter and Blattmann (1994)
0.5 g sample, N_2 distillation, NaCl/H_2SO_4	Preconc-HPLC–UV-ICP-MS (isotope dilution)	–	$MeHg^+$	0.015	Wilken and Falter (1998)
0.08 g sample, ultrasound, KOH/CH_3OH	Eth-CT-GC–AAS	On-line	Hg^{2+} $MeHg^+$	75 4	Fischer *et al*. (1993)

(*continued overleaf*)

Table 5.11 *(continued)*

Sample preparation	Separative method/detection	System type	Species	LOD (ng g^{-1})	References
0.25 g sample, ultrasound, KOH/CH_3OH	Eth-PTI-CT-GC–AFS	Off-line	$MeHg^+$	1	Bloom (1989)
0.2 g sample, MW digestion, $(CH_3)_4NOH$	Eth-Extraction-CGC–ICP-MS (isotope dilution)	Off-line	$MeHg^+$	0.3	Rodriguez Martin Doimeadios *et al*. (2002)
0.2 g sample, KCl/Cu^{2+}/H_2SO_4, extraction/back extraction	Preconc-HPLC–VG-AFS	–	$MeHg^+$	0.001	Shade and Hudson (2005)
2 g wet wt., ultrasound, HCl	HPLC–VG-AFS	–	Hg^{2+} $MeHg^+$	4.8 6.8	Li *et al*. (2005b)
0.3 g sample, ultrasound, Bu_4NBr/KOH, 60 °C	HG-CT-GC–AFS	On-line	Hg^{2+} $MeHg^+$	3.3 0.02	Heller and Weber (1998)
0.5 g sample, MW extraction, HNO_3	Eth-CGC–ICP-MS (isotope dilution)	Off-line	Hg^{2+} $MeHg^+$	0.2 0.2	Monperrus *et al*. (2008)

The need for the determination of individual chemical species occurs especially when these species are known to have a different toxicity, mobility or bioavailability. Great efforts have recently been performed in analytical methodologies and instrumentation, which has given rise to an increased awareness of organotin-related environmental issues. It is now well established that OTCs are ubiquitous contaminants of the marine ecosystem. Once in the water, OTCs are subjected to transformations and exchanges between the different environmental compartments (water, suspended matter, sediment) which make them available to filtering or sediment-feeding organisms. This leads to bioaccumulation in the food chain. Thus, TBT can be measured in tissues of organisms from superior levels of the marine trophic chain, such as tuna (Kannan *et al.*, 1996; Ueno *et al.*, 2004), whales, dolphins (Abel, 2000), seals (Belfroid *et al.*, 2000) and birds (Kannan *et al.*, 1998). This highlights the growing impact of this compound on large biological systems in acting as endocrine disrupters.

The speciation analysis of OTCs may be achieved in a number of different ways. Most analytical techniques developed for the speciation of organometallic compounds usually combine separation techniques such as HPLC (Inoue *et al.*, 1995; Hill *et al.*, 2000) or GC (Szpunar *et al.*, 1996a; Yang and Lam, 2001) with a specific detection mode such as a pulsed flame photometric detector (PFPD; Rodriguez Pereiro *et al.*, 1996; Szpunar *et al.*, 1996a), AAS (Cai *et al.*, 1993), AES (Jimenez and Sturgeon, 1997; Aguerre *et al.*, 2001a) or ICP-MS (Hintelmann *et al.*, 1995; Ruiz-Encinar *et al.*, 2000; Aguerre *et al.*, 2001b; Yang and Lam, 2001). The introduction of ICP-MS as a detector allows one to reach a wide choice of capabilities, including the simultaneous detection of several isotopes quasi-simultaneously. The coupling of GC to ICP-MS appears to be one of the more promising techniques for organotin analysis because of its multielement capability, the possibility of isotopic information and its high sensitivity (Aguerre *et al.*, 2001b; Yang and Lam, 2001; Ruiz-Encinar *et al.*, 2000). In addition, the quality of the results of speciation determination is closely associated with the quality and efficiency of the sample pretreatment stages. Routinely, for analysis using GC–ICP-MS, the analytes are extracted from the sample matrix and derivatized to yield volatile compounds. Once the extraction step is performed, the organometallic species are derivatized in order to obtain volatile species. This step is generally performed using Grignard reagent, HG or ethylation (Morabito *et al.*, 2000).

The application of isotope dilution analysis in elemental speciation has in recent years allowed the development of highly accurate and precise quantification approaches for the determination of a wide range of elemental species even when analysing complicated matrices. Rodríguez González *et al.* (2002) developed a methodology for the quantification of three butyltin species using two enriched isotopes for labelling able to correct for the sequential degradation of TBT. This methodology was later improved with the development of a triple spike methodology for the determination of three species (mono-, di- and tri-butyltin) using three isotopically enriched species able to correct for nine interconversion pathways (Rodríguez González *et al.*, 2005). More recently, isotope pattern deconvolution has been demonstrated to be a powerful and versatile approach not only to follow species degradation processes, but also for accurate quantification purposes (Rodríguez González *et al.*, 2007).

The quality of measurements necessary for monitoring programmes is directly linked to criteria such as sample location, type and treatment of the samples, sampling procedure, storage of the sample and quality control of the final determination (Quevauviller *et al.*, 1992). There are no particular strategies for OTC monitoring in environmental and biological matrices, but sample collection and storage have to ensure good representativity of samples with regard to OTCs. Indeed, despite the good performances of analytical techniques, several limitations concerning quality in species preservation and efficiency of recovery remain at the sample preparation level. Before sampling, a clear strategy should be planned with regard to the choice and the number of samples to be collected (depending on the site location and hydrodynamics of the area), the collection procedure (depending on whether water, sediment or biota are collected), the sampling period and the validity of the sampling site selected. After collection, sample preparation can require different and complementary steps before analysis. Inadequate sample preparation and analytical procedures may produce inaccurate data and inadequate interpretation about their biogeochemical cycle and ecological impact.

5.6.1 Seawater Samples

As organotin levels detected in marine water are most of the time extremely low (generally below the nanogram per litre level), the risk of contamination due to sample handling and treatments performed prior to the determination is high. Therefore, careful sampling and storage procedures become a critical step for the validity of the chemical species to be identified and quantified. For water samples, analyses are generally performed after filtration at 0.45 μm to determine the OTC concentration both in dissolved and particulate phases. Toxicity of OTCs is generally referred to their concentration in the dissolved phase, but it is also interesting to determine OTCs in the particulate phase because of the tendency of TBT to be adsorbed on particles. Analyses of nonfiltered water provide results that are not really representative of the real contamination owing to differences in the performances of the extraction techniques used. Moreover, these kinds of analysis do not provide results that are easily comparable. After sample collection and pretreatment, storage is the ultimate step before analysis. This point is one of the most critical aspects of the whole procedure, because physico-chemical degradations can occur in the sample and could then affect the quality of the final determination. OTCs are stable over a 4 months period in sea- and fresh-water samples, filtered and acidified at pH 2 with HCl, and stored in the dark at 4 °C (Bancon-Montigny, 2001). However, the stability of butyltin species is lower in nonfiltered samples containing high amount of suspended matter, probably due to possible interferences with particulates and/or microbial communities (Quevauviller, 1996); see Figure 5.4.

The analysis of OTCs at low concentration levels requires extremely sensitive and selective separation and detection techniques. This is particularly true for achieving the criteria of the WFD. Analytical methods of OTC determination in water are generally based on three steps: derivatization, separation and detection. The derivatization step providing volatile species (generally in isooctane or in hexane) requires the use of an alkylating agent such as sodium tetraethylborate or sodium *n*-propylborate. The organic phase is then directly injected into the GC-detector system. SPME can also be used as alternative method to liquid–liquid extraction (Bravo *et al.*, 2004) or SPE to extract OTCs from water without any organic solvent (Carvalho *et al.*, 2007). This technique is based on the equilibrium between the analyte concentration in the solution matrix and in the solid-phase fibre which is generally coated with a polymeric organic stationary phase. SPME sampling can be carried out in direct mode, immersed in liquid or in the headspace (Le Gac *et al.*, 2003).

Hyphenated techniques, which include species separation and detection, offer a most promising approach in environmental analysis nowadays (Prange and Jantzen, 1995; De Smaele *et al.*, 1996; Moens *et al.*, 1997; Bouyssiere *et al.*, 2002). In this sense, GC or LC coupled with highly selective and sensitive detectors, such as mass spectrometry (White *et al.*, 1998; Ikonomou *et al.*, 2002), atomic absorption spectrometry, flame photometry (Lespes *et al.*, 1998; Le Gac *et al.*, 2003) and flame ionization detection (Millán and Pawliszyn, 2000) are often used.

But among all these hyphenated techniques, ICP-MS in combination with GC provides excellent sensitivity, selectivity, multielemental (Prange and Jantzen, 1995; De Smaele *et al.*, 1996; Moens *et al.*, 1997; Aguerre *et al.*, 2001b; Bouyssiere *et al.*, 2002) and multi-isotope detection capabilities (Ruiz-Encinar *et al.*, 2000, 2001; Yang *et al.*, 2002; Monperrus *et al.*, 2005b). A few examples of this technique and the associated detection limits are presented in Table 5.12.

Table 5.12 Analytical methods used for tin speciation in seawater.[a]

Matrix	Sample preparation	Separative method/detection	System type	Species	LOD ($ng\,L^{-1}$)	References
Fresh water	Extraction: solvent Pr	GC–AED	On-line	MBT, DBT, TBT	7–12	Schubert *et al*. (2000)
Fresh water	Extraction: SPME Eth	GC–ICP-MS	On-line	MBT, DBT, TBT	0.3–2	Moens *et al*. (1997)
Natural water	Extraction: SPME Eth	GC–FPD	On-line	MBT, DBT, TBT	0.006–0.031	Aguerre *et al*. (1999)
Fresh and sea water	Extraction: HS-SPME Eth	GC–MIP-AES	On-line	MBT, DBT, TBT	5	Carpinteiro Botana *et al*. (2002)
Sea water	Extraction: HS-SPME Eth	GC–MS	On-line	TBT, DPhT, TPT, TET	0.025–1	Segovia-Martinez *et al*. (2010)
Sea water	Eth	GC–ICP-MS (IDMS)	On-line	MBT, DBT, TBT	0.04–0.27	Rodríguez González *et al*. (2002)
Sea water	Eth	GC–MS (IDMS)	On-line	MBT, DBT, TBT	0.18–0.25	Centineo *et al*. (2004)
Sea water	Eth	GC–ICP-MS (IDMS)	On-line	MBT, DBT, TBT	0.019–0.032	Monperrus *et al*. (2005b)

[a]MBT: monobutyltin; DBT: dibutyltin.

5.6.2 Marine Sediment Samples

Once entering in coastal waters, OTCs will be rapidly adsorbed onto suspended particulate matter (SPM), which will allow rapid settlement. This trapping in the sediment leads to long-term storage of the contaminants, ranging from 1 year to more than a decade (Craig, 1986). Concerning sampling, a careful approach needs to be taken into account in sediment collection, integrating the main geochemical parameters for meaningful comparative assessments. After samples collection, several different treatments for sediment storage (freezing, wet storage at 4 °C) and drying (air, freezing, oven) are suitable to preserve the stability of TBT over at least 4 months, but DBT and MBT are often subject to variation (mostly losses). In the case of long-term storage, freeze-drying is a better compromise (Quevauviller and Donard, 1990). Generally, the higher concentrations of OTCs are linked to the fine particles both for surface and deeper sediments, but sieving of sediment samples at 2 mm and correlation to grain size of particles may simplify comparison of results. Sampling in undisturbed sedimentary structure can highlight the OTC distribution profile in deeper sediment and, moreover, determine a sedimentation rate and to some extent a recording of the OTC contamination in the sediment (Benoit, 2005).

Sample preparation for tin speciation analysis requires extraction and derivatization steps. Most sample preparation techniques for butyltin speciation in solid matrix combine different complex preparation steps using stirring (Krone *et al*., 1989; Gomez-Ariza *et al*., 1995), sonication (Ceulemans *et al*., 1994; Gomez-Ariza *et al*., 1995; Pelligrino *et al*., 2000) or supercritical fluid extraction (Bayona and Cai, 1994). Open MW-assisted digestion has also gained wide acceptance as a rapid and efficient method for sample decomposition in inorganic analysis (Donard *et al*., 1995; Rodriguez Pereiro *et al*., 1996; Szpunar *et al*., 1996b; Schmitt *et al*., 1997). Investigations into the behaviour of OTCs under an MW field were carried out and the procedures were optimized to improve extraction recoveries and reduce the time for sample preparation (Ceulemans *et al*., 1994; Szpunar *et al*., 1996b).

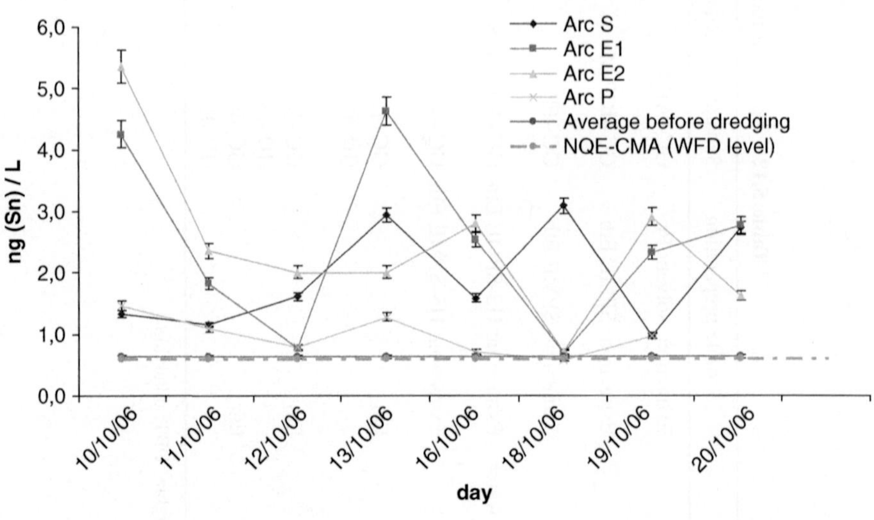

Figure 5.4 Environmental monitoring of TBT in a French harbour area during dredging operations. This graph presents the concentration of TBT in water samples (dissolved phase and SPM) which have been collected at different points of the port area (Arc S, Arc E1 and Arc E2). For each sampling station, TBT concentrations found are above the average concentration measured before dredging and also above the WFD level. See plate section for a colour version of this image.

The use of isotope dilution mass spectrometry (IDMS) analysis represents a good alternative method to avoid traditional problems related to nonquantitative recoveries and losses during the sample preparation procedure. If the equilibration of the spike and the analyte is fully achieved, IDMS is theoretically capable of compensating for nonquantitative sample preparation procedures, negating losses of analytes and instrument drift. Several applications of isotope dilution have been published for the analysis of butyltin species in sediments during the last decade (Rodríguez González *et al*., 2002, 2007; Monperrus *et al*., 2003a). Some applications for OTC determination in marine sediments are given in Table 5.13.

5.6.3 Marine Biota Samples

Knowledge of the concentrations of OTCs in biological tissues is required, particularly in benthic organisms, in order to understand the bioconcentration in the food web. Sampling strategies and suitable analytical methods are very important to determine OTCs in benthic populations. Indeed, sampling of the benthic population should represent as true as possible the real distribution of the different species and the analytical method should be sufficiently accurate and precise to measure low concentrations in very low mass samples; see Table 5.14.

For species extraction, it is obviously necessary to use gentle solubilization conditions to preserve their integrity before analysis. OTC extraction from biological samples is generally performed with acetic acid (Rodriguez Pereiro *et al*., 1996; Szpunar *et al*., 1996b; Schmitt *et al*., 1997). Alkaline hydrolysis with TMAH has also been successfully applied to biological tissues (Szpunar *et al*., 1996b). It was observed that MW extraction allows solubilization of biological tissues and the complete extraction of butyltin compounds in a short extraction time (3 min). Once the extraction step is performed, OTCs are derivatized to obtain volatile species. This step can be performed using Grignard reagent or HG, but aqueous ethylation remains the most commonly used derivatization method (Rapsomanikis, 1994; Tseng *et al*., 1999; García Fernandez *et al*., 2000). IDMS has also been developed for TBT analysis in biological tissues (Monperrus *et al*., 2003a,b).

5.7 QUALITY ASSURANCE/QUALITY CONTROL FOR SPECIATION ANALYSIS IN MARINE SAMPLES

5.7.1 Quality Assurance and Quality Control Requirements for Speciation Analysis

As demonstrated previously, the methods and techniques developed for speciation analysis involve many steps that must be fully mastered in order to prevent loss, contamination and the conversions of the species. The quality control of the measurements also means that calibrating solutions of the chemical species analysed are available as well as reference materials that meet all the requirements for validation, i.e. same matrix and same level of concentration (Quevauviller, 1998).

The choice of a method for speciation analysis will depend on various parameters, including its accuracy, its reliability (repeatability, reproducibility, etc.), its detection and quantification limits (sensitivity, blanks level), its selectivity or specificity, its resolving power and robustness. So far as there are very few standard methods for speciation analysis, validation should be performed rigorously. Following the selection procedures for sampling, sample processing and analysis technique, the first problem that arises for speciation analysis is the calibration. This calibration is usually performed with synthetic solutions containing the chemical forms to analyse. Although mono- or multi-elemental solutions provided with a certificate are commercially available for the determination of total element concentrations, this is very rarely the case for speciation analysis.

Table 5.13 Analytical methods used for tin speciation in sediments.

Sample preparation	Separative method/detection	System type	Species	LOD	References
Extraction: acetic acid and stirring, Eth	CT-GC–AAS	On-line	MMT, MBT, DBT, TBT	$1\,ng\,g^{-1}$	Martin *et al*. (1994)
Extraction: acetic acid and MW heating, Eth	GC–FPD	On-line	Methyl-, butyl-, phenyl- and octyl-tin	$10\,ng\,g^{-1}$	Lalère *et al*. (1995)
Extraction: acetic acid and MW heating, Eth	GC–AED	On-line	MBT, DBT, TBT	$2\,ng\,g^{-1}$	Szpunar *et al*. (1996a)
Extraction: acetic acid and ultrasound, Pr	GC–ICP-MS	On-line	MBT, DBT, TBT	52-$170\,fg\,g^{-1}$	De Smaele *et al*. (1996)
Extraction: acetic acid and ultrasound, Pr	GC–ICP-MS	On-line	TBT	93 fg	Monperrus *et al*. (2003a)
Eth	GC–ICP-MS (IDMS)	On-line	MBT, DBT, TBT	0.03–$0.06\,ng\,g^{-1}$	Rodriguez Gonzalez *et al*. (2005)
Extraction: acetic acid and ultrasound, Eth	GC-MS (IDMS)	On-line	MBT, DBT, TBT	–	Centineo *et al*. (2004)
Extraction: acetic acid and MW heating, Eth	GC-ICP-MS (IDMS)	On-line	MBT, DBT, TBT	–	Rodrı́guez González *et al*. (2007)

Table 5.14 Analytical methods used for tin speciation in biota.

Sample preparation	Separative method/Detection	System type	Species	LOD	References
Extraction: acetic acid and MW heating, Eth	GC–AED	On-line	MBT, DBT, TBT, TPhT	5–10 ng g^{-1}	Rodriguez Pereiro *et al*. (1996)
Extraction: TMAH and MW heating, Eth	GC–MIP-AED	On-line	MBT, DBT, TBT, TPhT	2–5 ng g^{-1}	Schmitt *et al*. (1997)
Extraction: TMAH and MW heating, Eth	GC–ICP-MS	On-line	MBT, DBT, TBT, MPhT, DPhT, TPhT	0.05 pg	Rodriguez *et al*. (1999)
Extraction: TMAH and MW heating, Eth	CGC–ICP-MS (IDMS)	On-line	MBT, DBT, TBT	0.15–0.30 ng g^{-1}	Monperrus *et al*. (2003b)
Extraction: TMAH and MW heating, Eth	CGC–ICP-MS (IDMS)	On-line	MBT, DBT, TBT	93 fg	Monperrus *et al*. (2003a)
Extraction: methanol–acetic acid and MW heating, Eth	GC–MS (IDMS)	On-line	MBT, DBT, TBT	–	Centineo *et al*. (2004)
Extraction: methanol–acetic acid and MW heating, Eth	GC–ICP-MS (IDMS)	On-line	MBT, DBT, TBT	–	Rodrı́guez González *et al*. (2007)

For the determination of chemical species of an element, it is common that each laboratory prepares its own calibrating solutions from solid salts. For the calibrating solutions used in speciation, a second parameter must be controlled, namely the stability of chemical species. For instance, if the stability of species is not controlled in the stock solution, the result obtained would be inaccurate (Quevauviller *et al*., 1995).

5.7.2 Calibration Methods Available for Speciation

The method of quantification of chemical species must also be chosen carefully. Indeed, the determination of chemical forms of elements can be achieved by different methods, such as external calibration, using an internal standard or standard additions method. They may all have drawbacks and, therefore, it is particularly difficult to recommend one of them. The method of the calibration curve is a rapid method for quantification, but it is essential to monitor any drift in sensitivity during the analytical session. In addition, the concentration of the compound to be assayed must be included in the concentration range of the calibration curve and it is often advisable to prepare standard solutions in a reconstituted matrix that contains all the reagents used to prepare samples. However, it is impossible to mimic the sample matrix perfectly. The changes in sensitivity and matrix effects can be corrected by using an internal standard. In the case of matrix effects, it is often preferable for speciation analysis to use the method of standard additions. Although tedious, especially if a large number of species must be determined, this method can resolve most of the matrix effects.

Specific quantification is usually performed by using standard substances of the species (internal calibration or standard additions). The calibration approach should comprise all analytical steps and can be greatly affected by matrix effects, losses and species degradation. If applicable, IDMS can be a powerful alternative and is considered as the optimal calibration technique offering the advantage of high accuracy of matrix-independent results. As discussed previously in this chapter, IDMS has been demonstrated to be very valuable for the analysis of chromium redox species, butyltin compounds and mercury species (IHg and MeHg) in environmental samples and for improving precision and accuracy of the results. It should be noted that IDMS does not provide a panacea for all problems encountered in speciation analysis; that is, it does not necessarily enable one to achieve traceability of measurements to SI units (Quevauviller, 2000). In all these methods, it is assumed that the added spike of enriched isotope and the endogenous element species behave in a similar way. If the spike is extracted from the solid more efficiently (e.g. because it was not allowed to equilibrate) than the analyte present in the sample, then incorrect data will be obtained. Furthermore, when several species need to be analysed simultaneously, each of them can be enriched in a different isotope. In this way, the concentration of the species can be quantified even when opposite or simultaneous degradation processes occur. It seems that these multi-isotope-labelled isotope dilution techniques will become a most powerful tool for optimizing and validating procedures for trace-metal speciation, going from species extraction from the solid to derivatization, separation and detection of those species for accurate final determination.

5.7.3 Reference Materials for Speciation Analysis

The best tool for the validation of a speciation method, particularly to highlight any problems associated with sample preparation, detection of species or to verify the accuracy of a method is the use of certified reference materials (CRMs). This is discussed extensively in Chapter 11 of this book and will not be developed here. In the field of speciation analysis, a number of matrix reference materials are available for species of Pb, Cr, As, Sn, Se or Hg in different types of samples; for example, see Quevauviller (1998). The selection of reference material depends primarily on two criteria. First, it must have a matrix similar to that of the sample and the same level of concentration. Although a significant effort has been made in the development of CRMs for speciation analysis, there is lack of available material. Indeed, in the environmental field, the concentrations assayed are often much lower than those of CRMs. CRMs for verifying the validity of results related to the introduction of new regulations, for which no standard analytical method is defined, also need to be developed. When the CRMs are not available, the only way to validate a method is to spike the sample with standard solutions and then calculate the recovery. However, this method is not as reliable as the endogenous compounds present naturally in the matrices, and compounds added display different binding capacities and, therefore, do not interact the same way with the matrix.

REFERENCES

Abasse G., Ouddane B. and Fischer J. C. (2002). *J. Anal. At. Spectrom.*, **17**, 1354.
Abasse G., Ouddane B. and Fischer J. C. (2003). *Anal. Sci.*, **19**, 529.
Abel P. D. (2000). *Sci. Total Environ.*, **258**, 1.
Abollino O., Aceto M., Sarzanini C. and Mentasti E. (2000). *Anal. Chim. Acta*, **411**, 223.
Abranko L., Yang L., Sturgeon R. E. *et al.* (2004). *J. Anal. At. Spectrom.*, **19**, 1098.
Ackley K. L., B'Hymer C., Sutton K. L. and Caruso J. A. (1999). *J. Anal. At. Spectrom.*, **14**, 845.
Aguerre S., Bancon-Montigny C., Lespes G. and Potin-Gautier M. (1999). *Analyst*, **125**, 263.
Aguerre S., Pécheyran C., Lespes G. *et al.* (2001a). *J. Anal. At. Spectrom.*, **16**, 1429.
Aguerre S., Lespes G., Desauziers V. and Potin-Gautier M. (2001b). *J. Anal. At. Spectrom.*, **16**, 263.
Ahern F., Eckert J. M., Payne N. C. and Williams K. L. (1985). *Anal. Chim. Acta*, **175**, 147.
Alberti G., Biesuz R. and Pesavento M. (2008). *Anal. Sci.*, **24**, 1605.
Álvarez-Llamas G., Fernández de la Campa M. R. and Sanz-Medel A. (2003). *J. Anal. At. Spectrom.*, **18**, 460.
Alzieu C. (1991). *Mar. Environ. Res.*, **32**, 7.
Alzieu C. (1998). *Ocean Coast. Management*, **40**, 23.
Amouroux D., Tessier E., Pecheyran C. and Donard O. F. X. (1998). *Anal. Chim. Acta*, **377**, 241.
Andrae M. O. (1978). *Deep Sea Res.*, **25**, 391.
Andrade S., Moffet J. and Correa J. A. (2006). *Mar. Chem.*, **101**, 203.
Andrle C. M., Jakubowski N. and Brokaert J. A. C. (1997). *Spectrochim. Acta Pt B*, **52**, 189.
Anninou P. and Cave R. R. (2009). *Estuar. Coast. Shelf Sci.*, **82**, 515.
Auger Y., Bodineau S., Leclercq S. and Wartel M. (1999). *Cont. Shelf Res.*, **19**, 2003.
Bacon J. R. and Davidson M. (2008). *Analyst*, **133**, 25.
Baeyens W. (1992). *Trends Anal. Chem.*, **11**, 245.
Baeyens W., Gao Y., De Galan S. *et al.* (2008). *Mol. Nutr. Food Res.*, **53**, 558.
Bancon-Montigny C. (2001). Thèse de l'Université de Pau et des Pays de l'Adour, Pau, France.
Barnowski C., Jakubowski N., Stuewer D. and Broekaert J. A. C. (1997). J. *Anal. At. Spectrom.*, **12**, 1155.
Bayen S., Worms I., Parthasarathy N. *et al.* (2006). *Anal. Chim. Acta*, **575**, 267.
Bayona J. M. and Cai M. (1994). *Trends Anal. Chem.*, **13**, 327.
Belfroid A. C., Purperhart M. and Ariese F. (2000). *Mar. Pollut. Bull.*, **40**, 226.
Benoit C. (2005). Thèse de l'Université de Bordeaux I, Bordeaux, France.
B'Hymer C. and Caruso J. A. (2004). *J. Chromatogr. A*, **1045**, 1.
Bloom N. (1989). *Can. J. Fish. Aquat. Sci.*, **46**, 1131.
Borak J. and Hosgood H. D. (2007). *Regul. Toxicol. Pharmacol.*, **47**, 204.
Bostick B. C., Chen C. and Fendorf S. (2004). *Environ. Sci. Technol.*, **38**, 3299.
Bouyssiere B., Szpunar J. and Lobinski R. (2002). *Spectrochim. Acta Part B*, **57**, 805.
Bowles K. C. and Apte S. C. (1998). *Anal. Chem.*, **70**, 395.
Bowles K. C. and Apte S. C. (2000). *Anal. Chim. Acta*, **419**, 145.
Braman R. S. and Foreback C. C. (1973). *Science*, **182**, 1247
Braungardt C. B., Achterberg E. P., Gledhill M. *et al.* (2007). *Environ. Sci. Technol.*, **41**, 4214.
Bravo M., Lespes G., De Gregori I. *et al.* (2004). *J. Chromatogr. A*, **1046**, 217.

Bravo-Sanchez L. R., Ruiz-Encinar J. R., Fidalgo Martinez J. I. and Sanz-Medel A. (2004). *Spectrochim. Acta Part B*, **59**, 59.
Brisbin J. A. and Caruso J. A. (2002). *Analyst*, **127**, 921.
Bryan G. W., Gibbs P. E., Burt G. R. and Hummerstone L. G. (1986). *J. Mar. Biol. Assoc. U.K.*, **66**, 611.
Buffle J. and Tercier-Waeber M. L. (2005). *Trends Anal. Chem.*, **24**, 172.
Cabañero A. I., Madrid Y. and Camara C. (2002). *J. Anal. At. Spectrom.*, **17**, 1595.
Cabon J. Y. and Cabon N. (2000). *Anal. Chim. Acta*, **418**, 19.
Cai Y., Rapsomanikis S. and Andreae M. O. (1993). *J. Anal. At. Spectrom.*, **8**, 119.
Cai Y., Jaffé R., Alli A. and Jones R. D. (1996). *Anal. Chim. Acta*, **334**, 251.
Campanella L., D'Orazio D., Petronio B. M. and Pietrantonio E. (1995). *Anal. Chim. Acta*, **309**, 387.
Campillo N., Peñalver R., Viñas P. *et al.* (2008). *Talanta*, **77**, 793.
Cao X., Hao C., Wang G. *et al.* (2009). *Food Chem.*, **113**, 720.
Capodaglio G., Coale K. H. and Bruland K. W. (1990). *Mar. Chem.*, **29**, 221.
Carpinteiro Botana J., Rodil Rodriguez R., Carro Diaz A. M. *et al.* (2002). *J.Anal. At. Spectrom.*, **17**, 904.
Carvalho P. N., Pinto L. F., Basto M. C. P. and Vasconcelos M. T. S. D. (2007). *Microchem. J.*, **87**, 147.
Centineo G., Rodríguez González P., Blanco E. *et al.* (2004). *J. Anal. At. Spectrom.*, **39**, 485.
Ceulemans M. and Adams F. C. (1996). *J. Anal. At. Spectrom.*, **11**, 201.
Ceulemans M., Witte C., Lobinski R. and Adams F. C. (1994). *Appl. Organometal. Chem.*, **8**, 451.
Connelly D. P., Statham P. J. and Knap A. H. (2006). *Deep-Sea Res.*, **53**, 1975.
Coquery M., Cossa D. and Sanjuan J. (1997). *Mar. Chem.*, **58**, 213.
Coquery M., Morin A., Bécue A. and Lepot B. (2005). *Trends Anal. Chem.*, **24**, 117.
Cornelis R., Caruso J. and Heumann K. G. (eds) (2003). *Handbook of Elemental Speciation: Techniques and Methodology*, John Wiley & Sons, Ltd, Chichester.
Cossa D., Martin J. M., Takayanagi K. and Sanjuan J. (1997). *Deep-Sea Res. Pt II*, **44**, 721.
Craig P. (1986). Occurrence and pathways of organometallic compounds in the environment – general overview, in *Organometallic Compounds in the Environment, Principles and Reactions* (ed. P. J. Craig), Longman, Essex.
Cranston R. E. and Murray J. W. (1980). *Limnol. Oceanogr.*, **25** (1104).
Cuong D. T. and Obbard J. P. (2006). *Appl. Geochem.*, **21**, 1335.
Davidson W. and Zang H. (1994). *Nature*, **367**, 546.
De Carvalho L. M., Do Nascimento P. C., Bohrer D. *et al.* (2008). *Electroanalysis*, **20**, 776.
De Diego A., Tseng C. M., Stoichev T. *et al.* (1998). *J. Anal. At. Spectrom.*, **13**, 623.
De Diego A., Pécheyran C., Tseng C. M. and Donard O. F. X. (1999). Cryofocusing for on-line metal and metalloid speciation in the environment, in *Flow Analysis with Atomic Spectrometric Detection* (ed. A. Sanz-Medel), Elsevier, Amsterdam, 375–406.
De Smaele T., Moens L., Dams R. and Sandra P. (1996). *Fresenius J. Anal. Chem.*, **355**, 778.
Demesmay D. and Ollé M. (1997). *Fresenius J. Anal. Chem.*, **357**, 1116.
Demuth N. and Heumann K. G. (2001). *Anal. Chem.*, **73**, 4020.
Dietz C., Madrid Y. and Camara C. (2001). *J. Anal. At. Spectrom.*, **16**, 1397.
Donard O. F. X., Lalère B., Martin F. M. and Lobinski R. (1995). *Anal. Chem.*, **67**, 4250.
Donard O. F. X., Lespes G., Amouroux D. and Morabito R. (2001). Organotin compounds in the environment: still a critical challenge, in *Trace Element Speciation for Environment, Food and Health* (eds R. Cornelis, L. Ebdon, O. F. X. Donard and Ph. Quevauviller), RSC, Cambridge, 142–175.
Dong L. M., Yan X. P., Li Y. *et al.* (2004). *J. Chromatogr. A*, **1036**, 119.
Dopp E., Hartmann L. M., Florea A. M. *et al.* (2004). *Toxicol. Appl. Pharmacol.*, **201**, 156.
Dufailly V., Noël L., Frémy J. M. *et al.* (2007). *J. Anal. At. Spectrom.*, **22**, 1168.
Dufailly V., Guérin T., Noël L. *et al.* (2008). *J. Anal. At. Spectrom.*, **23**, 1263.
Dugo G., La Pera L., Lo Turco V. and Di Bella G. (2005). *Chemosphere*, **61**, 1093.
Dunn R. K., Teasdale P. R., Warnken J. and Schleich R. R. (2003). *Environ. Sci. Technol.*, **37**, 2794.
Ellwood M. J. and Maher W. A. (2003). *Anal. Chim. Acta*, **477**, 279.
Ellwood M. J. and Van den Berg C. M. G. (2001). *Mar. Chem.*, **75**, 33.
Emteborg H., Baxter D. C. and Frech W. (1993). *Analyst*, **118**, 1007.
Emteborg H., Baxter D. C., Sharp M. and Frech W. (1995). *Analyst*, **120**, 69.
EN 15192 (2006). Characterization of waste and soil. Determination of chromium(VI) in solid material by alkaline digestion and ion chromatography with spectrophotometric detection.

European Commission (2004). Reports on tasks for scientific co-operation (SCOOP), Directorate General Health and Consumer Protection.
Evans S. M. (1999). *Mar. Pollut. Bull.*, **38**, 629.
Falter R. and Schöler H. F. (1995). *Fresenius J. Anal. Chem.*, **353**, 34.
Ferrarello C. N., Fernández de la Campa M. R., Sariego Muñiz C. and Sanz-Medel A. (2000). *Analyst*, **125**, 2223.
Filgueiras A. V., Lavilla I. and Bendicho C. (2002). *J. Environ. Monit.*, **4**, 823.
Filik H., Dogutan M. and Apak R. (2003). *Anal. Bioanal. Chem.*, **376**, 928.
Filippelli M., Baldi F., Brinckman F. E. and Olson G. J. (1992). *Environ. Sci. Technol.*, **26**, 1457.
Fischer J. C., Nganou R. and Wartel M. (1984). *Talanta*, **31**, 1057.
Fischer R., Rapsomanikis S. and Andreae M. O. (1993). *Anal. Chem.*, **65**, 763.
Fitzpatrick S., Ebdon L. and Foulkes M. (2002). *Int. J. Environ. An. Ch.*, **82**, 835.
Foster S., Maher W., Krikowa F. and Apte S. (2007). *Talanta*, **71**, 537.
Francesconi K. A. (2003). *Appl. Organomet. Chem.*, **17**, 682.
Francesconi K. A. and Edmonds J. S. (1996). *Adv. Inorg. Chem.*, **44**, 147.
Francesconi K. A. and Kuehnelt D. (2002). Arsenic compounds in the environment, in *Environmental Chemistry of Arsenic* (ed. W. Frankenberger), Marcel Dekker, New York, 51–94.
Francesconi K. A. and Kuehnelt D. (2004). *Analyst*, **129**, 373.
Gagnon C., Pelletier E., Mucci A. and Fitzgerald W. F. (1996). *Limnol. Oceanogr.*, **41**, 428.
Galceran J., Huidrobo C., Companys E. and Alberti G. (2007). *Talanta*, **71**, 1795.
Gao X., Chen S. and Long A. (2008). *Mar. Pollut. Bull.*, **56**, 770.
García Fernandez R. G., Bayon M. M., Alonso J. I. G. and Sanz-Medel A. (2000). *J. Mass Spectrom.*, **35**, 639.
Geng W., Komine R., Ohta T. *et al.* 2009. *Talanta*, **79**, 369.
Georgiadis M., Cai Y. and Solo-Gabriele H. M. (2006). *Environ. Pollut.*, **141**, 22.
Gerbersmann C., Heisterkamp M., Adams F. C. and Broekaert J. A. C. (1997). *Anal. Chim. Acta*, **350**, 273.
Gerringa L. J. A., Poortvliet T. C. W. and Hummel H. (1996). *Estuar. Coast. Shelf Sci.*, **42**, 629.
Gibbs P. E., Bryan G. W., Pascoe P. L. and Burt G. R. (1987). *J. Mar. Biol. Assoc. U.K.*, **67**, 507.
Gil S., Fragueiro S., Lavilla I. and Bendicho C. (2005). *Spectrochim. Acta B*, **60**, 145.
Gilmour C. C., Henry E. A. and Mitchel R. (1992). *Environ. Sci. Technol.*, **26**, 2281.
Goessler W. and Kuehnelt D. (2002). Analytical methods for the determination of arsenic and arsenic compounds in the environment, in *Environmental Chemistry of Arsenic* (ed. W. Frankenberger), Marcel Dekker, New York, 27–50.
Gomez-Ariza J. L., Morales E., Beltran R. *et al.* (1995). *Analyst*, **120**, 1171.
Goneaga Infante H., Van Campenhout K., Blust R. and Adams F. C. (2002). *J. Anal. At. Spectrom.*, **17**, 79.
Gong Z., Lu X., Ma M. *et al.* (2002). *Talanta*, **58**, 77.
Grabarczyk M. (2008). *Anal. Bioanal. Chem.*, **390**, 979.
Grabarczyk M., Korolczuk M. and Tyszczuk K. (2006). *Anal. Bioanal. Chem.*, **386**, 357.
Graham A. M., Wadhawan A. R. and Bouwer E. J. (2009). *Environ. Toxicol. Chem.*, **28**, 471.
Groeschner M. and Appriou P. (1994). *Anal. Chim. Acta*, **297**, 369.
Guérin T., Astruc A. and Astruc M. (1999). *Talanta*, **50**, 1.
Gunkel-Grillon P. and Buffle J. 2008. *Analyst*, **133**, 954.
Hagendorfer H. and Goessler W. (2008). *Talanta*, **76**, 656.
Hamza-Chaffai A., Amiard-Triquet C. and El Abed A. (1997). *Arch. Environ. Contam. Toxicol.*, **33**, 53.
Hamza-Chaffai A., Amiard C., Pellerin J. *et al.* (2000). *Comp. Biochem. Physiol. Pt C*, **127**, 185.
Hanaoka K., Gössler W., Irgolic K. J. *et al.* (1997). *Chemosphere*, **35**, 2463.
Hanaoka K., Tanaka Y., Nagata Y. *et al.* (2001). *Appl. Organomet. Chem.*, **15**, 299.
Heller A. A. and Weber J. H. (1998). *Sci. Total Environ.*, **221**, 181.
Hempel M., Chau Y. K., Dutka B. J. *et al.* (1995). *Analyst*, **120**, 721.
Hill S. J., Pitts L. J. and Fisher A. S. (2000). *Trends Anal. Chem.*, **19**, 97.
Hintelmann H. (1999). *Chemosphere*, **39**, 1093.
Hintelmann H. and Wilken R. D. (1993). *Appl. Organomet. Chem.*, **7**, 173.
Hintelmann H., Evans R. D. and Villeneuve J. Y. (1995). *J. Anal. At. Spectrom.*, **10**, 619.
Hirano S., Kobayashi Y., Cui X. *et al.* (2004). *Toxicol. Appl. Pharmacol.*, **198**, 458.
Hirata S., Honda K., Shikino O. *et al.* (2000). *Spectrochim. Acta Pt B*, **55**, 1089.

Hirata S., Toshimitsu H. and Aihara M. (2006). *Anal. Sci*., **22**, 39.
Horvat M., Bloom N. and Liang L. (1993). *Anal. Chim. Acta*, **281**, 135.
Hung J.-J., Lu C.-C., Huh C.-A. and Liu J. T. (2009). *J. Mar. Systems*, **76**, 479.
Huo D. and Kingston H. M. (2000). *Anal. Chem*., **72**, 5047.
Hursthouse A. S., Matthews J. M., Figures J. E. *et al*. (2003). *Environ. Geochem. Health*, **25**, 171.
Ikonomou M. G., Fernández M. P., He T. and Cullon D. (2002). *J. Chromatogr. A*, **975**, 319.
Inoue Y., Kawabata K. and Suzuki Y. (1995). *J. Anal. At. Spectrom*., 1995, **10**, 363.
Jackson T. A. (1988). *Can. J. Fish. Aquat. Sci*., **45**, 97.
Jain C. K. and Ali I. (2000). *Water Res*., **34**, 4304.
James B. R., Petura J. C., Vitale R. J. and Mussoline G. R. (1995). *Environ. Sci. Technol*., **29**, 2377.
Jimenez M. S. and Sturgeon R. E. (1997). *J. Anal. At. Spectrom*., **12**, 597.
Jitaru P. and Adams F. C. (2004). *J. Chromatogr. A*, **1055**, 197.
Jokai Z., Abranko L. and Fodor P. (2005). *J. Agric. Food Chem*., **53**, 5499.
Joksic A. S., Katz S. A., Horvat M. and Milacic R. (2005). *Water Air Soil Pollut*., **162**, 265.
Jones B. and Bolam T. (2007). *Mar. Pollut. Bull*., **54**, 1127.
Kaczynski S. E. and Kieber R. J. (1993). *Environ. Sci. Technol*., **27**, 1572.
Kaise T., Horiguchi Y., Fukui S. *et al*. (1992). *Appl. Organometal. Chem*., **6**, 369.
Kannan K., Corsolini S., Focardi F. *et al*. (1996). *Arch. Environ. Contam. Toxicol*., **31**, 19.
Kannan K., Guruge K. S., Thomas N. J. *et al*. (1998). *Environ. Sci. Technol*., **32**, 1169.
Karthikeyan S. H. and Hirata S. (2004). *Appl. Organometal. Chem*., **18**, 323.
Karthikeyan S., Hirata S. and Iyer C. S. P. (2004). *Int. J. Environ. Anal. Chem*., **84**, 573.
Kersten M. and Förstner U. (1986). *Water Sci. Technol*., **18**, 121.
Kirby J. and Maher W. (2002). *J. Anal. At. Spectrom*., **17**, 838.
Kneebone P. E., O'Day P. A., Jones N. and Hering J. G. (2002). *Environ. Sci. Technol*., **36**, 381.
Kohlmeyer U., Kuballa J. and Jantzen E. (2002). *Rapid Commun. Mass Spectrom*., **16**, 965.
Krivan V. and Haas H. F. (1988). *Fresenius J. Anal. Chem*., **332**, 1.
Krone C. A., Brown D. W., Burrows D. G. *et al*. (1989). *Mar. Environ. Res*., **27**, 1.
Kuehnelt D., Irgolic K. J. and Goessler W. (2001). *Appl. Organometal. Chem*., **15**, 445.
Kwokal Z. and Branica M. (2000). *Croat. Chem. Acta*, **73**, 97.
Lalère B., Szpunar J., Budzinski H. *et al*. (1995). *Analyst*, **120**, 2665.
Lambertsson L., Lundberg E., Nilsson M. and Frech W. (2001). *J. Anal. At. Spectrom*., **16**, 1296.
Landaluze J. S., De Diego A., Raposo J. C. and Madariaga J. M. (2004). *Anal. Chim. Acta*, **508**, 107.
Lansens P., Meuleman C., Leermakers M. and Baeyens W. (1990). *Anal. Chim. Acta*, **234**, 417.
La Pera L., Di Bella G., Rando R., *et al*. 2008. *Environ. Monit. Assess*., **145**, 119.
Larsen E. H., Engman J., Sloth J. J. *et al*. (2005). *Anal. Bioanal. Chem*., **381**, 339.
Le Gac M., Lespes G. and Potin-Gautier M. (2003). *J. Chromatogr. A*, **999**, 123.
Le X. C., Lu X. and Li X. F. (2004). *Anal. Chem*., **76**, 26A.
Leermakers M., Baeyens W., Quevauviller Ph. and Horvat M. (2005). *Trends Anal. Chem*., **24**, 383.
Leermakers M., Baeyens W., De Gietera M. *et al*. (2006). *Trends Anal. Chem*., **25**, 1.
Lespes G., Desauziers V., Montigny C. and Potin-Gautier M. (1998). *J. Chromatogr. A*, **826**, 67.
Li W., Zhao H., Teasdale P. R. *et al*. (2005a). *Anal. Chim. Acta*, **533**, 193.
Li Y., Yan X.-P., Dong L.-M. *et al*. (2005b). *J. Anal. At. Spectrom*., **20**, 467.
Liang L., Horvat M. and Bloom N. S. (1994). *Talanta*, **41**, 371.
Lobinski R., Schaumloffel D. and Szpunar J. (2006). *Mass Spectrom. Rev*., **25**, 255.
López-López J. A., Mendiguchía C., Pinto-Ganfornina J. J. and Moreno C. (2010). *Trends Anal. Chem*., **29**, 645.
López-Sánchez J. F., Sahuquillo A., Fiedler H. D. *et al*. (1998). *Analyst*, **123**, 1675.
Lorenzo J. L., Nieto O. and Beiras R. (2002). *Aquat. Toxicol*., **58**, 27.
Lorenzo R. A., Vazquez M. J., Carro A. M. and Cela R., 1999. *Trends Anal. Chem*., **18**, 410.
Macalady J. L., Mack E. E., Nelson D. C. and Scow K. M. (2000). *Appl. Environ. Microbiol*., **66**, 1479.
Magi E., Ianni C., Soggia F. *et al*. (2005). *J. Environ. Monit*., **7**, 1287.
Martin F. M., Tseng C. M., Belin C. *et al*. (1994). *Anal. Chim. Acta*, **286**, 343.
Martino M., Turner A. and Nimmo M. (2004). *Mar. Chem*., **88**, 161.
Mason A. Z. and Meraz R. F. (1998). *Mar. Environ. Res*., **46**, 573.
Mason R., Bloom N., Cappellino S. *et al*. (1998). *Environ. Sci. Technol*., **32**, 4031.

Mason R. P., Rolfhus K. R. and Fitzgerald W. F. (1995). *Water Air Soil Pollut.*, **80**, 665.
Mato Fernandez M. J., Otero Rey J. R., Moreda Pineiro J. *et al.* (2007). *Talanta*, **71**, 515.
McKiernan J. W., Creed J. T., Brockhov C. A. *et al.* (1999). *J. Anal. At. Spectrom.*, **14**, 607.
Mester Z., Lam J., Sturgeon R. and Pawliszyn J. (2000). *J. Anal. At. Spectrom.*, **15**, 837.
Meylan S., Behra R. and Sigg L. (2004). *Environ. Sci. Technol.*, **38**, 3104.
Millán E. and Pawliszyn J. (2000). *J. Chromatogr. A*, **873**, 63.
Moens L., De Smaele T., Dams R. *et al.* (1997). *Anal. Chem.*, **69**, 1604.
Monperrus M., Tessier E., Veschambre S. *et al.* (2005a). *Anal. Bioanal. Chim.*, **381**, 854.
Monperrus M., Point D., Graal J. *et al.* (2005b). *J. Environ. Monit.*, **7**, 693.
Monperrus M., Zuloaga O., Krupp E. *et al.* (2003a). *J. Anal. At. Spectrom.*, **18**, 247.
Monperrus M., Rodriguez Martin Doimeadios R., Scancar J. *et al.* (2003b). *Anal. Chem.*, **76**, 4095.
Monperrus M., Rodriguez Gonzalez P., Amouroux D. *et al.* (2008). *Anal. Bioanal. Chem.*, **390**, 655.
Morabito R., Massanisso P. and Quevauviller Ph. (2000). *Trends Anal. Chem.*, **19**, 113.
Morales-Munoz S., Luque-Garcia J. L. and Luque de Castro M. D. (2004). *Anal. Chim. Acta*, **515**, 343.
Morrison M. A. and Weber J. H. (1997). *Appl. Organomet. Chem.*, **11**, 761.
Mouneyrac C., Amiard-Triquet C., Amiard J. C. and Rainbow P. S. (2001). *Comp. Biochem. Physiol. Pt C*, **129**, 193.
Munksgaard N. C. and Parry D. L. (2003). *J. Environ. Monit.*, **5**, 145.
Nakazato T., Taniguchi T., Tao H. *et al.* (2000). *J. Anal. At. Spectrom.*, **15**, 1552.
Nakazato T., Tao H., Taniguchi T. and Isshiki K. (2002). *Talanta*, **58**, 121.
Nam S. H., Oh H. J., Min H. S. and Lee J. H. (2010). *Microchem. J.*, **95**, 20.
Ndungu K., Hurst M. P. and Bruland K. W. (2005). *Environ. Sci. Technol.*, **39**, 3166.
Neff J. M. (1997). *Environ. Toxicol. Chem.*, **16**, 917.
Nemati K., Bakar N. K. A. and Abas M. R. (2009). *Microchem. J.*, **91**, 227.
Ng J. C. (2005). *Environ. Chem.*, **2**, 146.
Niegel C. and Matysik F. M. (2010). *Anal. Chim. Acta.*, **657**, 83.
Nielsen S. C., Stürup S., Spliid H. and Hansen E. H., 1999. *Talanta*, **49**, 1027.
Nischwitz W. and Pergantis S. A. (2005). *Anal. Chem.*, **77**, 5551.
Nischwitz W. and Pergantis S. A. (2006). *J. Anal. At. Spectrom.*, **21**, 1277.
Nurnberg H. W., Valenta P., Mart L. *et al.* (1976). *Fresenius J. Anal. Chem.*, **282**, 357.
Orero Iserte L., Roig-Navarro A. F. and Hernandez F. (2004). *Anal. Chim. Acta*, **527**, 97.
Pandey A. K., Pandey S. D. and Misra V., 2000. *Ecotoxicol. Environ. Safety*, **47**, 95.
Parker J. L. and Bloom N. S. (2005). *Sci. Total Environ.*, **337**, 253.
Parthasarathy N., Pelletier M. and Buflle J. (2004). *J. Chromatogr. A*, **1025**, 33.
Pazos-Capeans P., Barciela-Alonso M. C., Bermejo-Barrera A. *et al.* (2006). *At. Spectroscopy*, **27**, 107.
Pelligrino C., Massanisso P. and Morabito R. (2000). *Trends Anal. Chem.*, **19**, 97.
Pempkowiak J., Sikora A. and Biernacka E. (1999). *Chemosphere*, **39**, 313.
Penrose W. R. and Woolson E. A. (1974). *Crit. Rev. Environ. Sci. Technol.*, **4**, 465.
Perna L., LaCroix-Fralish A. and Sturup S. (2005). *J. Anal. At. Spectrom.*, **20**, 236.
Pesavento M., Alberti G. and Biesuz R. (2009). *Anal. Chim. Acta*, **631**, 129.
Peshut P. J., Morrison R. J. and Brooks B. A. (2008). *Chemosphere*, **71**, 484.
Petrick J. S., Ayala-Fierro F., Cullen W. R. *et al.* (2000). *Toxicol. Appl. Pharmacol.*, **163**, 203.
Pettine M. and Capri S. (2005). *Anal. Chim. Acta*, **540**, 231.
Pettine M. and Millero F. J. (1990). *Limnol. Oceanogr.*, **35**, 730.
Pettine M., Millero F. J. and La Noce T. (1991). *Mar. Chem.*, **34**, 29.
Pettine M., Mastroianni D., Camusso M, *et al.* (1997). *Mar. Chem.*, **58**, 335.
Point D., Bareille G., Pinaly H. *et al.* (2007). *Talanta*, **72**, 1207.
Posta J., Alimonti A., Petrucci F. and Caroli S. (1996). *Anal. Chim. Acta*, **325**, 185.
Prange A. and Jantzen E. (1995). *J. Anal. At. Spectrom.*, **10**, 105.
Prange A. and Schaumlöffel D. (2002). *Anal. Bioanal. Chem.*, **373**, 441.
Prasada Rao T., Karthikeyan S., Vijayalekshmy B. and Iyer C. S. P. (1998). *Anal. Chim. Acta*, **369**, 69.
Price R. E. and Pichler T. (2005). *Chem. Geol.*, **224**, 122.
Price R. E., Amend J. P. and Pichler T. 2007. *Appl. Geochem.*, **22**, 2595.
Pueyo M., Rauret G., Bacon J. R. *et al.* (2001). *J. Environ. Monit.*, **3**, 238.
Puk R. and Weber J. H. (1994). *Appl. Organomet. Chem.*, **8**, 293.

Quevauviller Ph. (1996). The analysis of butylated tin compounds in the environment and in biological materials, in *Tributyltin: Case Study of an Environmental Contaminant* (ed. S. De Mora), Cambridge University Press, pp. 62–93.

Quevauviller Ph. (1998). *Method Performance Studies for Speciation Analysis*. The Royal Society of Chemistry, Cambridge, 271 pp. (ISBN: 0-85404-467-1).

Quevauviller Ph. (2000). *Environ. Sci. Pollut. Res*., **7**, 1.

Quevauviller Ph. (ed.) (2002). *Methodologies for Soil and Sediment Fractionation Studies*, The Royal Society of Chemistry, Cambridge, 180 pp. (ISBN: 085404-453-1).

Quevauviller Ph. and Donard O. F. X. (1990). *Appl. Organomet. Chem*., **4**, 353.

Quevauviller Ph., Donard O. F. X., Maier E. A. and Griepink B. (1992). *Mikrochim. Acta*, **109**, 169.

Quevauviller Ph., de la Calle-Guntiñas M. B., Maier E. A. and Cámara C. (1995). *Mikrochim. Acta*, **118**, 131.

Qvarnström J., Tu Q., Frech W. and Lüdke K. (2000). *Analyst*, **125**, 1193.

Rahman G. M. M. and Kingston H. M. (2005). *J. Anal. At. Spectrom*., **20**, 183.

Rahman G. M. M., Kingston H. M., Towns T. G. *et al*. (2005). *Anal. Bioanal. Chem*., **382**, 1111.

Ramalhosa E., Rio-Segade S., Pereira E. *et al*. (2001). *J. Anal. At. Spectrom*., **16**, 643.

Raml R., Goessler W. and Francesconi K. A. (2006). *J. Chromatogr. A*, **1128**, 164.

Rao C. R. M., Sahuquillo A. and López Sánchez J. F. (2008). *Water Air Soil Pollut*., **189**, 291.

Rapsomanikis S. (1994). *Analyst*, **119**, 1429.

Rattanachongkiat S., Millward G. E. and Foulkes M. E. (2004). *J. Environ. Monitor*., **6**, 254.

Rauret G., López-Sánchez J. F., Sahuquillo A. *et al*. (1999). *J. Environ. Monit*., **1**, 57.

Reuther R., Jaeger L. and Allard B. (1999). *Anal. Chim. Acta*, **394**, 259.

Reyes L. H., Guzmán Mar J. L., Mizanur Rahman G. M. *et al*. (2009). *Talanta*, **78**, 983.

Rio-Segade S. and Bendicho C. (1999). *J. Anal. At. Spectrom*., **14**, 263.

Ritsema R. and Donard O. F. X. (1994). *Appl. Organomet. Chem*., **8**, 571.

Rodil R., Carro A. M., Lorenzo R. A. *et al*. (2002). *J. Chromatogr. A*, **963**, 313.

Rodriguez I., Mounicou S., Lobinski R. *et al*. 1999. *Anal. Chem*., **71**, 4534.

Rodríguez Cea A., Linde A. R., Fernández de la Campa M. R., *et al*. (2006). *Talanta*, **69**, 963.

Rodríguez González P., Ruiz Encinar J., García Alonso J. I. and Sanz Medel A. (2002). *J. Anal. At. Spectrom*., **17**, 824.

Rodriguez Gonzalez P., Ignacio Garcia Alsonso J. and Sanz Medel A. (2005). *J. Anal. At. Spectrom*., **20**, 1076.

Rodríguez González P., Monperrus M., García Alonso J. I. *et al*. (2007). *J. Anal. At. Spectrom*., **22**, 1373.

Rodriguez Martin Doimeadios R. C., Krupp E., Amouroux D. and Donard O. F. X. (2002). *Anal. Chem*., **74**, 2505.

Rodriguez Martin Doimeadios R. C., Monperrus M., Krupp E. *et al*. (2003). *Anal. Chem*., **75**, 3202.

Rodriguez Martin Doimeadios R. C., Tessier E., Amouroux D. *et al*. (2004). *Mar. Chem*., **90**, 107.

Rodriguez Pereiro I., Schmitt V. O., Szpunar J. *et al*. (1996). *Anal. Chem*., **68**, 4135.

Rodriguez Pereiro I., Wasik A. and Lobinski R. (1998). *J. Chromatogr. A*, **795**, 359.

Romero R., Liu J., Mayer P. and Jönsson J. A. (2005). *Anal. Chem*., **77**, 7605.

Root R. A., Vlassopoulos D., Rivera N. A. *et al*. (2009). *Geochim. Cosmochim. Acta*., **73**, 5528.

Ruiz-Encinar J., Garcia-Alonso J. I. and Sanz-Medel A. (2000). *J. Anal. At. Spectrom*., **15**, 1233.

Ruiz-Encinar J., Monterde-Villar M. I., Santamaria V. G. *et al*. (2001). *Anal Chem*., **73**, 3174.

Sakai T., Inoue Y., Date Y. *et al*. (2001). *Appl. Orgamet. Chem*., **15**, 285.

Sander S. and Koschinsky A. (2000). *Mar. Chem*., **71**, 83.

Sanz E., Munoz-Olivas R. and Camara C. (2005). *J. Chromatogr A*., **1097**, 1.

Sanz-Medel A., Montes-Bayon M. and Luisa Fernandez Sanchez M. (2003). *Anal. Bioanal. Chem*., **377**, 236.

Saouter E. and Blattmann B. (1994). *Anal. Chem*., **66**, 2031.

Saverwyns, S., Zhang X., Vanhaecke F. *et al*. (1997). *J. Anal. At. Spectrom*., **12**, 1047.

Scarponi G., Capodaglio G., Toscano G. *et al*. (1995). *Microchem. J*., **51**, 214.

Schaeffer R., Fodor P. and Soeroes C. (2006). *Rapid Commun. Mass Spectrom*., **20**, 2979.

Schintu M., Durante L., Maccioni A. *et al*. (2008). *Mar. Pollut. Bull*., **57**, 832.

Schmeisser E., Goessler W., Kienzl N. and Francesconi K. A. (2004a). *Anal. Chem*., **76**, 418.

Schmeisser E., Raml R., Francesconi K. A. *et al*. (2004b). *Chem. Comm*., **16**, 1824.

Schmeisser E., Goessler W., Kienzl N. and Francesconi K. A. (2005). *Analyst*., **130**, 948.

Schmitt V. O., De Diego A., Cosnier A. *et al*. (1997). *Spectroscopy*, **13**, 99.

Schubert P., Rosenberg E. and Grasserbauer M. (2000). *Fresenius J. Anal. Chem*., **366**, 356.

Scindia Y. M., Pandey A. K., Reddy A. V. R. and Manohar S. B. (2004). *Anal. Chim. Acta*, **515**, 311.
Scoullos M. J. and Pavlidou A. S. (2003). *Water Air Soil Pollut.*, **147**, 203.
Séby F., Charles S., Gagean M. *et al.* (2003). *J. Anal. At. Spectrom.*, **18**, 1386.
Séby F., Benoit-Bonnemason C., Tessier E. *et al.* (2009). *Paralia*, **2**, s3.1.
Segovia-Martinez L., Bouzas-Blanco A., Campins-Falco P. and Seco-Torrecillas A. (2010). *Talanta*, **80**, 1888.
Shade C. W. and Hudson R. J. M. (2005). *Environ. Sci. Technol.*, **39**, 4974.
Simon S., Tran H., Pannier F. and Potin-Gautier M. (2004). *J. Chromatogr. A*, **1024**, 105.
Sirinawin W. and Westerlund S. (1997). *Anal.Chim. Acta*, **356**, 35.
Sirinawin W., Turner D. R. and Westerlund S. (2000). *Mar. Chem.*, **71**, 265.
Slaets S. and Adams F. C. (2000). *Anal. Chim. Acta*, **414**, 141.
Slaveykova V. I., Parthasarathy N., Buffle J. and Wilkinson K. J. (2004). *Sci. Total Environ.*, **328**, 55.
Slaveykova V. I., Karadjova I. B., Karadjov M. and Tsalev D. L. (2009). *Environ. Sci. Technol.*, **43**, 1798.
Sloth J. J., Larsen E. H. and Julshamn K. (2005). *J. Agric. Food Chem.*, **53**, 6011.
Som-Aum W., Liawruangrath S. and Hansen E. H. (2002). *Anal. Chim. Acta*, **463**, 99.
Som-Aum W., Threeprom J., Li H. and Lin J.-M. (2007). *Talanta*, **71**, 2062.
Stoichev T., Rodriguez Martin Doimeadios R. C., Amouroux D. *et al.* (2002). *J. Environ. Monit.*, **4**, 517.
Stoichev T., Amouroux D., Wasserman J. C. *et al.* (2004a). *Estuar. Coast. Shelf Sci.*, **59**, 511.
Stoichev T., Rodriguez Martin Doimeadios R. C., Tessier E. *et al.* (2004b). *Talanta*, **62**, 433.
Stoichev T., Amouroux D., Rodriguez Martin Doimeadios R. C. *et al.* (2006). *Appl. Spectrosc. Rev.*, **41**, 591.
Styblo M., Drobná Z., Jaspers I. *et al.* (2002). *Environ. Health Persp.*, **110**, 767.
Sumida T., Sabarudin A., Oshima M. and Motomizu S. (2006). *Anal. Sci.*, **22**, 161.
Suner M. A., Devesa V., Munoz O. *et al.* (2001). *J. Anal. Atom. Spectrom.*, **16**, 390.
Szpunar J. (2000). *Analyst*, **125**, 963.
Szpunar J. (2005). *Analyst*, **130**, 442.
Szpunar J., Schmitt V. O. and Lobinski R. (1996a). *J. Anal. At. Spectrom.*, **11**, 193.
Szpunar J., Ceulemans M., Schmitt V. O. *et al.* (1996b). *Anal. Chim. Acta*, **332**, 225.
Tessier A., Campbell P. G. C. and Bisson M. (1979). *Anal. Chem.*, **51**, 844.
Thomas P., Finnie J. K. and Williams J. G. (1997). *J. Anal. At. Spectrom.*, **12**, 1367.
Tirez K., Brusten W., Cluyts A. *et al.* (2003). *J. Anal. At. Spectrom.*, **18**, 922.
Tirez K., Scharf H., Calzorali D. *et al.* (2007). *J. Environ. Monit.*, **9**, 749.
Tseng C. M., De Diego A., Martin F. M. and Donard O. F. X. (1997a). *J. Anal. At. Spectrom.*, **12**, 629.
Tseng C. M., de Diego A., Martin F. M. *et al.* (1997b). *J. Anal. At. Spectrom.*, **12**, 743.
Tseng C. M., De Diego A., Pinaly H. *et al.* (1998). *J. Anal. At. Spectrom.*, **13**, 755.
Tseng C. M., De Diego A., Wasserman J. C. *et al.* (1999). *Chemosphere*, **39**, 1119.
Tseng C. M., Amouroux D., Brindle I. D. and Donard O. F. X. (2000). *J. Environ. Monit.*, **2**, 603.
Tseng C. M., Hammerschmidt C. R. and Fitzgerald W. F. (2004). *Anal. Chem.*, **76**, 7131.
Tu Q., Qian J. and Frech W. (2000). *J. Anal. At. Spectrom.*, **15**, 1583.
Tuzen M., Çıtak D., Mendil D. and Soylak M. (2009). *Talanta*, **78**, 52.
Ueno D., Inoue S., Takahashi S. *et al.* (2004). *Environ. Pollut.*, **127**, 1.
Ure A. M., Quevauviller Ph., Muntau H. and Griepink B. (1993). *J. Environ. Anal. Chem.*, **51**, 135.
USEPA (1995a). Alkaline digestion of hexavalent chromium, method 3060A, in *Test Methods for Evaluating Solid Waste – Physical/Chemical Methods*, SW-846, update 3, US Government Printing Office, Washington, DC.
USEPA (1995b). Chromium, hexavalent (colorimetric), method 7196A, in *Test Methods for Evaluating Solid Waste – Physical/Chemical Methods*, SW-846, update 3, US Government Printing Office, Washington, DC.
USEPA (1996). Determination of hexavalent chromium in drinking water, groundwater and industrial wastewater effluents by ion-chromatography, Method 7199A, in *Test Methods for Evaluating Solid Waste – Physical/Chemical Methods*, SW-846, update 3, US Government Printing Office, Washington, DC.
USEPA (1998). *Elemental and Speciated Isotope Dilution Mass Spectrometry, Method 6800*, US Government Printing Office, Washington, DC.
Usero J., Gamero M., Morillo J. and Gracia I. (1998). *Environ. Int.*, **24**, 487.
Välimäki I. and Perämäki P. (2001). *Microchim. Acta*, **137**, 191.
Van Campenhout K., Goenaga Infante H., Goemans G. *et al.* (2008). *Sci. Total Environ.*, **394**, 379.
Van den Berg C. M. G. (1991). *Mar. Chem.*, **34**, 211.

Vandenhecke J., Waeles M., Riso R. D. and Le Corre P. (2007). *Anal. Bioanal. Chem.*, **388**, 929
Vanhaecke F., Saverwyns S., De Wannemacker G. *et al*. (2000). *Anal. Chim. Acta.*, **419**, 55.
Varekamp J. C., Buchholtz ten Brink M. R., Mecray E. L. and Kreulen B. (2000). *J. Coast. Res.*, **16**, 613.
Vassileva E. (2000). *Analusis*, **28**, 878.
Vazquez M. J., Carro A. M., Lorenzo R. A. and Cela R. (1997). *Anal. Chem.*, **69**, 221.
Vergara Gallardo M., Bohari Y., Astruc A. *et al*. (2001). *Anal. Chim. Acta*, **441**, 257.
Vicente-Martorell J. J., Galindo-Riaño M. D., García-Vargas M. and Granado-Castro M. D. (2009). *J. Hazard. Mater.*, **162**, 823.
Waeles M., Riso R. D. and Le Corre P. (2005a). *Estuar. Coast. Shelf Sci.*, **65**, 143.
Waeles M., Riso R. D. and Le Corre P. (2005b). *Estuar. Coast. Shelf Sci.*, **62**, 313.
Wahlen R., McSheehy S., Scriver C. and Mester Z. (2004). *J. Anal. At. Spectrom.*, **19**, 876.
Wasik A., Rodriguez Pereiro I., Dietz C. *et al*. (1998). *Anal. Commun.*, **35**, 331.
White S., Catterick T., Fairman B. and Webb K. (1998). *J. Chromatogr. A*, **794**, 211.
Wilken R. D. and Falter R. (1998). *Appl. Organomet. Chem.*, **12**, 551.
Wuilloud R. G., Altamirano J. C., Smichowski P. N. and Heitkemper D. T. (2006). *J. Anal. At. Spectrom.*, **21**, 1214.
Yang G. D., Xu J. H., Zheng J. P. *et al*. (2009). *Talanta*, **78**, 471.
Yang L. and Lam J. W. H. (2001). *J. Anal. At. Spectrom.*, **16**, 724.
Yang L., Mester Z. and Sturgeon R. (2002). *Anal. Chem.*, **74**, 2968.
Yang L., Mester Z., Abranko L. and Sturgeon R. E. (2004). *Anal. Chem.*, **76**, 3510.
Yang L., Willie S. and Sturgeon R. E. (2009). *J. Anal. At. Spectrom.*, **24**, 958.
Yang Y.-J. and Huang H.-J. (2001). *Anal. Chem.*, **73**, 1377.
Yu L. P. and Yan X. P. (2003). *Trends Anal. Chem.*, **22**, 245.
Yuan C., Shi J., He B. *et al*. (2004). *Environ. Int.*, **30**, 769.
Zang H. and Davidson W. (1995). *Anal. Chem.*, **67**, 3391.
Zang H. and Davidson W. (2000). *Anal. Chem.*, **72**, 4447.
Zang H. and Davidson W. (2001). *Pure Appl. Chem.*, **73**, 9.
Zheng J., Hintelmann H., Dimock B. and Dzurko M. S. (2003). *Anal. Bioanal. Chem.*, **377**, 14.

6

Organic Micropollutants[1]

Robin J. Law[a], Lynda Webster[b], Norbert Theobald[c], Heather S. Rumney[a], and Jacob de Boer[d]

[a]*The Centre for Environment, Fisheries and Aquaculture Science, Cefas Lowestoft Laboratory, Suffolk UK*

[b]*Marine Scotland, Marine Laboratory, Aberdeen, UK*

[c]*Federal Maritime and Hydrographic Agency, Hamburg, Germany*

[d]*Vrije Universiteit Amsterdam, Institute for Environmental Studies Environmental Chemistry and Toxicology, Amsterdam, The Netherlands*

Chemical Marine Monitoring: Policy Framework and Analytical Trends, First Edition.
Edited by Philippe Quevauviller, Patrick Roose, and Gert Verreet.

6.1 GENERAL INTRODUCTION

Many organic compounds enter the marine environment as a result of man's activities, originating, for example, from discharges into rivers, estuaries and coastal waters. As they can be present at low (though not necessarily insignificant) concentrations, they are generally known as organic micropollutants. Improvements in the sensitivity of analytical techniques during past decades have meant that a wider range of these compounds can now be detected and quantified at lower concentrations than was possible hitherto. Organic micropollutants are important because, singly or in combination, they can exert adverse effects within the aquatic environment and, for compounds which are persistent and which bioaccumulate, potentially also on human consumers of seafood.

Many micropollutants of interest are organohalogen compounds. Most organohalogen pollutants are classical pollutants, which have been of environmental concern for nearly 50 years now. The insecticide DDT was identified as one of the first persistent pollutants found all over the world and causing harm to the environment (Carson, 1962). Other organochlorine pesticides, like hexachlorocyclohexane (HCH), hexachlorobenzene (HCB), toxaphene and many more, proved to be equally persistent. Many organohalogens are considered pollutants for priority action and have been banned, regulated and monitored on a regional or global scale as part of national and international monitoring programmes (HELCOM, OSPAR, EU, UN). Among the 'dirty dozen' of pollutants banned under the Stockholm Convention (aldrin, chlordane, dieldrin, endrin, heptachlor, HCB, mirex, toxaphene, DDT, polychlorinated dibenzodioxins (PCDDs), polychlorinated dibenzofurans (PCDFs) and polychlorinated biphenyls (PCBs)) the first nine compounds belong to the group of organochlorine pesticides. PCDDs, PCDFs and PCBs have different sources. PCBs have special uses and origins; they are described separately in Section 6.2.

In this chapter we summarize the current status of analytical methods for a number of organic micropollutants, PCBs, polybrominated diphenyl ethers (PBDEs), organochlorine pesticides and polycyclic aromatic hydrocarbons (PAHs). Analytical methods for other emerging contaminants are also described.

6.2 POLYCHLORINATED BIPHENYLS (PCBs)

6.2.1 Introduction

PCBs were originally produced at the end of the nineteenth century, although it was not until the late 1940s that they were produced and used in large volumes. PCBs were prepared industrially by the chlorination of biphenyl. The commercial PCB preparations (most well-known trade names: Aroclor, Clophen, Kanechlor and Phenoclor) were graded and marketed according to their chlorine content; for example, Aroclor 1232 contains 32% by weight of chlorine. During the 1960s and 1970s there were many international companies that produced Aroclors, with one of the biggest industrial producers of PCBs being the US business Monsanto Chemical Company. An estimated 1.3 million tonnes of PCBs were produced worldwide (Breivik *et al.*, 2002, 2007). Aroclors were widely used in transformers, capacitors, hydraulic fluids and as plasticizers in paints, plastics and sealants. However, owing to the concerns about the environmental impact of PCBs, production in western Europe and North America ceased in the late 1970s and in the early 1990s in eastern Europe and Russia.

On the basis of their persistence, toxicity and tendency to bioaccumulate, known as the PBT criteria, and following the OSPAR (Oslo and Paris Commission) Dynamic Selection and Prioritisation Mechanism for Hazardous Substances (DYNAMEC), PCBs are listed on the United Nations Environment Programme (UNEP) list of persistent organic pollutants (POPs) and they are classed as Chemicals for Priority Action under regional programmes such as OSPAR and HELCOM (Helsinki Commission). Theoretically, 209 individual PCB congeners could be produced, depending on the number and position of chlorine that is substituted onto the biphenyl moiety. However, only a limited number of PCBs were generally produced from the conditions involved during the industrial process. Individual congeners are generally named according to the short-hand system Ballschmiter *et al.* (1992) developed for PCB congeners. For this naming system a number, from 1 to 209, often prefixed with 'CB', was applied to each congener after the congeners have been sorted on the basis of their structural names. This system was summarized more recently by Mills *et al.* (2007).

Most monitoring programmes require the analysis of a limited number of the 209 PCB congeners. The seven 'Dutch' or ICES (International Council for the Exploration of the Seas) PCBs (CB28, 52, 101, 118, 153, 138 and 180) were recommended by the European Union Community Bureau of Reference (BCR); these PCBs were selected as indicators owing to their relatively high concentrations in technical mixtures and their wide chlorination range (three to seven chlorine atoms per molecule). The most toxic PCB congeners are the so-called 'dioxin-like' PCBs. The 'dioxin-like' PCBs are the four non-*ortho* (CB81, 77, 126 and 169) and eight mono-*ortho* PCBs (CB105, 114, 118, 123, 156, 157, 167 and 189) that also have chlorine atoms in both *para* and at least two *meta* positions. The non-*ortho* PCBs can obtain a planar configuration and the mono-*ortho* PCBs can obtain a near planar configuration. The 12 'dioxin-like' PCBs are stereochemically similar to 2,3,7,8-tetrachlorodibenzo-*p*-dioxin (TCDD) and, therefore, have similar toxic and biological responses to those of dioxins, but are normally found at much lower concentrations in the environment compared with the *ortho*-PCBs. However, of these 12 PCB congeners, only CB118 is one of the seven ICES PCBs and, therefore, included in routine monitoring programmes.

Historically, the main sources of PCBs to the marine environment included energy production, combustion industries, production processes and waste (landfill, incineration, waste treatment and disposal). PCBs will also enter the marine environment following the destruction and disposal of industrial plants and equipment or from emissions from old electrical equipment (for example, from landfill sites). Furthermore, run-off from urban areas and atmospheric deposition continue to be diffuse sources of hazardous substances to the marine environment. Owing to their persistent nature, high concentrations will still be found in the sediment, near industrialized and urbanized areas as a result of historical inputs. In these areas concentrations for the sum of the seven ICES PCBs (Σ7PCB) in sediment can be greater than $100\,\mu g\,kg^{-1}$ dry weight (dw). PCBs are hydrophobic and have the potential to bioaccumulate, particularly in lipid-rich tissue such as liver. Σ7PCB concentrations in fish liver are generally $<500\,\mu g\,kg^{-1}$ lipid weight (lw) at sites remote from industrial and urban activity and $>500\,\mu g\,kg^{-1}$ lw in more industrialized areas. Since the ban on the use of PCBs, concentrations in the marine environment have generally fallen. However, even low concentrations may give rise to adverse biological effects. Therefore, analytical methods need to be able to measure PCBs at these low concentrations ($<0.1\,\mu g\,kg^{-1}$).

6.2.2 Analytical Methods

Over the past 40 years there has been a considerable amount of research and monitoring on CBs in the environment and, therefore, validated analytical methods are well established. Analytical methods for PCBs in environmental monitoring have been reviewed (Smedes and De Boer, 1994, 1997; Wells and Hess, 2000; Muir and Sverko, 2006). Owing to the decreasing environmental concentrations of PCBs, analytical methods have had to improve to be able to detect PCBs at very low concentrations. Any analytical technique requires three main steps: extraction, clean-up and instrumental analysis.

6.2.2.1 Extraction Methods

A range of extraction methods has been used for the extraction of PCBs from environmental samples. These include the more traditional methods such as Soxhlet and the newer automated methods such as pressurized liquid extraction (PLE). Most extraction methods can be applied to sediment and biota samples. Historically, the extraction method of choice for PCBs in sediment and biota was Soxhlet extraction. Soxhlet requires relatively long periods for extraction (3–12 h) and large volumes of solvent (100–500 mL) compared with other, newer alternative techniques; however, it can be used for large sample sizes (up to 200 g of sediment and 100 g of biota) and it is not matrix dependent. Soxhlet extraction with nonpolar solvents is possible, but it takes considerably longer; more polar solvents such as dichloromethane are more efficient than polar/nonpolar mixtures, although for health and safety reasons dichloromethane should be avoided. An alternative solvent, which is as efficient as dichloromethane, is methyl-*tert*-butylether.

Although the use of binary non-polar/polar solvent mixtures and Soxhlet is still the benchmark for PCB extraction, there have been numerous attempts to find alternative procedures that are less time-consuming and use less solvent. Automated techniques include microwave-assisted extraction (MAE), supercritical fluid extraction (SFE) and PLE. These methods have the advantage of using less solvent, being fully automated and taking less time than traditional methods of extraction, such as Soxhlet.

For SFE, a supercritical fluid such as ammonia, methane or carbon dioxide is used for the extraction rather than a solvent, with carbon dioxide being the most common supercritical fluid (Lee and Peart, 1994; Tong and Imagawa, 1995). An advantage of SFE is that extracts are cleaner and additional clean-up steps are not required. However, only a small sample size (<5 g) may be used for SFE (which will increase the detection limits of the method) and reproducibility is poor compared with other extraction techniques, such as Soxhlet (Wells and Hess, 2000). Although SFE has been employed for environmental samples, it has not yet found a wide range of applications. MAE was first used for the extraction of organic contaminants in 1986. MAE uses microwaves to heat the sample and solvent; however, there are few publications using this technique for the extraction of PCBs in environmental samples and it is more commonly used for metal digests (Wells and Hess, 2000).

The most promising of the automated techniques in recent years is PLE (Sporring *et al.*, 2003; Gómez-Ariza *et al.*, 2002). This method is matrix independent and is used widely for the extraction of persistent organic pollutants (POPs) in environmental samples. The method further allows interesting modifications to enable partial sample clean-up by adding fat retainers (such as Florisil or alumina) to the extraction cell, as shown in Figure 6.1, and the use of a small carbon column in the extraction cell, which selectively adsorbs dioxin-like compounds (subsequently isolated by back-flushing with toluene) (Sporring *et al.*, 2003). The addition of the fat retainer to the extraction cell eradicates the need for additional lipid removal steps. PLE has the additional advantage of using less solvent, being fully automated and taking less time than the traditional Soxhlet method. There have been an increasing number of publications using this technique.

6.2.2.2 *Clean-up Procedures*

There is a critical need for clean-up steps when analysing environmental samples, especially if concentrations are low. The presence of higher levels of other substances that co-elute can cause interference in the final analysis and make accurate quantification difficult.

In biota samples, the removal of lipids from the extract needs to be carried out. A number of techniques are commonly used for lipid removal; these included gel permeation chromatography

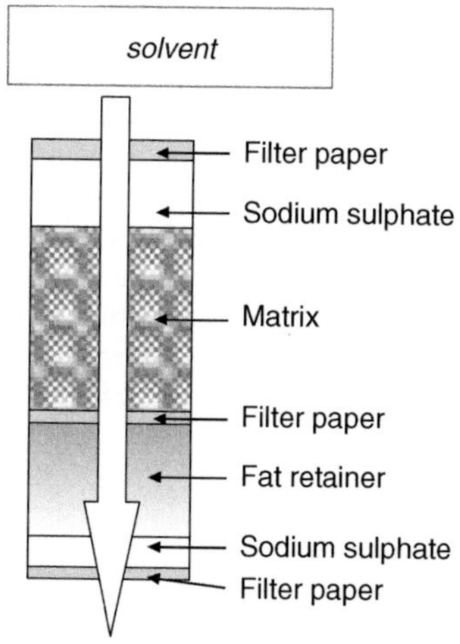

Figure 6.1 Schematic of PLE cell.

(GPC), adsorption column chromatography, dialysis, sulfuric acid treatment, saponification or a combination of these (Wells and Hess, 2000; Muir and Sverko, 2006). Removal of lipid using column chromatography can be achieved with adsorption on alumina columns; depending on the fat content, 5–15 g of alumina in glass columns can be used (van Leeuwen and de Boer, 2008). Silica columns and carbon columns can also be combined with alumina columns to achieve clean-up to a greater degree and to separate PCBs from organochlorine pesticides (Focant and De Pauw, 2002). Normally the silica is deactivated (1–5% water) and a nonpolar solvent such as hexane is used to elute the PCBs; PCBs will elute before most organochlorine pesticides. The activated carbon is used as a complementary tool to the alumina columns because the carbon can fractionate the non-*ortho* and mono-*ortho* PCBs from other classes of aromatic compounds and so improve the clean-up of the sample (Focant and De Pauw, 2002). GPC is an alternative method for clean-up; this method removes lipids through size-exclusion processes. GPC can be used with many solvents; however, one column is not able to remove all lipids and, thus, further clean-up steps are normally required. When applying GPC, two serial columns are often used for improved lipid separation. Solvent mixtures such as dichloromethane–hexane or cyclohexane–ethyl acetate can be used as eluents for GPC. GPC is the only technique which can be automated, but it requires large organic solvent volumes.

One advantage of PLE is that online clean-up and fractionation of dioxins, furans and PCBs (*ortho*, mono-*ortho* and non-*ortho*) with PLE for food, feed and environmental samples can be incorporated into the extraction method. The original method utilized a fat retainer for the online clean-up of lipid (Sporring *et al*., 2003). Silica impregnated with sulfuric acid, alumina and Florisil have all been used as fat retainers. To obtain fat-free extracts, sufficient fat retainer must be used. Björklund *et al*. (2006) investigated the relationship between the amount of fat retainer and the amount of fat in the extraction cells and proposed a 'fat to fat-retainer' ratio (FFR) of 0.025 (i.e. 40 times more fat retainer than fat is required to give a fat-free extract). Further developments combined the fat retainer method with an online carbon column for the fractionation of dioxins, furans and non-*ortho* PCBs (Björklund *et al*., 2006).

It is often necessary to apply to different clean-up methods in combination, such as silica gel and sulfuric acid. Concentrated sulfuric acid is often very useful to clean the extracts and ensures clean chromatograms. However, sulfuric acid is aggressive and will cause degradation of some organochlorine pesticides, such as dieldrin and endrin. However, for PCB clean-up it is very useful. It can also be used in column chromatography when mixed with silica gel.

All of the above clean-up procedures have also been applied to sediment to remove co-extractives. Fat retainer can also be used for sediment extractions by PLE, although less fat retainer is required than with biota samples. 10 g of fat retainer for the extraction of 20 g sediment will give a clean extract. In addition, sulfur should be removed from sediment extracts in order to reduce interferences and to protect the detectors. Sulfur clean-up is usually achieved by applying copper so that it reacts with the sulfur to form CuS or with tetrabutyl ammonium sulfite. This can be achieved by the addition of copper powder, wire or gauze during or after Soxhlet extraction. Copper can also be added to the PLE cell; however, this is not always sufficient, and further treatment with copper may be required following extraction. Mercury has also been used in the past for sulfur removal, but it should be avoided for environmental reasons. Ultrasonic treatment might improve the removal of sulphur. Using GPC as a clean-up step will also remove sulfur from sediment samples.

6.2.2.3 Clean-up for Non-ortho PCBs

Non-*ortho* PCBs are normally found at substantially lower concentrations than the *ortho* and mono-*ortho* PCBs. Therefore, it is necessary to separate the non-*ortho* CBs from the mono-*ortho* and *ortho* PCBs to allow for extract concentration. A number of methods have been reported for this, including activated carbon, adsorption columns, polyurethane foam with activated carbon, silica gel with activated carbon, Carbopack C mixed with Celite, charcoal, Florisil and porous graphic carbon, but the preferred method is by high-performance liquid chromatography (HPLC) using a PYE HPLC (2-(1-pyrenyl)ethyldimethylsilylated silica) column (Hess *et al.*, 1995). Using this technique, in addition to isolating the non-*ortho* PCBs from the *ortho* and mono-*ortho* PCBs, other co-extractives such as tetrachlorinated dioxins and furans can also be separated. The column packing separates similar molecules with different π-electron densities resulting from the configuration of the aryl rings (Wells and Echarri, 1994). However, there is one potential problem when using the PYE HPLC method for isolation of non-*ortho* PCBs. Polychlorinated naphthalenes (PCNs) have been shown to have the same capacity factor as the *ortho*, mono-*ortho* and non-*ortho* PCBs on the PYE column (Wells *et al.*, 1995). This is a particular problem when analysing by gas chromatography–mass spectrometry (GC-MS), because the PCNs produce ions with the same mass-to-charge ratios m/z as that of the ^{13}C-labelled PCB congeners for CB81 and CB77. PCNs have been found to co-elute with CB81 on columns used for analysis, such as CP Sil8 and HP5. As a consequence of PCN interference, there are currently no proficiency test schemes, for fish tissue or shellfish, which provide certified or assigned values for CB81.

6.2.2.4 Pre-concentration

Reducing the volumes at each step can improve the detection limit; however, this requires the clean-up steps to be rigorous to reduce possible interferences in the final analysis. Concentration steps will be required after extraction and prior to clean-up and after clean-up prior to GC analysis. No matter what technique is used, evaporation to dryness should be avoided. Up until recently the most common concentration method was with a rotary-film evaporator. Evaporation of solvents using this technique should be performed at low temperature (water bath temperature of $\leq$30 °C) and under controlled pressure conditions, in order to prevent losses of the more volatile PCBs. Rotary-film evaporators are more time consuming than other, newer concentration techniques, such as Turbo-Vap and Syncore. Turbo-Vap concentrators can be used to simultaneously reduce the solvent volume of a number of samples. Although a more rapid technique than rotary evaporation, Turbo-Vap needs to be carefully optimized and monitored to prevent both losses (both of volatiles and solvent aerosols) and cross-contamination. Syncore parallel evaporators (Büchi, Switzerland) can again be used to concentrate a number of samples at the same time. The Büchi Syncore Analyst also uses glass tubes, but the system is sealed, avoiding contamination from the laboratory air during evaporation, but again the water-bath temperatures should be minimized to prevent losses. When reducing the sample to the required final volume, solvents can be removed by a stream of clean nitrogen gas. Suitable solvents for injection into the gas chromatograph include *n*-hexane, *n*-heptane, toluene and *iso*-octane.

6.2.2.5 Detection and Quantification

GC is extremely suitable for the final analysis of PCBs. Therefore, the final stage of analysis of PCBs in environmental samples is to analyse the extracts by GC coupled to a suitable detector. The two GC injection techniques commonly used are splitless and on-column injection. Recently, other techniques, such as temperature-programmed or pressure-programmed injection, have become more prominent. They offer additional advantages such as an increased injection volume without the negative effects previously associated with that, but temperature and pressure programmes should be thoroughly optimized before use. Increasing the injection volume will allow either or both of the elimination of an extra evaporation step and a lowering of the detection limits.

A wide range of stationary phases can be used for PCB separation, such as HB-5 or CP-Sil8CB (Verenitch *et al*., 2007). Some of the capillary columns have been designed to prevent leaching of the stationary phase and, therefore, are suitable for MS. Adjustment of the column length and internal diameter may also help separation, but the majority of columns designed for PCB analysis are 30–60 m in length and 0.20–0.25 mm in diameter, although narrow bore columns at 0.10–0.15 mm have also been used. Wider bore columns should be avoided, as they may easily lead to co-elution problems. In recent years, new chromatographic phases have become available that result in an improved separation of critical PCB congener pairs. A good example is the HT-8 phase (1,7-dicarba-*closo*-dodecarborane phenylmethyl siloxane) (Larsen *et al*., 1995) that shows a remarkable selectivity for PCBs; see Figure 6.2. Separation of all 209 congeners is possible using this column (Dawes *et al*., 2004).

The detectors that have been used for PCB analysis include electron capture detection (ECD), low-resolution mass spectrometry (LRMS), ion trap mass spectrometry (ITMS/MS), time-of-flight mass spectrometry (TOF-MS) and high-resolution mass spectrometry (HRMS). MS techniques use either electron impact ionization (EI) or electron capture negative ionization (ECNI) (Wells and Hess, 2000; van Leeuwen and de Boer, 2008). Historically, ECD has been the most commonly used detector for the analysis of *ortho*-PCBs; however, this is changing, with MS being more commonly used. For analysis of non-*ortho* and mono-*ortho* CBs, EI-HRMS is the best method (van Leeuwen, and de Boer, 2008); however, ECD can be used for mono-*ortho* congeners but with co-elution risks and ECNI-LRMS has been used for non-*ortho* congeners. MS techniques are more selective

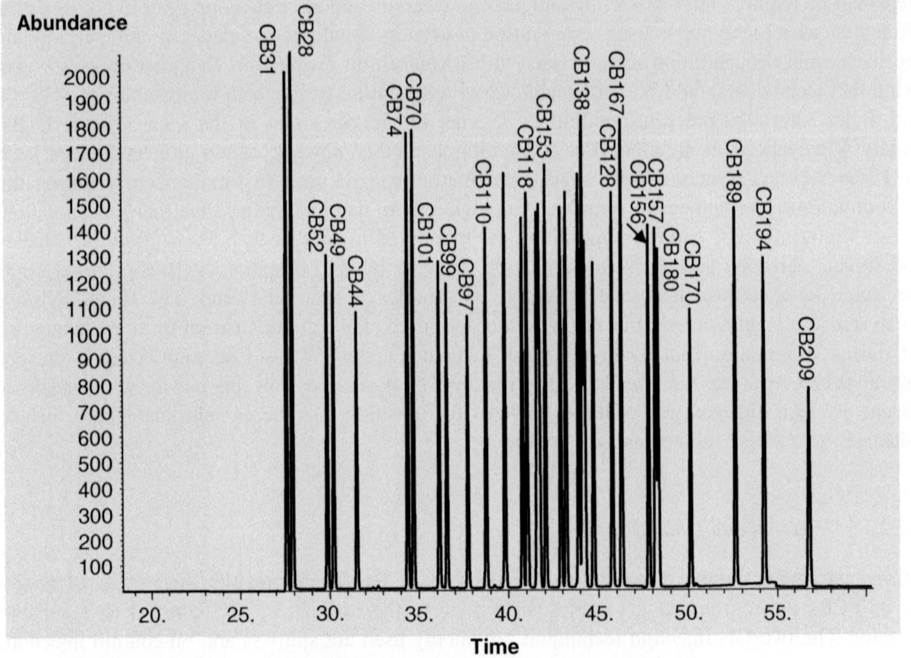

Figure 6.2 GC–MS chromatogram showing the separation of 32 PCB congeners using an HT-8 column (50 m × 0.22 mm × 0.25 μm; oven programme, 80 °C for 1 min; 20 °C min^{-1} to 170 °C hold 7.5 min; 3 °C min^{-1} to 290 °C hold 10 min).

and $^{13}C_{12}$-labelled internal standards can be used. Furthermore, ECNI is extremely sensitive for pentachlorinated to decachlorinated PCBs and is approximately tenfold more sensitive than ECD. For lower chlorinated congeners, the opposite is unfortunately true, although some exceptions are found for tetrachlorinated PCBs, such as CB77, which can be perfectly analysed by GC–ECNI-MS. Next to conventional GC–MS, the use of ion-trap with its MS^2 option (i.e. increased selectivity) is receiving increased attention. GC–ITMS is a less expensive alternative to HRMS, which is commonly used to determine PCDDs/PCDFs, and as such is also ideally suited for all CB groups (Eppe *et al.*, 2004).

A relatively new technique is the utilization of a second column of different stationary phase to help solve the problem of co-eluting compounds; this is referred to as multidimensional GC (or GC × GC) (Korytár *et al.*, 2002). By incorporating a second, smaller column with a different stationary phase the analytes are trapped for a short period of time on the second column before being released by heating the second column. The different times of elution from the columns and the responses give a three-dimensional chromatogram. This technique seems to be more complex because it involves proper selection of two columns as well as two temperature programmes, gas flow and timing release from the second column, but it has been used successfully for the separation of PCBs from complex samples.

6.2.3 Quality Assurance

In recent years there has been increased attention to quality assurance (QA), and there is a greater requirement for analytical methods to be accredited. As concentrations of some contaminants such as PCBs decrease in the environment, QA requirements will become more demanding. All analytical techniques are required to be fully validated and established; routine internal quality control (QC) programme should be applied to all analytical methods, to ensure the reliability of the analytical measurement. QC procedures should be well defined and include details on maintenance and calibration of equipment and conditions for sample storage and data handling and reporting. All methods and procedures used for an analytical technique should be well documented (standard operating procedures) to ensure reliable and comparable results are obtained. QC procedures should include system suitability checks, analysis of procedural blanks and analysis of appropriate reference materials (see Chapter 11). Procedural blanks should be measured with each batch of samples and should be prepared simultaneously using the same chemical reagents and solvents as for the samples. Their purpose is to indicate sample contamination by interfering compounds, which will result in errors in quantification. In addition, a laboratory reference material (LRM) or certified reference material (CRM) should be analysed within each sample batch. The analysis of an LRM is primarily intended as a check that the analytical method is under control and yields acceptable precision. A CRM may be analysed periodically in order to check the method bias; recoveries should be between 70 and 120%. The LRM must be homogeneous and well characterized for the determinands of interest within the analytical laboratory. Ideally, the LRM or CRM should be of the same matrix type (e.g. liver, muscle, mussel tissue) as the samples, and the determinand concentrations should be in the same range as those in the samples. The data produced for the LRM or CRM in successive sample batches should be used to prepare control charts. CRMs certified for PCBs are available from the Institute for Reference Methods and Materials, Geel, Belgium, and from the National Bureau of Standards and Technology (NIST, Washington, DC, USA) (de Boer and McGovern, 2001; Wise *et al.*, 2006). For analytical methods to be accredited, regular participation in intercomparison or proficiency exercises is required to provide an independent check on the performance of the analytical method (see further details in Chapter 11).

6.3 POLYBROMINATED DIPHENYL ETHERS

6.3.1 Introduction

PBDEs constitute a group of additive flame retardants (FRs) which were used in electrical equipment, textiles and furniture from the 1970s up until they were banned in Europe in the 2000s for reasons of environmental pollution. The use of penta- and octa-mix PBDE technical mixtures was banned in Europe (EU Directive 2002/95/EC) in 2004 and an EU ban on the use of the deca-mix PBDE product in electronics and electrical equipment became effective on 1 July 2008. The bromine industry has gradually terminated the production of all PBDE mixtures in North America. Between 1970 and 2000 it was estimated that approximately 3000 to 5000 t of the penta-mix PBDE product was produced in Europe, and between 9000 and 10 000 t imported to be used in products (Prevedouros *et al*., 2004). PBDEs are used as additives to polymers and resins and are thought to be more easily released to the environment than reactive FRs are. Commercial PBDE mixtures are classified according to the degree of bromination. The penta-mix contains mainly tetra- to hexa-BDEs, the octa-mix mainly hexa- to octa-BDEs and the deca-mix contains mainly BDE209. The penta-mix product was primarily used in furniture and upholstery, the octa-mix product in plastics and the deca-mix product in textiles and polymers.

Historically, PBDEs were released to the environment during their production and while manufacturing other products. Since the ban of PBDE formulations, the main sources will be from the disposal of products containing these chemicals. In addition, PBDEs may continue to leak out of treated material during everyday use of ordinary consumer products, which has been documented by high PBDE concentrations in the indoor environment (Jones-Otazo *et al*., 2005; Johnson-Restrepo and Kannan, 2009) and constitutes a diffuse source of these compounds to the environment. Atmospheric transportation is a major pathway for PBDEs into the marine environment; PBDEs have been found to concentrate in the Arctic and bioaccumulate in native animals and humans (de Wit *et al*., 2006). Volatilization from products still in use is thought to be the main source of PBDEs to the atmosphere today. Other possible pathways include direct discharge from point sources, such as storm waters and waste water.

PBDEs consist of two phenyl rings, connected by an ether bridge, each ring containing up to five bromine atoms. There are a possible 209 PBDE congeners, depending on the position and number of bromine atoms, with molecular weights ranging from 249 to 960 Da; see Figure 6.3. However, PBDE technical mixtures used as FRs contain only a limited number of these congeners (~20). PBDE congeners are named according to the Ballschmiter *et al*. (1992) short-hand system developed for PCB congeners. PBDE congeners currently analysed vary considerably; however, the congeners found in environmental samples are relatively consistent. Most laboratories analyse for the penta-mix compounds, containing the tetra- to hexa-PBDE congeners. These congeners are thought to be the most toxic and likely to bioaccumulate. In biota, the dominant congeners are normally BDE47, 99, 100, 153 and 154. BDE209 is less frequently measured, owing to the analytical difficulties, but can be the predominant congener in sediment samples. Law *et al*. (2006) proposed a minimum congener set for use when determining PBDEs to cover all three technical mixtures and

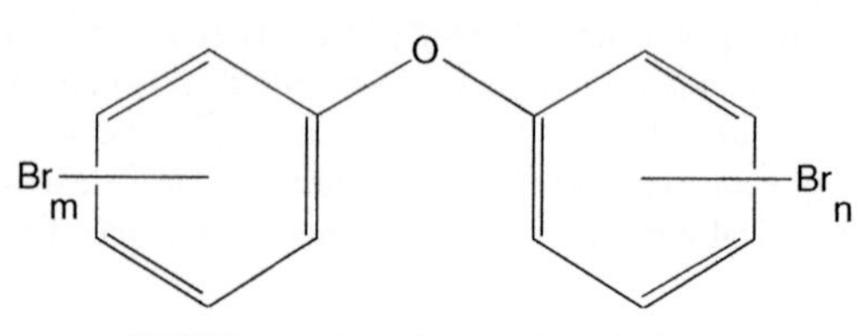

Figure 6.3 PBDE structure. PBDEs consist of two phenyl rings, connected by an ether bridge, each ring containing up to five bromine atoms.

those which are commonly found in biota and sediment samples. This list consisted of BDE28, 47, 99, 100, 153, 154, 183 and 209. In 2007, OSPAR (Working Group on concentrations, trends and effects of substances in the marine environment – SIME) recommended that nine PBDE congeners should be measured in environmental samples: BDE28, 47, 66, 85, 99, 100, 153, 154, 183 and that BDE209 should also be monitored in sediment (OSPAR Commission, 2007); see Figure 6.4.

PBDEs are hydrophobic and, therefore, tend to associate with particulate material and will accumulate in sediment particularly if it has a high organic carbon content. PBDEs are normally found at lower concentrations than PCBs, although high PBDE concentrations have been found close to industrialized areas and known sources (Morris *et al*., 2004). BDE209 is normally the dominant congener in sediment, though it can debrominate, producing lower brominated, more toxic and bioaccumulative PBDEs (Ross *et al*., 2009). PBDEs (except BDE209) are readily taken up by marine animals, both across gill surfaces and from their diet, and may bioaccumulate. Since the ban on PBDEs in Europe there is an indication that environmental concentrations are levelling off (Law *et al*., 2008).

6.3.2 Analytical Methods

Analytical methods for the analysis of PBDEs in environmental samples are well established and have been reviewed (de Boer *et al*., 2001; Covaci *et al*., 2003, 2007). The similarity in structure of the PBDEs to PCBs means that techniques used for the analysis of PCBs may also be applied to the analysis of PBDEs. More care is required for BDE209, which is often the major congener in sediment, as it is prone to photolytic and thermal degradation and is reactive. Blank problems can also be a major problem with BDE209, and detection limits are always poorer when compared with those for other congeners (approximately a factor of 10 higher) (de Boer and Wells, 2006). As concentrations can be very low and are likely to decrease, analytical techniques need to be able to detect PBDEs at very low levels.

6.3.2.1 Extraction Methods

Extraction methods used for other hydrophobic contaminants such as PCBs (Section 6.2.2.1) can often be applied to PBDEs, and it is possible to extract both PCBs and PBDEs together in the one

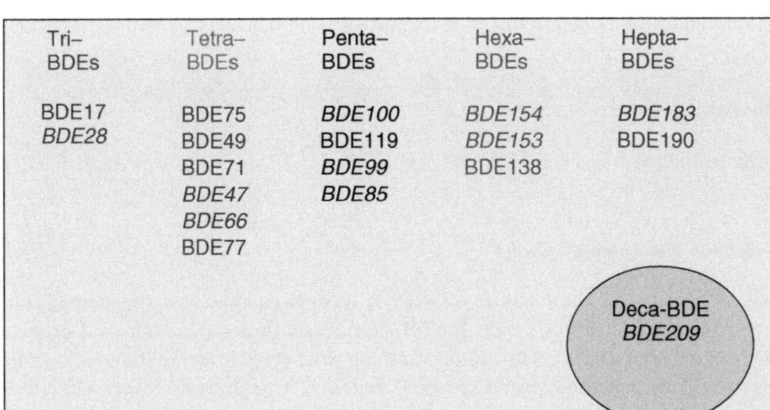

Figure 6.4 PBDEs commonly found in PBDE technical mixtures. Congeners in italics were recommended by OSPAR for analysis in environmental samples.

extraction. Similar methods are used for sediment and biota. Traditional extraction methods, such as Soxhlet, and the newer automated methods, such as PLE, have been used for the extraction of PBDEs from environmental samples (de Boer *et al.*, 2001; Covaci *et al.*, 2003, 2007). SFE has also been applied to PBDEs; however, the reproducibility was poor compared with Soxhlet (de Boer *et al.*, 2001; Covaci *et al.*, 2003). For Soxhlet extractions, hexane–acetone mixtures or toluene (particularly for BDE209) have been shown to give the best recoveries for the extraction of PBDEs combined with an extraction time of between 6 and 24 h (de Boer *et al.*, 2001). Hexane–acetone mixtures or toluene are also used with PLE (if no fat retainers are used) with an extraction time of ~10 min per sample. PLE or Soxhlet are, therefore, the preferred methods, with PLE having the advantage of using less solvent, being fully automated and taking less time than Soxhlet. However, problems have been highlighted with the analysis of BDE209, which can be lost during PLE through adsorption onto the extraction system tubing; therefore, recoveries tend to be poorer for this congener (Covaci *et al.*, 2003).

6.3.2.2 Clean-up Procedures

Environmental concentrations of PBDEs can be very low, especially in sediment; therefore, it is essential that a suitable clean-up is employed to remove compounds which may be present in the extract that may interfere with the subsequent analysis. Techniques used for PCBs can often be applied to PBDEs. The most commonly used clean-up methods involve the use of alumina or silica adsorption chromatography or GPC. Different techniques may be used, either singly or in combination. The choice will be influenced by the selectivity and sensitivity of the final measurement technique and also by the extraction method employed. PBDEs are stable under acid conditions; therefore, treatment with sulfuric acid or acid-impregnated silica columns may be used in the clean-up. As for PCBs, when applying GPC, two serial columns should be used for improved lipid separation with dichloromethane–hexane or cyclohexane–ethyl acetate as eluents. However, a second clean-up step is often required to separate the PBDEs from other organohalogenated compounds. For biota, destructive methods for lipid removal such as saponification have also been investigated; however, this method can result in the degradation of the higher brominated PBDEs and, therefore, is rarely used.

PLE methods with online clean-up previously described for PCBs (Section 6.2.2.1) can also be used for PBDEs (Björklund *et al.*, 2006). Silica impregnated with sulfuric acid, alumina and Florisil have all been used as fat retainers.

Sulfur should be removed from sediment extracts, in order to reduce interferences and to protect the detectors. Methods described in the PCB section (Section 6.2.2.2) can also be used for PBDEs.

6.3.2.3 Pre-concentration

Concentration techniques used for PCB analysis (Section 6.2.2.4) can also be used for PBDEs.

6.3.2.4 Detection and Quantification

Instrumental techniques for the analysis of PBDEs have been reviewed (Stapleton, 2006). The most common injection techniques used for PBDEs are splitless, on-column and programmable temperature vaporization (PTV). On-column has the advantage over splitless injection in that the potential for thermal degradation is reduced, although it requires a cleaner sample. Splitless injection needs to be combined with a pressure pulse, as otherwise degradation of BDE209 will occur due to the high injection temperature. For the same reason, the BDE209 analysis requires shorter columns to reduce the exposure of this compound to high temperatures as much as possible; columns with a length of 15 m or less are recommended (de Boer and Wells, 2006). For the other PBDEs, longer columns, as for PCBs, are recommended for a better separation. This means that samples need to be injected twice if BDE209 is to be determined in addition to the other BDEs. Mainly nonpolar columns are used; for example, DB5, HT-8, RTX1614 and STX-500. Korytár *et al.* (2005a) provide comprehensive information on various capillary columns used for PBDE analysis. One of the more difficult separations, even by GC × GC, is that of BDE154 with the brominated biphenyl BB153 (2,4,5,2′,4′,5′-hexabromobiphenyl). This co-elution can easily lead to false positive values for BDE154. Ackerman *et al.* (2005) used a 30 m DB-5 MS (Agilent Ltd) column to separate 39 PBDE congeners. Thinner films (0.1 μm) on the GC column have also been found to improve the determination of BDE209. The presence of a hump or rising baseline before BDE209 is an indication of degradation during injection. To minimize this degradation, the GC liners and injection syringe should be changed regularly and the liner silanized. The presence of lower brominated PBDE congeners (nona-, octa- and eventually other lower brominated BDEs) indicates possible degradation during extraction and clean-up (Korytár *et al.*, 2005b; Covaci *et al.*, 2003). The choice of guard column can also have an effect on the degradation of BDE209 during analysis. Siltek deactivated fused-silica retention gaps (Thames Restek, Buckinghamshire) have been shown to give the best and most consistent results for PBDEs, especially for BDE209.

GC–MS is the preferred technique for quantification of PBDEs in environmental samples, using either EI or ECNI (Stapleton, 2006). Although GC–HRMS with EI is the best method to unambiguously identify and quantify PBDEs in environmental standards, the expense and limited availability means that most laboratories have used GC–ECNI-MS. The majority of recent publications use this technique for the quantification of PBDEs in environmental samples. Ackerman *et al.* (2005) compared the detection of the PBDE congeners using low-resolution GC–MS in both EI and ECNI modes. The lower brominated PBDEs (mono- and di-BDEs) showed better sensitivity in EI mode. However, the higher brominated PBDEs (more than three bromine atoms) gave better sensitivity using ECNI mode. The base ions detected using ECNI are the bromine ions ($m/z = 79/81$) for the tri- to hepta-PBDEs. PBDE congeners show the typical ^{79}Br (50.5%) and ^{81}Br (49.5%) isotope distribution pattern. One of the drawbacks of the ECNI mode is that isotopically labelled standards (^{13}C) cannot be used as internal standards for quantification purposes as ECNI spectra of PBDE congeners normally only consist of the bromide ions. However, mono-fluorinated PBDEs, such as FBDE28, 100 and 160, are commercially available and may be used as internal standards, though they should be checked for the presence of BDEs (Korytár *et al.*, 2005b). Larger fragment ions, necessary for confirmation, are only found for BDE209. These are formed by the cleavage of the ether bond to give the pentabromo phenoxy ion ($m/z = 484/486$). An internal standard method including ^{13}C-labelled BDE209 standard should be used for the quantification of BDE209, which gives a base ion at m/z 494.

An illustrative chromatogram for PBDEs is shown in Figure 6.5.

Degradation of BDE209 during analysis can be avoided by using an liquid chromatography (LC) method instead of GC. A useful LC–APPI (atmospheric pressure photoionization)-MS method was introduced by Abdallah *et al.* (2009). This method offers a good sensitivity for BDE209, while degradation problems are completely avoided.

6.3.3 Quality Assurance

QA procedures described for PCBs will also apply for PBDEs. However, special precautions are required in the laboratory when analysing PBDEs owing to their sensitivity to UV light. PBDEs are prone to photolytic degradation; if exposed to UV light, debromination can occur, especially for BDE209 (Covaci *et al.*, 2003; de Boer and Wells, 2006). Therefore, incoming light to the

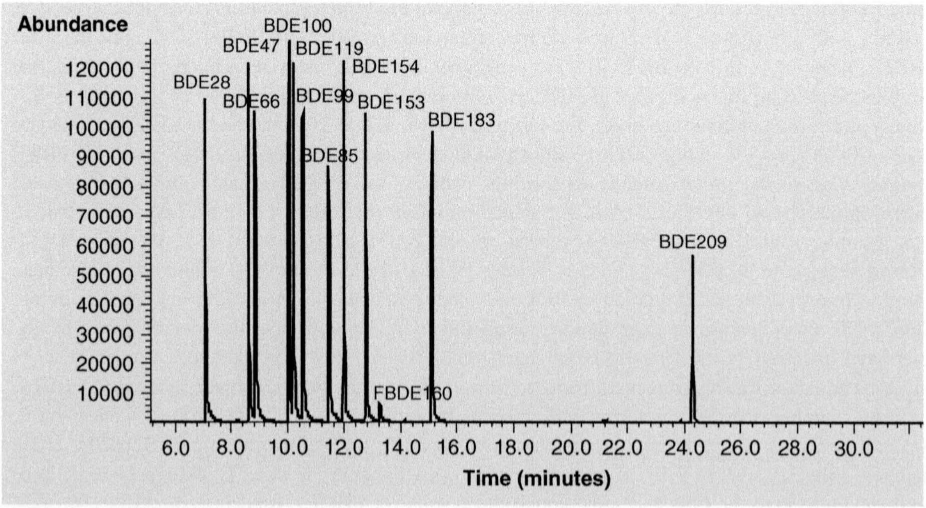

Figure 6.5 GC–MS chromatogram showing the separation of 10 PBDE congeners using an RTX-1614 column.

laboratory should be minimized by placing UV filters on the windows and over fluorescent lights. All calibration and spiking standards should be stored in amber glassware.

Plastics can contain PBDEs and should be avoided. Blank problems can be a major issue for PBDE analysis, particularly for BDE209, as it can stick to glassware (or any other chemically active sites), resulting in contamination of glassware. Heating of glassware in an oven (e.g. at 450 °C overnight) can also be useful for removing PBDE contamination; the use of older, scratched glassware should be avoided. Furthermore, BDE209 is associated with dust particles and can be a source of contamination in the laboratory. Therefore, it is essential that the laboratory is as dust free as possible; an ionizer placed in the laboratory may help.

6.4 EMERGING CONTAMINANTS

6.4.1 Introduction

Owing to the combination of an exploding world population and an ongoing industrial revolution, mankind is nowadays exposed to a large variety of chemicals. Many of those also occur in the marine environment. Some of these, such as the previously discussed PCBs and PBDEs, have been known already for several decades. However, while these are slowly disappearing due to bans and other regulations, others are emerging. Three classes have been selected for further discussion in this chapter: other brominated flame retardants (BFRs), phosphorus-based flame retardants (PFRs), chlorinated paraffins (CPs) and siloxanes.

6.4.2 Other Brominated Flame Retardants

In addition to the PBDEs, and partly due to the termination of their production, other BFRs are being used in increasing amounts. The highest production volumes of all BFRs are attributed to tetrabromobisphenol-A (TBBP-A). In comparison with this substantial production, environmental concentrations are relatively low. This is, on the one hand, due to the reactive binding of this BFR, which makes direct release from products less easy. On the other hand, its analysis is more difficult, as, owing to its polar character, derivatization followed by GC–MS or LC–MS analysis is required. Finally, the polar character of this BFR makes it less bioaccumulating and the fat–water equilibrium will be more to the aquatic side in comparison with most PBDEs. Nevertheless, there are reports on the presence of TBBP-A in biota, although mainly in freshwater organisms. TBBP-A was found in sediments and biota from some polluted estuaries along the North Sea (Morris *et al.*, 2004).

Hexabromocyclododecane (HBCD) is another BFR that is produced in major volumes. This BFR is strongly bioaccumulating and persistent. HBCD has initially been analysed by GC–ECD or GC–MS. However, HBCD consists of three diastereomers that cannot be separated by GC. On top of that these diastereomers can be converted into each other above 160 °C. This means that GC data are less reliable, as the response factors in GC–ECNIMS are also different. Results of interlaboratory studies normally show larger coefficients of variation (CVs) for this compound when analysed by GC in comparison with most PBDEs. Consequently, LC–MS is recommended for the analysis of HBCD (Morris *et al.*, 2006). This can be done in a diastereomer-specific way, as all three diastereomers can be separated by LC. There are even more enantiomers of HBCD which can also be separated by using chiral stationary phases (Gómara *et al.*, 2007, Hühnerfuss and Sah, 2009). Interestingly, the LC analysis of HBCD can be combined with that of TBBP-A (Morris *et al.*, 2004). Both electrospray ionization (ESI) and atmospheric pressure chemical ionization (APCI) can be used as ionization techniques, but ESI provides the highest sensitivity both for TBBP-A and HBCD. An alternative technique is atmospheric pressure photoionization mass spectrometry (APPI-MS), which offers a comparable sensitivity (Abdallah *et al.*, 2009). This technique can also be used for deca-BDE detection, which takes away the difficulties with regard to thermal degradation of BFRs. Extraction and clean-up techniques for HBCD and TBBP-A are more or less similar to those used for the PBDEs. Only for TBBP-A is one extra partitioning step needed to ensure that TBBP-A remains in the non-dissociated state.

There are a number of other BFRs that are being introduced to serve as alternatives for the PBDEs. These may not have come yet to detectable environmental concentrations, but some may have the potential to do so. Lopez *et al.* (2011) detected other BFRs, such as bromophenols, intermediates in FR formulation like bromoanilines, and their brominated and non-brominated by-products, such as bromoanisoles, bromotoluenes, bromoalkanes and 1,5,9-cyclododecatriene. These compounds have very low odour and taste thresholds, below sub-nanogram per litre. 2,6-Dibromophenol and 2,4,6-tribromoanisole could be detected at levels of 1.6 and 2.1 ng L^{-1}, which is higher than their respective odour thresholds. All aforementioned target compounds were found in sediment from the Western Scheldt estuary in the Netherlands at concentrations between 0.3 ng L^{-1} for 2,4,6-tribromotoluene and 15 ng L^{-1} for 2,5-dibromotoluene. Lopez *et al.* (2011) tested a number of methods for the analysis of these compounds. The combination of solid-phase extraction (SPE) and GC–ECNI-MS was the most sensitive and reliable method for the determination of most of the brominated compounds, whereas SPE in combination with GC–EI-MS was the optimal method to quantify the three isomers of 1,5,9-cyclododecatriene.

Kierkegaard *et al.* (2004) identified and quantified decabromodiphenylethane (DBDPE) in various matrices. DBDPE can be used as an alternative for deca-BDE. It was also found in sediment from the Baltic, Sweden (Ricklund *et al.*, 2010) and in sediment from the Western Scheldt (Kierkegaard *et al.*, 2004), the Netherlands. The method for the determination of DBDPE is comparable to that of deca-BDE. Similar problems due to thermal instability are met during GC analysis. Other BFRs that have been found in the environment include pentabromotoluene (De Wit *et al.*, 2009), 2,3,4,5,6-pentabromoethyl-benzene (PEB) and 1,2-*bis*-(2,4,6-tribromophenoxy)ethane (Hoh and Hites, 2005). Combined brominated/chlorinated FRs, such as tetrabromo-*o*-chlorotoluene (TB*o*CT) and pentabromochlorocyclohexane (PBCCH), and brominated/phosphorus-based FRs, such as *tris*(2,3-dibromopropyl)phosphate (TDBPP), are also reported. Many technical FR mixtures contain combinations of phosphorus-based and brominated FRs, such as SAYTEX® RX-8500 (Albemarle, Baton Rouge, Louisianan, USA), a mix of a reactive BFR, aryl phosphate and triphenyl phosphate (TPP), SAYTEX® RZ-243 (Albemarle), containing tetrabromophthalate, an aryl phosphate and TPP, and Firemaster® 552 (Great Lakes Chemicals, UK), a mix of halogenated aryl esters, triaryl phosphate, isopropylated triarylphosphate and TPP. Obviously, methods for most of these compounds have either not been developed or still need proper validation. The analysis of these novel BFRs, and their environmental fate and behaviour, has been reviewed recently by Covaci *et al*. (2011).

6.4.3 Phosphorus-Based Flame Retardants

Phosphorus flame retardants (PFRs), which have already been used for over 150 years and which were responsible for 20% of the FR consumption in 2006 in Europe, are often proposed as alternatives for BFRs. PFRs can be divided into three main groups: inorganic, organic (non-halogen) and halogen-containing PFRs. The latter two groups contain hundreds of compounds. We have selected three compounds for further discussion, these being the most frequently reported ones in the environment, with relatively high production figures: TPP, *tris*(2-chloroethyl)phosphate (TCEP) and *tris*(chloropropyl) phosphate (TCPP). Data on the occurrence of these compounds in air, water and sediment is available, but there are less data on the marine environment and less concerning the occurrence in biota. In all matrices the dominant PFR was TCPP. TPP is not considered to meet the persistence or very persistent criteria (half-life >40 days and >60 days in freshwater respectively and half-life >120 days and >180 days in freshwater sediment (Pakalin *et al*., 2007). TPP concentrations in the River Ruhr (Germany) ranged up to 40 ng L^{-1} (Andresen *et al*., 2004). In general, the non-halogenated PFRs will be less persistent and mineralize relatively easily in comparison with the chlorinated PFRs.

TCPP is able to accumulate in food chains (Leisewitz *et al*., 2000). Kawagoshi *et al*. (2002) tested the degradation of some PFRs in leachates from a sea-based solid waste disposal site and found that TCPP showed low degradability. TCPP concentrations of 80–100 ng L^{-1} were found in River Rhine (Germany) water and 100 ng L^{-1} in the River Lippe (Germany), while TCPP concentrations in the River Ruhr (Germany) varied between 20 and 200 ng L^{-1} (Andresen *et al*., 2004). River sediments show concentrations of up to 180 µg kg^{-1} dw. Eight different PFRs have been found in sediments from Liverpool Bay, the rivers Mersey and Tees, all in the UK (Kim *et al*., 2007).

TCEP concentrations in the water samples from the River Danube were 13–23 ng L^{-1}. In the River Schwechat (Austria) and the River Liesing, concentrations of 130 and 37 ng L^{-1} were found (Martínez-Carballo *et al*., 2007). In the River Schwechat sediment the TCEP concentration was 160 µg kg^{-1} dw.

6.4.3.1 Analysis

Andresen *et al*. (2004) analysed TPP, TCPP and TCEP in surface waters. SPE was used for TCEP extraction, using methyl-*tert*-butylether and toluene. D_{27}-TBP was added as internal standard. The final volume of the extract was 1 mL. The extracts were analysed by GC–MS. The injection mode used was PTV-splitless and the injection volume was 1 µL. The column used was a DB-5 MS column (30 m × 0.25 mm × 0.25 µm). Recoveries of TCEP were 67%, with residual standard deviation RSD = 15%.

Rodriguez *et al*. (2006) tested the feasibility of solid phase micro-extraction (SPME) for the determination of several PFRs, including TCEP, TCPP and TPP in water with GC-NPD (nitrogen–phosphorus detection). Maximum efficiencies were achieved using PDMS-DVB and PDMS-CAR-DVB fibres. PDMS-DVB showed no carryover effect, whereas important memory effects (~40 % of the peak area in the first desorption) were noticed for TPP with PDMS-CAR-DVB. With the SPME method the RSDs were 3–13%. Limits of quantification (LOQs) were 0.010–0.025 µg L^{-1}.

Results of the SPME method were compared with an SPE method with OASIS HLB cartridges (60 mg), which were eluted with ethyl acetate. Recoveries for the SPE method were 83–105%. LOQs of the method ranged from 0.005 to 0.010 µg L^{-1} with RSDs of 1.9–16.7%. Both extraction methods were used in combination with a GC-NPD. Two capillary columns, DB-5 (30 m × 0.25 mm × 0.25 µm) and SPB-1701 (30 m × 0.25 mm × 0.25 µm), were compared. TBEP and TPP could be separated on an SPB-1701 column but not on a DB-5 column. However, the retention time of TCEP at the SPB-1701 column matches exactly that of an isomer of TCPP. LOQ values were 5–10 µg L^{-1} and the repeatability of the injection for standards at different concentration levels was around 2–3%.

Ishikawa *et al*. (1985) developed an analytical method for determination of TPP, TCPP and TCEP in sediment. Sediment was collected with an Ekman–Berge dredge and collected in 3 L glass bottles that had been cleaned with acetone. Extraction was performed with acetone. A Florisil column (10cm × 1 cm) was used for clean-up. Analyses were performed with GC-FPD (flame photometric detector). Recoveries were 78–95% and the limits of detection (LODs) were 2–10 µg kg^{-1}.

Martínez-Carballo *et al*. (2007) analysed TCEP, TCPP and TPP in sediment from Austria. All glassware was precleaned with acetone and baked at 250 °C for at least 10 h. The sediment samples were dried at 30 °C and sieved (0.63 µm). Extraction was performed with ultrasound-assisted solvent extraction with ethyl acetate–acetonitrile (30:70) (v/v). Analysis was performed with LC–MS/MS. The analytical column used was a Luna C8 column (150 mm × 2 mm × 5 µm). Eluents were water and methanol, each modified with 0.1% (v/v) formic acid and 10 mM ammonium acetate, and the injection volume was 10 µL. A triple quadrupole MS in the positive ion mode was used. Quantification of all compounds was made by multiple reaction monitoring (MRM). LOQs were 0.48–11 µg kg^{-1}, and recoveries ranged from 74 to 104%.

García-López *et al*. (2009) analysed PFRs in sediments from rivers located in Spain and the USA. Ten PFRs were analysed, including TCEP, TCPP and TPP. MAE was used for the extraction. Ethyl acetate, dichloromethane and acetone were tested as extraction solvents. Extraction with acetone followed by extraction with acetonitrile provided the best results for the extraction of PFRs with MAE. Silica cartridges of 50 mg were used for clean-up of the extracts. The final extracts were concentrated to 200 µl, of which 2 µl was injected in the gas chromatograph. Analyses were performed using GC–ICP-MS equipped with an HP-5 capillary column (30 m × 0.32 mm × 0.25 µm). Conventional and pulsed splitless were compared as injection modes. Pulsed splitless reduces the chance of analyte decomposition and adsorption on the internal surface of the liner. In addition, it provides a narrow injection band in the head of the GC column. A two- to three-fold improvement in the peak height of target compounds was achieved compared with conventional splitless.

Nitrogen was used as additional gas in the central argon plasma channel to enhance phosphorus ionization. The addition of small amounts of alternative gases (N_2, O_2, He) is usually employed to improve the detection limits of elements with high ionization potentials, such as phosphorus (García-López *et al*., 2009). Recoveries for the MAE-ICP-MS method were between 78 and 105% and the LOQs varied from 2 to 4 µg kg^{-1}.

According to Björklund *et al*. (2004), advantages of using GC-NPD are low LODs and high sensitivity for phosphorus-containing compounds. García-López *et al*. (2009) mention as a disadvantage of NPD an unsatisfactory selectivity for PFRs. Another disadvantage is that it does not offer positive identification. MS is a more powerful identification tool (Björklund *et al*., 2004). It has a higher selectivity and the option of using isotope-labelled compounds for quantification.

A disadvantage of GC–MS-EI is the extensive fragmentation of alkylated phosphates that disables a proper identification. GC–MS-positive ion chemical ionization (PICI) is needed for identification of the alkylated phosphates (Carlsson *et al*., 1997; Björklund *et al*., 2004). However, GC–MS-PICI has a limited sensitivity (García-López *et al*., 2009). In the study of Hartmann *et al*. (2004) on GC–MS in single-ion monitoring (SIM) mode, interfering peaks occurred often in samples, although these were not present in standards. This resulted in the need to use the second, third or fourth most abundant peaks for identification and quantification of most analytes, which again resulted in higher LODs.

When using LC–MS detection for phosphorus analysis, the formation of stable complexes with metal cations such as Na^+ (e.g. $[M + Na]^+$ and $[2M + Na]^+$) that may be present in the samples is disadvantageous. The relative abundance of these complexes is influenced by the concentration of metal cations and the pH. By adjusting the pH the complexes formed with metal cations are more efficiently suppressed, which results in lower LODs (0.4–19 pg m^{-3}) (Tollbäck *et al*., 2006).

6.4.3.2 Precautions

PFRs may occur in laboratory air. Laboratory equipment can be contaminated if no precautions are taken. Soaking all glassware in a solution of ethanol containing 5% (w/w) NaOH and rinsing afterwards with H_2O, ethanol and acetone can reduce blank problems. To avoid blanks from the sampling devices, glass-fibre filters can be ultrasonicated in methanol, acetone and dichloromethane, cellulose support discs can be ultrasonicated in dichloromethane (Staaf and Östman, 2005), polyurethane foams can be washed with H_2O, acetone, and dichloromethane, and Soxhlet-extracted for 12 h in dichloromethane (Carlsson *et al*., 1997; Staaf and Östman, 2005). The use of glassware should be avoided when extracting or cleaning samples for PFR analysis.

6.4.4 Chlorinated Paraffins

CPs have not received much attention of the environmental analyst, while the worldwide production surpasses by far that of PCBs and substantial environmental levels are being reported. The enormous complexity of the mixture with tens of thousands of congeners may be one of the reasons for that. The environmental analysts have been deterred by this complexity, which demands qualitative and quantitative methods at a level beyond the performance characteristics of their instruments. However, nowadays, selectivity and sensitivity of analytical instruments are rapidly improving and have reached performance levels that were previously unthought of. That brings, though still not easy, a reliable CP analysis within reach.

6.4.4.1 Environmental Occurrence

CPs are normally divided into three sub-groups: short-chain (SCCPs), medium-chain (MCCPs) and long-chain (LCCPs) chlorinated paraffins. SCCPs have carbon chains of C_{10}–C_{13}, MCCPs of C_{14}–C_{17} and LCCPs of C_{20}–C_{30}. CPs are sometimes also called polychlorinated alkanes (PCAs). Kraemer and Ballschmiter (1987) were the first to report the presence of CPs in the marine environment. They found concentrations of ~50 µg L^{-1} in surface water of the Sargasso Sea near Bermuda, and concentrations decreasing to 3 ng L^{-1} at depths of 1200 m. More recently, Castells *et al*. (2008) reported SCCP concentrations in coastal sediments along the Catalonia coast near Barcelona (Spain) and under the influence of wastewater treatment plant effluents and sewage sludges of up to 2 mg kg^{-1}. Reth *et al*. (2005, 2006) reported SCCP–MCCP concentrations and homologue patterns in a range of fish species from the North Sea, the Baltic Sea and the North Atlantic. The SCCP and MCCP concentrations were higher in Atlantic cod (*Gadus morhua*) from the Baltic (19–143 µg kg^{-1} ww and 25–106 µg kg^{-1} ww respectively) than in the same species from Iceland (11–70 µg kg^{-1} ww and 7–47 µg kg^{-1} ww respectively). North Sea dab (*Limanda limanda*) had higher concentrations of SCCP–MCCPs than cod or flounder (*Platichthys flesus*). SCCP–MCCP concentrations in Arctic char from Ellasjøen (11–27 µg kg^{-1} ww and 13–43 µg kg^{-1} ww in liver respectively) were much lower than total PCBs, which averaged from about 500–1000 µg kg^{-1} ww in separate collections (Evenset *et al*., 2004). Two species of seabird, little auks (*Alle alle*) and kittiwakes (*Rissa tridactyla*), collected at Bjørnøya had a similar range of SCCP–MCCPs concentrations in liver and muscle as the char liver (Muir, 2010).

6.4.4.2 Analytical Methods

Tomy (2010) recently reviewed analytical methods for the determination of CPs in environmental matrices. Soxhlet extraction and PLE are the most frequently used methods for CP extraction from biota and sediments. Three methods have been used for extracting or pre-concentrating CPs in water (i) SPE (Castells *et al*., 2004; Dick *et al*., 2010), (ii) SPME (Castells *et al*., 2004) and (iii) liquid–liquid extraction (LLE; Moore *et al*., 2004). GPC (Tomy *et al*., 2000; Fridén *et al*., 2005; Thomas *et al*., 2006; Ismail *et al*., 2009), concentrated sulfuric acid treatment, or sulfuric acid–silica gel column chromatography (Iino *et al*., 2005) have all been used to remove co-extracted lipids from extracts. Adsorption chromatography on Florisil, silica and alumina have been used to separate CPs from interfering compounds (Tomy *et al*., 2000; Moore *et al*., 2004; Thomas *et al*., 2006), but Rieger and Ballschmiter (1995) reported dehydrochlorination of CPs on alumina. Recently, a carbon-skeleton method, in which CPs are also dehydrochlorinated to linear alkanes by a palladium catalyst in the liner, was presented by Ricci and Pellizzato (2010). This method results in much simpler chromatograms, but may lead to false positive when alkanes are present in the sample.

Korytár *et al*. (2005c) applied comprehensive two-dimensional GC (GC × GC) to improve CP separations. While the separation of the congeners was not complete, an ordered structure with four parallel groups of peaks corresponding to chlorine content was observed. Both EI-MS and ECNI-TOF-MS were used. ECD has not been widely adopted as the detector of choice for single-column GC analysis of CPs, but its application in combination with GC × GC could be useful.

The first report of the use of ECNI HRMS was in 1997 (Tomy *et al*., 1997). This method was based on measuring the $[M - Cl]^-$ ions of each CP congener in the SIM mode at a resolving power of 12 000. Interferences from other co-extracted organochlorine compounds, like PCBs, chlordane, toxaphene and other organochlorine pesticides, were not observed. One of the big advantages of this approach is the ability to measure the relative molar amounts of each formula group. This is important because the molecular composition of samples can be very different to that of commercial mixtures that are used as external standards for quantification purposes (Tomy *et al*., 1997, 2000; Tomy and Stern, 1999). Coelhan *et al*. (2000) analysed SCCPs in fish by low-resolution ECNI-MS using a technical formulation and several individual CP standards, of varying carbon chain length and chlorine content. The results varied by a factor of 10 depending on the standard used. Coelhan *et al*. (2000) quantified the extent to which errors can arise when using SCCP standards whose chlorine content differed from that of the sample. Using three standards of varying chlorine content, concentration differences of >1000% were observed when quantifying the same sample. Tomy *et al*. (1997) have shown that it is possible to correct for these differences. They introduced a numerical correction factor, average molar mass (AMM), to correct for differences between sample and standard. Using the AMM, the error in quantifying one technical mixture using another as an external standard was less than 40% (Tomy *et al.*, 1999).

Moore *et al.* (2004) reported on the use of metastable atomic bombardment (MAB) ionization and HRMS in the positive ion mode for the detection of CPs. With inert gases like argon, MAB ionization generates ions via an electrophilic reaction of a metastable atom with the analyte of interest. Using MAB, the two most intense ions in the $[M - Cl]^+$ ion cluster of CP congeners are monitored. A major advantage of MAB over ECNI is the detection of congeners of a low chlorination degree. More recently, Reth *et al.* (2005) reported a method that compensates for the difference in the profiles between sample and standard. The method makes use of the linear relationship between the ECNI response factor and the chlorine content, which has only been validated on a quadrupole MS.

Only two interlaboratory studies were carried out for SCCPs, and none for MCCPs or LCCPs (Tomy *et al.*, 1999; Pellizzato *et al.*, 2009). Pellizzato *et al.* (2009) distributed an industrial soil extract. Commercially available single CP congener solutions were used by the six participating laboratories. With the exception of a single outlier laboratory, the results of four laboratories that employed MS were in generally good agreement with a range of only 10 ± 1 to 15 ± 2 ng L^{-1}. The one laboratory that employed an atomic emission detector reported on a value that was fourfold greater than the mean of the laboratories that used MS detection. Part of the discrepancy found in this study was thought to be due to the choice of detection and quantitative approaches used. Obviously, the number of participants is too low to draw firm conclusions. In an earlier interlaboratory study on SCCPs by Tomy *et al.* (1999), seven participants were supplied with an injection-ready extract (biota) along with two technical solutions. All laboratories used single-capillary GC coupled with ECNI-MS; two laboratories used HRMS at a resolving power of 10 000–12 000. One laboratory also used ECD. Surprisingly, the results of the synthetic C_{10} solution whose carbon and chlorine profile differed quite markedly to that of the supplied technical mixtures were better (CVs: 2–74%) than results of the technical mixture that was treated as a sample (error range: 30–310%) (Tomy *et al.*, 1999; Tomy, 2010). It is clear from both interlaboratory studies that differences in measurement data on SCCPs can be substantial. The choice of external standard employed can substantially contribute to the overall error. Recently, impurity-free synthetic C_{10}–C_{13} solutions have become available (Tomy, 2010). However, other confounding variables, like choice of extraction and clean-up procedures, have not been quantitatively assessed and are likely to contribute to the uncertainty in CP data.

6.4.5 Siloxanes

Siloxanes are building blocks for silicone products. Siloxanes contain a silicone–oxygen backbone and their structure can be linear or cyclic; see Figure 6.6. Siloxanes have some useful properties, like high thermal stability, water repellency, good dielectric properties, radiation resistance, physiological inertness, good damping behaviour, low surface tension and smooth texture. Therefore, siloxanes are widely applied in paper and mould-release agents, medical and cosmetic products, polishes, coolants and dielectric fluids for electric systems and personal care products (Lassen *et al.*, 2005). The commonest forms of siloxane polymers are fluids, gels, elastomers and resins. Each group is classified by the length of the backbone, the extent of cross-linking and the type and number of organic groups attached to the silicon atoms. Silicone fluids are usually straight chains of polydimethylsiloxanes (PDMSs). Silicone gels are lightly cross-linked PDMS fluids (Silicon Technology Dow-Corning, 2008). Low molecular weight (MW) siloxanes have an MW < 1000 and high MW siloxanes such as PMDS an MW > 1000. The estimated worldwide annual production volume of silicone materials (fluids, elastomers and resins) in 2002 was 0.85 million tonnes, with a growth rate of 2% per year from 1998 to 2002. D4 (octamethylcyclotetrasiloxane) is commercially

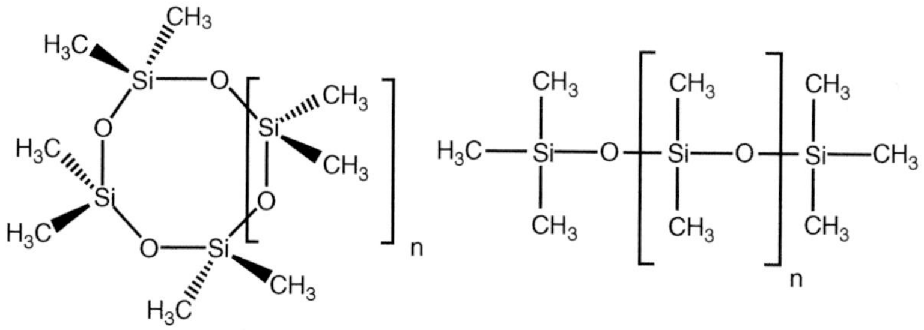

Figure 6.6 The cyclic and linear structures of siloxane.

the most important cyclic siloxane, with a production volume of 136×10^6 kg in 1993, and still growing. Some 80% of the D4 produced is used as an intermediate for the production of polymeric dimethylsiloxanes and the other 20% is used in personal care products and other formulations (Varaprath *et al.*, 2009).

Although not halogenated, siloxanes may show a persistent and bioaccumulating behaviour in the environment. PMDS is intensively used as an antifouling agent on ships. Rough calculations show that with a production volume of 4000 t in 2007 the major shipping lanes will be coated with 2 mm of PDMS within 10 years. This will have a severe effect on sediment-dwelling organisms. The PDMS that is tightly bound to the sediment can change the physical characteristics of sediment, which could influence the bioavailability of energy sources, fat-soluble vitamins and organic contaminants.

6.4.5.1 *Environmental Occurrence*

The highest siloxane concentration, 2300 $\mu g\,kg^{-1}$ dw (sum of D4, D5 (decamethylcyclopentasiloxane) and D6 (dodecamethylcyclohexasiloxane)) was observed in sediment from Roskilde (Denmark). For the sum of the linear siloxanes L4 (decamethyltetrasiloxane) and L5 (dodecamethylpentasiloxane) this was 87 ng g^{-1}. In sediment, D5 also predominates. The siloxane concentrations (sum D4, D5, D6) in fish livers from northern European locations varied from <0.5 to 100 $\mu g\,kg^{-1}$ ww. No significant differences in concentrations were observed between marine and freshwater livers. D5 concentration found in seals and pilot whales ranged from 10 to 42 $\mu g\,kg^{-1}$ ww. The toxic effects found for D4 and the relatively high concentrations observed in the environment ask for more information about bioaccumulation. Stevens *et al.* (2001) produced a review on the fate and effects of PMDS in the marine environment.

6.4.5.2 *Analytical Methods*

The analytical methods used to measure the PDMS, low MW siloxanes and the polar metabolites are summarized by Varaprath *et al.* (2006). PDMS is most commonly analysed with ICP atomic emission spectroscopy (ICP-AES) by measuring the silicon (Si) concentration in the sample (Carpenter *et al.*, 1995; Fendinger *et al.*, 1997). By coupling GPC to ICP-AES, not only can the Si concentration be determined, but also the MW. Another advantage of using GPC is that inorganic silicates that are naturally present in the environment can be separated from the PDMS, which minimizes the interference during the ICP-AES analysis (Varaprath *et al.*, 2006). In most cases 350 cSt PDMS is used for the calibration curve and octaphenylcyclotetrasiloxane as a marker to correct for retention time differences (Dorn and Skelly Fram, 1994).

The more polar siloxanes can only be separated by GPC and measured by GC if they are derivatized. Another option is measuring them with HPLC (Dorn and Skelly Fram, 1994). The advantage of supercritical fluid chromatography is that it extends the range of GC and yields nearly identical resolution and selectivity, while it is also possible to measure volatile and thermally labile compounds (Varaprath *et al*., 2006). Other detection methods used for qualitative analysis are infrared and nuclear magnetic resonance (NMR). NMR, in particular, is widely used for the characterization of silicon compounds (Varaprath *et al*., 2006). More detailed information about different detection systems was presented by Varaprath *et al*. (2006).

6.4.5.3 Background Problems

Some parts of the gas chromatograph, like the septum and the column, normally contain silicone compound which can easily cause interferences. Septa are normally made of cross-linked PDMS. During use septa are exposed to high temperatures and can start to bleed (De Zeeuw, 2005). Bleed products found are cyclic siloxanes from D3 up to D12. Those bleed products produce ghost peaks, which interfere in the analysis of low MW siloxanes. Even bigger background problems will appear when septum particles are deposited in the liner. Therefore, De Zeeuw (2005) recommended using pre-pierced septa in combination with tapered needles, or to avoid septa and use valve injection or Merlin MicroSeal® (made from high-temperature-resistant fluorocarbon elastomers) instead.

Horii and Kannan (2008) tested the release of siloxanes from several types of septum at various inlet temperatures with GC–EIMS. They reported decreasing background levels of D4 to D6 with decreasing inlet temperatures. An inlet temperature of 200 C showed the best compromise between low background and good sensitivity. The decrease of the inlet temperature from 250 °C to 200 °C caused a background level decrease of 79%, while the response of the standard cyclic siloxanes mixture decreased only by 10%; see Table 6.1.

GC columns also contain siloxane compounds, which can bleed and give background problems. Horrii and Kannan (2008) tested the siloxane background levels for two low-bleed capillary columns (DB-1 and DB-5 ms) and observed that the release of siloxanes from a conditioned,

Table 6.1 Background levels and relative peak areas of D4–D6 at different inlet temperatures and background levels of different septa at an inlet temperature of 200 °C (Horii and Kannan, 2008).

Inlet temperature (°C)	Background[a] (pg)			Standard peak area (%)
	D4	D5	D6	
250	6.0	1.6	1.6	100
200	1.7	0.6	0.3	90
150	0.8	0.3	0.9	80
100	0.6	nd	nd	72
Septa at 200 °C				
Restek BTO septum	0.8	0.3	0.2	
Restek thermolite septum	1.6	0.6	0.3	
Agilent advanced green septum	1.7	0.6	0.3	
General purpose red septum	1.9	1.0	0.8	
General purpose grey septum	4.1	6.0	2.1	

[a]nd: not detected.

low-bleed column is minimal in comparison with that from septa. Varaprath *et al.* (2006) emphasized that extracts must be dried before GC analysis.

Other materials that may contribute to siloxane background levels are extraction solvents, liners, syringe lubricants, gloves, hand lotions and even diffusion pump fluids (De Zeeuw, 2005; Varaprath *et al.*, 2006; Loretz *et al.*, 2008). Flassbeck *et al.* (2001) tested different solvents like tetrahydrofuran and ethyl acetate, which are suggested for successful extraction of siloxanes. Trace amounts of D3–D6 were detected in the blanks of these two solvents in the SIM mode, while no contamination was found in hexane. The analyst should not use hand lotion during sample preparation (Horii and Kannan, 2008). Also, talc-powdered gloves must be avoided during sample preparation. Silanized glass wool in the liner and rubber o-rings may also contribute to siloxane background levels (Varaprath *et al.*, 2006).

6.4.5.4 Pre-treatment and Clean-up

Low MW siloxanes are volatile and can therefore easily escape from the matrix during sample pre-treatment or concentration (Varaprath *et al.*, 2006). For siloxanes containing the Si-H functionality, the glassware used must be pretreated with concentrated hydrochloric acid, washed with water and dried before use to avoid decomposition on the alkaline surface (Varaprath *et al.*, 2006).

For the sludge, water, soil and sediment samples, Kaj *et al.* (2005) used a purge-and-trap apparatus. Samples were diluted with water and a 0.5 mL buffer solution (2 M K_2HPO_4, 0.4 M HCl, 80 g $Na_2EDTA \cdot 2H_2O$ per litre) was added and purged at 70 °C with a gas stream (nitrogen, 50 mL min^{-1}) passing through an adsorbent trap (0.25 g Tenax) from which the analytes were later thermally desorbed and analysed by GC–EIMS. Dewil *et al.* (2007) developed a method for extracting siloxanes from sludge samples: 50 mL of sludge sample was extracted with 10 mL hexane by mixing the sludge–hexane mixture on a vortex mixer at high speed for 10 min. The extracts were then centrifuged at 4400 rpm for 5 min and the top phase consisting of hexane and the extracted siloxanes was removed and analysed by GC-flame ionization detection (FID). The extraction efficiency found by Dewil *et al.* (2007) for spiked sludge samples was 88 ± 10 % for D4 and 100 ± 1 % for D5.

Sample preparation of biota samples can be performed by extraction with hexane. Kaj *et al.* (2005) extracted 0.3 g of a biota sample with 1 mL hexane by vortex-mixing for 5 min. The extract was centrifuged at 10 000 rpm and the clear solution was carefully removed and transferred to a GC vial. No further treatment was applied and the extract was analysed by GC–HRMS.

Kaj *et al.* (2005) did not observe any differences between results from sample extracts that were dried with sodium sulfate according to Varaprath *et al.* (2006) and results from undried extracts. Kaj *et al.* (2005) used GC–HRMS. Therefore, lower sample amounts (0.3 g) did not create a sensitivity problem.

6.4.5.5 GC Methods

In most cases the separation was performed on a 30 m low-bleed capillary column with 0.25 mm inner diameter and a film thickness of 0.25 μm. The oven programmes used all started at low temperatures (35–70 °C). In most cases splitless injection was performed at 200 °C, which was the best compromise between low background of the septum and good sensitivity (see discussion on background) (Horii and Kannan, 2008). The best way to characterize the compounds is by MS. However, with proper separation, identification based on retention times is also possible with GC-FID. The LODs found for sediment, biota and air are given in Table 6.2.

No CRMs are available that are certified for siloxanes. The results from interlaboratory studies were also not found in the literature.

Table 6.2 LODs for the low MW siloxanes in different matrices (Kaj *et al.*, 2005).

Matrix	LOD
Water	0.013–0.027 μg L^{-1} (D4–D6) 0.0003 μg L^{-1} (L2–L5)
Sediment	1–2 μg kg^{-1} (D4–D6) 0.02 μg kg^{-1} (L2–L5)
Biota	6.7–13 μg kg^{-1} (D4–D6) 0.13 μg kg^{-1} (L2–L5)
Air	0.006–0.03 μg m^{-3} (D4–D6) 0.002–0.008 μg m^{-3} (L2–L5)

6.4.6 Miscellaneous

There is a multitude of other emerging organic contaminants that may be found in lower or higher concentrations in marine environmental matrices. Many of them will concentrate in sediments and marine organisms. Occurrences in rivers and estuaries are normally first indications of their occurrence in the marine environment. Many compounds can be analysed by GC–MS or LC–MS, but all have their specific difficulties, and methods have often not been developed or validated (Lepom, 2009). A higher polarity generally causes problems in the clean-up, as the separation from lipids will be more difficult. This also implies that the window related to the usual clean-up methods for organic contaminants such as column chromatography (silica gel, alumina) and GPC may exclude relevant contaminants from detection. Alternative approaches, such as effect-directed analysis (EDA), in which bio-assays and fractionation techniques are combined, should, therefore, be considered for future detection of emerging contaminants. Hoogenboom *et al*. (2009) showed the power of accurate mass screening with an LC–hybrid linear ion trap-Fourier transform (LTQ-FT)-Orbitrap MS–MS in EDA. They were able to identify a number of pharmaceuticals such as metropolol, diclofenac, sotalol, sulfamethoxazole and tramadol, and illicit drugs such as benzdazepines, cocaine and δ-9-tetrahydrocannabinol in surface water samples. Finally, the use of various types of passive sampler may also help in the identification of emerging pollutants in marine waters (Greenwood *et al*., 2009; Söderström *et al*., 2009).

6.5 ORGANOHALOGENS IN WATER

6.5.1 Introduction

The determination of concentrations of organohalogen compounds in seawater are important, as transport of these compounds to ocean areas remote from sources, supplemented by deposition from the atmosphere, can represent a significant route by which biota are exposed to these contaminants. It also allows declining trends in concentration to be studied and the efficacy of controls to be assessed.

In the absence of specific local sources, HCH isomers have generally been found among the organohalogens having the highest concentrations in seawater (lower nanogram per litre range).

Concentrations of other halogenated POPs are one to three orders of magnitude lower (lower picogram per litre to sub-picogram per litre range). The causes of the relatively high HCH concentrations are their large production amounts and higher water solubility compared with other POPs. Restrictions in the production and use of HCH have led to a 90% decline of HCH levels in European sea areas to 20–200 pg L^{-1} over the last decade.

Data on chlordane, heptachlor, mirex and toxaphene concentrations in marine waters are not available. Because of their high lipophilicity (and existing production and application restrictions, at least in Europe) water is not likely to be a primary monitoring matrix for these compounds. The same is true for PCDD and PCDF, which are expected to be found only in the low femtogram per litre range (using environmental concentrations of PCB for a coarse extrapolation estimate).

The water phase is the favoured matrix for more polar compounds having log K_{ow} values below about 5. With increasing polarity of the pollutants, the portion which is freely dissolved in the water phase increases. On the other hand, nonpolar pollutants adsorb to solid particles (suspended particulate matter (SPM) or sediments) and thereby are removed by sedimentation faster from the water column. The water phase has some specific advantages over sediments or biota as an investigation matrix when the focus of the survey is on the spatial distribution of pollutants, especially on regional and global scales. When investigating input sources, transport paths and the ultimate fate of pollutants, water concentrations often yield a clearer picture than concentrations in sediments or biota. This is due to the fact that the water phase generally is a 'primary' compartment and does not require normalization (SPM content should only be considered with more lipophilic compounds). In contrast, in sediments and biota, enrichment and other factors have to be considered which may depend on intrinsic parameters of the matrices themselves. The water phase, unlike sediments and biota, has a short response time, and there is hardly any time integration. This may be of advantage in detecting peak events but, on the other hand, it may yield higher natural variability, which has drawbacks in long-term monitoring and may require high sampling frequencies, causing high costs. However, this is often less problematic in open oceans owing to the slow dynamics of large water bodies. In coastal areas and estuaries, a correlation to salinity may 'normalize' for variability caused by freshwater mixing. A different approach to overcome variability by high dynamics in the water and to increase time integration is to use passive samplers (see Chapter 13).

6.5.2 Analytical Methods

Because of the low concentrations to be expected in the marine environment (low picograms per litre to low nanograms per litre range) the analytical procedure always consists of an extraction and enrichment step and a very selective chromatographic–mass spectrometric determination step. Depending on the analytes chosen, the water body studied and expected pollutant concentrations, an additional clean-up step may be necessary prior to the instrumental analysis.

The organohalogenated POPs under consideration belong to different chemical compound classes that have different structures and differ in their chemical and physical–chemical properties. Among the physical–chemical parameters influencing the analytical procedures, polarity and boiling points are the most important ones. All 'classical' pollutants are relatively nonpolar with log K_{ow} values between 3.5 and 8. They are quite stable and can be evaporated without destruction. Therefore, they can be analysed by gas chromatographic techniques. LLE using a nonpolar solvent such as hexane or pentane is a suitable extraction and enrichment method. For more polar and nonvolatile compounds like PFOS, LC by means of tandem MS is the method of choice. SPE is the preferred extraction and enrichment technique.

This repertoire of two extraction options (LLE, SPE) and two determination techniques (GC–MS and HPLC–MS/MS) allows a wide range of different compounds to be analysed.

The special challenge in determining organohalogen pollutants in seawater is their very low concentration range compared with surface or drinking water. In addition, concentrations in coastal waters may have a high dynamic range (low or sub-picogram per litre to low nanogram per litre).

6.5.2.1 Extraction Methods

In order to reach the projected LOQs in the low picogram per litre range, large sample volumes of 10 to 100 L (or more) have to be sampled and extracted. With modern MS instruments, ultra-low LOQs often are not limited by the absolute signal intensity but by blank problems or insufficient selectivity due to matrix background. Blank problems often can only be overcome by using a large sample volume (for selectivity improvement, see Sections 6.5.2.2 and 6.5.2.4).

'Classical' organohalogen pesticides (such as DDT, HCH, HCB, dieldrin, endrin and many more) can be extracted and enriched from seawater by means of LLE using hexane or pentane. Up to 100 L of water can be extracted by LLE, while larger volumes are generally extracted by SPE. Gaul and Ziebarth (1983) described a 10 L glass bowl sampler allowing extraction in the sampling bowl itself, thus minimizing problems with sample handling, blanks and adsorption. In 1998, the same principle was expanded to a 100 L bowl, thus increasing the sample volume and lowering the LOQ by a factor of 10 (Theobald *et al.*, 1990). Figure 6.7 shows a picture of the sampling bowl. Extraction is done by agitating the sample with 1 L of pentane using a stirrer.

For larger volumes of 200 to 1000 L, Schulz *et al.* (1995) described an SPE procedure using large extraction cartridges filled with XAD resins. With this adsorbent, they obtained good extraction results for PCBs, DDT and PAHs, but not for HCH.

LLE and SPE do not yield exactly the same concentrations, as they use different extraction principles. While SPE effectively extracts only freely dissolved compounds, LLE extracts freely dissolved humic acids complexed and (in part) particle-bound pollutants (Sturm *et al.*, 1989). For the analysis of polar compounds by HPLC–MS/MS, smaller sampling volumes (1–30 L) and

Figure 6.7 A 100 L glass bowl sampler for sampling seawater for organic pollutant analysis.

smaller SPE cartridges (1–3 g of resin) are used. The reason for this downscaling is the fact that polar compounds often occur at higher concentrations.

6.5.2.2 *Clean-up Procedures*

There are two main objectives for a clean-up step prior to GC–MS analysis. The first one is to protect the GC injector and column from polar, nonvolatile matrix compounds such as fatty acids, esters, alcohols and others. Without such a cleaning step, the GC column would be ruined rather quickly and residues deposited in the injector would lead to adsorption of sensitive compounds like DDT. The second objective for a clean-up step is to remove matrix compounds that can overlap with the signals of the analytes. In general, clean-up techniques similar to those described for PCBs or PBDEs can also be used for organohalogen pesticides (Sections 6.2.2.3 and 6.3.2.3). As the matrix load of seawater is generally smaller than that of biological tissues or sediments, a simpler clean-up step may be used. Therefore, chromatography using a small silica gel column (with about 200 mg of deactivated SiO_2) is sufficient for most purposes. The main effect of such a step is to remove nonvolatiles. The effectiveness needed for the clean-up depends on the detection technique used and on the concentration range. Less-selective techniques like ECD require a more stringent clean-up than MS/MS detection, and with an LOQ of 100 pg L^{-1} less clean-up may be necessary than with a range below 5 pg L^{-1}. Larger amounts of silica gel may also be required for estuarine water samples or samples from a region with high biological activity (algal bloom). For special applications, even an HPLC pre-separation step may be necessary (Petrick *et al.*, 1988).

The activity of the silica gel and the elution step have to be carefully controlled. In particular, medium-polar compounds like HCH or dieldrin are susceptible to losses or co-elution with matrix compounds. Samples analysed by HPLC–MS/MS mostly do not require a clean-up step because the technique is quite robust and highly selective.

6.5.2.3 *Pre-concentration*

Pre-concentration is carried out using the techniques described in Section 6.2.2.4.

6.5.2.4 *Detection and Quantification*

GC–MS All 'classical' organohalogen pesticides can be analysed using the procedures described for PCBs in Section 6.2.2.5. They can be analysed by GC using the same stationary phase, such as HB-5 or HT-8. Owing to their good response to the selective ECD, they ensure fairly selective separation of interfering matrix compounds from the samples. Thus, in environmental investigations during the 1970s to 1990s, GC with ECD was the method of choice for the analysis of organohalogen pesticides (Gaul and Ziebarth, 1983). Using this method and with 10 L samples, LOQs in the range 0.1–0.05 ng L^{-1} were obtained for HCH and DDT metabolites.

MS initially was used mainly for identification and structure elucidation. Later, in the 1990s, when GC–MS systems became more affordable and more sensitive, it was used for quantitative determination as well. SIM using quadrupole mass spectrometers proved to be more selective than ECDs and finally led to the same limits of quantification. The main advantage of GC–MS is its better selectivity and the suppression of background 'noise' from matrix constituents. The higher selectivity of GC–MS often allows simpler clean-up steps. Further, the range of multicomponent analysis has been extended. For example, with GC–MS in SIM mode, simultaneous analysis of PAHs and chlorinated hydrocarbons is possible in one sample extract and one GC–MS analysis.

With special mass spectrometric techniques like HRMS, ECNI or reaction monitoring in tandem mass spectrometers (MS/MS), a considerable improvement in the selectivity and sensitivity of MS detection has been achieved in the past few years. While general use of HRMS is still hampered by the high price of the instrumentation, ECNI has some practical limitations resulting from its high molecular structure dependency for sensitivity (i.e. compounds with fewer than five halogen atoms show poor detection response), poor long-term stability and higher demands for ion source cleaning. MS–MS techniques were first used with ion trap instruments because of their attractive price. With ion traps, MS–MS (in time) works very well for relatively 'clean' samples without a large matrix burden, where only a few compounds have to be determined and few MS transitions are necessary. These limitations are overcome by using triple quadrupole instruments. Though more expensive than ion trap mass spectrometers, they have become quite affordable and are now the most universal and sensitive MS instruments available for ultrasensitive analysis of organohalogen pollutants (and, of course, other compounds as well).

For ultimate sensitivity, the combination of large-volume injection (by a temperature-programmed injector) and a highly selective triple quadrupole mass spectrometer is a most powerful approach. The increased selectivity and sensitivity of reaction monitoring (MRM) in comparison with normal SIM is shown in the examples in Figures 6.8 and 6.9. In Figure 6.8, the ion chromatogram of fragment m/z 219 of the HCH isomers is shown; in Figure 6.9 the fragmentation step from m/z 255 to 219. The total amount injected for GC–MS was a few

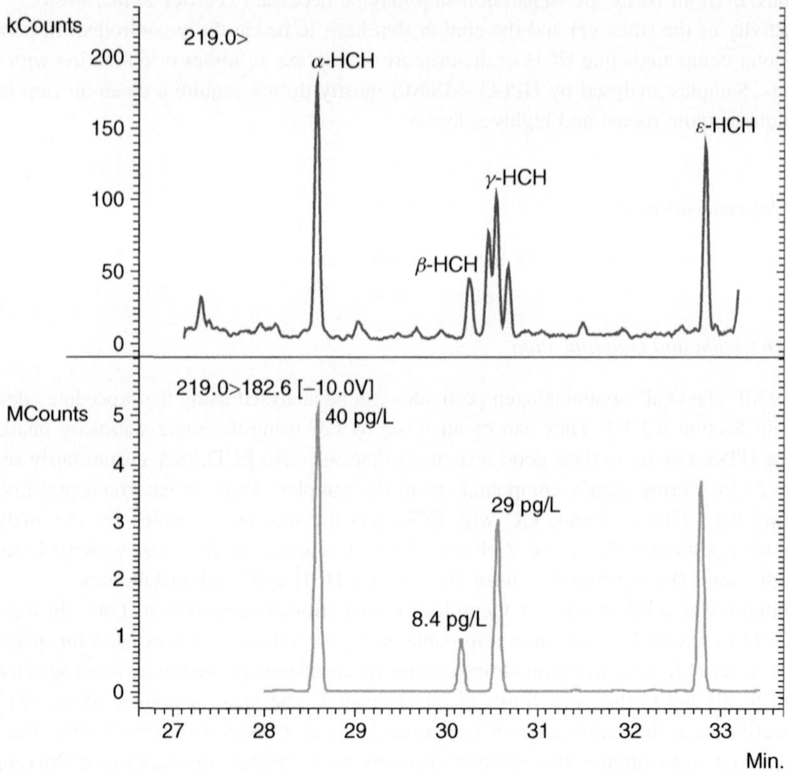

Figure 6.8 Analysis of HCH isomers of a 100 L sample extract from the central North Sea (August 2008); GC–MS (upper trace, m/z 219) and GC–MS/MS (lower trace, m/z 219 to 182.6).

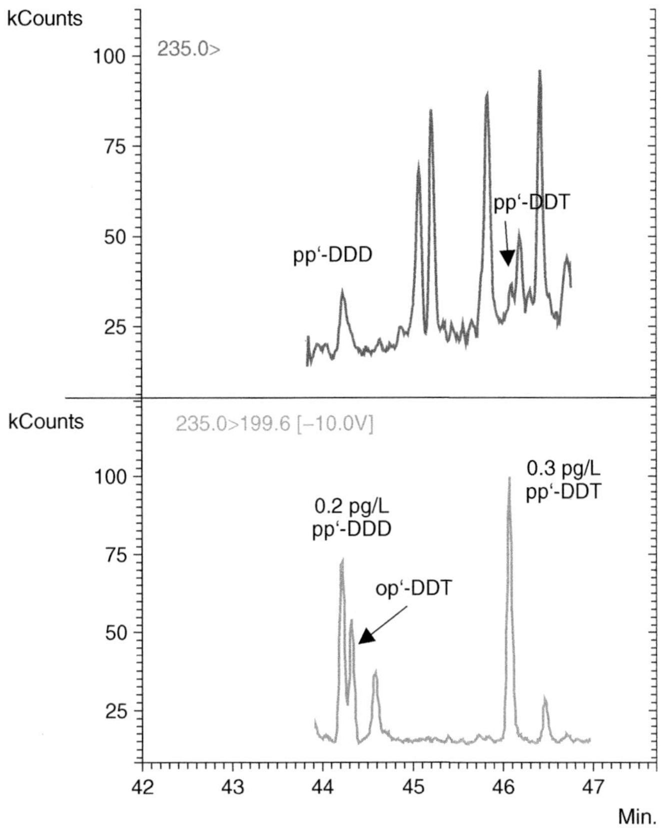

Figure 6.9 Analysis of DDD and DDT of a 100 L sample extract from the central North Sea (August 2008); GC–MS (upper trace, m/z 235) and GC–MS/MS (lower trace, m/z 235 to 199.6).

picograms; the calculated water concentration was in the range of 10 to 40 pg L^{-1} (100 L water sample, 6 µL of 100 µL extract was injected during GC).

A similar example with p, p'-DDD and p, p'-DDT is shown in Figure 6.9. Water concentrations in this case are two orders of magnitude lower, in the sub-picogram per litre range. The figures clearly demonstrate a more than 10-fold better sensitivity of the MS–MS technique, with signals that are easily processed by automatic integration.

Table 6.3 summarizes the LOQs of a selection of other organohalogen compounds determined under the same conditions. In addition, their typical concentration ranges are listed for North Sea surface water (in August 2008). Of course, other (e.g. non-halogenated) pollutants of similar polarity and volatility can be analysed by this procedure as well.

It should be noted that analyses in such a low concentration range require extensive experience. The whole procedure and the GC–MS/MS must be carefully controlled for optimal performance; otherwise an order of magnitude or more in sensitivity is quickly lost.

6.5.3 Quality Control

Because of the ultralow concentrations of pollutants in seawater, QA is a highly challenging task. All aspects mentioned with respect to PCBs in Section 6.1.3 also apply to the determination of

Table 6.3 Typical LOQs of selected organohalogen pollutants and observed concentrations in sea water of the southern North Sea (pg L^{-1}) (August 2008, 100 L samples, GC–MS/MS).

	LOQ	River Elbe		Coastal			Open sea		
		Min.	Max.	Min.	Median	Max.	Min.	Median	Max.
Salinity		0.5	0.5	20	32.11	32.93	34.26	34.47	34.83
α-HCH	0.1	967	1 353	39.2	49.5	264	33.8	37.8	56.9
β-HCH	0.1	1589	2 868	29.5	53.3	689	8.3	8.8	10.3
γ-HCH	0.3	374	562	75.9	87.9	209	28.9	51.8	60.9
δ-HCH	0.1	64.2	653.0	8.9	10.3	131	0.3	0.8	2.9
HCB	0.1	19.8	196	2.2	3.2	18.2	1.8	2.3	3.9
QCB	0.1	8.2	39.9	0.8	1.2	6.3	0.5	0.8	1.5
Trifluralin	0.1	5.9	8.8	0.4	0.5	2.5	<0.1	0.6	1.1
Aldrin	4.0	<4	13.0	<4	<4	<4	<4	<4	<4
Isodrin	5.0	<5	26.8	<5	<5	<5	<5	<5	<5
Dieldrin	1.0	<1	40.4	13.5	20.2	22.2	5.6	12.5	20.9
Endrin	2.0	<2	49.5	<2	<2	6.9	<2	<2	<2
p,p'-DDE	0.1	26.3	165.9	1.2	3.0	22.5	0.2	0.6	1.1
p,p'-DDD	0.1	44.8	442.9	1.2	7.1	100.6	0.2	0.9	1.9
p,p'-DDT	0.1	6.1	15.0	0.4	0.7	4.2	0.2	0.3	0.4
CB28	0.1	6.7	30.8	1.1	1.9	5.0	0.3	0.4	0.7
CB52	0.2	7.3	28.6	0.7	1.4	5.5	0.4	0.5	0.8
CB153	0.5	26.5	154.8	2.3	6.1	24.4	1.3	1.9	3.1
CB138	0.4	22.2	123.3	3.8	5.1	18.8	1.4	2.6	5.1

organohalogens in water. Some aspects (e.g. blank problems) are even more relevant and more difficult to control because values are even lower. Even the ultrapure chemicals and solvents used sometimes have to be pre-cleaned additionally. All equipment has to be thoroughly cleaned and continuously checked for blanks. The same is true for the whole procedure. Several blank samples thus have to be processed with each analytical sample series. Concentrations often are close to the LOQs, which means difficult calibration and integration, and reduced analytical precision. Although GC–MS/MS and HPLC–MS/MS have very high selectivities, it is good practice to use a second MS transition as a qualifier.

In addition, the following special problems are encountered in seawater analyses:

- Because of the large sample volumes, it is not possible to analyse replicate samples routinely or to take samples for back-up analysis. However, it is often possible to make a plausibility check by comparing the results with those of samples taken from a homogeneous water body at adjacent stations. Homogeneity can be derived from oceanographic parameters like salinity.
- No CRMs are available for water analysis. Therefore, at least an LRM has to be used, which should preferably be a natural or spiked extract from a typical monitoring station. Extraction efficiencies should be controlled by standard addition tests.
- Standard performance tests (e.g. by QUASIMEME) are difficult to perform and to evaluate because sample volumes in the tests (maximum 1 L) differ from those used in real analysis (>10 L). Thus, concentration ranges in the tests are often higher than in real-life samples.

Thorough internal quality control is thus of crucial importance. When the monitoring objective is long time series, it is of particular importance to perform reliable and reproducible high-quality analyses over decades. Therefore, such analyses require well-documented procedures and highly experienced analysts.

6.6 POLYCYCLIC AROMATIC HYDROCARBONS

De Boer and Law (2003) summarized and assessed the development of the use of chromatographic techniques for the determination of organohalogenated contaminants and PAHs in samples from the marine environment. In that paper, the development of PAH analysis from the determination of a small number of PAHs using column, paper and thin-layer chromatography (e.g. the six of Borneff) to the determination of the 16 USEPA priority parent PAHs using HPLC and of a wider range of both parent and alkylated PAHs using GC–MS was described. The main findings and summaries of that review in respect of PAH are still up to date and relevant, and so we propose, in this chapter, only to review advances from that point. At that time, our view was that ion-trap benchtop GC–MS systems operated in full-scan mode were particularly suited to the analysis of PAHs and other hydrocarbons as, unlike in quadrupole MS systems, there is no sensitivity advantage to be gained by limiting the number of ions detected (e.g. using single/multiple ion monitoring). Other GC–MS methodologies may, of course, be employed effectively. This is still the case for PAHs up to about molecular weights of 302 Da; above that level LC–MS may be more appropriate and effective due to the lower volatility of the higher molecular weight PAH compounds, although GC–MS methods may still be applicable if due attention is paid to the details of the methodology (de Boer and Law, 2003).

Poster *et al.* (2006) published a review of GC methods for PAH analysis. This review recognized that the major drivers for innovation were the needs for increased sensitivity, selectivity and speed of analysis. As the approaches described in de Boer and Law (2003) were entirely adequate for the determination of two- to six-ring PAHs, further developments were necessarily directed towards either extending the range of PAH compounds to be quantified or towards more effective (i.e. cheaper) methods and methodology. Poster *et al.* (2006) identified three basic enhancements to GC methods applied to PAH analysis: large-volume injection GC; thermal desorption techniques with GC; and fast GC. Large-volume injections have a number of advantages: improved sensitivity and so lowered limits of detection due to the greater sample volume injected, and a reduction in the time needed for sample extract concentration. Thermal desorption techniques (such as SPME) in conjunction with GC–MS analyses have been applied effectively to PAH analyses in a variety of matrices. Fast GC generally involves the use of short microbore columns which reduce analysis time, but risks overloading of the stationary phase if analytes are highly concentrated (Poster *et al.*, 2006). The application of fast GC is currently limited, owing to the necessary need for rapid detector response, mainly to GC-FID, GC-ECD and GC–TOF-MS (high-resolution GC coupled to TOF-MS). Column length is typically ~10 m and analysis times may be ~10 min or less for the USEPA 16 priority parent PAHs. Using a liquid-crystal stationary phase in combination with GC–MS, complete separation of the USEPA 16 priority parent PAHs was achieved in less than 5 min on a 6 m × 0.1 mm ID column, and in less than 7 min on a 15 m × 0.25 mm ID column (Britten and Naikwadi, 2008). Fast GC–HRMS was used to separate 16 PAH compounds of 216–302 Da, from benzo[*c*]fluorene to dibenzo[*a*, *h*]pyrene (Ziegenhals *et al.*, 2008). These compounds represent priority PAHs within the EU as defined by the Scientific Committee on Food (SCF-PAH), which could include shellfish. Using a 10 m × 0.1 mm TR-50 ms column a runtime of 25 min was obtained.

Two-dimensional gas chromatography (GC × GC) has been applied successfully to a number of complex mixtures of isomers and congeners of environmental contaminants, including dibenzo-*p*-dioxins and furans, PCBs and PBDEs. Although some problems remain to be resolved, GC × GC–TOF-MS has been applied with reasonable success to analysis of an urban dust standard reference material (SRM1649a) (Amador-Muñoz *et al.*, 2008) and polluted sediments from the River Elbe in the Czech Republic (Skoczyñska *et al.*, 2008). In addition, oil fingerprinting studies have utilized both GC × GC-FID (Arey *et al*., 2007; Nelson *et al*., 2006) and offline multidimensional HPLC with diode array detection (Saravanabhavan *et al*., 2007) in order to study the complex mixtures of hydrocarbons present in weathering heavy fuel oil from the *Bouchard 120* heavy fuel oil spill and the heavy gas oil fraction of Alaskan North Slope crude oil respectively. Unresolved complex mixtures (the complex 'hump' observed in weathered petroleum and petroleum-contaminated sediments) have traditionally been impenetrable using GC methods; however, their composition is of interest and their toxicity of possible concern (Gough and Rowland, 1990). Recent advances (GC × GC) using a variety of stationary phases, including chiral columns, have opened a window into their composition (Frysinger *et al*., 2003).

Within the EU, the implementation of the Water Framework Directive (WFD; 2000/60/EC) has generated interest in a wider range of organic contaminants than hitherto determined, and so a desire for improved multi-residue analytical methods. A method for the determination of 36 priority substances, including eight PAH compounds from naphthalene to indeno[1,2,3-*cd*]pyrene has been developed, using SPE with Strata X cartridges (Barrek *et al*., 2009). Naphthalene was determined using GC–MS, and all eight PAHs using HPLC with fluorescence detection. Other compounds were determined using either GC–MS or HPLC–MS/MS. Using a new extraction technique known as stir bar sorptive extraction (SBSE), Huertas *et al*. (2007) developed a method for 24 priority substances within the WFD. This included four PAHs from benzo[*b*]fluoranthene to benzo[*ghi*]perylene. The detection and quantification limits were less than 5 and 10 ng L^{-1} for all four PAHs, with repeatabilities around 10%. Pitarch *et al*. (2007) developed a method for the quantification of more than 50 organic micro-pollutants in water using SPE with C_{18} cartridges followed by GC–EIMS/MS. This included 16 PAHs from naphthalene to dibenz[*a*, *h*]anthracene. The method did not prove adequate for the determination of naphthalene and acenaphthylene; for the other 14 PAHs, LODs and LOQ were 1–5 ng L^{-1} and 25 ng L^{-1} respectively. A similar range of PAHs were determined, along with other classes of compounds, in a multiresidue method developed by Zhang *et al*. (2007). SPE using two-layer discs of C18 and SDB-XC was followed by GC–MS. LODs were approximately 0.4–2.0 ng L^{-1} per compound.

One recent paper reports the use of a molecularly imprinted polymer as a solid-phase adsorbent for the quantitative enrichment of PAHs in coastal sediments (Krupadam *et al*., 2009). Recoveries from 10 g of coastal sediment samples ranged from 85 to 96%. Determination of PAHs (eight compounds from acenaphthene to benzo[*a*]pyrene) was by fluorescence spectrometry, but this technique could usefully be applied ahead of GC–MS. Tondeur and Hart (2009) reviewed the field of ultratrace extraction of POPs, and identified mismatches between high-volume extractions and analytical capabilities. Simply increasing the volume of water extracted (to 1000 L, say) may not lead to improvements in LODs and other performance criteria without other methodological changes. The general application of passive sampling devices for sample collection has a number of advantages which will be further exploited in the future. Unlike biomonitoring techniques (e.g. using mussels), these devices concentrate contaminants by diffusion and partitioning processes or by sorption, and offer time-integrated sampling that compensates for fluctuating discharge levels (Harman *et al*., 2009) (i.e. offering time-weighted average concentrations). Deployed within sediment samples, these devices can also indicate concentrations of contaminants within sediment pore waters; that is, those bioavailable to sediment-dwelling organisms that uptake from that matrix rather than by ingesting sediment particles.

6.6.1 Certified Reference Materials for PAH Analysis

When assessed by de Boer and Law (2003), nine CRMs were available for PAHs in marine samples. These comprised four mussel samples, four sediment samples and one sample of seaweed. They also noted that there was a need for CRMs certified for a wider range of PAH compounds than at

Table 6.4 CRMs for PAHs in marine samples.[a]

CRM	Material	Form	Producer	Country	No. of certified PAHs
SRM 1974b	Mussel	Frozen	NIST	USA	22
SRM 2977	Mussel	Freeze-dried	NIST	USA	22
IAEA-140-OC	Seaweed (*Fucus* sp.)	Freeze-dried	IAEA	Monaco	21
IAEA-406	Fish	Freeze-dried	IAEA	Monaco	22
IAEA-432	Mussel	Freeze-dried	IAEA	Monaco	22
IAEA-435	Tuna	Freeze-dried	IAEA	Monaco	22
EDF-2524	Pacific herring	Slurry	CIL	Monaco	11
SRM 1941b	Sediment	Freeze-dried	NIST	USA	24
SRM 1944	Sediment	Freeze-dried	NIST	USA	24
IAEA-159	Sediment	Freeze-dried	IAEA	Monaco	22
IAEA-408	Sediment	Freeze-dried	IAEA	Monaco	22
IAEA-417	Sediment	Freeze-dried	IAEA	Monaco	22

[a]The mussel tissue CRM SRM 2 977 from NIST is listed, but is shown on their web site as out of stock. NIST: National Institute for Standards and Technology; IAEA: International Atomic Energy Agency; CIL: Cambridge Isotopes Limited.

present, and for the provision of homogeneous but uncharacterized materials for use as LRMs, as scientists in many laboratories are uncertain of their ability to prepare these themselves (Law and Klungsøyr, 2000). Currently, there are 12 CRMs available, listed in Table 6.4. A wider range of CRMs is still needed. Further details of the CRMs available, including certified values, are given in Chapter 11.

ACKNOWLEDGEMENTS

Work on PAHs undertaken by Cefas is funded by the Department of Environment, Food and Rural Affairs as part of its programme of marine environmental research. We also wish to acknowledge Ike van der Veen and Sicco Brandsma (MSc students at the VU University, Amsterdam), for their studies on PFRs and siloxanes respectively.

REFERENCES

Abdallah M. A.-E., Covaci A. and Harrad S. (2009). *Anal. Chem.*, **81**, 7460.
Ackerman L. K., Wilson G. R. and Simonich S. L. (2005). *Anal. Chem*, **77**, 1979.
Amador-Muñoz O., Villabolos-Pietrini R., Aragón-Piña A. *et al.* (2008). *J. Chromatogr. A*, **1201**, 161.
Andresen J. A., Grundmann A. and Bester K. (2004). *Sci. Total Environ.*, **332**, 155.
Arey J. S., Nelson R. K. and Reddy C. M. (2007). *Environ. Sci. Technol.*, **41**, 5738.
Ballschmiter K., Bacher R., Mennel A. *et al.* (1992). *J. High Resolut. Chromatogr.*, **15**, 260.
Barrek S., Cren-Olivé C., Wiest L. *et al.* (2009). *Talanta*, **79**, 712.
Björklund J., Isetun S. and Nilsson U. (2004). *Rapid Commun. Mass Spectrom.*, **18**, 3079.
Björklund E., Sporring S., Wiberg K. *et al.* (2006). *Trends Anal. Chem.*, **25**, 318.
Breivik K., Sweetman A., Pacyna J. M. and Jones K. C. (2002). *Sci. Total Environ.*, **290**, 199.
Breivik K., Sweetman A., Pacyna J. M. and Jones K. C. (2007). *Sci. Total Environ.*, **377**, 296.

Britten A. J. and Naikwadi K. P. (2008). *Polycyclic Arom. Cpds.*, **28**, 165.

Carlsson H., Nilsson U., Becker G. and Östman C. (1997). *Environ. Sci.Technol.*, **31**, 2931.

Carpenter J. C., Celia J. A. and Dorn S. B. (1995) . *Environ. Sci. Technol.*, **29**, 864.

Carson R. (1962). *Silent Spring* (2002 edn). Houghton Mifflin, Mariner Books (ISBN 0-618-24906-0).

Castells P., Santos F. J. and Galceran M. T. (2004). *J. Chromatogr. A*, **1025**, 157.

Castells P., Parera J., Santos F. J. and Galceran M. T. (2008). *Chemosphere*, **70**, 1552.

Coelhan M., Saraci M. and Parlar H. (2000). *Chemosphere*, **40**, 685.

Covaci A., Voorspoels S. and de Boer J. (2003). *Environ. Int.*, **29**, 735.

Covaci A., Voorspoels S., Ramos L. *et al*. (2007). *J. Chromatogr. A*, **1153**, 145.

Covaci A., Harrad S., Abdallah M. A.-E. *et al*. (2011). *Environ. Int.*, **37**, 532.

Dawes P., Hibberd A., Sharp G. and Pfeffer B. (2004). The analysis of PCB congeners using a selective PCB capillary column, http://www.sge.com/uploads/44/a8/a837a2ff82208c46c2dc2b5ceb0072/TP-0105-C.pdf.

De Boer J. and Law R. J. (2003). *J. Chromatogr. A*, **1000**, 223

De Boer J. and McGovern E. (2001). *Trends Anal. Chem.*, **20**, 140.

De Boer J. and Wells D. E. (2006). *Trends Anal. Chem.*, **25**, 364.

De Boer J., Allchin C. R., Zegers B. and Boon J. P. (2001). *Trends Anal. Chem.*, **20**, 591.

Dewil R., Appels L., Baeyens J. *et al*. (2007). *Talanta*, **74**, 14.

De Wit C., Herzke D. and Vorkamp K. (2009). *Sci. Total Environ.*, **408**, 2885. DOI: 10.1016/j.scitotenv.2009.08 .037, PubMed ID: 19815253.

De Wit C. A., Alaee M. and Muir D. C. G. (2006). *Chemosphere*, **64**, 209.

De Zeeuw J. (2005). *Am. Lab.*, **37**, 18.

Dick T. A., Gallagher C. P. and Tomy G. T. (2010). *World Rev. Sci. Technol. Sustain. Develop*. **7**, 387.

Dorn S. B. and Skelly Fram E. M. (1994). *Analyst*, **199**, 1687.

Eppe G., Focant J. F., Pirard C. and De Pauw E. (2004). *Talanta*, **63**, 1135.

Evenset A., Christensen G. N., Skotvold T. *et al*. (2004). *Sci. Total Environ.*, **318**, 125.

Fendinger N. J., McAvoy D. C., Eckhoff W. S. and Price B. B. (1997). *Environ. Sci. Technol.*, **31**, 1555.

Flassbeck D., Pfiederer B., Grumping R. and Hirner V. (2001). *Anal. Chem.*, **73**, 606.

Focant J.-F. and De Pauw E. (2002). *J. Chromatogr. B*, **776**, 199.

Fridén U., Jansson B. and Parlar H. (2005). *Chemosphere*, **54**, 1083.

Frysinger G. S., Gaines R. B., Xu L. and Reddy C. M. (2003). *Environ. Sci. Technol.*, **37**, 1653.

García-López M., Rodríguez I., Cela R. *et al*. (2009). *Talanta*, **79**, 824.

Gaul H., and Ziebarth U. (1983). *Dtsch. Hydrogr. Zeitschr.*, **36**, 191–212.

Gómara B., Lebron-Aguilra R., Quantanilla-Lopez J. E. and González M. J. (2007). *Anal. Chim. Acta*, **605**, 53.

Gómez-Ariza J. L., Bujalance M., Giraldez I. *et al*. (2002). *J. Chromatogr. A*, **946**, 209.

Gough M. A. and Rowland S. J. (1990). *Nature*, **344**, 648.

Greenwood R., Mills G. A. and Vrana B. (2009). *J. Chromatogr. A*, **1216**, 631.

Harman C., Thomas K. V., Tollefsen K. E. and Meier S. (2009). *Mar. Pollut. Bull.*, **58**, 1671.

Hartmann P. C., Burgi D. and Giger W. (2004). *Chemosphere*, **57**, 781.

Hess P., de Boer J., Cofino W. P. *et al*. (1995). *J. Chromatogr. A*, **703**, 417.

Hoh E. and Hites R. A. (2005). *Environ. Sci. Technol.*, **39**, 2472.

Hoogenboom A., Van Leerdam J. A. and de Voogt P. (2009). *J. Chromatogr. A*, **1216**, 510.

Horii Y. and Kannan K. (2008). *Environ. Contam. Toxicol.*, **55**, 701.

Huertas C., Morillo J., Usero J. and Gracia-Manarillo I. (2007). *Talanta*, **72**, 1149.

Hühnerfuss H. and Sah M. R. (2009). *J. Chromtaogr. A*, **1216**, 481.

Iino F., Takasuga T., Senthilkumar K. *et al*. (2005). *Environ. Sci. Technol.*, **39**, 859.

Ishikawa S., Taketomi M. and Shinohara R. (1985). *Water Res.*, **19**, 119.

Ismail N., Gewurtz S., Pleskach K. *et al*. (2009). *Environ. Toxicol. Chem.*, **28**, 910.

Johnson-Restrepo B. and Kannan K. (2009). *Chemosphere*, **76**, 542.

Jones-Otazo H, Clarke J. P., Diamond M. L. *et al*. (2005). *Environ. Sci. Technol.*, **39**, 5121.

Kaj L., Schlabach M., Andersson J. *et al*. (2005). TemaNord 593, Nordic Council of Ministers, Copenhagen, Denmark (ISBN 92-893-1268-8).

Kawagoshi Y., Nakamura S. and Fukunaga I. (2002). *Chemosphere*, **48**, 219.

Kierkegaard A., Björklund J. and Fridén U. (2004). *Environ. Sci. Technol.*, **38**, 3247.

Kim S. D., Cho J. W., Kim I. S. *et al*. (2007). *Water Res.*, **41**, 1013.

Korytár P., Covaci A., de Boer J. *et al.* (2005a). *J. Chromatogr. A*, **1065**, 239.
Korytár P., Leonards P. E. G., de Boer J. and Brinkman U. A. Th. (2005b). *J. Chromatogr. A*, **1100**, 200.
Korytár P., Parera J., Leonards P. E. G. *et al.* (2005c). *J. Chromatogr. A*, **1086**, 71.
Korytár P., Leonards P. E. G., de Boer J. and Brinkman, U. A. Th. (2002). *J. Chromatogr. A*, **958**, 203.
Kraemer W. and Ballschmiter, K. (1987). *Fresenius Z. Anal. Chem.*, **327**, 47.
Krupadam R. J., Bhagat B., Wate S. R. *et al.* (2009). *Environ. Sci. Technol.*, **43**, 2871.
Larsen B., Cont M., Montanarella L. and Platzner N. (1995). *J. Chromatogr. A*, **708**, 115.
Lassen C., Libak Hansen, C., Hagen Mikkelson S. and Maag J. (2005). Environ, Project no. 1031, Danish Ministry of the Environment, Copenhagen, Denmark.
Law R. J. and Klungsøyr J. (2000). *Int. J. Environ. Pollut.*, **13**, 262.
Law R. J., Allchin C. R., de Boer J. *et al.* (2006). *Chemosphere*, **64**, 187.
Law R. J., Herzke D., Harrad S. *et al.* (2008). *Chemosphere*, **73**, 223.
Lee H. B. and Peart T. E. (1994). *J. Chromatogr. A*, **663**, 87.
Leisewitz A., Kruse H. and Schramm E. (2000). Environmental research plan of the German Federal Ministry for the Environment, Nature Conservation and Nuclear Safety, Research Report 297 44 542, Berlin, Germany.
Lepom P. (2009). *J. Chromatogr. A*, **1216**, 470.
Lopez P., Brandsma S., Leonards P., de Boer J. (2011). *Anal. Bioanal. Chem.*, in press.
Loretz L. J., Api A. M., Babcock L. *et al.* (2008). *Fd. Chem. Toxicol.*, **46**, 1516.
Martínez-Carballo E., González-Barreiro C., Sitka A. *et al.* (2007). *Sci. Total Environ.*, **388**, 290.
Mills S. A., Thal D. I. and Barney J. (2007). *Chemosphere*, **68**, 1603.
Morris S., Allchin C., Zegers B. *et al.* (2004). *Environ. Sci. Technol.*, **38**, 5497.
Moore S., Vromet L. and Rondeau B. (2004). *Chemosphere*, **54**, 453.
Morris S., Bersuder P., Allchin C. R. *et al.* (2006). *Trends Anal. Chem.*, **25**, 343.
Muir D. and Sverko E. (2006). *Anal. Bioanal. Chem*, **386**, 769.
Muir D. C. G. (2010). Environmental levels and fate, in *Chlorinated Paraffins* (ed. J. De Boer), Handbook of Environmental Chemistry, Springer, Heidelberg, Germany.
Nelson R. K., Kile B. M., Plata D. L. *et al.* (2006). *Environmental Forensics*, **7**, 33.
OSPAR Commission (2007). Which brominated flame retardants should be monitored as part of the CEMP? SIME 07/5/1-E, London, UK.
Pakalin S., Cole T., Steinkellner J. *et al.* (2007). European Report EUR 22693 EN, Brussels, Belgium, http://ecb.jrc.ec.europa.eu/documents/Existing-Chemicals/Review_on_production_process_of_deca BDE.pdf.
Pellizzato F., Ricci M., Held A. *et al.* (2009). *Trends Anal. Chem.*, **28**, 1029.
Petrick G., Schulz D. and Duinker J. C. (1988). *J. Chromatogr,.* **435**, 241.
Pitarch E., Medina C., Portolés T. *et al.* (2007). *Anal. Chim. Acta*, **583**, 246.
Prevedouros K., Jones K. C. and Sweetman A. J. (2004). *Environ. Sci. Technol.*, **38**, 3224.
Poster D. L., Schantz M. M., Sander L. C. and Wise S. A. (2006). *Anal. Bioanal. Chem.*, **386**, 859.
Reth M., Zencak Z. and Oehme M. (2005). *Chemosphere*, **58**, 847.
Reth M., Ciric A., Christensen G. N. *et al.* (2006). *Sci. Total Environ.*, **367**, 252.
Ricci M. and Pellizzato F. (2010). Reliable determination of SCCPs. A dream? in *QUASIMEME Workshop on the Analysis of Chlorinated Paraffins*, Ostend, Belgium, 18–19 March 2010, www.quasimeme.org.
Ricklund N., Kierkegaard A. and MacLachlan M. S. (2010). *Environ. Sci. Technol.*, **44**, 1987. DOI: 10.1021/es903701q.
Rieger R. and Ballschmiter K. (1995). *Fresenius J. Anal. Chem.*, **352**, 715.
Rodriguez I., Calvo F., Quintana J. B. *et al.* (2006). *J. Chromatogr. A*, **1108**, 158.
Silicon Technology-Dow Corning (2008). www.dowcorning.com.
Ross P. S., Couillard M., Ikonomou M. G. *et al.* (2009). *Mar. Pollut. Bull.*, **58**, 7.
Saravanabhavan G., Helferty A., Hodson P. V. and Brown R. S. (2007). *J. Chromatog. A*, **1156**, 124.
Schulz D. E., Petrick G., Kannan N. and Duinker J. C. (1995). *Mar. Chem.*, **48**, 245.
Skoczy*n*ska E., Korytár P. and de Boer J. (2008). *Environ. Sci. Technol.*, **42**, 6611.
Smedes F. and de Boer J. (1994). *Quim. Anal.*, **13**, 100.
Smedes F. and de Boer J. (1997). *Trends Anal. Chem.*, **16**, 503–517.
Söderström H., Lindberg R. H. and Fick J. (2009). *J. Chromatogr. A*, **1216**, 623.
Sporring S., Wiberg K., Bjorklund E. and Haglund P. (2003). Dioxin 2003, Boston, MA, USA. *Organohalogen Cpds.*, **60–65**.

Staaf T. and Östman C. (2005). *J. Environ. Monit.*, **7**, 344.
Stapleton H. M. (2006). *Anal. Bioanal. Chem.*, **386**, 807.
Stevens C., Powell D. E., Makela P. and Karman C. (2001). *Mar. Pollut. Bull.*, **42**, 536.
Sturm B., Knauth H. D., Theobald N. and Wünsch G. (1998). *Fresenius J. Anal. Chem.*, **361**, 803.
Theobald N., Lange W., Rave A. *et al.* (1990). *Dtsch. Hydrogr. Zeitschr.*, **43**, 311.
Thomas G. O., Farrar D. G., Braekevelt E. *et al.* (2006). *Environ. Int.*, **32**, 34.
Tomy G. T. (2010). Analysis of chlorinated paraffins in environmental matrices: the ultimate challenge for the analytical chemist, in *Chlorinated Paraffins* (ed. J. De Boer), Handbook of Environmental Chemistry, Springer, Heidelberg, Germany.
Tomy G. T. and Stern G. A. (1999). *Anal. Chem.*, **71**, 4860.
Tomy G. T., Stern G. A., Muir D. C. G. *et al.* (1997). *Anal. Chem.*, **69**, 2762.
Tomy G. T., Westmore J. B., Stern G. A. *et al.* (1999). *Anal. Chem.*, **71**, 446.
Tomy G. T., Muir D. C. G., Stern G. A. and Westmore J. B. (2000). *Environ. Sci. Technol.*, **34**, 1615.
Tondeur Y. and Hart J. (2009). *Trends Anal. Chem.*, **28**, 1137.
Tong P. and Imagawa T. (1995). *Anal. Chim. Acta*, **310**, 93.
Van Leeuwen S. P. J. and de Boer J. (2008). *J. Chromatogr. A*, **1186**, 161.
Varaprath S., Stutts D. H. and Kozerski G. E. (2006). *Silicon Chem.*, **3**, 79.
Verenitch S. S., de Bruyn A. M. H., Ikonomou M. G. and Mazumder A. (2007). *J. Chromatogr. A*, **1142**, 199.
Wells D. E. and Echarri I. (1994). *Anal. Chim. Acta*, **286**, 431.
Wells D. E. and Hess P. (2000). Separation, clean-up and recoveries of persistent trace organic contaminants from soils, sediments and biological matrices, in *Sample Handling and Trace Analysis of Pollutants, Techniques, Applications and Quality Assurance* (ed. D. Barceló), Elsevier, Amsterdam, pp. 73–116.
Wells D. E., Echarri I. and McKenzie C. (1995). *J. Chromatogr. A*, **718**, 107.
Wise S. A., Poster D. L., Kucklick J. R. *et al.* (2006). *Anal. Bioanal. Chem.*, **386**, 1153.
Zhang S., Zhang Q., Darisaw S. *et al.* (2007). *Chemosphere*, **66**, 1057.
Ziegenhals K., Hübschmann H.-J., Speer K. and Jira W. (2008). *J. Sep. Sci.*, **31**, 1779.

7

Nutrients

Carlos Rocha[a] and Malcolm Woodward[b]

[a]*Geography Department, School of Natural Sciences, Trinity College Dublin, Dublin, Ireland*
[b]*Plymouth Marine Laboratory, Plymouth UK*

7.1 INTRODUCTION

The term 'monitoring' could suggest in popular understanding the process by which the wellbeing of a patient in hospital is closely followed by a team of medical staff. This image is reinforced by repeated, regular evaluation of a medical chart designed to show the progress and evolution of the patient's vital signals. However, without any pre-available standard by which to compare any temporal changes on the medical chart, this becomes a meaningless exercise for any specific patient, since there would be no way to know whether the person was improving, responding to medication or indeed regressing. The a priori availability of this comparative health standard will thus rely on sound background knowledge of the pre-monitoring state of the patient's health and the main drivers of its change, which then provide a rational framework for preventive or palliative care. Indeed, the definition of what healthy status is in the first instance constitutes the foundation on which health management stands, and this status may be different for different patients. While the comparison between the medical field and the environmental one is not original

Chemical Marine Monitoring: Policy Framework and Analytical Trends, First Edition.
Edited by Philippe Quevauviller, Patrick Roose, and Gert Verreet.

(Elliott and Cutts, 2004), the parallels that can be drawn between health and environmental monitoring are obvious. In fact, whilst an assessment of any ecosystem provides an overview of its main characteristics and also the identification of potential risks, it will also highlight the drivers and influences that threaten its health (US EPA, 1998). Monitoring of an environmental system requires measuring and recording, like the medical chart, of a set of chosen characteristics against some sort of standard, set previously (Elliott and De Jonge, 1996), in order to understand whether the particular ecosystem is deviating significantly from its previous state, and to then, where possible, give advice and implement preventive or remedial actions.

Environmental management, therefore, stems from the study of the main drivers of any ecosystem change, and in order to define boundaries within which different measures of environmental quality status might be identified, it then progresses through a system of constant evaluation of its responses to further forcing, including the result of implemented corrective measures (Grumbine, 1994). Hence, a rational approach to monitoring the environment, either to ascertain its evolving status (against a predefined standard) or to provide forecasting on its potential reaction to external and internal forcing (including management policies), will require the study of environmental change feedbacks and also a strong definition of what the monitoring should specifically entail (Boesch, 2006; Foden *et al*., 2008). This will include the level of expectation that is required in meeting the primary goal that is established, and aimed for, by any stated conservation policies.

Within a limited view, environmental monitoring is deemed necessary to judge the trade-off between an acceptable loss of environmental quality and the derived socio-economic benefits caused by this loss. However, there is the notion whereby acceptable degradation limits to an ecosystem may be established as the compliance targets for environmental policy, as long as the corresponding socio-economic gain is deemed advantageous (Preston and Wood, 1971). This has evolved by incorporating the knowledge of the interdependent nature of ecosystem components, including human occupancy, and then integrating the ecological, social, economic and stakeholder perspectives, which stem from the recognition of the fundamental importance of ecosystem structure and function for sustainability (WCED, 1987). This outlook then provides a management framework by which environmental impacts may be minimized and/or kept within predefined acceptable limits (Gray *et al*., 1991), aiming ultimately to provide safe and acceptable parameters for human existence (Rockstrom *et al*., 2009).

As such, environmental monitoring programs can actually be seen as a means to an end. They can be viewed as a quantitative foundation underpinning management programmes that are aiming to conserve whole-ecosystem health, inclusive of humans (Grumbine, 1994), by preserving their ecological integrity, functionality and services rendered (McLeod *et al*., 2005); however, this latter aim is usually taken for granted (Costanza *et al*., 1997). The current ecosystem-based approach (EBA) to environmental management has evolved from the recognition that compartmentalization of impacts by sector and driver, pervasive in previous management strategies, was ineffective (Arkema *et al*., 2006).

Within this rationale, it is then trivial to take the understanding that environmental monitoring is more than just the temporal collection and analysis of data. It must have the stated objectives of: (1) understanding the evolution of environmental quality of a system against a set of standards or reference guidelines that are set before monitoring commences, (2) in the case where a management strategy is in place, evaluating the effects of its implementation, and (3) providing robust temporal data trends from which the forecasting of environmental quality status and the subsequent modelling of potential impacts of human activity can help to inform policy decisions (McLeod *et al*., 2005). As such, monitoring requires physically meaningful benchmarks for comparison and reference, which in turn are drawn from established environmental quality standards. These follow on from the scientific understanding of the main drivers of ecosystem health. The touchstone here (Murawski, 2007; Murawski *et al*., 2010) is as to whether the system is understood sufficiently for the determinants of environmental quality to be actually stated in the first place. These determinants include the precision of forecasting tools used to identify tipping points beyond which the ecosystem might not recover back to its previous state. This might then induce the collapse of its services, and thus compound the failure of management's ultimate goal of sustainability (Botsford *et al*., 1997). It may be impossible to predict to a high degree of accuracy the moment when these tipping points might be reached; therefore, applying a precautionary approach may be the most prudent course of action whenever increasing deviation from the benchmark is observed. This capability to identify deviations from pre-existing states requires that the confidence with which data are collected and recorded sustains the statistical significance of discriminatory temporal trend analysis. Thus, an important feature of any monitoring programme will be the precision with which data are measured and reported, the quality control systems and the intercomparability of data sets, particularly when different laboratories are executing monitoring programmes, and where different states share management responsibilities over whole ecosystems, hence linking monitoring with the science of metrology (Quevauviller, 2004).

The identification of the determinants of marine environmental health requires sufficient background information and rigorous functional analysis in order to be confidently established. These determinants can only be drawn from previous appropriate research programmes, designed with well-defined research questions and aims. If there is substantial doubt on the causal links between functional aspects of the ecosystem and its determinants, the rational foundation of any monitoring program that informs a management strategy may be greatly undermined, thus leading to a waste of time, effort and financial resources, something managers would be rightly concerned about (Elliott and De Jonge, 1996). Hence, within the Ecosystem Based Approach (EBA) to management, monitoring plays a vitally important role. It ideally works as a continuous feedback loop informing both the nature of any causal links between the monitored variable trends and the health of the ecosystem, plus also our ability to discern these with confidence. Likewise, to continue to be effective in sustaining the ultimate goal of protecting and restoring the healthy function of marine ecosystems, monitoring should be adaptive, and should also incorporate any new knowledge on ecosystem function, underlying processes and methodological advances to metrology, as they become available.

These general principles are inscribed in recent European environmental policy via the Marine Strategy Framework Directive (MSFD) (European Commission, 2008), and are likely to be inserted into the legal marine environment framework in the USA soon (PEW Oceans Commission, 2003; USCOP, 2004; Interagency Ocean Policy Task Force, 2009). For example, the EU guidelines for environmental monitoring programmes under the Water Framework Directive (WFD) (European Commission, 2000), which will provide similar guiding principles for the MSFD, will aim to ensure that member states are able to provide an 'assessment on the deviation of observed conditions to those that would normally be found under reference conditions' (see point (1) above). This is whilst still providing for natural habitat variation, accounting for the range of natural and anthropogenic variability of all quality elements in all water types, and accounting for their interactions, in order to provide detection of a full range of potential impacts, and to enable a robust classification of ecological status criteria (points (2) and (3) above). The guideline document (European Commission, 2003) also adds that it does not provide guidance on how to (a) establish aforementioned reference conditions, (b) develop assessment and classification systems for different water types or (c) provide constraints into data analysis and reporting formats. This apparent lack of definition arises from the fact that the drivers of ecosystem function and the nature and extent of pressures over the same environment are geographically variable, a crucial tenet of the EBA to marine environmental management (Arkema *et al*., 2006). In addition, the need to balance different societal ways in which resources are allocated or prioritized also have a bearing on the trade-off between the acceptable environmental loss and perceived socio-economic benefits mentioned before. This will tend to be an important issue during the paradigmatic shift from traditional sectorial management approaches and the full implementation of the EBA as discussed (Borja, 2006; Barnes and McFadden, 2008).

Whilst both former points are recognized in the basic tenets of the ecosystem approach to management (Apitz *et al*., 2006; Arkema *et al*., 2006; Barnes and McFadden, 2008; McFadden and

Barnes, 2009), their potential conflict in a scenario of multi-stakeholder participation in ecosystem management remains the fulcrum of ongoing discussions over the way to implement the conceptual paradigm shift in environmental management worldwide (Guerry, 2005; de Jonge *et al.*, 2006; Borja *et al.*, 2008; Brack *et al.*, 2009). This is particularly true with regard to marine ecosystems, which, because of their geographical extent and trans-boundary nature, uphold the need to integrate multinational policies and management strategies. It is also because oceans tend to be the ultimate convergence point of cumulative stressors imposed on the health of the global ecosystem, thus requiring a much larger knowledge basis about the interdependence of processes, functions and drivers of environmental health. This knowledge, which is increasing at a very rapid pace, is far from being comprehensive to provide sufficient information.

The overlapping areas within the discussion of these latter points are often, perhaps understandably, at the heart of the apparent conflict between managers and scientists regarding the appropriateness and timeliness of EBA implementation (Murawski, 2007) to the management of marine ecosystems. Nevertheless, it is reasonably consensual that, in order for a monitoring strategy to be useful for management, the following questions should be addressed: (1) What is the health status of marine and coastal waters (definition of state and functionality)? (2) Where, how and why are water quality conditions changing (spatial and temporal coverage of processes, functions and drivers)? (3) If observed, are any of the changes relevant to ecosystem health and would then this imply adaptation of the monitoring programme to extended or even different priorities, in view of the new methodological advances or improved knowledge about the function of the ecosystem?

Notably, a workable response to these queries implies the necessity of long-term, scientifically informed monitoring programmes in order to understand if the observed trends are independent from the natural variability of the underlying processes relevant to ecosystem function, which will cover a wide range of temporal and spatial scales. Ancillary questions include what data should be collected, coupled with the understanding that collected data might not resolve temporal variability of functional processes. Also, where should data be collected and how much data is enough in order to provide spatial and temporal coverage of processes? Finally, it is important to consider what the quality of the data is, including its precision and accuracy, with a view to ensuring its traceability and comparability across different laboratories and stakeholders.

7.2 MARINE NUTRIENT MONITORING

It is well established that humankind relies on ecosystem goods and services provided by the healthy function of the marine ecosystem as a whole (UNEP, 2006). These include food, water, fuel and minerals, medicine, water purification, storm and flood protection, moderation of climate through the control of atmospheric composition, weather modulation, transport, recreation and, importantly, the key supporting service of nutrient cycling. Generation and availability of these different provisioning, regulating and supporting services depends on the interconnections of different characteristics and functional processes within marine ecosystems, including trophic structure and biodiversity, the biogeochemical cycling of elements between the land, atmosphere and sea and, naturally, the modulation of these by human activity. Anthropogenic influence over marine ecosystem functionality occurs via multiple stressors, most derived from cumulative impacts and feedbacks of different drivers of environmental change, and largely interconnected to varying degrees. These include climate change (Barnett *et al.*, 2001; Walther *et al.*, 2002), the alteration of coastal ecosystems, habitat destruction and loss of biodiversity (Lotze *et al.*, 2006), nutrification (Mackenzie *et al.*, 2002), either from land sources (Howarth *et al.*, 1996; Seitzinger *et al.*, 2002) or atmospheric deposition (Duce *et al.*, 2008), the dumping of sewage (UNEP and GPA, 2006), changes in sediment loading due to the constraint of hydrological flow (Freeman *et al.*, 2007), increasing levels of eutrophication (Nixon, 1995; Smith *et al.*, 2006) and the decline of fish populations (Myers and Worm, 2003).

However, the state of health of the oceans has been assessed with rising urgency and been found increasingly to be of concern (GESAMP, 2001a; UNEP, 2006; UNEP and GPA, 2006; Halpern *et al.*, 2008). Visible signs of collapsing ecosystem integrity and, hence, the ability to provide goods and services are observed in different ways, including increases in the size and spread of hypoxic or oxygen-deficient areas (dead zones) in the oceans, the increasing frequency and extent of harmful algae blooms (HABs), greater global-scale bleaching of coral reefs due to ocean acidification, and the collapse of fish stocks and massive habitat loss, particularly in the coastal zone (Nelleman *et al.*, 2008).

Eutrophication has been identified as one of the major threats to marine ecosystem health since the early 1970s (Ryther and Dunstan, 1971). Both hypoxic phenomena (Diaz and Rosenberg, 2008) and the increase in HABs (Heisler *et al.*, 2008) are causally linked to eutrophication, mostly driven by increasing organic matter and nutrient loading (Cloern, 2001; Rabalais *et al.*, 2003, 2009; Smith, 2003; Smith *et al.*, 2006). The European Union, through the WFD (European Commission, 2000), the USA, through the implementation of the Clean Water Act (McLusky and Elliott, 2004; Kapp, 2005), and other nations are pursuing the implementation of an EBA to the management of the marine ecosystem. These include Canada (O'Boyle and Jamieson, 2006), Australia (Foster *et al.*, 2005; Fletcher, 2006) the Republic of South Africa (Taljaard *et al.*, 2006) and others who have introduced comprehensive legislative framework, including nutrient monitoring guidelines in the North American and European cases (US EPA, 2001; European Commission, 2003), that address, inter alia, eutrophication as a process that affects ecosystem health in brackish and coastal waters.

Our conceptual understanding of eutrophication merits mention at this stage, since it reflects the evolution of this particular approach to environmental management, especially with regard to marine ecosystems. The evolving nature of the definition of eutrophication in the context of monitoring of coastal waters and the assessment of their ecological health has been previously discussed at length (Nixon, 1995; Cloern, 2001; Andersen *et al.*, 2006). In Europe, management strategies have focused on measures that target the sources of nutrients that cause eutrophication (Elliott *et al.*, 1999; de Jonge *et al.*, 2002), as the result of this excessive nutrient enrichment of the water column causes accelerated growth of algae and higher forms of plant life (OSPAR Commission, 2003b). However, eutrophication as a process is almost exclusively confined to the coastal or near coastal zone, since accelerated nutrient loading can be traced back to the intensification of land-based activities (Matson *et al.*, 1997). Indeed, the European policy frameworks addressing the issue are grounded in other legislative directives addressing land-based inputs of nutrients or organic matter (GESAMP, 2001b), such as the Urban Wastewater Treatment Directive (European Commission, 1991) and the Nitrates Directive (European Commission, 1998). However, Cloern (2001) noted the evolution of the conceptual understanding of eutrophication and postulated that its future outlook within environmental management strategies should be organized around five main questions: (1) How do system-specific attributes constrain or amplify the responses of coastal ecosystems to nutrient enrichment? (2) How does nutrient enrichment interact with other stressors (toxic contaminants, fish stock exploitation, aquaculture, human influences, presence of non-indigenous species, habitat loss, climate change, hydrologic manipulations) to change coastal ecosystems? (3) How are responses to multiple stressors linked? (4) How does human-induced change in the coastal zone impact the Earth system not only as a habitat for humanity, but also for other species? And finally, (5) How can a deeper scientific understanding of the coastal eutrophication problem be applied to develop tools for building strategies for ecosystem restoration or rehabilitation? This perspective entails not only the assessment of the health status of an ecosystem, but also an understanding of the processes underpinning it, which might be closer to the longer term goal of following potential shifts in ecosystem function with a view to protect services rendered on a global scale.

Since the emphasis of the new EBA to management is going to be placed on ecosystem function (Guerry, 2005), then, for example, shifts in phytoplankton community composition in the oligotrophic areas of the ocean in response to changes in nutrient ratio availability (Tyrrell, 1999; Klausmeier *et al.*, 2004; Mackey *et al.*, 2009; Moore *et al.*, 2009) potentially caused by enhanced atmospheric deposition of nitrogen (Duce *et al*., 2008), amongst other stressors, should be considered as well. Indeed, the most rapidly growing inputs of anthropogenic N to the marine ecosystem, both in terms of loading and geographical scale, are atmospheric deposition and groundwater inputs (Paerl, 1999). As the understanding of eutrophication grows, and additional links are drawn between differential drivers of nutrient enrichment and their impacts on the global health of the marine ecosystem (Vaquer-Sunyer and Duarte, 2008; Rabalais *et al*., 2009; Davis *et al*., 2010), so too grows the realization of the critical importance in developing strategies for marine nutrient monitoring into the future. This includes the choice of which parameters are monitored and what sampling and analytical techniques are used, all of which are a function of the target matrices and quality guidelines that are crucial to underpin the traceability of the gathered data.

7.3 NUTRIENT MONITORING GUIDELINES

Notwithstanding the debate surrounding the legislative implementation and effective use of an EBA to marine ecosystem management globally (VanderZwaag and Powers, 2008; Curtin and Prellezo, 2010), facilitating the overall aim of maintaining or recovering a healthy ecological status for marine waters requires a set of quality objectives to be agreed upon (Gray *et al*., 1991). These have to reflect the underlying processes and function of the monitored ecosystem, and must conceptually include the acceptance of geographical uniqueness and context (Guerry, 2005; Apitz *et al*., 2006). In Europe, regional implementation of the ecosystem approach to marine ecosystem management draws fundamentally on the experience of the Convention on the Protection of the Marine Environment of the Baltic Sea Area, known as the Helsinki Convention (HELCOM, 1974), signed on 22 March 1974 by the seven Baltic states, Denmark, Finland, Germany (split at the time between GDR and FRG), Poland, Sweden and the USSR, and the Convention for the Protection of the Marine Environment of the Northeast Atlantic (OSPAR, 1992), signed on 22 September 1992 in Paris, by Belgium, Denmark, Finland, France, Germany, Iceland, Ireland, Luxembourg, Netherlands, Norway, Portugal, Spain, Sweden, Switzerland and the UK, and accepted by the European Community.

The subsequent amendments to both conventions have included more countries, expanded their tenets to include habitat and species protection and drawn on significant, multilateral experience in the monitoring of regional seas, particularly in the case of eutrophication (OSPAR Commission, 2003a,b; Painting *et al*., 2005; Backer *et al*., 2009; Borja and Collins, 2009). This experience has been recognized by the European Union in the WFD (European Commission, 2000) and will be determinant to the implementation of the MSFD (European Commission, 2008). For example, key features of chemical and physicochemical quality elements for coastal waters listed in the monitoring guidelines for nutrients under the WFD (European Commission, 2003) are consistent throughout the current OSPAR and HELCOM countries, which by itself already entails a significant degree of trans-boundary integration of monitoring programmes and environmental policy, drawn from common interests. In addition, although some differences persist in the definition of the best biological indicators of eutrophication, for example (OSPAR Commission, 2005; Backer *et al*., 2010), monitoring guidelines applied by both regional sea monitoring programmes (OSPAR Commission, 1997, and subsequent updates; HELCOM, 2008) are included in the international standards to be followed in the implementation phase of the WFD (European Commission, 2003: 153). Hence the accumulated experience in integrating the role of science in policy making (de Jong, 2006) will certainly underscore the future implementation of the MSFD.

In the case of eutrophication, a particular set of ecological quality objectives (EcoQO's) have been developed in Europe to assess the biological response to nutrient enrichment (Anon, 2002). According to the outcome of the Schveningen Workshop (Skjoldal *et al*., 1999), ecological quality (EcoQ) is defined as an overall expression of the structure and function of the aquatic system, thereby establishing EcoQO's as the desired level of EcoQ relative to the reference level at which anthropogenic influence over the ecological system is minimal (Heslenfeld and Enserink, 2008; Johnson, 2008). In the spirit of EBA to marine ecosystem management, EcoQs suggested by the Bergen declaration include both biological and physicochemical indicators of marine ecological health, as follows:

1. winter nutrient concentrations – dissolved inorganic nitrogen (DIN) and dissolved inorganic phosphate (DIP);
2. oxygen;
3. phytoplankton chlorophyll *a*;
4. phytoplankton indicator species for eutrophication (system-specific); and
5. changes/kills in zoobenthos in relation to eutrophication.

Nutrient concentrations are direct evidence of existing chemical support for phytoplankton growth and, hence, causal variables of eutrophication, whereas chlorophyll *a* and oxygen are response variables, depicting the reaction of a system impacted by eutrophication through the increase in water column primary production and eventual decrease in oxygen levels through enhanced mineralization of the organic matter created in the process. Related EcoQO's are hence defined (OSPAR Commission, 2003) as: winter DIN (the sum of the concentrations of ammonium (NH_4^+), nitrite (NO_2^-) and nitrate (NO_3^-)) and/or DIP (soluble reactive orthophosphate concentrations, PO_4^{3-}) which should remain below stated elevated levels, concurrently defined as the concentration $> 50\%$ above salinity-related and/or region-specific natural background concentrations. Winter values are taken here because they correspond to the season of lowest algal activity due to light and temperature limitation, thus providing an assessment of background nutrient levels in the absence of significant primary production within the monitored area.

Unresolved issues regarding the applicability of such a definition of EcoQOs to the marine environment remain specifically in offshore areas, because part of the salinity harmonization standard refers to what would be considered to be the marine component in the fresh/marine water continuum characteristic of the coastal ocean, and is taken from relevant nutrient background levels in offshore areas close to the site in question, thus compounding the mixing gradient paradigm underpinning coastal oceanography. For example, Atlantic water entering the shelf seas of northern Europe provides the background conditions from which to derive EcoQ standards for the North Sea (Heslenfeld and Enserink, 2008).

Nevertheless, whilst the effort to develop a common monitoring strategy for the open ocean increases, some operational determinants of marine nutrient monitoring, namely the concentrations of NO_3^-, NO_2^-, NH_4^+ and PO_4^{3-} are directly applied in establishing the EcoQ and as the causal variables of eutrophication in regional seas. These compounds, along with silicate ($Si(OH)_4$), are commonly designated by the generic term 'nutrients' and are an integral part of nutrient monitoring programmes, since both their quantities and relative availability (nutrient ratios) have a direct bearing on phytoplankton speciation and growth. Total phosphorus (TP), defined as DIP + DOP (dissolved organic phosphorus), and total nitrogen (TN), defined as DIN + DON (dissolved organic nitrogen) +PON (particulate organic nitrogen), are also included as measurements because of their importance to ecosystem function analysis and nutrient budgets. However, in the context of ocean monitoring, the absolute quantities and temporal variability might not be relevant in the same way as is observed in shelf seas that have a continental land-mass influence, but they are rather indicative of continuous and sustained pressure (Scheffer and Carpenter, 2003; Scheffer and van Nes, 2004). This sort of analysis requires long-term monitoring and the consequent provision of high-quality data sets, particularly in oligotrophic regions.

Essentially the same suite of parameters is recommended in the monitoring guidelines of ongoing management programs. Some are required (European legislative framework) or prioritized (Tier 1) in the North American proposed tiered system (Caffrey *et al*., 2007), whilst others are seen as supporting parameters needed to understand functional links between nutrient availability and biological primary production. Hence, the knowledge acquired on the function of marine ecosystems and the drivers of eutrophication has helped to improve the analytical effort in monitoring programmes through more informative lists of target compounds. For example, in the USA, the EPA suggested in 2001 (US EPA, 2001: Table 4-1, Ch. 4, pp. 10–12) that TN and TP were sufficient as preferred parameters to indicate water quality status for coastal waters, while after the publication of the Ocean Blueprint for the 21st Century (USCOP, 2004), the Nutrients Workgroup (Caffrey *et al*., 2007: 5, Table 1) suggested that there should be instead a tiered system of parameters to be monitored in marine environments, including offshore waters, listing TN, NH_4^+, $NO_3^- + NO_2^-$, TP, DRP and DRSi (dissolved reactive silicate) as priority targets of quantitative monitoring (Tier 1). Then adding TDN, TDP, PN (particulate nitrogen (PIN + PON)) and PP (particulate phosphorus (PIP + POP)) as supporting parameters (Tier 2). Similarly, the European legislation framework in the WFD (European Commission, 2000) required (European Commission, 2003: 72, Table 3.12), TN, NH_4^+, $NO_3^- + NO_2^-$, TP, PO_4^{3-} and DRSi, while recommending POC (particulate organic carbon), PON and POP as variables that would support the functional understanding of eutrophication. In Annex III of the MSFD (European Commission, 2008: L 164/36, Table 1), the spatial and temporal distribution of nutrients (DIN, TN, DIP, TP), TOC (total organic carbon) and oxygen are listed as chemical features to be monitored.

However, local or salinity-specific reference levels will be essential to the rational application of EcoQOs, and this point should be strongly emphasized, since the intent to extend nutrient monitoring programmes to cover a continuum of aquatic environments effectively targets the whole hydrological cycle, instead of just the link between land-based nutrient sources and the aquatic basins directly interconnected in the vicinity of the coastal zone. For example, whilst according to the OSPAR comprehensive procedure DIN and DIP concentrations (and their ratios (N/P)) may be compared with local EcoQ standards, these might be harder to establish in offshore areas due to the lack of spatial coverage, or indeed the limited data availability (e.g. Painting *et al*., 2005). These factors alone condition the evaluation of the statistical significance of long-term trends, particularly in derived parameters such as nutrient ratios. This is because of the intrinsic precision and accuracy needed for the determination of each component of the ratios, and also because the robust linkages between the parameter values and ecosystem response might actually be difficult to establish further out at sea and away from land (Scheffer and van Nes, 2004).

This fact makes the comparability and traceability of nutrient data in the global ocean a matter of crucial importance, particularly when these data underscore knowledge of global environmental change. The Intergovernmental Panel on Climate Change, for instance, recognizes the importance of the issue by stating in its 2007 report (Bindoff *et al*., 2007):

> Uncertainties in deep ocean nutrient observations may be responsible for the lack of coherence in the [observed] nutrient changes. Sources of inaccuracy include the limited number of observations and the lack of compatibility between measurements from different laboratories at different times

This state of affairs is not novel. In fact, the Brundtland report (WCED, 1987: 315, item 2.2.2, point 49) already stated that:

> Although more is known about the state of the global environment now than a decade ago, there are still major gaps and limited international capability for monitoring, collecting and combining basic and comparable data needed for authoritative overviews of key environmental issues and trends. Without these, the information needed to help set priorities and develop effective policies will remain limited.

The need for intercomparability in order to comprehend ecosystem functional resilience to human pressure is made more urgent since the global biogeochemical cycles of nitrogen and phosphorus closely link food security, sustainability, biodiversity and ecosystem health (Vitousek *et al.*, 1997; UNEP and WHRC, 2007; Cordell *et al.*, 2009).

However, the issues of data quality, comparability and traceability are insufficiently discussed (Quevauviller *et al.*, 2008) in both the WFD and, by extension, in the MSFD, and it is generally accepted as a serious shortcoming of available nutrient monitoring programmes, in spite of the consensual agreement that monitoring has to be designed, necessarily, in a way that permits long-term trends or shifts in concentrations to be detected over vast geographical areas. Hence, the demonstration of data comparability and its traceability is not only central to the trans-boundary and multi-stakeholder nature of marine nutrient monitoring in the future, it is the vital point on which the coherent feasibility of an EBA to marine ecosystem management hinges. According to the Vocabulary in Metrology (VIM), traceability is defined as (ISO, 2007):

> the property of the result of a measurement or the value of a standard whereby it can be related to stated references, usually national and international standards, through an unbroken chain of comparisons all having stated uncertainties.

Because the definition of traceability inherently implies that the smaller the chain of comparison the smaller will be the uncertainty of the final reported result, it becomes clear that, within the process of quantization of nutrient concentrations, the common implementation of all steps of the procedure from sampling to final analysis and reporting, has to be carried out if possible under the utmost precautions so as to limit error and inconsistency. Accordingly, each step of the process should ideally be followed by a rigorous quality assurance protocol referring to a standard methodology in order to keep the chain of metrological comparisons intact within any prestated boundaries of uncertainty. The latter would ensure, for instance, that even in the cases where different analytical methods or operational procedures are applied to each step, intercomparability of the final result could be achieved within statistically robust scenarios.

The process of nutrient analysis in seawater includes a succession of actions, namely: (1) sampling, (2) calibration of analytical system in preparation to receive samples (can be carried out whilst sampling occurs to minimize any storage), (3) analysis of samples; however, if this is not possible, then (4) storage and/or preservation of representative samples for later analysis, and finally (5) calculation and reporting of results. Whilst the notion of how traceability may be understood within chemical monitoring programmes in the WFD has been examined before (Quevauviller *et al.*, 2008), the analytical determination of nutrient concentrations in saline matrices is affected by a suite of specific difficulties that have to be understood in the context of a future guarantee of the quality of results, particularly in the context of monitoring programmes designed to cover a wide range of aqueous matrices, from dystrophic waters to the vast oligotrophic regions of the global ocean.

One observes immediately that even the way in which each of the required parameters is determined may influence the outcome of a comparability exercise, through the accumulation of quantification error and operational inconsistencies leading to different (and oft untraceable) uncertainty levels in the final result. For example, DIN is the sum of the concentrations of three different solutes. Even disregarding the known issues affecting the determination of ammonium in seawater (e.g. Aminot *et al.*, 1997), both nitrate and nitrite are normally determined as a total ($NO_3^- + NO_2^-$), which must be carried out by ensuring complete reduction of nitrate to nitrite. Nitrite must also be determined in isolation, sometimes after the removal of the cadmium reduction process from the system, but normally with different methodological precision and accuracy (e.g. Pai and Riley, 1994). This makes the process intrinsically dependent on (a) the technical and operational proficiency of the analyst, a rather subjective and undeterminable influence on the result, and (b) the efficiency of the reduction step of nitrate to nitrite, which is more objective and can be traced if internal quality control (IQC) protocols are followed and are properly documented. However, if no effective certified standard reference material for nutrients is used, hence providing a guide to evaluate the consistency of the whole chain of procedures from sampling to final reporting, then the traceability of the measurement will always have the potential to be compromised, or at the very least affect the accuracy of the final reported DIN concentration. This use of a reference material in the context of the traceability and accuracy issues mentioned in the example above is, nevertheless, relevant whatever the nutrient species under investigation.

7.4 QUALITY ASSURANCE AND CONTROL IN MARINE NUTRIENT MONITORING

The success of a chemical monitoring programme in meeting the aims of a defined management strategy depends on the implementation of a robust quality assurance programme. The International Union of Pure and Applied Chemistry (IUPAC) defines quality assurance in analytical laboratories as 'the essential organizational infrastructure that underlies all reliable analytical measurements' (IUPAC, 1997). More generally, this implies that there needs to be a definition of all planned and systematic actions necessary to provide adequate confidence in that a product or service satisfies the required standard of analysis (Thompson and Wood, 1995). In other words, any part of an analytical procedure that can influence the quality of the measurement so it meets the needs of end users has to be considered and optimized in a way that allows its outputs to be treated statistically, providing results of known accuracy within a known degree of probability (e.g. Dux, 1990).

Measurement quality in the aforementioned sense can, however, be influenced by a series of factors that may be divided into three separate categories: (a) the training and experience of personnel, (b) laboratory management practices (e.g. adequacy of the laboratory environment, record keeping, maintenance and calibration of instruments, etc.), and finally (c) the execution of the whole sample analytical process to produce the final results according to defined protocols. Whilst it is essential that all these factors are optimized individually in order to improve the quality of the final product, which is a reliable measurement, we will concentrate on the last category in this chapter.

For nutrient analysis, technical operations comprise an analytical chain that progresses stepwise through sampling, storage and preservation (if required), calibration of the analyser, analysis of the samples and, finally, calculation and reporting of measured values. Yet, because nutrients are present throughout the marine environment in a wide range of concentrations, quality assurance procedures should aim at guaranteeing that, regardless of the specific choice of methodology, the measured quantity should represent the concentration at the instant of sampling (the true value); that is, the measurement should be accurate. However, accuracy is a qualitative concept that describes the closeness of agreement between the result of a measurement and the true value of the measured quantity (IUPAC, 1997), and should not be confused with precision, which depicts the nearness of agreement between a series of independent measurements of the same quantity carried out under prescribed conditions. Hence, a result, however reproducible, may in fact be far from accurate. As pointed out by Quevauviller *et al*. (1995), many wrong but highly reproducible (i.e. precise) results have led in the past to misinterpretation of environmental processes. So, the absolute fundamental requirement within collaborative monitoring programmes worldwide is to accomplish the intercomparability of data obtained by different laboratories, irrespective of country of origin (Topping, 1992; Thompson and Wood, 1993). This has not been achieved to date, with significant bias being found and reported in the literature dealing with large-basin or global studies (e.g. Gouretski and Jancke, 2000). These observations have ultimately led to the acceptance that the inability to provide a fully comprehensive understanding of shifts in nutrient availability in the global ocean and to reconcile sectorial mass balances, particularly in studies involving deep ocean areas into a coherent global picture may be largely due to the inaccuracy of the different available datasets (Bindoff *et al*., 2007). However, the situation has a positive outcome in the foreseeable future. Within the Japanese-led INSS programme there have been a series of repeat hydrographical sections on WOCE and CLIVAR lines that are starting to allow intercomparability to be achieved. As more teams use the reference materials and these become more widely available, so this intercomparability will grow, hopefully exponentially, once the reference materials are finally certified.

Hence, improving the accuracy of nutrient monitoring programmes worldwide necessarily implies assessing the accuracy of nutrient measurements across different countries and laboratories and evaluating their comparability through laboratory comparison studies as part of a concerted effort. Comparison studies can be divided into three major types (IUPAC, 1997): (1) method-performance studies, in which several laboratories follow the same procedure and analytical methodology to measure performance parameters of the analytical method, (2) laboratory performance studies (LPSs), where a group of laboratories use their chosen methodology to measure a concentration of an analyte in order to compare the performance of different laboratories or different analytical technicians, and finally (3) material-certification studies, in which a reference value is assigned, with stated uncertainty, to a common test sample using the best-performing laboratories and the less-biased methodologies.

The quality assurance framework, so defined, puts the question of the delay in providing a quality assurance and control scheme to nutrient determinations into perspective, since the three steps above are necessarily interdependent and should progress in a sequential, focused manner, in which no progress beyond each of the stages can be made while guaranteeing an unbroken chain of traceability without the completion to the intended quality standard of the previous one.

7.4.1 Methodology

Various methods for nutrient analysis in seawater have been established (e.g. Wurl, 2009 and references cited therein), these being as a result of early research into the quantitative analytical chemistry of freshwater and then subsequently adapted to the different matrices while limnology as a research area naturally expanded from rivers into coastal waters and out to deep oceans; this was the expansion and development of oceanography and marine chemistry through the 1960s, 1970s and 1980s. Prior to the 1970s, some nutrients were regularly determined by manual colorimetric methods (Strickland and Parsons, 1968, 1972), but the introduction of automated batch analysis, in the form of continuous flow analysis (CFA) systems, predicated the expansion of marine chemistry and chemical oceanography worldwide. By adapting the manual sequence of colorimetric analytic procedures to a sequential, automated batch analysis procedure, the gain was not so much in the ability to process a larger number of samples or the possibility of determining simultaneously the full suite of nutrients from a single aliquot, but rather rested in the stabilization of the conditions in which analysis of several samples progressed, by guaranteeing the exact same prescribed and maintained operational conditions for each sample, which limited random error and increased precision. Unsegmented flow injection analysis (FIA) developed very quickly for several applications (Ruzicka and Hansen, 1986, 2008), but while FIA definitely increases sample throughput and lowers overall reagent and sample consumption, the lack of flow segmentation makes the technique susceptible to transverse sample contamination or carryover within the flowstream in the analytical manifolds, decreasing its sensitivity for the analysis of batches of very low-level concentrations. For seawater samples with very low nutrient concentrations, the technique may thus be insufficient for the required performance goals. However, FIA is still a technique used for porewater nutrient analysis, for instance, where the maximization of speed and the crucial need to use very small sample volumes carrying substantial concentrations of all the nutrients allows for the possibility of sacrificing some of the sensitivity in favour of overall accuracy.

The methods of choice for nutrient analysis are based on long-established manual procedures that have been continuously streamlined and refined over the years (e.g. Strickland and Parsons, 1968, 1972; Grasshoff, 1976; Grasshoff *et al.*, 1983, 1999; Parsons *et al.*, 1984); see Wurl (2009) for an overall summary of these. Their use and operational limitations as studied in marine monitoring programmes are well documented (Kerouel and Aminot, 1987; Kirkwood, 1994, 1996; Lysiak-Pastuszak and Krysell, 2004; ICES, 2005; HELCOM, 2008), whilst their evolving adaptation to CFA systems until the present day (Mee, 1986; Aminot and Kerouel, 2007; Aminot *et al.*, 2009) has been concurrently processed with a basis on frequent laboratory performance studies (UNESCO, 1965, 1967; Jones and Folkard, 1968; ICES, 1977; Kirkwood *et al.*, 1991; Kirkwood, 1994; Glanville *et al.*, 1997; Aminot *et al.*, 1997; Willie and Clancy, 2000; Clancy and Willie, 2003).

However, with few exceptions (e.g. US EPA, 2001), there are no pre-prescribed analytical methods in the regulatory framework for nutrient monitoring programmes. There are several reasons for this, including the fact that the choice of methods by different laboratories in different countries is often made according to the specific nature of the seawater matrix of the area (salinity, pH, quantity of suspended solids, presence of different quantities of organic matter, particularly humic or fulvic acids derived from land-based organic exports, and amino acids). It can also be dependent on the availability of laboratory equipment, technicians and/or the sampling and laboratory conditions.

The disadvantages of prescriptive methods in the case of marine nutrient monitoring programmes are due largely to some of the assumptions made regarding method performance characteristics. When developed and published in the literature, the prescription methods usually only define a detection level, in effect directly dependent on the sample solution matrix and analytical chain performance, which includes the proficiency of the technician, the sampling conditions (when applicable) and also any laboratory quality assurance procedures. However, these often do not provide sufficient information on the precision and quantization limits of the measurements needed to comply with regulatory requirements expressed or implied in monitoring programme guidelines, since samples are often purposefully chosen within uniform matrices or indeed prepared specifically for method development and testing. In addition, there is the question of transferring the successfully tested method from the laboratory environment to 'real-world' conditions. The transfer will again be affected by matrix differences, but more so by sampling and storage conditions or, in other words, by the pre-analysis steps in the analytical chain. Moreover, laboratory capabilities to detect and quantify a specific analyte vary on a daily basis (laboratory temperature, air quality in the sampling processing area, operator changes, etc.). This variability is naturally set to augment when different laboratories are compared. To compensate for this uncertainty, too subjective to evaluate without some sort of overview of all relevant steps of the quality assurance process, regulatory instances opt to define minimum quantization standards for the target species rather than prescribing a benchmark method for a specific parameter. Also, an important point in this context is the inertia to be overcome if a certain suite of methods that have become engrained into the laboratory methodology, tweaked and optimized over the years, have to be changed radically for some reason. These will not change unless proven to be exceedingly better than the previous ones, and the burden of proof without a certified reference material (CRM) is much larger.

For example, OSPAR (OSPAR Commission, 1997: 4, Table 1) delineates detection limits for phosphate, nitrate + nitrite, nitrite, ammonium and silicate, whereas the EPA (US EPA, 2001: 4–10, Table 4-1) suggests not only the range or detection limits, but also references particular analytical protocols, offering several alternatives in each case. The prescription of analytical protocols under some regulatory frameworks has often more to do with the fact that previous IQCs have been developed for a standard methodology, and this way their direct application by stakeholder institutions or individuals would benefit from existing quality assurance. However, unless each laboratory conducts a rigorous quality control study with the method of choice on the matrix it is analysing, with the analytical equipment available and the laboratory environment in which the analysis is effectively to be conducted, rarely can it be assumed that published performance metrics have been achieved. The alternative is not good practice, and can induce a false sense of security, compounding error and lack of traceability with the additional nefarious effect of encouraging less-rigorous IQC studies by stakeholder laboratories.

There are other perceived disadvantages of prescriptive methods. These include the difficulty of determining degrees of comparability between data gathered by different prescribed methodologies, particularly if IQCs are tentative; a lower incentive for the development of new, potentially better, analytical methods or monitoring technologies, which is a significant constraint within the new EBA to marine monitoring, which predicates an evolving analytical framework and continuous feedback of new methodology to overcome deficiencies of data coverage and functional understanding of the system, as well as reduce the cost. There is also a lack of any extensive interlaboratory comparability studies actually made between different methodologies. It is trivial to understand that, even if well documented, individual method refinements that include particular adaptations to specific matrices are difficult to implement under a restrictive administrative framework empowered by any prescriptive paradigm. Nevertheless, the fundamental point is that the actual method performance often remains unknown, since the use of a particular method does not by itself guarantee achievement of the published performance characteristics, as these are established under a particular set of circumstances which, taken as whole, constitute the complete analytical system (Thompson and Wood, 1995). The end result is that data quality can be questionable, particularly so for the more established methods.

Under the justified assumption that comprehensive marine nutrient monitoring programmes now join a continuum of monitoring requirements that effectively target the hydrological cycle as whole under a holistic approach to ecosystem management, the issues above take an even more poignant note. Effectively, nutrients are now expected to be monitored under a vast range of different, albeit aqueous, solutions. This fact compels analysts to consider the matrix of the sample as a major determinant of the analytical system, defined by Thompson and Wood (1995) as 'the particular set of circumstances surrounding the routine application of a method of choice to the analysis of a single determinant'. The analytical matrix includes all the components of the sample other than the analyte (IUPAC, 2009). Unsurprisingly, in the marine environment, salinity is the major matrix interference in marine nutrient analysis, particularly if CFA is employed (e.g. Mantoura and Woodward, 1983; Stewart and Elliott, 1996; Aminot *et al.*, 1997; Woodward *et al.*, 2010)

The so-called 'salt effect' is caused by the geometry of flow cells in CFA systems. By having rounded edges to maximize uniform plug flow, the flow-cell design creates diffraction effects in the colorimeters that are read as absorbance signals. This is especially counterproductive when analysing samples in the very low concentration range typical of offshore or oligotrophic seawater (Kirkwood, 1996). However, seawater creates other problems affecting the batch analysis of samples. Indeed, the variable ionic strength and buffer capacity found along optical density gradients (from estuaries to coastal oceans and beyond) affect the pH of the mixed reagent plug flow to an unpredictable degree, particularly if no prior knowledge of the main characteristics of the analysed samples is held (Aminot *et al.*, 1997; Aminot and Kérouel, 2001; Woodward *et al.*, 2010). This usually affects the sensitivity of the analysis, and has to be tightly monitored by the analyst. The same effect, since the majority of the methodologies are colorimetric, extends to the wash and blank solutions. Indeed, the optical density of the water carried through the flow cell ideally should be constant at all times, since even small changes cause matrix effects when using a freshwater wash and a saline sample being analysed (for example, when using ultrapure 18.2 MΩ water for the wash solution in a marine-designated analytical manifold). This is a common practice in many marine laboratories and to be recommended, as it provides the only reliable way of ensuring a 'true' zero from which to measure sample peaks against. However, this salinity effect must be measured and corrected when calculating the final results. In addition, fluctuation of reaction rates due to changing ionic strength, variable laboratory, sample or standard solution temperatures, uneven flow rates induced by worn pump tubes or defective pumps, for example, will all affect the colour-forming reaction process. And since CFA (or FIA) techniques are predicated on the notion that colour development (in other words, the extension of the reaction) does not necessarily need to go to completion (Aminot and Kerouel, 2007; Aminot *et al*., 2009), but must progress at a constant rate, then any variability of those factors has to be accounted for to guarantee proportional signals in the detector for different samples and standards.

The disruptive effect of these variations does not necessarily lead to a poor method performance if the training of the operator is comprehensive enough to be aware of these factors, allowing for the adaptation of the operational conditions of the CFA system in use or the chemical composition of the reagents to overcome the issue – but then again, this requires that CFA systems not be used indiscriminately by operators who treat the equipment as a 'black box' producing a precise and accurate result, thus underscoring the importance of including focused operator training in any quality assurance protocols, and then the need for the absolute adherence to the set analytical protocols. Because during the calibration and quantification stage of the analytical chain all these potential adverse effects over the quality of the data may occur, the output has to be validated. This implies verifying that all the accuracy and precision criteria for method performance have been complied with (Ibe and Kullenberg, 1995). However, in reporting the data, it is essential to couple both analytical results and quality assessment parameters that will allow external users to verify the accuracy level in the first instance, and that in order to maintain performance reliability, the laboratory must participate frequently in comparison studies and allow methodology to be scrutinized and vetted by external end users, lest nutrient databases be contaminated by poor results (Goldberg and Taylor, 1985). The first result of laboratory performance studies will be an effective determination of analytical precision and bias, and will evaluate the overall accuracy in determining the concentration of the target analytes in reference samples.

7.4.2 Laboratory Performance Studies

The previous observation about the almost complete lack of comparability and traceability for seawater nutrient analysis worldwide, coupled with the perceived notion of variable accuracy and lack of validation of data, prompted a number of laboratory performance studies executed under ICES and UNESCO sponsorship (UNESCO, 1965, 1967; Jones and Folkard, 1968; ICES, 1977; Kirkwood *et al*., 1991), and eventually led to the launch, in Europe, of the Quality Assurance of Information for Marine Environmental Monitoring (QUASIMEME) programmes in the early 1990s (Wells *et al*., 1993; Cofino and Wells, 1994). These were established to assist European laboratories to improve the data quality they provided in monitoring programmes. For nutrients, assessment of data quality was carried out using Z- and P- scores to describe bias and precision (Thompson and Wood, 1993; Cofino and Wells, 1994; Wells and Cofino, 1997). The results of the early performance study, involving 50 laboratories across Europe, indicated that, in general, systematic rather than random errors were the major cause of poor analytical performance, hence justifying the need to efficiently adapt the analytical range of the instrumentation to the matrix and analyte concentrations in the samples (Aminot *et al*., 1997).

The same programme confirmed overall below-par performance in the determination of ammonium in seawater, a particular problem at low concentration levels generally attributed to contamination of the analytical chain. Whilst for nitrate + nitrite, the analytical system was judged to be under control by the end of the experiment in the vast majority of participating laboratories, poorer performance with phosphate was noted and attributed to both systematic and random (point-source contamination) errors. The strong point made by the QUASIMEME project was in showing that, irrespective of the method employed in the determination of nutrients, basic IQC procedures and analytical chemistry principles remained the central foundation upon which metrological quality lies. The overarching conclusion was that analysts first had to choose and apply the most appropriate methodology for their seawater analysis, then adjust the analytical system to the appropriate range of concentrations, and finally systematically control analytical sensitivity and assess and correct for the blank and calibration procedures with appropriate care for any matrix effects; only then could potential contamination issues be addressed.

However, QUASIMEME and similar performance studies are intercomparison exercises and, as such, are unable to guarantee traceability; thus, even though useful as a process to understand and pinpoint performance bias, they are inadequate to address the main issue of improving day-to-day precision in the laboratory or to achieve a known level of comparability between laboratories. In fact, because such studies focus only on a small part of the analytical chain, from calibration onwards, they overlook the fundamental absence of any quality assurance of the previous links in the analytical chain, namely sampling and sample handling, including in the event of absolute need the preservation and storage of samples in order to maintain unadulterated in-situ (true) concentrations. However, following the success of the QUASIMEME I (1993–1996) programme in Europe, the logical step forward was also to implement quality assurance protocols to sampling and sample handling procedures up to the quantization step in the analytical chain, while further extending the support provided by QUASIMEME to the participating laboratories by following a continuous programme of LPSs in order to maintain the standard attained before, and hence to be able to validate the data produced by marine monitoring activities in which the laboratories were involved. This role was played by the Quality Assurance for Sample Handling (QUASH) project funded by the EU to satisfy one need and QUASIMEME II, now co-funded by the participating laboratories and acting to support quality assurance/quality control for the ongoing OSPAR, HELCOM, ICES and the Arctic Monitoring Programme (AMAP) monitoring programmes (Wells and Cofino, 1997).

Although rarely considered, but oft mentioned as a fundamental component of the analytical chain, the quality of the actual sampling pre-empts the overall accuracy of the determination. In fact, the maxim that establishes that the final measurement is only as good as the initial sample is central to this point. While the issue might be obscured by the perceived methodological performance and 'natural variability', if the ambient concentrations are high, however (such as in the upper reaches of estuaries, dystrophic or eutrophic waters and some coastal lagoons), then it is much more important when confident quality analysis of oligotrophic or offshore waters is required. Given that geographical diversity will be the norm in the EBA to marine management under the recent worldwide regulatory framework (see Sections 7.4.1 and 7.4.2), it is important that this issue is taken into account in any quality assurance procedures, lest it be ignored or overlooked by undemonstrated local circumstances.

From the point of view of improving analytical accuracy, determination of nutrients in seawater samples should be carried out immediately following collection, thus avoiding any intermediate preservation or storage. Once any increase in sample handling or any manipulation occurs, then that will necessarily amplify the risk of random errors, which will then propagate throughout the chain and compromise the accuracy of the final result. However, there are other reasons for avoiding the deferral of nutrient analysis after sample collection. One is the serious possibility of sample alteration occurring during storage and handling, since nutrients are unstable in seawater and their in-vitro reactivity is high. It is generally considered that NH_4^+, NO_2^- and PO_4^{3-} would be the most unstable under storage conditions; however, silicate with elevated concentrations above 40 µm must, if frozen, be defrosted according to certain protocols (MacDonald *et al.*, 1986), otherwise results are severely compromised. However, NH_4^+ is considered the determinand that is most affected also by contamination arising from the careless manipulation of the samples and is prone to many sources of contamination. Indeed, the human skin contains significant amounts of PO_4^{3-}, NH_4^+ and urea, and these vary from person to person; the slightest contact with water samples would promote serious random contamination. However, simply wearing gloves is not an answer, as the type and material of the gloves must be considered, with some types of plastic gloves being a significant source of contamination in themselves. Ideally, the gloves should also be tested as a further quality control check. This is a particularly major factor when sampling oligotrophic areas, where it would cause serious issues with the determinations of DIP, DIN, TN and TOC, where, for example, the actual surface seawater concentrations can be as low as 1–2 nmol L^{-1} (Krom *et al.*, 2005). Similarly, the air we breathe in confined spaces (such as ship laboratories) may contain appreciable amounts of ammonium and nitrogen oxides, depending on the number of people, the meteorological conditions like humidity and temperature, cleanliness of the laboratory space and the presence or absence of concurrent samples and methodologies that may have reagents that could contaminate the samples. Therefore, standard operational procedures should always account for these potential contamination sources. Systematic study of contamination sources and their effects on analytical performance (Kerouel and Aminot, 1987) ranked nutrients in decreasing order of contamination risk as follows: nitrate, ammonia, phosphate and finally nitrite. This contrasts with the reactivity/lability ranking that suggests ammonium, nitrite and phosphate as the most risky to store. However, the analytical procedures for ammonium in particular have changed since the 1987 paper and new analytical techniques have highlighted the alarming potential for ammonium contamination from a myriad of sources. Silica seems both the least reactive in vitro and one of the least prone to frozen storage liability, but only if precautions regarding the thawing of stored samples are followed closely (MacDonald and McLaughlin, 1982; MacDonald *et al.*, 1986). By contrast, it is consensual that ammonium cannot be reliably determined after storage.

It is fairly straightforward to understand that there is a lack of any Certified Reference Materials (CRMs) that can be used to identify these random errors by tracking the composition of the aliquot through the whole process, including sampling and subsequent manipulations, hence denying the possibility of effectively identifying and quantifying these sources of error, and thus undermining any subsequent quality control of the upper levels of the sampling chain.

If immediate analysis is not feasible, then the possibility of preserving an unadulterated sample has to be considered. Apart from the increase in the number of links in the traceability and analytical chains, it is never to be recommended, as mentioned before, unless absolutely necessary, and there are other concerns that might make us want to avoid this. In effect, a successful preservation technique must maintain the original composition of the sample (true value) during storage, to the level of accuracy required by the sampling programme (Dore *et al.*, 1996). Also, the prepping of samples for storage or quantization when the water carries high particulate loads usually involves filtering, but the type of filter used may affect the concentration of a number of the nutrients in different ways, as with phosphates for example (Li and Hansell, 2008). The two main methods of preservation, both of which are not devoid of consequences for the subsequent quality performance of the analysis, at least for some nutrients, are freezing and the addition of a chemical preservative (Kirkwood, 1992).

Dore *et al.* (1996) reviewed 41 previous studies of preservation by low-temperature storage in the context of quality assurance of quantitative nutrient analysis in the period 1953–1995. The study convincingly shows that frozen storage (minimum −20 °C) of unfiltered open ocean waters in clear polyethylene bottles reasonably preserves, in decreasing order of efficiency, the in-situ concentrations of nitrate + nitrite, silicate (DRSi) and phosphate (DIP). In addition, samples for the determination of DRSi should be allowed to thaw for a minimum of 12–24 h or more to allow depolymerization to occur prior to analysis (MacDonald *et al.*, 1986). The addition of chemical preservatives, of which mercuric chloride (Kirkwood, 1992; Kattner, 1999) is the most effective, presents other issues as well, since it systematically lowers the NH_4^+ concentrations in stored samples by ∼10 % (Kattner, 1999), and can poison the nitrate channel's copper–cadmium reduction column, hence affecting its reducing capacity.

The highest variability in the determination of nutrients after frozen storage or preservation may mostly affect the analytical accuracy of low-level concentration samples taken in oligotrophic areas.

These marine regions include more than three-quarters of the global ocean area and sustain more than 80% of the global marine productivity (Karl *et al.*, 1996). When this knowledge is combined with the need to quantify low-level concentrations of nutrients, the necessity for accurate, testable and precise analysis of nutrient concentration to feed into process studies and global mass budgets becomes clear. Again, the introduction of quality control of the sample handling process and the subsequent analytical determination, by including a stable solution of a CRM with the stored material and trace its alteration (a tracking certified standard), would allow direct assessment of potential systematic bias during storage, and subsequent analysis, thus preserving the traceability of analytical performance and, consequently, the coherence of the analytical chain.

7.4.3 The International Nutrients Scale System (INSS)

It has been well established that, despite continuing efforts, adequate comparability and traceability of marine nutrient data worldwide have not been achieved so far. However, suitable reference materials, defined as 'stable substances for which one or more properties have been established sufficiently well to calibrate or validate a measurement process' (IUPAC, 2009), should ideally be available in large quantities; and, besides being prepared in a natural seawater matrix, they should also have the potential to achieve comparability of nutrient analysis to a quantifiably high level. In combination with good analytical laboratory practice and the use of tracking standards through measurement campaigns, these should be a crucial part in accomplishing the level of traceability necessary to reconcile different datasets collected globally. The paradigm shift reversing the unfortunate prior situation occurred when Michio Aoyama (Meteorological Research Institute, Japan) and Hidezaku Ota (KANSO Technos, Japan) organized in 2006 a global performance study involving 55 different laboratories to test proposed reference marine nutrient solutions (RMNS) produced in Japan. These were natural seawater samples prepared by following the lessons learned by prior performance comparison studies and stabilization procedures for nutrient standard solutions (Aminot and Kerouel, 1991, 1995, 1997a,b) and subsequently tested in internal laboratory performance studies. The results so far (Aoyama *et al.*, 2008, 2010) show that the use of these materials combined with best practice make it possible to achieve comparability of nutrient analysis to a level better than 1%. As a consequence, an international workshop on Chemical Reference Materials in Ocean Science, focusing on the measurement of nutrients and CO_2, but particularly on the current availability of certified nutrient references in ocean science, was organized in Tsukuba, Japan, from 29 October to 1 November 2007. The result of the workshop was the launch of a collaborative programme called the International Nutrient Scale System (INSS) aiming to establish global comparability and traceability of nutrient data. An international steering committee led this programme forward and promoted the next international nutrient intercomparability exercise in 2008 (Aoyama *et al*, 2010). Following an INSS workshop held in Paris under the auspices of UNESCO in February 2009, a joint ICES–IOC study group on nutrient standards (SGONS) was proposed, and then accepted and established by the ICES–IOC in July 2009, with the aims to develop international standards for nutrients to establish comparability and traceability of nutrient data in the worlds oceans (ICES–IOC, 2009).

The rapid evolution observed since the Tsukuba workshop in 2007 in the testing of these materials suggests, therefore, that the crucial aim to have a common, certified reference seawater standard for dissolved nutrients that would allow (1) traceability to be achieved within the full analytical chain and which subsequently would permit (2) day-to-day quality control procedures to be established – and on this basis, performance improved and validated on a continuous basis, which in the whole would secure the ultimate goal of ensuring that a known level of comparability is achieved between laboratories anywhere – might be at hand.

7.5 CONCLUSIONS

This chapter focuses on the notion that marine nutrient monitoring is much more than the temporal collection of data reflecting the levels of nutrients in seawater. Rather than comparing measurable levels to predefined thresholds in order to gauge the health status of a particular system, the current EBA to management aims to understand and follow the response of the ecosystem to external drivers of environmental change, in order to, with the hindsight of forecasting ability, constrain shifts in health status within boundaries that ensure long-term provision of services essential to the sustainability of the human population through scientifically informed management policy. Because temporal and spatial patterns of nutrient concentrations directly link the pace of carbon cycling to the ability of ecosystems to support human activities, nutrient monitoring has a responsibility to deliver data that will ensure not only compliance to predefined thresholds, but also insight into the internal response of ecosystems to environmental change drivers and internal feedbacks at different timescales. The ability to forecast response is thus inextricably linked to the quality of data ensuring measurable statistical confidence in the observed trends, while also dependent on the available understanding of how systems respond to drivers of change across different pressure ranges from short to long timescales as a way to inform monitoring strategies and field and laboratory operational procedures. Historical progress in the area has thus evolved from a self-informed cyclic process with an embedded evolving trend after much iteration.

Over the last few decades, environment managing directives have evolved in pace with metrology advances and support provided by increased understanding of ecosystem function within limited geographical and geomorphological boundaries. At the brink of extension of those directives to a global, holistic perspective of nutrient behaviour, from the freshwater end-members to the deepest reaches of the ocean, unresolved issues still remain regarding the applicability of currently used EcoQOs, because wide gaps in knowledge still persist in what relates to the global interactions of different nutrient reservoirs of the Earth system across a gradient of temporal and spatial scales. Since the issues of data quality, comparability and traceability are as yet insufficiently established, serious shortcoming of nutrient monitoring programmes going forward are to be expected, in spite of the consensual agreement that monitoring has to be designed in a way that permits long-term trends or shifts in concentrations to be detected over vast geographical areas. Hence, the demonstration of data comparability and its traceability is not only central to the trans-boundary and multi-stakeholder nature of marine nutrient monitoring in the future, it is the vital point on which the coherent feasibility of an EBA to marine ecosystem management hinges.

Recognition of this central point coupled with the rapid advances leading to the current development of the INSS have provided the solace of a credible framework within which data quality, comparability and traceability might be achieved worldwide. This framework will hopefully be the starting point from which the new EBA to marine monitoring, which predicates an evolving analytical framework and continuous feedback of new methodology to overcome deficiencies of data coverage and functional understanding of the system, as well as reduce the cost, might develop on solid scientific grounds, thus improving our chances of efficient and objective managing of our seas.

REFERENCES

Aminot A. and Kerouel R. (1991). Autoclaved sea water as a reference material for the determination of nitrate and phosphate in sea water. *Anal. Chim. Acta*, **248**, 277–283.

Aminot A. and Kérouel R. (1995). Reference material for nutrients in seawater: stability of nitrate, nitrite, ammonia and phosphate in autoclaved samples. *Mar. Chem.*, **49**, 221–232.

Aminot A. and Kérouel R. (1997a). Assessment of heat treatment for nutrient preservation in seawater samples. *Anal. Chim. Acta*, **351**, 299–309.

Aminot A. and Kérouel R. (1997b). Reference material for nutrients for the QUASIMEME laboratory performance studies 1993–1996. *Mar. Poll. Bull.*, **35**, 78–83.

Aminot A. and Kérouel R. (2001). An automated photo-oxidation method for the determination of dissolved organic phosphorus in marine and fresh water. *Mar. Chem.*, **76**, 113–126.

Aminot A. and Kerouel R. (2007). *Dosage Automatique des Nutriments dans les Eaux Marines, Méthodes en Flux Continu*, 1st edn, IFREMER.

Aminot A., Kirkwood D. S. and Kérouel R. (1997). Determination of ammonia in seawater by the indophenol-blue method: evaluation of the ICES NUTS I/C 5 questionnaire. *Mar. Chem.*, **56**, 59–75.

Aminot A., Kerouel R. and Coverly S. (2009). Nutrients in seawater using segmented flow analysis, in *Practical Guidelines for the Analysis of Seawater* (ed. O. Wurl), CRC Press, Taylor and Francis Ltd, Boca Raton, FL, pp. 143–178.

Andersen J. H., Schluter L. and Aertebjerg G. (2006). Coastal eutrophication: recent developments in definitions and implications for monitoring strategies. *J. Plankton Res.*, **28**, 621–628.

Anon (2002). Ministerial Declaration of the Fifth International Conference on the Protection of the North Sea (Bergen Declaration), http://www.ospar.org/html_documents/ospar/html/bergen_declaration_final.pdf.

Aoyama M., Barwell-Clarke J., Becker S. *et al.* (2008). 2006 Inter-laboratory comparison study for reference material for nutrients in seawater. Technical Reports of the Meteorological Research Institute, No. 58, 99 pp.

Aoyama M., Anstey C., Barwell-Clarke J. *et al.* (2010). 2008 Inter-laboratory comparison study for reference material for nutrients in seawater. Technical Reports of the Meteorological Research Institute, No. 60, 134 pp.

Apitz S. E., Elliott M., Fountain M. and Galloway T. S. (2006). European environmental management: moving to an ecosystem approach. *Integrated Environ. Assess. Manag.*, **2**, 80–85.

Arkema K. K., Abramson S. C. and Dewsbury B. M. (2006). Marine ecosystem-based management: from characterization to implementation. *Front. Ecol. Environ.*, **4**, 525–532.

Backer H, Leppänen J. M., Brusendorff A. C. *et al.* (2010). HELCOM Baltic Sea action plan – a regional programme of measures for the marine environment based on the ecosystem approach. *Mar. Poll. Bull.* **60**, 642–649.

Barnes C. and McFadden K. W. (2008). Marine ecosystem approaches to management: challenges and lessons in the United States. *Mar. Policy*, **32**, 387–392.

Barnett T. P., Pierce D. W. and Schnur R. (2001). Detection of anthropogenic climate change in the world's oceans. *Science*, **292**, 270–274.

Bindoff N. L., Willebrand J., Artale V. *et al.* (2007). Observations: oceanic climate change and sea level, in *Climate Change 2007: The Physical Science Basis, Contribution of Working Group I to the Fourth Assessment Report of the Intergovernmental Panel on Climate Change* (eds S. Solomon, D. Qin, M. Manning *et al.*). Cambridge University Press, Cambridge, UK, p. 1009.

Boesch D. F. (2006). Scientific requirements for ecosystem-based management in the restoration of Chesapeake Bay and coastal Louisiana. *Ecol. Eng.*, **26**, 6–26.

Borja Á. (2006). The new European Marine Strategy Directive: difficulties, opportunities, and challenges. *Mar. Poll. Bull.*, **52**, 239–242.

Borja A. and Collins M. (2009). Regional Seas integrative studies, as a basis for an ecosystem-based approach to management: the case of the Bay of Biscay. *Cont. Shelf Res.*, **29**, 951–956.

Borja A., Bricker S. B., Dauer D. M. *et al.* (2008). Overview of integrative tools and methods in assessing ecological integrity in estuarine and coastal systems worldwide. *Mar. Poll. Bull.*, **56**, 1519–1537.

Botsford L. W., Castilla J. C. and Peterson C. H. (1997). The management of fisheries and marine ecosystems. *Science*, **277**, 509–515.

Brack W., Apitz S. E., Borchardt D. *et al.* (2009). Towards a holistic and risk-based management of European river basins. *Integrated Environ. Assess. Manag.*, **5**, 5.

Caffrey J., Younos T., Connor M. *et al.* (2007). Nutrient requirements for the national water quality monitoring network for U.S. coastal waters and their tributaries – Recommendations of the Nutrients Workgroup regarding the specific constituents to be monitored and performance requirements for analytical methods within the National Water Quality Monitoring network for U.S. coastal waters and their tributaries, under the US Ocean Action Plan, http://acwi.gov/monitoring/network/.

Clancy V. and Willie S. (2003). NOAA/NRC Inter-comparison for Nutrients in Seawater, NOAA Technical Memorandum NOS NCCOS CCMA 158, 32 pp.

Cloern J. E. (2001). Our evolving conceptual model of the coastal eutrophication problem. *Mar. Ecol. Prog. Ser.*, **210**, 223–253.

Cofino W. P. and Wells D. E. (1994). Design and evaluation of the QUASIMEME inter-laboratory performance studies: a test case for robust statistics. *Mar. Poll. Bull.*, **29**, 149–158.

Cordell D., Drangert J. and White S. (2009). The story of phosphorus: global food security and food for thought. *Global Environ. Change*, **19**, 292–305.

Costanza R. d'Arge R., de Groot R. *et al.* (1997). The value of the world's ecosystem services and natural capital. *Nature*, **387**, 253–260.

Curtin R. and Prellezo R. (2010). Understanding marine ecosystem based management: a literature review. *Mar. Policy*, **34**, 821–830.

Davis J. M., Rosemond A. D., Eggert S. L. *et al.* http://www.pnas.org/content/107/1/121.full-aff-3 (2010). Long-term nutrient enrichment decouples predator and prey production. *Proc. Natl. Acad. Sci. U. S. A.*, **107**, 121–126.

Diaz R. J. and Rosenberg R. (2008). Spreading dead zones and consequences for marine ecosystems. *Science*, **321**, 926–929.

Dore J. E., Houlihan T., Hebel D. V. *et al.* 1996. Freezing as a method of sample preservation for the analysis of dissolved inorganic nutrients in seawater. *Mar. Chem.*, **53**, 173–185.

Duce R. A., LaRoche J., Altieri K. *et al.* (2008). Impacts of atmospheric anthropogenic nitrogen on the open ocean. *Science*, **320**, 893–897.

Dux J. P. (1990). *Handbook of Quality Assurance for the Analytical Chemical Laboratory*, 2nd edn, Van Nostrand Reinhold, New York, 203 pp.

Elliott M. and Cutts N. D. (2004). Marine habitats: loss and gain, mitigation and compensation. *Mar. Poll. Bull.*, **49**, 671–674.

Elliott M. and De Jonge V. N. (1996). The need for monitoring the monitors and their monitoring. *Mar. Poll. Bull.*, **32**, 248–249.

Elliott M., Fernandes T. and De Jonge V. (1999). The impact of European Directives on estuarine and coastal science and management. *Aquat. Ecol.*, **33**, 311–321.

European Commission (1991). Council Directive 91/271/EEC of 21 May 1991 concerning urban waste-water treatment. *OJ* L 135, 30.5.1991, pp. 40–52.

European Commission (1998). Council Directive 91/676/EEC of 12 December 1991 concerning the protection of waters against pollution caused by nitrates from agricultural sources. *OJ*, L 375, 31.12.1998, pp. 1–8.

European Commission (2000). Directive 2000/60/EC of the European Parliament and of the Council of 23 October 2000 establishing a framework for Community action in the field of water policy. *OJ*, L 327, 22.12.2000, pp. 1–73.

European Commission (2003). Monitoring under the Water Framework Directive, CIS Guidance Document No. 7.

European Commission (2008). Directive 2008/56/EC of the European Parliament and the Council of 17 June 2008 establishing a framework for community action in the field of marine environmental policy (Marine Strategy Framework Directive). OJ, L 164, 25.6.2008, pp. 19–40.

Fletcher W. (2006). Frameworks for managing marine resources in Australia through ecosystem approaches: do they fit together and are they useful? *Bull. Mar. Sci.*, **78**, 691–704.

Foden J., Rogers S. I. and Jones A. P. (2008). A critical review of approaches to aquatic environmental assessment. *Mar. Poll. Bull.*, **56**, 1825–1833.

Foster E., Haward M. and Coffen-Smout S. (2005). Implementing integrated oceans management: Australia's south east regional marine plan (SERMP) and Canada's eastern Scotian shelf integrated management (ESSIM) initiative. *Mar. Policy*, **29**, 391–405.

Freeman M. C., Pringle C. M. and Jackson C. R. (2007). Hydrologic Connectivity and the Contribution of Stream Headwaters to Ecological Integrity at Regional Scales. *J. Am. Water Resour. Assoc.*, **43**, 5–14.

GESAMP (IMO/FAO/UNESCO/WMO/WHO/IAEA/UN/UNEP Joint Group of Experts on the Scientific Aspects of Marine Environmental Protection) (2001a). A sea of troubles, UNEP/GRID-Arendal.

GESAMP (IMO/FAO/UNESCO/WMO/WHO/IAEA/UN/UNEP Joint Group of Experts on the Scientific Aspects of Marine Environmental Protection) (2001b). Protecting the oceans from land-based activities: land-based sources and activities affecting the quality and uses of the marine, coastal and associated freshwater environment, UNEP/GRID-Arendal.

Glanville S. D., Clark A. G., Wells D. E. *et al.* (1997). A review of the achievements of the EU project 'QUASIMEME' 1993–1996. *Mar. Poll. Bull.*, **35**, 3–17.

Goldberg E. D. and Taylor J. K. (1985). The VD conspiracy. *Mar. Poll. Bull.*, **16**, 1.

Gouretski V. V. and Jancke K. (2000). Systematic errors as the cause for an apparent deep water property variability: global analysis of the WOCE and historical hydrographic data. *Prog. Oceanogr.*, **48**, 337–402.

Grasshoff K. (ed.) (1976). *Methods of Seawater Analysis*, 1st edn, Verlag Chemie, Weinheim, Germany, 317 pp.

Grasshoff K., Ehrhardt M. and Kremling K. (eds) (1983). *Methods of Seawater Analysis*, 2nd edn, Verlag Chemie, Weinheim, Germany, 419 pp.

Grasshoff K., Ehrhardt M. and Kremling K. (eds) (1999). *Methods of Seawater Analysis*, 3rd edn, Verlag Chemie/Wiley–VCH, Weinheim/New York, 600 pp.

Gray J. S., Calamari D., Duce R. *et al.* (1991). Scientifically based strategies for marine environmental protection and management. *Mar. Poll. Bull.*, **22**, 432–440.

Grumbine R. (1994). What is ecosystem management? *Conserv. Biol.*, **8**, 27–38.

Guerry A. D. (2005). Icarus and Daedalus: conceptual and tactical lessons for marine ecosystem-based management. *Front. Ecol. Environ.*, **3**, 202–211.

Halpern B. S., Walbridge S., Selkoe K. A. *et al.* (2008). A global map of human impact on marine ecosystems. *Science*, **319**, 948–952.

Heisler J., Glibert J. P., Burkholder J. *et al.* (2008). Eutrophication and harmful algal blooms: a scientific consensus. *Harmful Algae*, **8**, 3–13.

HELCOM (1974). Convention on the protection of the Marine Environment of the Baltic Sea Area, http://www.helcom.fi/stc/files/Convention/convention1974.pdf.

HELCOM (2008). Manual for marine monitoring in the COMBINE programme of HELCOM, http://www.helcom.fi/groups/monas/CombineManual/en_GB/main/.

Heslenfeld P. and Enserink E. L. (2008). OSPAR ecological quality objectives: the utility of health indicators for the North Sea. *ICES J. Mar. Sci.*, **65**, 1392–1397.

Howarth R. W., Billen G., Swaney D., *et al.* (1996). Regional nitrogen budgets and riverine N and P fluxes for the drainages to the North Atlantic Ocean: natural and human influences. *Biogeochemistry*, **35**, 75–139.

Ibe A. C. and Kullenberg G. (1995). Quality assurance/quality control (QA/QC) regime in marine pollution monitoring programmes: the GIPME perspective. *Mar. Poll. Bull.*, **31**, 209–213.

ICES (1977). The International Intercalibration Exercise for Nutrient Methods, ICES (International Council for the Exploration of the Sea), ICES Cooperative Research Report No. 67. Copenhagen, Denmark, 44 pp.

ICES (2005). ICES/HELCOM Steering Group on Quality Assurance of Chemical Measurements in the Baltic Sea, Copenhagen: ICES (International Council for the Exploration of the Sea).

ICES–IOC (2009). Proposal for a joint ICES–IOC study group on nutrient standards (SGONS), IOC/INF-1260, Paris, 11 May 2009, http://unesdoc.unesco.org/images/0018/001827/182777e.pdf.

Interagency Ocean Policy Task Force (2009). Interim Report of the Interagency Ocean Policy Task Force, http://www.whitehouse.gov/assets/documents/09_17_09_Interim_Report_of_Task_Force_FINAL2.pdf.

ISO (International Organization for Standardization) (2007). ISO/IEC Guide 99:2007, International vocabulary of metrology – basic and general concepts and associated terms (VIM), ISO.

IUPAC (1997). Compendium of Analytical Nomenclature – Definitive Rules 1997 3rd edn, IUPAC (International Union of Pure and Applied Chemistry), http://old.iupac.org/publications/analytical_compendium/.

IUPAC (2009). Compendium of Chemical *Terminology*, 2nd edn (the 'Gold Book'), compiled by A. D. McNaught and A. Wilkinson, Blackwell Scientific Publications, Oxford (1997). XML on-line corrected version: http://goldbook.iupac.org (2006–) created by M. Nic, J. Jirat, B. Kosata; updates compiled by A. Jenkins. (ISBN 0-9678550-9-8. DOI: 10.1351/goldbook. Last update: 2009-09-07; version: 2.1.5. DOI: doi:10.1351/goldbook.M03758.)

Johnson D. (2008). Environmental indicators: their utility in meeting the OSPAR Convention's regulatory needs. *ICES J. Mar. Sci.*, **65**(8), 1387–1391.

Jones P. G. W. and Folkard A. R. (1968). Report on the analysis of phosphate at the ICES intercalibration trials of chemical methods held in Copenhagen, 1966, organized by ICES. UNESCO Tech. Pap. Mar. Sci., p. 9.

De Jong F. (2006). *Marine Eutrophication in Perspective: On the Relevance of Ecology for Environmental Policy*, 1st edn, Springer-Verlag, Berlin, 335 pp.

De Jonge V., Elliott M. and Orive E. (2002). Causes, historical development, effects and future challenges of a common environmental problem: eutrophication. *Hydrobiologia*, **475–476**, 1–19.

De Jonge V., Elliott M. and Brauer V. (2006). Marine monitoring: its shortcomings and mismatch with the EU Water Framework Directive's objectives. *Mar. Poll. Bull.*, **53**, 5–19.

Kapp R. (2005). Clean Water Act (CWA), US. In *Encyclopedia of Toxicology*. Elsevier, New York, pp. 618–620.

Karl D. M., Christian J. R., Dore J. E., *et al*. (1996). Seasonal and interannual variability in primary production and particle flux at Station ALOHA. *Deep-Sea Res. (2 Top. Stud. Oceanogr.)*, **43**, 539–568.

Kattner G. (1999). Storage of dissolved inorganic nutrients in seawater: poisoning with mercuric chloride. *Mar. Chem*., **67**, 61–66.

Kerouel R. and Aminot A. (1987). Procédure optimisee hors-contaminations pour l'analyse des eléments nutritifs dissous dans l'eau de mer. *Mar. Environ. Res*., **22**, 19–32.

Kirkwood D. (1994). Nutrients: practical notes on their determination in seawater, in *ICES/HELCOM Workshop on Quality Assurance of Chemical Analytical Procedures for the Baltic Monitoring Programme. Baltic Sea Environment Proceedings* (eds G. Topping and U. Harms), pp. 23–47.

Kirkwood, D. (1996). Nutrients: Practical notes on their determination in seawater, ICES Techniques in Marine Environmental Sciences, No 17. Copenhagen, Denmark.

Kirkwood D., Aminot A. and Perttila M. (1991). Report on the results of the fourth ICES Inter-comparison study for Nutrients in Seawater, ICES Cooperative Research Report No. 174, Copenhagen, Denmark.

Kirkwood D. S. (1992). Stability of solutions of nutrient salts during storage. *Mar. Chem*., **38**, 151–164.

Klausmeier C. A., Litchman E., Daufresne T. and Levin S. A. (2004). Optimal nitrogen-to-phosphorus stoichiometry of phytoplankton. *Nature*, **429**, 171–174.

Krom M., Woodward E. M. S., Herut B. *et al*. 2005. Nutrient cycling in the south east Levantine basin of the eastern Mediterranean – results from a P starved system. *Deep-Sea Res. (2 Top. Stud. Oceanogr.)*, **22–23**, 2879–2896.

Li Q. P. and Hansell D. A. (2008). Intercomparison and coupling of magnesium-induced co-precipitation and long-path liquid-waveguide capillary cell techniques for trace analysis of phosphate in seawater. *Anal. Chim. Acta*, **611**, 68–72.

Lotze H. K., Lenihan H. S., Bourque B. J. *et al*. (2006). Depletion, degradation, and recovery potential of estuaries and coastal seas. *Science*, **312**, 1806–1809.

Lysiak-Pastuszak E. and Krysell M. (eds) (2004). Chemical measurements in the Baltic Sea: guidelines on quality assurance, ICES Techniques in Marine Environmental Sciences, No. 35, Copenhagen, Denmark, 149 pp.

MacDonald R. and McLaughlin F. (1982). The effect of storage by freezing on dissolved inorganic phosphate, nitrate and reactive silicate for samples from coastal and estuarine waters. *Water Res*., **16**, 95–104.

MacDonald R., McLaughlin F., and Wong C. S. (1986). The storage of reactive silicate samples by freezing. *Limnol. Oceanogr*., **31**, 1139–1142.

Mackenzie F. T., Ver L. M. and Lerman A. (2002). Century-scale nitrogen and phosphorus controls of the carbon cycle. *Chem. Geol*., **190**, 13–32.

Mackey K. R., Rivlin T., Grossman A. R. *et al*. (2009). Picophytoplankton responses to changing nutrient and light regimes during a bloom. *Mar. Biol*., **156**, 1531–1546.

Mantoura R. F. C. and Woodward E. M. S. (1983). Optimization of the indophenol blue method for the automated determination of ammonia in estuarine waters. *Estuar. Coast. Shelf Sci*., **17**, 219–224.

Matson, P. A. Parton W. J., Power A. G. and Swift M. J. (1997). Agricultural intensification and ecosystem properties. *Science*, **277**, 504–509.

McFadden K. W. and Barnes C. (2009). The implementation of an ecosystem approach to management within a federal government agency. *Mar. Policy*, **33**, 156–163.

McLeod K. L., Lubchenco J., Palumbi S. R. *et al*. (2005). *Scientific Consensus Statement on Marine Ecosystem-Based Management*, signed by 221 academic scientists and policy experts with relevant expertise and published by the Communication Partnership for Science and the Sea, http://www.compassonline.org/sites/all/files/document_files/EBM_Consensus_Statement_v12.pdf.

McLusky D. S. and Elliott M. (2004). *The Estuarine Ecosystem: Ecology, Threats, and Management*, 3rd edn, Oxford University Press, Oxford, UK, 240 pp.

Mee L. D. (1986). Continuous flow analysis in chemical oceanography: principles, applications and perspectives. *Sci. Total Environ*., **49**, 27–87.

Moore M. C., Mills M. M., Achterberg E. P. *et al*. (2009). Large-scale distribution of Atlantic nitrogen fixation controlled by iron supply. *Nature Geosci*., **2**, 867–871.

Murawski S. A. (2007). Ten myths concerning ecosystem approaches to marine resource management. *Mar. Policy*, **31**, 681–690.

Murawski S. A., Steele J. H., Taylor P. *et al*. (2010). Why compare marine ecosystems? *ICES J. Mar. Sci*., **67**, 1–9.

Myers R. A. and Worm B. (2003). Rapid worldwide depletion of predatory fish communities. *Nature*, **423**, 280–283.

Nelleman C., Hain S. and Alder J. (2008). In dead water – merging of climate change with pollution, over-harvest, and infestations in the world's fishing grounds, Norway, United Nations Environment Programme, GRID-Arendal.

Nixon S. W. (1995). Coastal marine eutrophication: a definition, social causes and future concerns. *Ophelia*, **41**, 199–219.

O'Boyle R. and Jamieson G. (2006). Observations on the implementation of ecosystem-based management: experiences on Canada's east and west coasts. *Fish. Res*., **79**, 1–12.

OSPAR (1992). Convention for the Protection of the Marine Environment of the North-East Atlantic. EUR-Lex reference 21998A0403(01). http://eur-lex.europa.eu/LexUriServ/LexUriServ.do?uri=CELEX:21998A0403(01):EN:NOT.

OSPAR Commission (1997). JAMP Eutrophication Monitoring Guidelines: Nutrients, OSPAR.

OSPAR Commission (2003a). 2003 Strategies of the OSPAR Commission for the Protection of the Marine Environment of the North-East Atlantic, Reference 2003-21, http://www.ospar.org/html_documents/ospar/html/Revised_OSPAR_Strategies_2003.pdf#nameddest=eutrophication.

OSPAR Commission (2003b). OSPAR Integrated Report 2003 on the eutrophication status of the OSPAR Maritime Area based upon the first application of the comprehensive procedure.

OSPAR Commission (2005). Synergies between the OSPAR Comprehensive Procedure, the integrated set of OSPAR EcoQOs for eutrophication and the EC Water Framework Directive, http://www.ospar.org/documents%5Cdbase%5Cpublications%5Cp00231_Synergies%20ComProc-EcoQO-WFD.pdf.

Paerl H. W. (1999). Cultural eutrophication of shallow coastal waters: coupling changing anthropogenic nutrient inputs to regional management approaches. *Limnol. Ecol. Manag. Inland Waters*, **29**, 249–254.

Pai S. and Riley J. P. (1994). Determination of nitrate in the presence of nitrite in natural waters by flow injection analysis with a non-quantitative on-line cadmium reductor. *Int. J. Environ. Anal. Chem*., **57**, 263–277.

Painting S., Devlin M. J., Rogers S. I. *et al*. (2005). Assessing the suitability of OSPAR EcoQOs for eutrophication vs ICES criteria for England and Wales. *Mar. Poll. Bull*., **50**, 1569–1584.

Parsons T. R., Maita Y. and Lalli C. M. (1984). *A Manual of Chemical and Biological Methods for Seawater Analysis*, Pergamon Press, New York, 173 pp.

PEW Oceans Commission (2003). America's living oceans – charting a course for sea change. A Report to the Nation, Recommendations for a New Ocean Policy, http://www.pewtrusts.org/our_work_report_detail.aspx?id=30009andcategory=130.

Preston A. and Wood P. C. (1971). Monitoring the marine environment. *Proc. R. Soc. Lond. [Biol.]*, **177**, 451–462.

Quevauviller Ph. (2004). Traceability of environmental chemical measurements. *Trends Anal. Chem*., **23**, 171–177.

Quevauviller Ph., Maier E. and Griepink B. (1995). 1. Quality assurance for environmental analysis, in *Quality Assurance for Environmental Analysis – Method Evaluation within the Measurements and Testing Programme (BCR)*, Elsevier, Amsterdam, pp. 1–25.

Quevauviller Ph., Borchers U., Thompson C. and Simonart T. (eds) (2008). *The Water Framework Directive – Ecological and Chemical Status Monitoring*, John Wiley & Sons, Ltd, Chichester, 496 pp.

Rabalais N. N., Turner R. E. and Wiseman W. J. (2003). Gulf of Mexico hypoxia, a.k.a. 'the dead zone'. *Annu. Rev. Ecol. Syst*. **33**, 235–263.

Rabalais N. N., Turner R. E., Diaz R. J. and Justic D. (2009). Global change and eutrophication of coastal waters. *ICES J. Mar. Sci*., **66**, 1528–1537.

Rockstrom J., Steffen W., Noone K. *et al*. (2009). A safe operating space for humanity. *Nature*, **461**, 472–475.

Ruzicka J. and Hansen E. H. (1986). The first decade of flow injection analysis: from serial assay to diagnostic tool. *Anal. Chim. Acta*, **179**, 1–58.

Ruzicka J. and Hansen E. H. (2008). Retro-review of flow-injection analysis. *Trends Anal. Chem*., **27**, 390–393.

Ryther J. H. and Dunstan W. M. (1971). Nitrogen, phosphorus, and eutrophication in the coastal marine environment. *Science*, **171**, 1008–1013.

Scheffer M. and Carpenter S. R. (2003). Catastrophic regime shifts in ecosystems: linking theory to observation. *Trends Ecol. Evol.*, **18**, 648–656.

Scheffer M. and van Nes E. H. (2004). Mechanisms for marine regime shifts: can we use lakes as microcosms for oceans? *Prog. Oceanogr.*, **60**, 303–319.

Seitzinger S., Kroeze C., Bouwman A. F. *et al*. (2002). Global patterns of dissolved inorganic and particulate nitrogen inputs to coastal systems: recent conditions and future projections. *Estuaries*, **25**, 640–655.

Skjoldal H. R., Gool S. V., Offringa H. *et al*. (1999). Report on the Workshop on Ecological Quality Objectives (EcoQOs) for The North Sea. Scheveningen – the Netherlands, 1–3 September 1999, TemaNord, The Hague, 591 pp.

Smith V. (2003). Eutrophication of freshwater and coastal marine ecosystems a global problem. *Environ. Sci. Pollut. Res.*, **10**, 126–139.

Smith V. H., Joye S. B. and Howarth R. W. (2006). Eutrophication of freshwater and marine ecosystems. *Limnol. Oceanogr.*, **51**, 351–255.

Stewart B. M. and Elliott P. A. W. (1996). Systematic salt effects in the automated determination of nutrients in seawater. *Water Res.*, **30**, 869–874.

Strickland J. D. H. and Parsons T. R. (1968). *A Practical Handbook of Seawater Analysis*, 1st edn, Fish. Res. Bd. Can. Bull., Ottawa, Bulletin 167, 293 pp.

Strickland J. D. H. and Parsons T. R. (1972). *A Practical Handbook of Seawater Analysis*, 2nd edn, Fish. Res. Bd. Can. Bull., Ottawa, Canada, 122 pp.

Taljaard S., Monteiro P. M. S. and Botes W. A. M. (2006). A structured ecosystem-scale approach to marine water quality management. *Water SA*, **32**, 535–542.

Thompson M. and Wood R. (1993). The international harmonized protocol for the proficiency testing of (chemical) analytical laboratories (technical report). *Pure Appl. Chem.*, **65**, 2123–2144.

Thompson M. and Wood R. (1995). Harmonized guidelines for internal quality control in analytical chemistry laboratories (technical report). *Pure Appl. Chem.*, **67**, 649–666.

Topping G. (1992). The role and application of quality assurance in marine environmental protection. *Mar. Poll. Bull.*, **25**, 61–66.

Tyrrell T. (1999). The relative influences of nitrogen and phosphorus on oceanic primary production. *Nature*, **400**, 525–531.

UNEP (United Nations Environment Programme) (2006). Marine and coastal ecosystems and human well-being: a synthesis report based on the findings of the Millennium Ecosystems Assessment, Nairobi, Kenya, United Nations Environmental Programme (UNEP), www.MAweb.org.

UNEP (United Nations Environment Programme) and GPA (Global Programme of Action for the Protection of the marine Environment from Land-based Activities) (2006). The state of the marine environment: trends and processes, http://www.gpa.unep.org/documents/soe_-_trends_and_english.pdf.

UNEP (United Nations Environment Programme) and WHRC (Woods Hole Research Center) (2007). Reactive nitrogen in the environment: too much or too little of a good thing, http://www.gpa.unep.org/documents/non-technical_review_on_reactive_english.pdf.

UNESCO (1965). Report on the intercalibration measurements in Copenhagen, 9–13 June 1965, UNESCO.

UNESCO (1967). Report on intercalibration measurements, Leningrad, 24–28 May 1966 and Copenhagen, September 1966.

USCOP (2004). An ocean blueprint for the 21st century. Final report, http://oceancommission.gov/documents/full_color_rpt/welcome.html.

US EPA (1998). Guidelines for Ecological Risk Assessment.

US EPA (2001). Nutrient Criteria Technical Guidance Manual: Estuarine and Coastal Marine Waters, CDG Document Ref. EPA-822-B-01-003, October 2001, http://www.epa.gov/waterscience/criteria/nutrient/guidance/marine/coastal.pdf.

VanderZwaag D. L. and Powers A. (2008). The protection of the marine environment from land-based pollution and activities: gauging the tides of global and regional governance. *Int. J. Mar. Coast. Law*, **23**, 423–452.

Vaquer-Sunyer R. and Duarte C. M. (2008). Thresholds of hypoxia for marine biodiversity. *Proc. Natl. Acad. Sci. U. S. A.*, **105**, 15452–15457.

Vitousek P. M., Mooney H. A., Lubchenco J. and Melillo J. M. (1997). Human domination of Earth's ecosystems. *Science*, **277**, 494–499.

Walther G., Post E., Convey P. *et al*. (2002). Ecological responses to recent climate change. *Nature*, **416**, 389–395.

WCED (United Nations World Commission on Environment and Development) (1987). Report of the World Commission on Environment and Development: Our Common Future, transmitted to the General Assembly as an Annex to document A/42/437 – Development and International Cooperation: Environment.

Wells D. E. and Cofino W. P. (1997). The assessment of the QUASIMEME laboratory performance studies data: techniques and approach. *Mar. Poll. Bull.*, **35**, 18–27.

Wells D. E., Cofino W. P., Quevauviller Ph. and Griepink B. (1993). Quality assurance of information in marine monitoring – a holistic approach. *Mar. Pollut. Bull.* **26**, 368–375.

Willie S. and Clancy V. (2000). NOAA/NRC Inter-comparison for nutrients in seawater, NOAA Technical Memorandum NOS NCCOS CCMA 143, 176 pp.

Woodward M, Coverly S., Kerouel R. *et al*. 2010. Some specific analytical considerations for the continuous flow analysis of nutrients, INSS book.

Wurl O. (ed.) (2009). *Practical Guidelines for the Analysis of Seawater*, 1st edn, CRC Press, Taylor and Francis Ltd, Boca Raton, FL, 408 pp.

Part 4

Types of Monitoring

8

Classical Chemical Monitoring of the Marine Environment

Colin F. Moffat, Lynda Webster, and Rob Fryer

Marine Scotland, Marine Laboratory, Aberdeen, UK

8.1 INTRODUCTION

The ultimate objective of marine monitoring is to provide data which can be used to make an assessment, spatial and/or temporal, of some aspect of the marine environment, be it nutrient concentrations in water, metals in biota, polycyclic aromatic hydrocarbons (PAHs) in shellfish or halogenated hydrocarbons in fish. This may require the isolation of trace compounds from a complex matrix which, in itself, is a very small component of a population. On the other hand, it can involve the regular determination of nutrient concentrations in water samples from a designated location

Chemical Marine Monitoring: Policy Framework and Analytical Trends, First Edition.
Edited by Philippe Quevauviller, Patrick Roose, and Gert Verreet.

over many years. There may be a need to determine additional physical or chemical characteristics which will be used, for example, to normalize contaminant concentrations. The various monitoring procedures present challenges in respect of sampling, analysis and, ultimately, interpretation of the numbers. In an ideal situation, the numbers would be compared with appropriate assessment criteria and the spatial implications of any result would be clear; see Box 8.1. There would also be a temporal series of sufficient duration to provide high-confidence data on trends. However, sampling in the marine environment is expensive and the area/volume that is being assessed is often large. As such, the situation regarding sufficiency of data to make an assessment in which there is a high degree of confidence is often far from ideal. That said, in recent years, technical guidelines have been written, quality assurance procedures devised and assessment criteria developed. This has allowed assessments of the status of marine regions to be undertaken with reasonable confidence. Although there are undoubted gaps, the underlying classical chemical techniques which result in the production of numbers have reached a stage of reliability such that, with appropriate care in terms of sample collection, sample storage, and the actual generation of numbers, the data can be compared with the relevant assessment criteria and conclusions drawn. These conclusions can then be used in the development of strategies and the further management of the marine environment.

BOX 8.1

When undertaking classical chemical monitoring, whether it is for hazardous substances, determination of salinity or assessment of nutrients, there is a need to establish the monitoring requirements, including sample selection (mottled grey box). This is followed by sample collection, storage and, where appropriate, subsampling (black boxes). Quantification of the analyte (chemical entity being investigated; Prichard *et al.*, 2001) may be undertaken directly following sample collection, following a period of storage or directly from a subsample. However, there may be a need to extract the analyte from the matrix (all components other than the analyte present in the sample) and undertake a clean-up procedure (separation of analyte(s) from other sample components that might otherwise make the analysis more difficult) prior to either direct quantification through instrumental analysis or else a chromatographic separation prior to the quantification (dark grey boxes). Following a quality check there is a need for assessment, reporting and the drawing of conclusions (white box) which should ultimately result in a consideration of the requirement for further sampling and another round of the 'classical chemical analytical wheel'. All aspects of the procedure should incorporate quality assurance processes and procedures (light grey box).

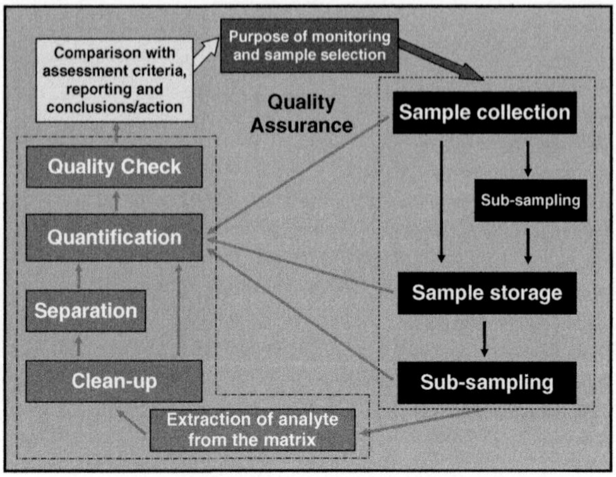

8.2 THE BASIS AND PURPOSE OF MARINE MONITORING

Marine monitoring is required to meet local, national and international commitments. As such, the basis under which marine monitoring is conducted is complex, but there is a considerable drive at the research institute level to design monitoring programmes so that the data are fit for as many purposes as possible. For simplicity, we have selected to focus on the main drivers which require classical chemical monitoring in UK seas and the wider North-East Atlantic.

The basis and purpose of classical chemical monitoring in the marine environment have been described in Chapters 1–3. However, it is worth reiterating that classical chemical monitoring is undertaken for a number of reasons. These include:

- Assessment of status or condition, often across a specified spatial area.
- Assessment of temporal trends.
- Assessment of the impact of legislation and implemented measures.
- Compliance with various international agreements. These include agreements from Regional Seas Conventions and Action Plans (see http://www.unep.ch/regionalseas/programmes/actions/default.asp), for members of the European Union, relevant European Directives such as the Water Framework Directive (WFD; Directive 2000/60/EC; see http://ec.europa.eu/environment/water/water-framework/index_en.html) and agreements that cover wider geographical areas (e.g. The Stockholm Convention on Persistent Organic Pollutants, adopted on 22 May 2001; see www.pops.int).

The OSPAR Convention (Annex IV, Article 1) defines monitoring as 'the repeated measurement of:

1. the quality of the marine environment and each of its components, i.e. water, sediments and biota;
2. activities of natural and anthropogenic inputs which may affect the quality of the marine environment; and
3. the effects of such activities and inputs.'

The above definition requires that chemical monitoring be undertaken and, for OSPAR, there are a number of programmes where chemical measurements are made. These include:

- *The Co-ordinated Environmental Monitoring Programme (CEMP).* This is currently focused on monitoring the concentrations and effects of selected contaminants and nutrients in the marine environment as follows:
  - mercury, cadmium and lead in biota and sediments;
  - polychlorinated biphenyls (PCBs; ICES[1] seven congeners) in biota and sediments;
  - PAHs (nine parent compounds) in biota and sediment;
  - hexabromocyclododecane and polybrominated diphenyl ethers (PBDEs; nine congeners) in biota and sediments and BDE 209 in sediment;
  - tributyltin (TBT) in sediments and TBT-specific effects;

[1] ICES, International Council for the Exploration of the Sea.

- direct and indirect eutrophication effects;
- nutrients in seawater and eutrophication effects.

The CEMP also includes a pre-CEMP covering components which the OSPAR Contracting Parties are preparing to monitor in a co-ordinated manner through the development of monitoring guidance, quality assurance procedures and/or assessment tools. Currently, the pre-CEMP includes the following components:

- planar PCBs in biota;
- alkylated PAHs in biota and sediment;
- TBT in biota;
- perfluorooctane sulfonates (PFOS) in sediment, biota and water;
- dioxins and furans in biota and sediment;
- PAH- and metal-specific biological effects;
- general biological effects.

- *The Comprehensive Atmospheric Monitoring Programme (CAMP).* This aims to assess the input of selected contaminants to the OSPAR maritime area and its regions via atmospheric deposition. This is based on annual monitoring and reporting of concentrations of mandatory and voluntary components in precipitation and air. The CAMP is currently focused on mandatory monitoring of metals (arsenic, cadmium, chromium, copper, lead, mercury, nickel and zinc), lindane and nitrogen species (ammonium and nitrate) in precipitation, and nitrogen species (NO_2, HNO_3, ammonia, ammonium and nitrate) in air. Monitoring of PCBs and PAHs in precipitation and air and of metals as well as lindane in air are voluntary.
- *The Comprehensive Study on Riverine Inputs and Direct Discharges (RID).* This aims to assess the input of selected contaminants to the OSPAR maritime area and its regions which are carried via rivers into tidal waters or are discharged directly into the sea (e.g. through sewage pipelines or activities such as aquaculture applying substances directly in the sea). The RID study currently focuses on mandatory monitoring and reporting of the concentrations and loads of the metals cadmium, copper, lead, mercury and zinc, the organic pollutant lindane, nitrogen and phosphorus species and suspended particulate matter. Monitoring of PAHs, mineral oil, PCBs and other hazardous substances, especially organohalogens, is recommended for voluntary monitoring.

To assist the coordination of these monitoring programmes across the contracting parties, OSPAR has published guidelines for the monitoring of contaminants in biota and sediments. Initially agreed in 1999, and subsequently updated in 2008 and 2009, the Joint Assessment and Monitoring Programme (JAMP) Guidelines for Monitoring Contaminants in Biota provide sampling strategies, procedures for sample storage and analytical methodologies. They also provide recommendations for the species of fish, shellfish and seabirds that should be used. They detail first-choice species (dab (*Limanda limanda*), cod (*Gadus morhua*) and *Mytilus* sp.) and second-choice species (whiting (*Merlangius merlangus*), flounder (*Platichthys flesus*) and Pacific oyster (*Crassostrea gigas*)). The seabirds recommended for use in the monitoring programme include the guillemot (*Uria aalge*), oystercatcher (*Haematopus ostralegus*) and the common tern (*Sterna hirundo*). The determinands covered by the guidelines for biota include PCBs, PAHs, PBDEs, hexabromocyclocdodecane, perfluorinated compounds, organotin compounds, cadmium, mercury, lead and copper.

The output from the three OSPAR monitoring programmes is annual theme-specific assessments as well as more occasional holistic assessments, including the OSPAR Quality Status Reports.

These are innovative, comprehensive reports on the quality of the marine environment for the whole North-East Atlantic, the most recent of which is the QSR 2010 (OSPAR, 2010).

HELCOM, the governing body of the Helsinki Convention, works to protect the marine environment of the Baltic Sea from all sources of pollution through intergovernmental co-operation between Denmark, Estonia, the European Community, Finland, Germany, Latvia, Lithuania, Poland, Russia and Sweden. The Cooperative Monitoring in the Baltic Marine Environment (COMBINE) programme was instituted in 1992. The aims of COMBINE are:

- to identify and quantify the effects of anthropogenic discharges/activities in the Baltic Sea, in the context of the natural variations in the system, and
- to identify and quantify the changes in the environment as a result of regulatory actions.

COMBINE includes the collection of data on core hydrographic variables, including temperature, salinity, oxygen and hydrogen sulfide. This is done so as to set the background for all other measurements related to the identification and quantification of the effects of anthropogenic discharges/activities, the parameters providing an indication of natural fluctuations in the hydrographic regime of the Baltic Sea. COMBINE details the data required to investigate problems related to eutrophication; more specifically, to determine the extent and the effects of anthropogenic inputs of nutrients on marine biota. This requires the determination of the concentrations of nutrients and the response of the different biological compartments. COMBINE also details the requirements for monitoring the concentrations and effects of contaminants with the objectives of:

- Comparing the concentrations of contaminants in selected species of biota (including different parts of their tissues) from different geographical regions of the Baltic Sea in order to detect possible contamination patterns, including areas of special concern (or 'hot spots').
- Measuring concentrations of contaminants in selected species of biota at specific locations over time in order to detect whether levels are changing in response to the changes in inputs of contaminants to the Baltic Sea.
- Measuring concentrations of contaminants in selected species of biota at different locations within the Baltic Sea, particularly in areas of special concern, in order to assess whether the levels pose a threat to these species and/or to higher trophic levels, including marine mammals and seabirds.
- Carrying out biological effects measurements at selected locations in the Baltic Sea, particularly at sites of special concern, in order to assess whether the concentrations of contaminants in sea water and/or suspended particulate matter and/or sediments and/or in the organisms themselves are causing detrimental effects on biota (e.g. changes in community structure).

As with OSPAR, guidance is provided by HELCOM on the selection of appropriate species (e.g. herring (*Clupea harengus*), guillemot (*Uria aalge*), blue mussel (*Mytilus edulis*), bladder wrack (*Fucus vesiculosus*), flounder (*Platichhys flesus*), perch (*Perca fluviatilis*), viviparous blenny (*Zoarces viviparus*), common seal (*Phoca vitulina*)), tissues (e.g. liver, muscle) and deteminands (e.g. copper, lead, PCB congeners).

The two examples of 'Regional Seas' co-operation illustrated above give an indication of the types of chemical for which concentrations are measured using classical chemical analysis and the components in which they are measured. They also provide some insight into the need for international cooperation and the requirements of such programmes; the OSPAR pre-CEMP is a direct consequence of the need for monitoring guidance, quality assurance procedures and/or assessment tools to be available so that data produced by a number of laboratories across a range of countries can be combined so as to provide an appropriate geographical description of the status of a sea area.

The OSPAR Convention, covering the North-East Atlantic, and the Convention on the Protection of the Marine Environment of the Baltic Sea' (the Helsinki Convention) are both independent programmes. Globally, there are 19 'Regional Seas' with cooperative agreements covering the marine environment of more than 140 of the world's coastal countries. The Regional Seas Programme was launched in 1974 in the wake of the 1972 United Nations Conference on the Human Environment held in Stockholm. Within European waters the only United Nations Environment Programme (UNEP)-administered programme is The Convention for the Protection of the Marine Environment and the Coastal Region of the Mediterranean (Barcelona Convention) while the Convention on the Protection of the Black Sea Against Pollution (Bucharest Convention) is a non-UNEP administered programme.

As part of the WFD there is a requirement to monitor both ecological status and chemical status of surface waters, including coastal marine waters. The monitoring network is designed to provide a coherent and comprehensive overview of chemical status. Member states of the European Union are required to monitor parameters which are indicative of the status of relevant quality elements. Within this directive there are several types of monitoring described. These include:

- surveillance monitoring,
- operational monitoring, and
- investigative monitoring.

The differences between these types of monitoring are described in the Directive. In brief, the objective of the surveillance monitoring is to provide information on long-term changes in natural conditions and those resulting from widespread anthropogenic activity as well as providing information on the design of future monitoring programmes. Operational monitoring is undertaken to establish the status of WFD water bodies that have been identified as being at risk of failing to meet their environmental objectives and to assess the changes in status of those water bodies as a result of programmes and measures. Finally, investigative monitoring is carried out to ascertain the cause and effects of a failure when either the reason for exceedance is unknown or the magnitude of accidental pollution is unknown. Guidance on monitoring design related to surveillance, operational and investigative monitoring is provided by the European Commission (European Communities, 2009). The guidance covers selection of monitoring points and monitoring parameters as well as detailing the procedures for the actual sampling of, for example, water, sediments and biota. Furthermore, the guidance highlights that the standards for monitoring of quality elements for physico-chemical parameters shall be 'any relevant CEN/ISO standards or such other national or international standards which will ensure the provision of data of an equivalent scientific quality and comparability'. The methods have to be validated and for chemical monitoring certain minimum performance criteria have to be met. It is heartening to note that this drive for minimum standards in classical chemical monitoring is basically uniformly recognized, and where there is information already provided this is utilized in compiling guidance. For example, the WFD Guidance directs the reader to the OSPAR Joint Assessment and Monitoring Programme (JAMP) Guidelines for Monitoring Contaminants in Sediments. Since many countries undertake monitoring for a variety of reasons, it is helpful if the data can be gathered once and used many times. By insisting on recognized standards for classical chemical monitoring, as well as using tried and tested methods, it is possible to optimize the monitoring process so that the data can be applied in a range of programmes as well as used in an international context.

At a national level, the UK Marine Monitoring and Assessment Strategy (UKMMAS) was established following the publication, in 2005, of the first integrated assessment of UK Seas, Charting Progress (Defra, 2005). The thrust of the UKMMAS is to make the efficient use of UK resources in terms of existing obligations and to be prepared for emerging requirements. It represented a step-change in the way the UK monitors and assesses UK seas. From the perspective of classical chemical monitoring there are two primary areas of interest:

- The Clean and Safe Seas Evidence Group (CSSEG). This group focuses particularly on the direct effect on the marine environment of chemicals and pathogens, and their likely impact on human health. As such, it relies heavily on monitoring data for hazardous substances, nutrients (in respect of eutrophication), phycotoxins, radioactivity, oil and chemical spills and microbiological contamination.
- The Ocean Processes Evidence Group (OPEG). This evidence group focuses on sedimentary processes, sea level, circulation, temperature, salinity and waves, as well as CO_2 and acidification. Thus, there is a requirement for measurements of nutrient concentrations, salinity and the parameters used to investigate ocean acidification: total alkalinity (TA) and dissolved inorganic carbon (DIC).

The key to this process is that the outputs permit an assessment of the status of UK seas as well as allowing trends to be investigated. This requires a substantial amount of classical chemical analysis to be carried out by a number of organizations across the UK. This analysis has to be carried out using documented procedures and to recognized standards such that the data from the various laboratories can be compared with nationally agreed assessment criteria. The most recent assessment of this classical chemical monitoring is presented in Charting Progress 2 (UKMMAS, 2010).

The EU Marine Strategy Framework Directive (MSFD), adopted in June 2008, has brought further requirements for the monitoring of European seas (see Chapter 3). Under Article 11 of the MSFD, member states are required to:

> ... establish a monitoring programme for the on-going assessment of environmental status of their marine waters on the basis of the indicative lists of elements set out in Annex III and the list set out in Annex V, and by reference to the environmental targets pursuant to Article 10.

Amongst the indicative lists of pressures and impacts detailed in Annex III are 'contamination by hazardous substances', 'systematic and/or intentional release of substances' and 'nutrient and organic enrichment'. Characteristics include 'physical and chemical features'. These cover:

- spatial and temporal distribution of salinity;
- spatial and temporal distribution of nutrients (DIN,[2] TN, DIP, TP, TOC) and oxygen; and
- pH, pCO_2 profiles or equivalent information used to measure marine acidification.

In addition, there is a category under the heading of 'Characteristics' which covers:

- A description of the situation with regard to chemicals, including chemicals giving rise to concern, sediment contamination, hotspots, health issues and contamination of biota (especially biota meant for human consumption).

Thus, there will be a need for classical chemical monitoring to be included in support of the MSFD. This is especially the case when it is noted that the qualitative descriptors of good

[2] DIN, dissolved inorganic nitrogen (nitrite, nitrate and ammonia); TN, total nitrogen; DIP, dissolved inorganic phosphorus; TP, total phosphorus, TOC, total organic carbon.

environmental status (GES[3]), as detailed in Annex I of the MSFD, include the issues of hazardous substances and eutrophication.

An interesting aspect of the MSFD is the list of the features of a monitoring programme. These are detailed in Annex V of Directive 2008/56/EC and are reproduced in Box 8.2 because, although they are specific for the requirements of the MSFD, there are within the 12 requirements aspects which can be applied to marine monitoring in general. Specifically, this highlights the need for such programmes to provide information on environmental status, allow the identification of appropriate indicators and ensure that corrective measures are delivering the desired outcome. Classical chemical monitoring will contribute significantly to achieving these objectives of monitoring programmes.

BOX 8.2

Features of a monitoring programme as detailed in the European Marine Strategy Framework Directive (Directive 2008/56/EC).

Monitoring programmes
(referred to in Articles 11(1) and 24)

1. Need to provide information for an assessment of the environmental status and for an estimate of the distance from, and progress towards, good environmental status in accordance with Annex III and with the criteria and methodological standards to be defined pursuant to Article 9(3).
2. Need to ensure the generation of information enabling the identification of suitable indicators for the environmental targets provided for in Article 10.
3. Need to ensure the generation of information allowing the assessment of the impact of the measures referred to in Article 13.
4. Need to include activities to identify the cause of the change and hence the possible corrective measures that would need to be taken to restore the good environmental status, when deviations from the desired status range have been identified.
5. Need to provide information on chemical contaminants in species for human consumption from commercial fishing areas.
6. Need to include activities to confirm that the corrective measures deliver the desired changes and not any unwanted side effects.
7. Need to aggregate the information on the basis of marine regions or subregions in accordance with Article 4.
8. Need to ensure comparability of assessment approaches and methods within and between marine regions and/or subregions.
9. Need to develop technical specifications and standardised methods for monitoring at Community level, so as to allow comparability of information.

[3] 'Good environmental status' means the environmental status of marine waters where these provide ecologically diverse and dynamic oceans and seas which are clean, healthy and productive within their intrinsic conditions, and the use of the marine environment is at a level that is sustainable, thus safeguarding the potential for uses and activities by current and future generations.

10. Need to ensure, as far as possible, compatibility with existing programmes developed at regional and international level with a view to fostering consistency between these programmes and avoiding duplication of effort, making use of those monitoring guidelines that are the most relevant for the marine region or subregion concerned.
11. Need to include, as part of the initial assessment provided for in Article 8, an assessment of major changes in the environmental conditions as well as, where necessary, new and emerging issues.
12. Need to address, as part of the initial assessment provided for in Article 8, the relevant elements listed in Annex III including their natural variability and to evaluate the trends towards the achievement of the environmental targets laid down pursuant to Article 10(1), using, as appropriate, the indicators established and their limit or target reference points.

A requirement of the MSFD is that member states sharing a marine region or subregion ensure that their monitoring programmes are coherent and coordinated. This requires that monitoring methods are consistent across regions or subregions, thereby allowing comparability of monitoring results produced by different member states and ensuring that relevant transboundary impacts and transboundary features can be taken into account. This highlights the specific need of marine monitoring that arises because our seas do not recognize administrative boundaries and are a fluid and flowing medium. As highlighted earlier, within the UK there are several government-based organizations which contribute to a joint assessment of UK seas. On an OSPAR-wide scale there are 16 contracting parties[4] with responsibility for an oceanic area of $\sim 13.5 \times 10^6\,\text{km}^2$ with a volume of $30 \times 10^6\,\text{km}^3$. As such, the data arising from the classical chemical monitoring must be produced in accordance with specific guidelines and be of a known quality so that comparisons of data produced in many countries can be made, thereby providing an assessment of status and trends across vast geographical areas.

It is apparent from the above that classical chemical monitoring has, and will continue to have, a pivotal position in determining the status and trends in the global seas and oceans. That is not to say that this particular type of monitoring will not be enhanced, or in some cases replaced, by modern technologies such as remote sensing or automated in-situ analyses. Furthermore, there is a need to understand the impacts of hazardous substances on biota, and so 'biological effects monitoring' may replace classical chemical monitoring in some instances. However, the more recent technologies and effects methodologies will generally require validation through classical chemical monitoring and, in some instances, it is unlikely that classical chemical monitoring will be superseded by alternatives for many years to come. It is, though, worth briefly considering and contrasting current classical chemical monitoring with the more recent technologies and methodologies.

8.3 SOME CONSIDERATIONS AROUND CLASSICAL CHEMICAL MONITORING AND OTHER METHODOLOGIES

Classical chemical monitoring covers a vast array of measurements using many different techniques; see Table 8.1. However, the ultimate aim is the production of a quantitative output. The data derived

[4] Fifteen states in the North East Atlantic catchments (Belgium, Denmark, Finland, France, Germany, Iceland, Ireland, Luxembourg, The Netherlands, Norway, Portugal, Spain, Sweden, Switzerland, and UK) and the European Community.

Table 8.1 Typical analytes used to assess different components of the marine environment through classical chemical monitoring.

Component	Analyte(s)	Typical methodology			Reference
		Extraction/isolation	Clean-up	Instrumental determination	
Water					
	Nutrients (nitrate, nitrite, phosphate, silicate) Nitrite (low concentrations in oceanic waters)	Sample used as collected, if analysed within a few hours. Samples should be protected from light. If samples are to be stored and are from turbid waters they should be filtered. A number of storage methods are possible, including refrigeration, freezing and poisoning by the addition of mercuric chloride	Sample used as collected. Homogenized by shaking	Continuous-flow auto-analyser with absorption spectroscopy	Grasshoff *et al*. (1999)
	Ammonia (low concentrations in oceanic waters)	Sample used as collected, if analysed within a few hours. Samples should be protected from light. If samples are to be stored and are from turbid waters they should be filtered. A number of storage methods are possible, including freezing and poisoning	Sample used as collected. Homogenized by shaking	Continuous-flow auto-analyser with absorption spectroscopy or fluorometric determination	Grasshoff *et al*. (1999) Kerouel and Aminot (1997)

Salinity	Sample used as collected	Sample used as collected	Conductivity	UNESCO (1981, 1983)
Chlorophylls (a, b, c_1 and c_2)	Samples are extracted with solvent by either grinding, soaking or sonication	Clarification of the sample, if required, is performed by centrifugation or filtration	Spectrophotometric determination	Jeffrey and Humphrey (1975)
Chlorophyll a and Pheophytin a			Fluorometric determination	Lorenzen (1967) Welschmeyer (1994)
Chlorophylls, cartenoids and chlorophyll degradation products (including chlorophylls a, b, c_1, c_2, c_3, diadinoxanthin, fucoxanthin, lutein, zeaxanthin, pheophytin a)			HPLC with either UV, diode array or fluorescence detection	Jeffrey *et al*. (1997)
Total alkalinity (TA)	Treat with mercuric chloride to prevent biological activity		Potentiometric titration (open or closed cell)	Dickson *e al*. (2007)
Dissolved inorganic carbon (DIC)			Acidification followed by IR or coulometric titration	
Carbon dioxide fugacity (fCO_2)			Sea water in equilibration with air and IR detection	

(*continued overleaf*)

Table 8.1 *(continued)*

Component	Analyte(s)	Typical methodology			Reference
		Extraction/isolation	Clean-up	Instrumental determination	
	pH			Spectrophotometric detection or glass electrode (Cl ref electrode)	
	Metals (e.g. Cd, Pb, Hg)	None		Inductively coupled plasma mass spectrometry (ICP-MS)	Epov *et al*. (2004)
		Chelex resin preconcentration		Cold vapour atomic fluorescence spectrometry (CV-AFS) for Hg	
Sediment					
	Water content	Not applicable	Not applicable	Oven drying at $80 \pm 5\,°C$ for 22 ± 2 h	Webster *et al*. (2007)
	Particle size	Wet slurry or as a freeze-dried powder	Not applicable	Light-scattering based particle size	Russell *et al*. (2008)
	Total organic carbon	Freeze-dried and ground	Acid treatment to remove carbonate	Elemental analyser	Russell *et al*. (2008)
	Organic contaminants (e.g. PAHs, PCBs, PBDEs)	Soxhlet extraction	Liquid–liquid extraction	GC–MS or variants thereof	Webster *et al*. (2008, 2009c,d)

		Liquid–solid extraction	Adsorption chromatography		
		Pressurized liquid extraction (PLE)	Gel permeation chromatography (GPC)		
		Supercritical fluid extraction (SFE)	HPLC		
	Metals (e.g. Pd, Cd, Hg, Al)	Microwave digestion with hydrofluoric acid		ICPMS	Loring and Rantala (1990)
		Microwave digestion with other acids		Graphite furnace atomic absorption spectrometry (GF-AAS)	Robinson and Devalla (2006)
		Sequential extractions		CV-AFS (for Hg)	Rauret *et al*. (1999)
Fish					
	Length	Not applicable	Not applicable	Standard ruler	
	Lipid content	Liquid–liquid extraction	Removal of solvent	Weighing	Bligh and Dyer (1959) Hanson and Olley (1963) Smedes (1999)
	Metals (e.g. Pd, Cd, Hg)	Microwave digestion with nitric acid, sometimes with H_2O_2 or other acids in addition		ICPMS GF-AAS CV-AFS (for Hg)	CEN (2009)
	Organic contaminants (e.g. PBDEs, PCBs)	Saponification	Liquid–liquid extraction	GC–MS or variants thereof.	Webster *et al*. (2009c)
		Soxhlet	Adsorption chromatography		
		Pressurized liquid extraction (PLE)	Gel permeation chromatography (GPC) HPLC		

(*continued overleaf*)

Table 8.1 *(continued)*

Component	Analyte(s)	Typical methodology			Reference
		Extraction/isolation	Clean-up	Instrumental determination	
Shellfish					
	Shell length (mussels)	Not applicable	Not applicable	Standard ruler	
	Lipid content	Liquid–liquid extraction	Removal of solvent	Weighing	Bligh and Dyer (1959) Hanson and Olley (1963) Smedes (1999) CEN (2009)
	Metals (e.g. Pd, Cd, Hg)	Microwave digestion with nitric acid sometimes with H_2O_2 or other acids in addition		ICPMS GF-AAS CV-AFS (for Hg)	
	Organic contaminants (e.g. PAHs, PBDEs, PCBs)	Saponification	Liquid–liquid extraction	GC–MS or variants thereof	Webster *et al*. (2009c,d)
		Soxhlet	Adsorption chromatography		

		Pressurized liquid extraction (PLE)	Gel permeation chromatography (GPC) HPLC		
Biotoxins:					
	Paralytic shellfish poisoning (PSP) Diarrhetic shellfish poisoning (DSP) and Lipophilic shellfish toxin (LST)	Liquid–solid extraction	C18 SPE and COOH SPE (PSP) Extract filtration (DSP and LST)	HPLC-FLD & LC-MS (PSP) LC-MS & LC-MS (DSP)	AOAC (2006) (PSP) Dell'Aversano *et al*. (2005) McNabb *et al*. (2005) Stobo *et al*. (2005)
	ASP		SAX-SPE (ASP)	HPLC-DAD (ASP)	AOAC International (2000)

from classical chemical monitoring are used to assess status and trends with respect to water chemistry, eutrophication, hazardous substances and phycotoxins. The components investigated cover water, sediment and biota (primarily fish and shellfish, but also marine mammals (Pierce *et al*., 2008; Law *et al*., 2010), seabird eggs (Becker, 1989; Pusch *et al*., 2005) and seaweed (Caliceti *et al*., 2002)), while the techniques incorporate methods that allow normalization of data as well as those that produce concentrations of compound X in component Y. Sample preparation (extraction/isolation of the analyte and associated clean-up) is, for some methodologies, relatively simple. An example is the determination of salinity. In this case the sample is analysed directly using a salinometer with no specific sample preparation (UNESCO, 1981, 1983). In contrast, other techniques, such as the determination of PAHs in shellfish, involve a complex and prolonged procedure to extract the PAHs and provide a sample that can be injected onto a gas chromatograph. The methodologies used to measure the various compounds of interest to those undertaking marine assessments take in absorption spectroscopy (nutrients), conductivity (salinity), liquid chromatography–mass spectrometry (LC–MS; shellfish toxins) and gas chromatography–electron capture negative ionization mass spectrometry (GC–ECNIMS, PBDEs). This range of instrumental analytical methods, as indicated in Table 8.1, uses a number of chemistries and techniques, but all are characterized by the production of a numerical value which is generally both precise and accurate and can often be compared with clearly defined assessment criteria.

Classical chemical monitoring is generally characterized by discrete sampling and compound-specific analysis. This results in very specific information being generated but does not provide access to the full story, since the methodologies do not generally isolate only the bioavailable fraction, nor do they provide information on the range of compounds to which an animal may have been exposed. Rather, classical chemical analysis of hazardous substances provides very precise information on the concentration of a specific compound or group of compounds. There are 209 individual PCB congeners, and often the analysis of this group of compounds is limited to a recognized group (e.g. ICES seven PCBs). This does provide valuable data, but there are limitations. Some of the limitations can be overcome through the use of new technologies such as passive sampling (see Chapter 10) or by making measurements which provide some indication of the biological effects of hazardous substances (see Chapter 9). Indeed, the impact of the (now banned) hull antifoulant TBT is best investigated through its biological effect; the imposition of male sexual characteristics on female marine snails such as the dog whelk (*Nucella lapillus* L.) or the common whelk (*Buccinum undatum*). In recent years, assessment criteria have been established and used to assess the impact of TBT on coastal regions (Gubbins *et al*., 2004; OSPAR, 2004; Gubbins, 2010). However, this particular biological effect is very specific and the high sensitivity of marine snails to TBT means that the biological effect is evident at concentrations that are difficult to detect by classical chemical methods. Development of assessment criteria for other biological effects measurements has proved more elusive, and the linking of 'cause and effect' has also been a challenge. However, there have been some recent developments through a joint OSPAR/ICES Study Group on Integrated Monitoring of Contaminants and Biological Effects (ICES 2010).

A characteristic of the marine environment is the variability which exists both spatially and over time. Some of the variation is well documented with relatively long time scales. For example, nutrient concentrations are high in the winter months and low in the summer months. Such gross differences are readily measured with weekly sampling and classical chemical analysis in the laboratory; see Figure 8.1. However, more detailed changes in nutrient concentrations may not be observed with such a sampling programme. The lack of temporal data has made it difficult to assess the performance of nutrient monitoring programmes in the context of eutrophication assessments. Recently, 'in situ' Smart Buoy technologies have become available. These monitor a range of chemical and environmental variables as frequently as every 15 min. The temporal resolution of Smart Buoy data can be used to inform the design of ship-based nutrient monitoring programmes (Heffernan *et al*., 2010).

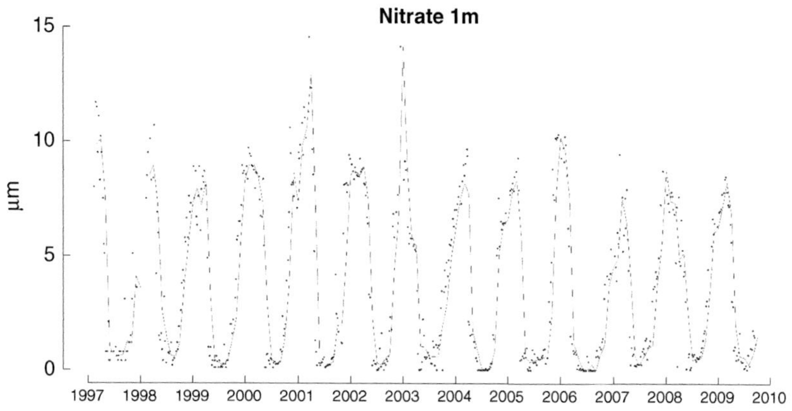

Figure 8.1 Temporal variation in the surface (1 m) nitrate concentrations at the Stonehaven ecosystem monitoring site (56°57.8′N 02°06.2′W) operated by Marine Scotland Science. Water samples have been collected on a weekly basis (weather permitting) and the nitrate determined using segmented flow analysis with detection by absorption spectroscopy. Specifically, seawater (from a 20 mL sample) is mixed with sulfanilamide under acidic conditions to form a diazo compound. This is coupled with *N*-1-naphthylethylenediamine hydrochloride to form a purple azo dye. The absorption is measured at 520 nm. The analysis time is 15 min. The dots are the weekly observation points and the line is monthly means. (Figure provided by Jens Rasmussen, Marine Scotland Science).

In conclusion, classical chemical monitoring, biological effects methodologies and 'in situ' technologies all have their own benefits and limitations; see Table 8.2. The viability of any methodology is dependent on many factors, including cost, availability of the technology, delivery of data of appropriate precision and accuracy and the availability of relevant assessment criteria. Ultimately, any methodology must be 'fit for purpose', and this means providing data that is interpretable such that it is possible to come to conclusions on status and trends. It is likely that future monitoring programmes will be a mixture of classical chemical monitoring, 'in situ' monitoring and biological effects determinations. The challenge will be integrating the different methodologies to provide truly holistic assessments.

8.4 DESIGNING A SAMPLING PROGRAMME

In order to be able to fulfil the vision of clean, healthy, safe, productive and biologically diverse seas and oceans there is a need to detect trends, to be able to describe water bodies and areas of the seabed, to be able to compare one sea area with another and describe any difference with respect to, for example, concentrations of persistent organic pollutants (POPs) and to be in a position to say if changes in practices and procedures have resulted in improved environmental status for marine waters. Ultimately, this requires that samples of water, sediment and biota are collected and analysed. Classical chemical analysis is characterized by collecting discrete samples. Obtaining sediment from the seabed and water samples or biota that are representative of an area is a costly and time-consuming process. There is often the need for a ship. This means that the sampling process can be hindered by bad weather and that sampling is being undertaken from a moving platform. Collecting a sediment sample from the seabed that is under 130 m of water requires

Table 8.2 Comparison of classical chemical monitoring with 'effects' monitoring and continuous monitoring processes.

Characteristic	Classical chemical monitoring	'In-situ' chemical monitoring[a]	Biological effects monitoring
Type of sampling	Spot sampling, often from a ship	Single site using, e.g., Smart Buoy	Spot sampling, often from a ship or 30 min trawl
Frequency of sampling	Medium (weekly) or low (monthly or annual)	Very high (15 min) or high (hourly)	Low (monthly or annual)
Spatial coverage	Good	Limited by the availability of 'in situ' monitoring devices	Good
Temporal resolution	Limited	Good	Limited
Components	Water, sediment, biota	Water	Biota
Analytes	Hazardous substances, nutrients, salinity, chlorophyll, biotoxins,	Nutrients, salinity, chlorophyll	Indicators of exposure to hazardous substances, e.g. imposex, EROD, vitellogenin, acetylcholinesterase activity, scope for growth
Benefit	Large number of analytes in water, sediment and biota	High frequency of monitoring giving excellent temporal resolution	Can provide information on exposure to a specific contaminant (e.g. TBT)
	Technical guidelines available		Provides information on the impact of exposure to the environment in which the animal lives

	Well-established methodologies and quality assurance Assessment criteria available for many analytes		Takes account of what is bioavailable
Limitation/problems	Single compound or group of compounds Limited temporal resolution Requires a sampling platform, often a ship	Analytes currently limited to nutrients, salinity and chlorophyll Subject to biofouling, which limits operation Limited spatial resolution	Many techniques still being developed Limited number of assessment criteria available Cause and effect can be unclear

[a]Passive sampling is not included in this summary. Although it provides ‘in situ’ sampling, the determination of the contaminant concentration relies on classical chemical analysis.

specialized samplers, but to date you are, in effect, sampling blind, since it is generally not possible to observe the seabed as it is being sampled; nor is it readily possible to assess how characteristic the specific sample is of the surrounding sediment. Constraints on sampling require that optimal sampling regimes, with a sound statistical basis, are developed such that the optimal number of samples is collected in a way that ultimately allows description of a designated area to be made and also permits comparisons with other areas. Many institutes across the world have, over many years, sampled the seabed so as to assess the status of marine sediments with regards to hazardous substances (e.g. Kelly and Campbell, 1996; Webster *et al*., 1997; Gutiérrez-galindo *et al*., 1998; Yang *et al*., 1998; Zanardi *et al*., 1999; Midorikawa *et al*., 2004; Ünlü and Alpar, 2005; Edgar *et al*., 2006; El Deeb *et al*., 2007; Russell *et al*., 2008; Ameur *et al*., 2010). During this time there has been an improved understanding of how to sample the marine environment with statistical tools being used to develop appropriate sampling protocols so as to ensure the production of data which can be usefully used to describe the status of the seabed and to detect trends (Moffat *et al*., 2007). Such studies on the sampling of sediments have been extended to included biota. The designs of many large marine environmental monitoring programmes, such as those of OSPAR, HELCOM and the Barcelona Convention, have their origins in the monitoring programmes of contaminant concentrations in biota begun in the late 1970s and early 1980s. These were set up primarily to detect temporal trends and had a network of monitoring stations, typically placed close to industrial or estuarine sources, where the effect of any reductions in inputs might be detected. The sampling protocol was essentially to sample in the same place at the same time each year, as indicated in Figure 8.2, the idea being to control the sources of variation not of interest, such as seasonal variation, to focus on long-term trends (Phillips, 1980). As monitoring of sediment contaminant concentrations, benthic communities and biological effects was introduced, the concept of monitoring a network of fixed stations annually was repeated, with sediment networks sometimes more densely populated so that spatial trends could also be investigated. However, over time it has become clear that sampling in the same place at the same time each year gives less control on the unwanted sources of variation than had been envisaged. In particular, many time series were shown to have significant random between-year variation, changes in, for example, concentration from year to year with no particular pattern. This, often dramatically, reduced the power (or probability) of long-term trend detection (Fryer and Nicholson, 1993). Random between-year variation and its effect on power is still an issue. Detecting a doubling or halving of contaminant concentrations by annual monitoring over 10 years has often been used as a convenient target for monitoring programmes (Fryer and Nicholson, 1997). Yet the assessments leading up to the OSPAR QSR 2010 found that, for example, only one-third of mercury time series and one-quarter of CB153 time series in biota met this criterion.

It has been difficult to identify the main causes of random between-year variation, although this is fundamental to improving the power of trend monitoring programmes. If driven by a mixture of events on a yearly scale, such as annual variation in inputs or summer temperatures, or cohort effects in biota, then there is little that can be done and good power will often require several decades of monitoring. However, there is a strong suggestion that the random between-year variation has often been artificially induced by cluster sampling, also known as pseudo-replication. Ironically, this arises from the 'same place, same time' sampling philosophy. For example, in sediment monitoring, if several grabs are taken each year from effectively the same spot, but that spot varies from year to year due to navigational error, then any short-scale spatial differences in concentration between the spots (e.g. due to differences in substrate, currents or deposition) will be modelled as between-year variation. A similar effect arises if all samples from a station are analysed in the same batch, but there is variation in analytical performance between batches. A simple way to reduce the cluster-sampling effect is to take samples over a larger area, or at several times, and to analyse them in several batches. This should reduce the between-year variation and lead to designs for monitoring programmes with better power. For example, the French mussel

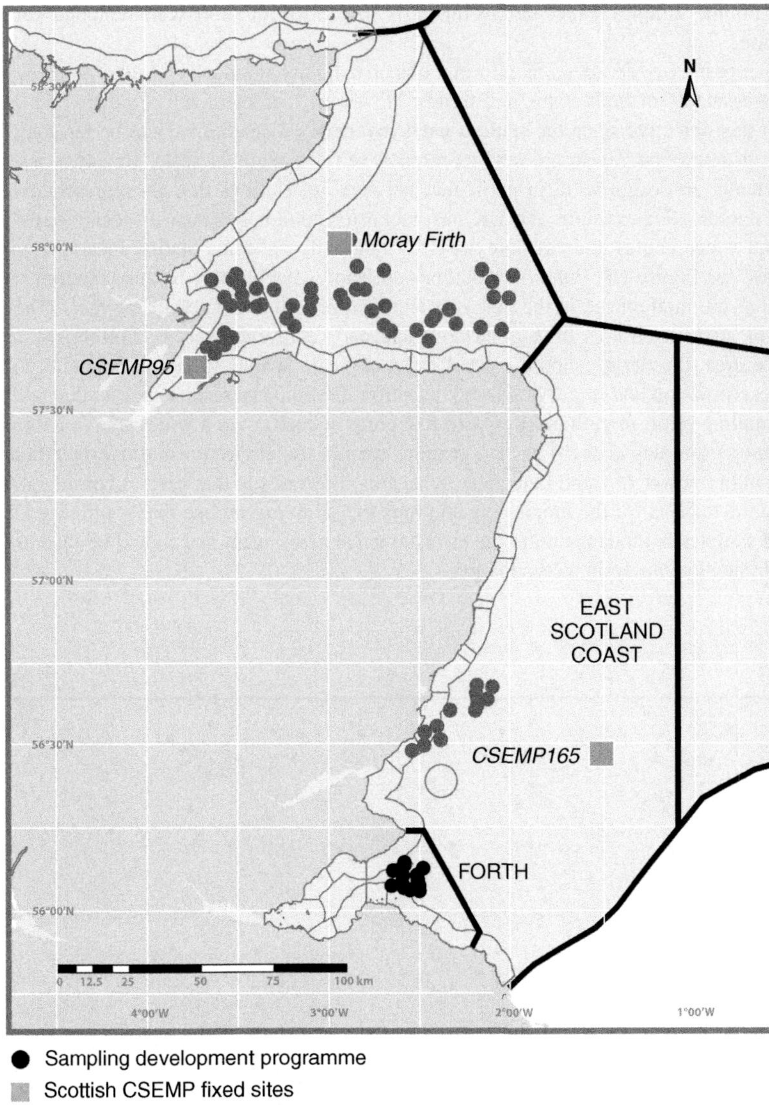

Figure 8.2 Assessing trends in the contaminant concentration of sediments along the east coast of Scotland has traditionally been done through repeated collection of five discrete samples at designated sites (squares) at a specified time of year. More recently, the sampling programme has been revised and now uses stratified random sampling of the muddy sediments (dots). Sites within a specific stratum are selected at random on an annual basis. This facilitates reporting on a regional basis. The figure covers the years 2006–2009. See plate section for a colour version of this image.

watch programme samples some stations quarterly to control for short-scale temporal variation in concentration.

Over the past two decades, monitoring objectives have gradually shifted towards making regional assessments of both status and trends. However, current monitoring networks are often unsuited to this objective since the stations each cover only a small area, can be far apart and are often close to point sources so are not 'representative' at a regional scale. Regional assessments require a change in design with, in particular, networks of stations that are representative of the regions of interest. For example, the UK has begun to monitor sediment concentrations in some large coastal water bodies and offshore areas by randomly choosing station locations within the water bodies; see Figure 8.3. Either new stations are chosen each year (stratified random sampling) or the stations are randomized in the first year and then revisited on subsequent years. The random allocation of stations ensures both good coverage and representativeness. Further, by sampling over a wide area, the design should control for short-scale spatial variation, potentially reducing between-year variation and improving trend detection. Preliminary results suggest that reallocating existing sampling effort (e.g. three sites with five grabs at each) over a wider area (e.g. three water bodies with five samples in each) has led to more meaningful environmental interpretation with at least comparable power of trend detection. What this illustrates is that careful consideration must be given to the reasons for the monitoring and thus the sampling regime that is utilized. Obtaining appropriate samples is fundamental to the environmental assessment and should be seen as integral to classical chemical analysis methodology.

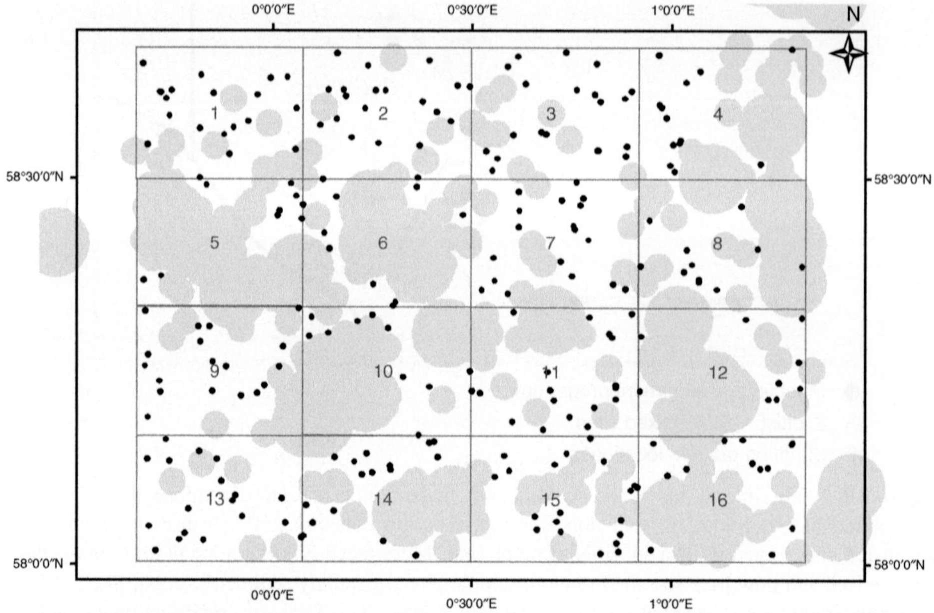

Figure 8.3 An example of a stratified random sampling scheme used to investigate the possible impacts of the offshore oil and gas industry in the Fladen Ground of the North Sea. In this case the Fladen Ground was divided into 16 zones. Within each zone, stratified random sampling sites (black dots) in the far-field areas (defined as areas more than 5 km away from multiple oil wells or 2 km from single oil wells) were assigned. Grey circles are near-field sites (big circles are multiple oil wells and small circles are single wells).

8.5 SAMPLE COLLECTION AND IMMEDIATE HANDLING

The objective of environmental sampling is, as has been stated, to provide the relevant information such that appropriate assessment can be made in respect of status and trends. The selection and collection of appropriate materials (whether water, sediment or biota) is an integral part in assessing the marine environment. A proper assessment can only be achieved if relevant samples are collected and appropriately handled immediately following collection. This includes the collection of samples for sensory assessment, which is often used in conjunction with classical chemical monitoring following an oil spill (Davis *et al.*, 2002). Sample collection is only the first stage in a process which involves a number of steps that ultimately result in the production of scientific data and provision of information; see Box 8.1. If this aspect of the process is not conducted appropriately then there is no point in progressing to subsequent stages, as any assessment process would be of little value. Thus, the handling, processing and storage of samples immediately following collection in the field should be strictly controlled.

As an example, hydrocarbon compounds are ubiquitous within the environment and also abound on sampling vessels. Thus, the opportunities for contamination at any stage, including initial collection, during subsampling, from the storage container and during the transfer to the laboratory, are significant. Adventitious contamination can make the assessment of PAH concentrations in a sample impossible, thereby wasting considerable time, effort and money. This means that considerable care must be taken, especially when collecting material, to provide background or reference data. Samples should not come in contact with plastic, rubber or similar materials (such as gloves, bags, etc.), solvents, cleaning fluids or oils. Other preparative procedures are essential. These include rinsing sample containers with an appropriate organic solvent (e.g. pentane, hexane or dichloromethane) as well as rinsing any aluminium foil used in the storage of fish samples with an appropriate solvent. Furthermore, such solvents should be of an appropriate quality and should be tested to ensure their suitability.

Although some of the procedures which have to be followed may appear to be unnecessary and time consuming, they are essential if sample integrity is to be maintained and assured.

8.6 SAMPLE STORAGE (SHORT AND LONG TERM)

Classical chemical analysis is characterized by the collection of discrete samples, whether it be sediment, biota or water. In some cases the subsequent analysis will involve the taking of a subsample, while in other cases the analysis will require the combination of, for example, several fish livers. On occasions, such as when investigating the effects of an oil spill, fish will be subsampled with one fillet undergoing sensory assessment while the other fillet will be subject to chemical analysis. When this is the case, sampling has to take account of the fact that these fish will be tasted by a sensory panel. Subsampling or sample combination may be done at the time of obtaining the sample in the field. However, it may be the case that subsampling is performed on return to the laboratory. Either way, sample integrity can only be assured by appropriate storage, whether it is short or long term. The key to suitable storage is appropriate:

- Packaging,
- labelling and
- storage temperature.

In the case of storage temperature the objective should be to attain the apposite temperature as rapidly as possible following collection of the sample; see Table 8.3. For small sample sizes,

Table 8.3 Storage conditions for marine samples to be analysed by classical methods, for biological effects measurements and sensory assessment.

Type of sample	Treatment and storage
Sediment	Frozen and stored at ≤−20 °C
Fish flesh, shellfish tissue and liver tissue for chemical analysis (e.g. PCBs, PBDEs, PAHs)	Frozen and stored at ≤−20 °C
Fish liver for biological effects measurement	Immediately frozen in liquid nitrogen and stored at ≤−80 °C
Fish flesh for sensory assessment	If tested within 72 h, refrigerate at 4 ± 2 °C If there is a delay in testing beyond 72 h then samples should be frozen and stored at ≤−20 °C for a maximum of 3 months
Shellfish tissue for sensory assessment (excluding mussels)	If tested within 72 h, refrigerate at 4 ± 2 °C If there is a delay in testing beyond 72 h then samples should be frozen and stored at ≤−20 °C for a maximum of 3 months
Mussels for sensory assessment	If tested within 72 h, refrigerate at 4 ± 2 °C If there is a delay in testing beyond 72 h then samples should be cooked and stored at ≤−20 °C
Shellfish tissue for biological effects measurements	Immediately frozen in liquid nitrogen and stored at ≤−80 °C
Seawater for salinity determination	Room temperature, out of direct sun light
Seawater for nutrient analysis (nitrite, phosphate, total oxidized nitrogen, silicate and ammonia)	If tested within 10 h of sampling, refrigerate at 4 ± 2 °C If there is a delay in testing beyond 10 h then samples should be frozen and stored at ≤−20 °C except for determination of silicate, in which case the samples should be refrigerated at 4 ± 2 °C

snap-freezing in liquid nitrogen is appropriate and cryogenic storage in dry containers is also very useful when undertaking field work. For larger sample sizes (e.g. bulk sediment or fish muscle) it is important that the sample is packaged so as to allow rapid cooling to the core of the sample. Once the sample has attained its storage temperature, every effort should be made to maintain that temperature throughout the lifetime of that sample. Repeated freezing and thawing of samples should be avoided; this means that the further processing of any samples should be planned with a life history of every sample being maintained.

Samples must be packaged so as to maintain their integrity. Furthermore, the purpose of the analysis has to be considered when deciding how to package the sample and what material should be used. When determining metals in sediment you want to store the sample in a plastic container, which is something you must avoid when determining PAHs in sediment. Similarly, determination of silicate in seawater necessitates the use of plastic bottles rather than the standard glass bottles in which samples for the determination of salinity or nitrate concentration are packaged. Biological samples can be small and must be wrapped carefully in, for example, aluminium foil before being labelled and transferred to an appropriate container. The key here is simply to identify that there is a need to consider the packaging of samples, since their loss or damage means that time has been wasted, a time series may have been broken and an assessment is not possible.

It is not unknown for a sample to lose its 'label', at which point it can only be discarded. Labelling with 'indelible ink pens' does not guarantee a permanent label, especially when samples are stored frozen and then defrosted or when you are working with wet samples. Under such conditions an alternative should be considered, such as tie-on labels. Another problem is that a name or a code may make perfect sense at the time of writing, but some months later could well be unintelligible. The key to good labelling of samples and subsamples, especially when operating in the field, is to prepare well in advance and to include the labelling within a recognized quality system. That way there is a structure to the numbering system and a prescribed method for recording the data which is then maintained in an appropriate data archive. Another methodology which can assist in ensuring accurate labelling and tracking of samples is through the use of bar coding. This can be incorporated into a laboratory information management system. To facilitate this, preprinted sample sheets may be used to ensure that all the required information is collected and recorded at the time of sampling. Such information should include, for example, a sample identification code, date and time when the sample was collected, depth from which the sample was collected (if pertinent) and location, and so on. If care is taken, and attention to detail maintained during the early stages of sample collection, then the risks of errors occurring and samples becoming misidentified will be reduced. Furthermore, the correct recording of all relevant details will enable a concise fully accountable, audit trail to be undertaken, including the necessity of 'chain of custody' statements.

8.7 LABORATORY ANALYSIS

By the time samples have reached the laboratory they will have already underdone a certain amount of processing, which may have included subsampling. However, it will often be the case that the actual sample analysed (the test sample) is a subsample of the total material available. This allows, if required, the analysis to be repeated. As detailed earlier, for some classical chemical methods the primary sample or a subsample is analysed directly. This includes determination of salinity or nutrients. However, in the majority of cases, there is a need to extract the analyte of interest, be it a metal, organic contaminant, biotoxin or the pigments present in phytoplankton, from the matrix; see Table 8.1. The resulting extract is then subject to a clean-up process prior to instrumental analysis. The instrumental reading is then processed so as to provide a quantitative output. Finally, the data are quality checked; see Box 8.1. Specifics regarding the laboratory analysis of various analytes are covered in Chapters 4–7. However, there are several common key features which should ensure data from classical chemical methods are both accurate and precise.[5] These are presented below.

8.7.1 Quality Assurance Procedures and Method Validation

Quality assurance represents the procedures undertaken by laboratory staff to ensure that data of the appropriate quality is obtained to meet the defined aims of the laboratory. There are two main components:

Quality Control – This is achieved by performing a number of procedures which maintain measurements within an acceptable level of accuracy and precision. This includes the determination

---

[5] Accuracy is the closeness of an experimental measurement or result to the true or accepted value. Precision is the random or indeterminate error associated with the result and is measured by the standard deviation. See Fifield and Haines (2000) and Prichard *et al*. (2001).

of recoveries for the analyte in question, an evaluation of the limit of detection (LoD) and limit of quantification (LoQ),[6] the use of laboratory reference materials (LRMs) and certified reference materials (CRMs) (see Chapter 11), participation in an external proficiency testing scheme (e.g. QUASIMEME; see www.quasimeme.org) and, if appropriate, accreditation by an external organization such as the United Kingdom Accreditation Service (UKAS; www.ukas.com).

Quality Assessment – This is delivered through procedures which provide documented evidence that quality control is achieved. This includes controls which are used to:

- ensure a method is under control,
- check for trends, and
- highlight problems.

Integral to quality assessment is the use of control charts (Mullins, 2003), which are a form of internal quality control. An example of such a chart is the Shewart chart, which is constructed on the basis of the repeated analysis of an LRM. This produces a mean value for the analyte with

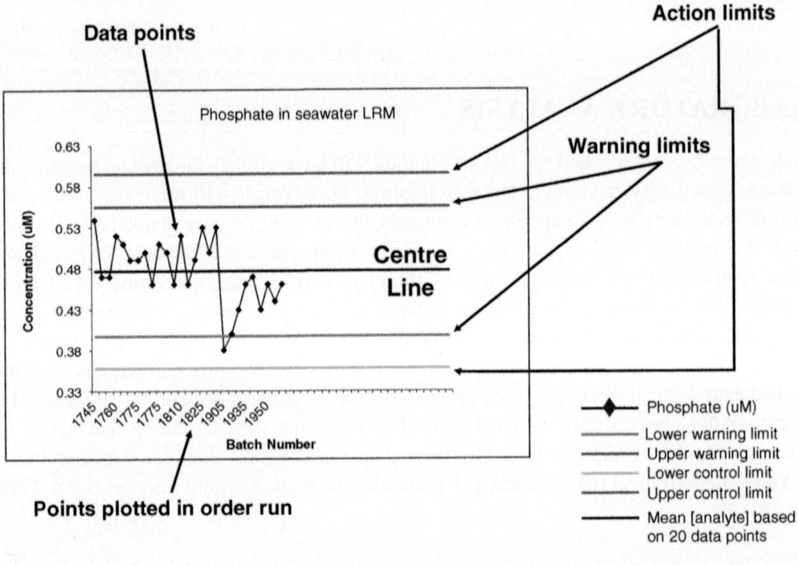

Figure 8.4 Example of a Shewart chart used as part of the internal quality assurance procedures when conducting classical chemical monitoring. A mean value (the centre line) is determined for the specific analyte using an LRM. Warning limits and action limits are usually defined at ±2 standard deviations and ±3 standard deviations respectively. See plate section for a colour version of this image.

[6] The LoD for an analytical procedure is defined as the minimum single result which, with a stated probability, can be distinguished from a suitable blank value (see Prichard *et al*. (2001)). For an analytical method the determination of the LoD is based on a multiplier of the sample standard deviation of a blank determination or the repeated analysis of a sample spiked with a low concentration of the analyte. In the authors' laboratory the LoD is calculated using a numerical factor of 4.65.The LoQ is the lowest concentration of an analyte that can be determined with acceptable uncertainty. For most trace analysis the LoQ is taken as 10 times the sample standard deviation.

warning and action limits being set at 2× and 3× the standard deviation; see Figure 8.4. A set of rules are put in place to ascertain whether or not a test point passes or fails. Failure can occur when:

- one point is outside either the upper or lower *action* limit;
- two consecutive points are outside either the upper or lower *warning* limit;
- seven points in a row are either *above or below* the mean.

Ultimately, the laboratory method being used should be fully validated. Method validation checks should be carried out on all methods before they are brought into use, in other words used to analyse actual test samples. This applies to new methods, when new instruments are installed or if existing methods are significantly changed. For chemical methods, the minimum required for validation is calculation of LoD, precision (reproducibility) and spiking recovery (bias). In the authors' laboratory, the method LoD is determined by analyzing at least seven reagent blanks (or low matrix samples if the blank is not detectable) extracted and run over several days, with the resulting standard deviation expressed as a concentration. The LoD is then expressed as 4.65 times the blank standard deviation. The LoQ values can normally be calculated using the same information (e.g. 10 times standard deviation). However, as the LoD is an idealized value, when reporting 'less than' values in test reports, it may be more appropriate for some methods to run a low calibration standard or reference material at or around the level of the theoretical LoD. This demonstrates that the idealized LoD value can be achieved on the day of analysis.

To determine reproducibility, the authors undertake at least seven replicate determinations of a standard at the upper (90%) and lower (10%) end of the working range of the analyte(s), run 'as a sample'. These are run on separate days (reproducibility conditions) and expressed as concentrations. The mean, standard deviation and percentage variance are calculated and then compared with a target value.

To determine bias/recovery the preference is to analyse at least seven replicate determinations of an appropriate CRM. If such a material is not available, then at least seven replicate determinations of an external quality assessment sample, with a known assigned value or at least seven replicate determinations of a sample and a spiked sample can be used. The samples are extracted and run on separate days and expressed as concentrations. The resulting data are averaged and the sample standard deviation calculated, and expressed as a percentage. This is then compared with a target value.

Such procedures, as outlined above, contribute to ensuring that the data used in assessments are fit for purpose. However, there is a need for other 'common features' of the chemical analysis which contribute to ensuring accurate and precise data.

8.7.2 The Preparation and Use of Standard Operating Procedures

Standard operating procedures (SOPs) are written procedures which, put simply, describe the processes and procedures undertaken by staff in delivery of the laboratory services and cover such topics as:

- routine inspection, cleaning, maintenance, testing, calibration and standardization of instruments;
- actions to be taken in response to equipment failure;
- analytical methods;

- definition of raw data;
- data handling, storage and retrieval;
- health and safety precautions;
- receipt, identification, storage, mixing and method sampling of test and control articles;
- record keeping, reporting, storage and retrieval of data;
- coding of studies, handling of data, including the use of computerized data systems;
- operation of quality assurance personnel in performing and reporting study audits, inspections and final study report reviews.

SOPs should be written by those undertaking the analytical procedures, but checked, and ultimately signed off, by the laboratory manager. SOPs should not be written to explain how procedures are supposed to work, but how they work. This ensures that the information is adequate and that the document invites rather than discourages routine use. There should be SOP on writing SOPs. This is important for consistency and efficiency. For example, it should be defined who is responsible for initiating, authoring and approving SOPs and how procedures are distributed and how the use of SOPs is enforced. Deviation from SOPs can be allowed under certain circumstances, but deviations should be approved and documented.

8.7.3 Looking after Your Equipment

To some extent, this links in with the SOPs. However, the need for calibration of laboratory balances, appropriate maintenance of pipettes and regular servicing of instrumentation has first to be identified. All equipment should be adequately inspected, cleaned and maintained. Equipment used for generation, measurement or assessment of data should also be adequately tested, calibrated and/or standardized. The frequency for calibration, revalidation and testing (performance verification) depends on the instrument itself, the recommendations from manufacturers of equipment, laboratory experience and the extent of use. For instance, a pH meter or a balance should be calibrated before each use and the wavelength of a high-performance liquid chromatograph variable-wavelength detector should be calibrated about every month or whenever the cell is removed and reinstalled. However, the whole process of ensuring the generation of appropriate data is very much intertwined with the different processes and procedures outlined; all must to be performed.

Chemical reactions are temperature sensitive and analytes can be light sensitive. As such, an appropriate laboratory environment is critical. In this context, protection from light should be an integral part of the analysis of PBDEs; this could include the use of amber glassware and the placing of blinds over windows. Temperature regulation can be an issue, especially when operating on-board a research vessel. Drift in silicate determinations at sea was found to be due to the change in air temperature in the instrument room on the ship. This was remedied by the installation of air conditioning. Staying with nutrient analysis, the use of a tungsten filament as the light source in the nutrient analyser proved problematic due to the vibration of the ship. This was resolved by using a light-emitting diode. These examples illustrate how crucial it is to look at all aspects of the analytical procedure.

8.7.4 Solutions and Reagents

Many high-quality, high-purity reagents are available to purchase. However, the quality of the products should never be simply accepted; reagents should be tested and laboratory procedures

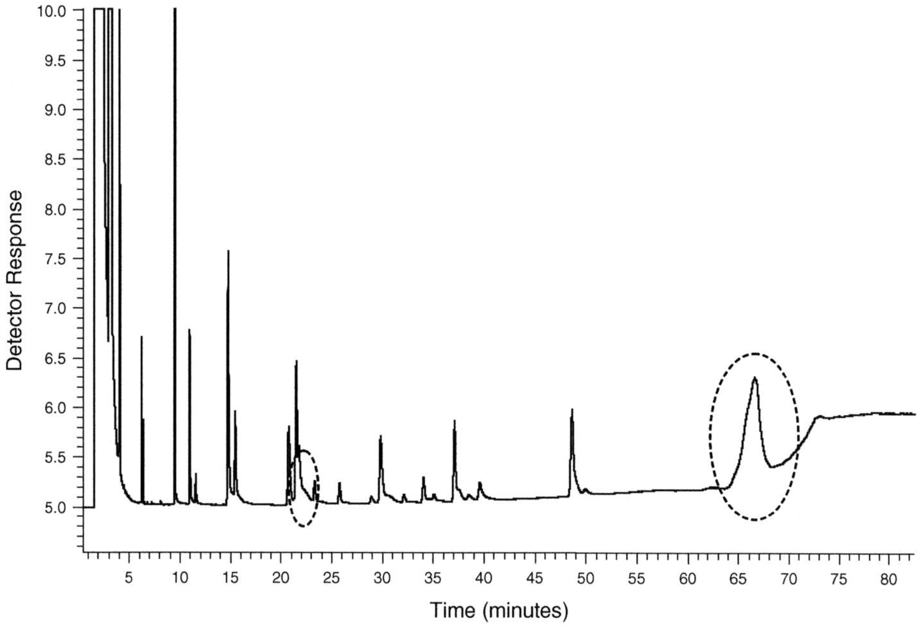

Figure 8.5 Example of a gas chromatography–flame ionization detection (GC–FID) chromatogram of fatty acid methyl esters where there is excessive peak tailing and a substantial contaminant peak at approximately 66 min. The peak tailing is often related to contamination of the retention gap, a 30 cm × 0.53 mm i.d. silanized pre-column placed between the on-column injector and the analytical column.

established to ensure that the products being used do not introduce unwelcome contaminations or result in poor chromatography; see Figure 8.5. The nature of the insert within the cap on the bottle of solvent, impurities in the sodium sulfate used to dry extracts or the cleaning product used to clean the glassware can all influence the analytical result in an adverse manner. All reagents and solutions should be labelled to indicate identity, titre or concentration, storage requirements and expiration date. The expiration date depends on the nature of the chemical. Sodium chloride has practically no expiration date. In these cases it is generally acceptable to indicate 'none' or 'not applicable' on the label for expiration date. However, such a designation must be justifiable. Formal studies are not always required to justify assigned expiration dates. It can be sufficient to assign expiration dates based on literature references and/or laboratory experience. Deteriorated or outdated reagents and solutions should be discarded. Finally, it is also good practice to include the date opened; this can be critical for some chemicals, such as diethyl ether.

8.7.5 Effective Document Control

Integral to any quality system should be effective documentation. This can include the use of controlled laboratory notebooks as well as SOPs. All controlled documents (SOPs and worksheets) should include, as a minimum, the title, version number or date, owner and pagination. Where appropriate, standard forms should be prepared. Worksheets should be used for recording information on the analytical batch, such as the date of extraction, the weight of sample extracted and the amount of internal or recovery standard added. Worksheets can also be used for calculating the final concentration in the sample. These can be set up to automatically blank correct and adjust for the weight of the sample used. In addition, sheets can be protected such that staff are only able to insert the relevant raw data required and cannot accidently change formulae used in the calculations. All completed records should be uniquely identified. This is especially useful when operating within a quality system, when this can be linked with the issuing of sample numbers.

8.8 THE FINAL ASSESSMENT

Determining the concentration of PCBs in fish liver, nitrate in seawater or cadmium in sediment is fundamental to the processes of assessing the status of our seas. However, there is a need to understand the relevance of the concentration that has been determined. Does the measured concentration raise concerns in terms of human health, environmental impact or the risk of eutrophication? Ultimately, the concentration has to be compared with some criteria such that a conclusion can be drawn about the significance of the data in terms of risk to the plants and animals that live in our seas. The gas chromatography with electron capture detection (GC–ECD) analysis of a sample of fish liver provides an image from which the various peaks can clearly be identified and quantified; see Figure 8.6. However, the significance of the concentrations determined can only be fully understood by comparison with assessment criteria; see Figure 8.6. In recent years, working groups and committees of the OSPAR Commission have devoted a considerable amount of effort to developing assessment criteria that can be used to interpret the hazardous substances data produced by Contracting Parties and to assess delivery of the Hazardous Substances Strategy.[7] This requires the derivation of background or reference concentrations. As mentioned earlier, historical sampling sites for hazardous substances were often associated with point sources and so contain elevated concentrations. The derivation of background concentrations requires samples to be collected from pristine sites. The ICES Marine Chemistry Working Group has defined such sites as being remote from industry or large populations, subject to limited atmospheric transportation (i.e. currents and prevailing wind direction) and not appreciably influenced by major riverine discharges (ICES, 2008). In this context, OSPAR has derived background concentrations (BCs) and background assessment concentrations (BACs) (Webster *et al.*, 2009a). The BAC is a value for testing whether concentrations at a site are close to background with observed concentrations being 'near background' if the mean concentration is statistically significantly below the corresponding BAC. BACs were adopted by OSPAR as the transition value differentiating between those concentrations that were near background and those that were above background: transition T_0, in Figure 8.6. A second transition point (T_1) provides an assessment between acceptable status (green in a colour-based traffic light system) and unacceptable status (red in a colour-based traffic light system). This transition point is described by the environmental assessment criteria (EAC). The EAC is the contaminant concentration in the environment below which it can be assumed that no chronic effects will occur in marine species, including the most sensitive species (Webster *et al.*, 2009b). In the specific case of PCBs, a derivative of the EAC, the $EAC^{passive}$, is used. The output from the assessment, Figure 8.6, is a spatial representation of contaminant concentrations which presents a clear picture of the status with respect to the contaminant group. Where there is sufficient temporal data, trend information can also be presented; see Figure 8.6. The OSPAR objective was to have EACs for all the various determinands in the Coordinated Environmental

[7] The OSPAR Strategy objectives for hazardous substances are: move towards the cessation of discharges, emission and losses of hazardous substances by 2020. The ultimate aim is to achieve concentrations of hazardous substances in the marine environment near background values for naturally occurring substances and close to zero for man-made substances.

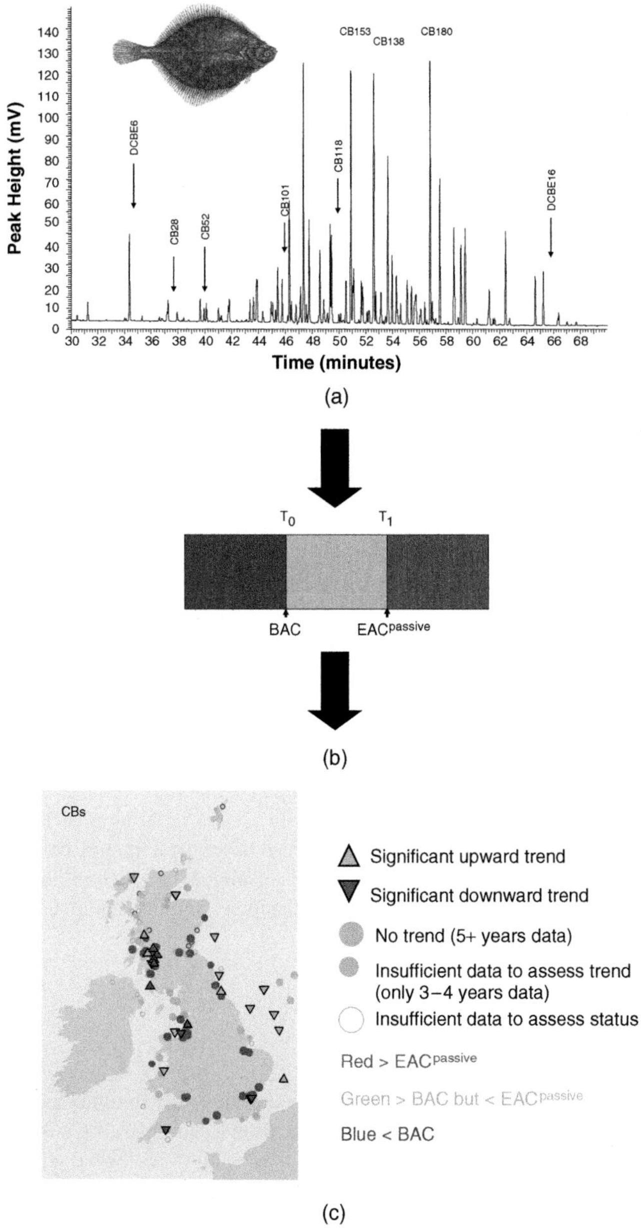

Figure 8.6 Determination of PCBs in plaice liver results in a chromatogram (a) from which a quantitative result can be determined. However, this must be compared with assessment criteria (b) in order that the data can be visualized in a clearly understandable manner (c). T_1 and T_2 are the transition points between the categories. BAC: background assessment concentration; $EAC^{passive}$: environmental assessment criteria calculated from the EAC for sediments based on partitioning theory. See plate section for a colour version of this image.

Monitoring Programme (Section 8.2). However, this has not yet been achieved. Consequently, the second transition point (T_1) in the current OSPAR assessment scheme utilizes a variety of criteria including the EACs (PAHs and TBT in biota, PCBs in sediment), $EAC^{passive}$ (CBs in biota), EU food limits (metals in biota) and the United States National Oceanic and Atmospheric Administration (NOAA) effects range-low[8] (ERL; PAHs and metals in sediments). The latter category, the effects range-low, is a set of numerical sediment quality guidelines that were developed as informal, interpretive tools for the US National Status and Trends Program (NOAA, 1999).

HELCOM uses the Hazardous Substances Status Assessment Tool (CHASE) for the integrated assessments of hazardous substances in the Baltic Sea (HELCOM, 2010). Contamination ratios (CRs) are calculated by dividing individual contaminant concentrations (or biological effects measurements) by the relevant assessment criteria. Internationally agreed assessment criteria are used, such as EC environmental quality standards or OSPAR EACs. A contamination sum can then be calculated by summing the available CRs and dividing by the square root of the number of indicators. An overall assessment level can then be assigned to one of five possible classes: High (0–0.5), Good (0.5–1), Moderate (1–5), Poor (5–10) and Bad (>10).

HELCOM has used a similar assessment tool as CHASE to assess the status of eutrophication in the Baltic. In this case the HELCOM Eutrophication Assessment Tool (HEAT) derives an ecological quality ratio (EQR), which is the ratio of reference conditions to the actual conditions. This is compared with what is deemed to be an acceptable deviation, the definition of which can vary (HELCOM, 2009). From a classical chemical monitoring perspective this requires the determination of nutrient concentrations, total phosphorus, total nitrogen, dissolved inorganic nitrogen and dissolved inorganic phosphorus, as well as chlorophyll-*a* concentrations.

Assessment criteria have also been developed within the OSPAR framework for the determination of the status in respect of nutrient concentrations and eutrophication. As with HELCOM, there are a range of parameters measured. However, it includes the measurement of nutrient and chlorophyll-*a* concentrations. From the perspective of assessment, OSPAR has set winter nutrient background concentrations with assessment criteria being 50% above the background.

The above are only examples of the assessment criteria that are available to assess status and trends in the marine environment and have recently been used in status reports. The key point is that development of such criteria must continue and such development should be cooperative across Regional Sea conventions so that the best available techniques are shared and, where possible, common assessment tools are used across the international frameworks that have developed to manage our global marine environment.

8.9 CONCLUSION

In conclusion, it is apparent that classical chemical monitoring will continue as an integral part of the assessment of our marine ecosystems well into the future. However, there is a need to develop relevant assessment criteria, and this will require the integration of biological effects measurements with data from classical chemical monitoring. Furthermore, regardless of whether it is classical chemical monitoring, biological effects monitoring or else the use if in-situ monitoring, there are common requirements around ensuring that all processes are appropriately quality controlled. Some of the challenges are different in respect of the different methodologies; however, without a coming together of the various monitoring processes and methodologies, the advice that scientists

[8] Effects range low (ERL): concentration below which effects are rarely observed or predicted among sensitive life stages and (or) species of biota for sediment effect concentrations used to evaluate sediment concentrations of trace elements and synthetic organic compounds.

can provide to those developing marine policy will be inappropriately limited. This requires that we all think outside the box, learn from each other and are prepared to grasp opportunities afforded by the new technologies while maintaining the high-quality classical monitoring tools where they are the best solution.

REFERENCES

Ameur W. B., Trabelsi S. and Driss M. R. (2010). Polycyclic aromatic hydrocarbons in superficial sediments from Ghar El Melh Lagoon, Tunisia. *Bull. Environ. Contam. Toxicol*., **85**, 184–189.

AOAC (2006). Paralytic shellfish poisoning toxins in shellfish, prechromatographic oxidation and liquid chromatography with fluorescence detection, first action 2005, method 2005.06, in *Methods of Analysis of the Association of Official Analytical Chemists*, AOAC International, Gaithersburg, MD, p. 83.

AOAC International (2000). AOAC Official Method 991.26. Domoic acid in mussels, liquid chromatographic method, in *Official Methods of Analysis of AOAC International*, (ed. W. Horowitz), AOAC International, Gaithersburg, MD.

Becker P. H. (1989). Seabirds as monitor organisms of contaminants along the German North Sea coast. *Helgoland Mar. Res*., **43**, 395–403.

Bligh E. G. and Dyer W. J. (1959). A rapid method of total lipid extraction and purification. *Can. J. Biochem. Physiol*., **37**, 911–917.

Caliceti M., Argese E., Sfriso A. and Pavoni B. (2002). Heavy metal contamination in the seaweeds of the Venice lagoon. *Chemosphere*, **47**, 443–454.

CEN (2009). Foodstuffs – Determination of trace elements – Determination of arsenic, cadmium, mercury and lead in foodstuffs by inductively coupled plasma mass spectrometry (ICP-MS) after pressure digestion, EN 15763.

Davis H. K., Moffat C. F. and Shepherd N. (2002). Experimental tainting of marine fish by chemically dispersed petroleum products, with comparisons to the *Braer* oil spill. *Spill Sci. Technol. Bull*., **7(5–6)**, 257–278.

Dell'Aversano C., Hess P. and Quilliam M. A. (2005). Hydrophilic interaction liquid chromatography–mass spectrometry analysis of paralytic shellfish poisoning (PSP) toxins, *J. Chromatogr. A*, **1081**, 190–201.

Defra (2005). Charting progress – an integrated assessment of the state of UK seas, 120 pp.

Dickson A. G., Sabine C. L. and Christian J. R. (eds) (2007). *Guide to Best Practices for Ocean CO_2 Measurements*, PICES Special Publication 3, 191 pp.

Edgar P. J., Hursthouse A. S., Matthews J. E. *et al*. (2006). Sediment influence on congener-specific PCB bioaccumulation by *Mytilus edulis*: a case study from an intertidal hot spot, Clyde Estuary, UK. *J. Environ. Monit*., **8**, 887–403.

El Deeb K. Z., Said T. O., El Naggar M. H. and Shreadah M. A. (2007). Distribution and sources of aliphatic and polycyclic aromatic hydrocarbons in surface sediments, fish and bivalves of Abu Qir Bay (Egyptian Mediterranean Sea). *Bull. Environ. Contam. Toxicol*., **78**, 373–379.

Epov V. N., Lariviere D., Epova E. N. and Evand R. D. (2004). Polyatomic interferences produced by macroelements during direct multi-elemental ICP-MS hydrochemical analysis. *Geostand. Geoanal. Res*., **28**(2), 213–224.

European Communities (2009). *Common Implementation Strategy for the Water Framework Directive (2000/60/EC), Guidance Document No. 19, Guidance on surface water chemical monitoring under the Water Framework Directive*, European Commission, Luxembourg (ISBN 978-79-11297-3).

Fifield F. W. and Haines P. J. (eds) (2000). *Environmental Analytical Chemistry*, Blackwell Science Ltd, Oxford, UK, 490 pp.

Fryer R. J. and Nicholson M. D. (1993). The power of a contaminant monitoring programme to detect linear trends and incidents. *ICES J. Mar. Sci*., **50**, 161–168.

Fryer R. J. and Nicholson M. D. (1997). *Throwaway Remark Now Set in Stone*. Atlas Bar, Copenhagen.

Grasshoff K., Ehrhardt M. and Kremling K. (1999). *Methods of Seawater Analysis*. Wiley–VCH.

Gubbins M. J. (2010). SOTEAG rocky shore monitoring programme. TBT contamination in Sullom Voe, Shetland. (2009) dogwhelk survey. *Scottish Mar. Freshwater Sci*., **1**(1), 15 pp.

Gubbins M. J., O'Reilly M., McIlroy L., *et al*. (2004). A decade of UK organo-tin specific biological effects monitoring – trends, ecological quality assessments and future monitoring requirements. ICES CM 2004/Z:06, 17 pp.

Gutiérrez-galindo E. A., Mendoza L. M. R., Munoz G. F. and Celaya J. A. V. (1998). Chlorinated hydrocarbons in marine sediments of the Baja California (Mexico)–California (USA) border zone. *Mar. Pollut. Bull*., **36**(1), 27–31.

Hanson S. W. F. and Olley J. (1963). Application of the Bligh and Dyer method of lipid extraction to tissue homogenates. *Biochem. J*., **89**, 101–102.

Heffernan J., Barry J., Devlin M. and Fryer R. (2010). A simulation tool for designing nutrient monitoring programmes for eutrophication assessments. *Environmetrics*, **21**, 3–20.

HELCOM (2009). Eutrophication in the Baltic Sea – an integrated thematic assessment of the effects of nutrient enrichment in the Baltic Sea region, Balt. Sea Environ. Proc. No. 115B, 148 pp.

HELCOM (2010). Hazardous substances in the Baltic Sea – An integrated thematic assessment of hazardous substances in the Baltic Sea, Balt. Sea Environ. Proc. No. 120B, 116 pp.

ICES (2008). ICES Marine Chemistry Working Group Report, http://www.ices.dk/reports/MHC/2007/MCWG07.pdf.

ICES (2010). Report of the Joint ICES/OSPAR Study Group on Integrated Monitoring of Contaminants and Biological Effects (SGIMC), 25–29 January, Copenhagen, Denmark. ICES CM 2010/ACOM:30,. 211 pp.

Jeffrey S. W. and Humphrey G. F. (1975). New spectrophotometric equations for determining chlorophylls *a*, *b* and *c*, in higher plants, algae and natural phytoplankton. *Biochem. Physiol. Pflanzen*, **167**, 191–194.

Jeffrey S. W., Mantoura R. F. C. and Wright S. W. (1997). *Phytoplankton Pigments in Oceanography*, Monographs on oceanographic methodology, UNESCO.

Kelly A. G. and Campbell L. A. (1996). Persistent organo-chorine contaminants in the Firth of Clyde in relation to sewage sludge input. *Mar. Environ. Res*., **41**, 99–132.

Kerouel R. and Aminot A. (1997). Fluorometric determination of ammonia in sea and estuarine waters by direct segmented flow analysis. *Mar. Chem*., **57**, 265–275.

Law R. J., Barry J., Bersuder P. *et al*. (2010). Levels and trends of brominated diphenyl ethers in blubber of harbor porpoises (*Phocoena phocoena*) from the UK, 1992–2008. *Environ. Sci. Technol*., **44**, 4447–4451.

Lorenzen C. J. (1967). Chlorophyll and pheo-pigments, determination of spectrophtometric equations. *Limnol. Oceanogr*., **12**, 343–346.

Loring, D. H. and Rantala, R. T. T. (1990). Sediments and suspended particulate matter: total and partial methods of digestion. ICES Techniques in Marine Science (TIMES) no. 9, 14 pp.

McNabb P., Selwood A. I. and Holland P. T. (2005). Multiresidue method for determination of algal toxins in shellfish: single-laboratory validation and interlaboratory study. *J. AOAC Int*., **88**, 761–772.

Midorikawa S., Arai T., Harino H. *et al*. (2004). Concentrations of organtin compounds in sediment and clams collected from coastal areas in Vietnam. *Environ. Poll*., **131**, 410–408.

Mullins E. (2003). *Statistics for the Quality Control Chemistry Laboratory*, Royal Society of Chemistry, Cambridge, UK, 455 pp.

Moffat C. F., Webster L., Pollard P. *et al*. (2007). Sampling sediments in the marine environment, in *The Encyclopedia of Statistics in Quality and Reliability*, (eds F. Ruggeri, R. S. Kenett, and F. W. Faltin), John Wiley & Sons, Ltd, Chichester, UK, pp. 1756–1761.

NOAA (1999). Sediment quality guidelines developed for the National Status and Trends program. http://response.restoration.noaa.gov/book_shelf/121_sedi_qual_guide.pdf.

OSPAR (2004). Proposal for assessment criteria for TBT-specific biological effects, ASMO 04/3/3, OSPAR Environmental Assessment and Monitoring Committee, Stockholm, 29 March–2 April 2004.

OSPAR (2010). Quality Status Report 2010, OSPAR Commission, London, 176 pp.

Phillips D. J. H. (1980). *Quantitative Aquatic Biological Indicators*, Applied Science, London.

Pierce G. J., Santos M. B., Murphy S. *et al*. (2008). Bioaccumulation of persistent organic pollutants in female common dolphins (*Delphinus delphis*) and harbour porpoises (*Phocoena phocoena*) from western European seas: geographical trends, causal factors and effects on reproduction and mortality. *Environ. Pollut*., **153**(2), 401–415.

Pusch K., Schlabach M., Prinzinger R. and Gabrielsen G. W. (2005). Gull eggs – food of high organic pollutant content? *J. Environ. Monit*., **7**, 635–639.

Prichard E., Green J., Houlgate P. *et al*. (2001). *Analytical Measurement Terminology. Handbook of Terms Used in Quality Assurance of Analytical Measurement*, Royal Society of Chemistry, Cambridge, UK, 74 pp.

Smedes F. (1999). Determination of total lipid using non-chlorinated solvents. *Analyst*, **124**, 1711–1718.

Rauret G., Lopez-Sanchez J. F., Sahuquillo A. *et al*. (1999). Improvement of the BCR three step sequential extraction procedure prior to the certification of new sediment and soil reference materials. *J. Environ. Monit*., **1**, 57–61.

Robinson C. D. and Devalla S. (2005). An improved method for total digestion and ICP-MS determination of metals in sediment. Fisheries Research Services Internal Report No. 09/05, 13 pp.

Russell M., Webster L., Walsham P. *et al*. (2008). Composition and concentration of hydrocarbons in sediment samples from the oil producing area of the East Shetland Basin, Scotland *J. Environ. Monit*., **10**, 559–569.

Stobo L. A., Lacaze J.-P. C. L., Scott A. C. *et al*. (2005). Liquid chromatography with mass spectrometry detection of lipophilic shellfish toxins. *J. AOAC Int*., **88**(5), 1371–1382.

UKMMAS (2010). *Charting Progress 2 The State of UK Seas*, United Kingdom Marine Monitoring and Assessment Strategy, Department for Environment Food and Rural Affairs, UK, 166 pp.

UNESCO (1981). The Practical Salinity Scale 1978 and the International Equation of State of Seawater 1980, UNESCO Technical Papers in Marine Science, 36, 25 pp.

UNESCO (1983). Algorithms for computation of fundamental properties of seawater, UNESCO Technical Papers in Marine Science, 44, 53 pp.

Ünlü S. and Alpar B. (2005). Polycyclic aromatic hydrocarbon pollution in sediments of the Çanakkale (Dardanelles) Strait, Turkey. *Bull. Environ. Contam. Toxicol*., **74**, 946–953.

Webster L., Mackie P. R., Hird S. J. *et al*. (1997). Development of analytical methods for the determination of synthetic mud base fluids in marine sediments. *Analyst*, **122**, 1485–1490.

Webster L., Russell M., Philips L. *et al*. (2007). Measurements of organic contaminants and biological effects in Scottish waters between 1999 and 2005. *J. Environ. Monit*., **9**, 616–629.

Webster L., Russell M., Adefehinti F. *et al*. (2008). Preliminary assessment of polybrominated diphenyl ethers (PBDEs) in the Scottish aquatic environment, including the Firth of Clyde. *J. Environ. Monit*., **10**, 463–473.

Webster L., Fryer R., Davies I. M. *et al*. (2009a). A proposal for the assessment criteria to be used for the assessment of monitoring data for the concentrations of hazardous substances in marine sediments and biota in the context of the OSPAR QSR 2010, FRS Collaborative Report No. 02/09, 24 pp.

Webster L., Russell M., Walsham P. *et al*. (2009b). An assessment of persistent organic pollutants (POPs) in wild and rope grown blue mussels (*Mytilus edulis*) from Scottish coastal waters. *J. Environ. Monit*., **11**, 1169–1184.

Webster L., Tronczynski J., Bersuder P. *et al*. (2009c). Determination of polybrominated diphenyl ethers (PBDEs) in sediment and biota, ICES Techniques in Marine Environmental Sciences, No. 4.

Webster L., Tronczynski J., Korytar P., *et al*. (2009d). Determination of polycyclic aromatic hydrocarbons (PAHs) in sediment and biota, ICES Techniques in Marine Environmental Sciences, No. 45.

Welschmeyer N. A. (1994). Fluorometric analysis of chlorophyll *a* in the presence of chlorophyll *b* and phaeopigments. *Limnol. Oceanogr*., **39**, 1985–1992.

Yang G-P., Liu X-L. and Zhang, J-W. (1998). Distribution of dibenzothiophene in the sediments of the South China Sea. *Environ. Pollut*., **101**, 405–414.

Zanardi E., Bícego M. C., Bruner L. *et al*. (1999). Distribution and origin of hydrocarbons in water and sediment in São Sebastião, SP, Brazil. *Mar. Pollut. Bull*., **38**(4), 261–267.

9
Biomonitoring

Michael Haarich

Johann Heinrich von Thünen-Institut, Federal Research Institute for Rural Areas, Forestry and Fisheries, Institute of Fishery Ecology, Analytical Department, Hamburg, Germany

9.1 INTRODUCTION

Biomonitoring is associated with different fields of investigations on biological material. One main task within biological monitoring programmes is the investigation of organisms with respect to

Chemical Marine Monitoring: Policy Framework and Analytical Trends, First Edition.
Edited by Philippe Quevauviller, Patrick Roose, and Gert Verreet.

abundance, condition, age distribution and other biological parameters related to biodiversity, like the composition of species to detect changes in marine habitats. The chemical analysis of marine organisms is another and well-established part of marine monitoring programmes, using biota as suitable matrix to detect and quantify presence and concentration of hazardous substances. A more recent kind of biomonitoring is the investigation of biological effects (hereafter, bioeffects) using cellular and molecular tests determining diseases or using histopathological methods, with the aim to detect effects of natural and anthropogenic impacts on the health status of biota. The ideal case would be a definite correlation between a hazardous substance, its biological effect and implications on the reproduction of marine organisms, so that all mentioned kinds of marine biomonitoring will form a closed cause–effect chain.

In about the last 15 years, the co-operation between scientists performing chemical monitoring and those investigating biological effects in marine biota has been developed to integrate chemical and bioeffect monitoring. Therefore, this chapter will focus on the chemical monitoring of hazardous substances in marine biota, but will also include aspects of the integration of chemical and bioeffect methods.

One of the reasons to use biota for environmental chemical monitoring is that organisms are accumulating some metals, metalloids and many persistent lipophilic substances which are known to be toxic, carcinogenic or possessing endocrine-disrupting properties. As concentrations of hazardous substances are generally relatively low in the marine environment due to the large dilution when entering the sea from the rivers, local sources and the atmosphere, it is more likely to be able to detect and quantify them in biological tissues due to the bioaccumulation. Less polar, hydrophobic substances, therefore, are mainly analysed in biological matrices, even if they are still detectable in sediment or water, because of the higher concentrations due to bioaccumulation.

Another advantage is that the accumulation period is restricted to the age of the organism. Particularly for young individuals and females, which are releasing part of the contaminants by spawning, the monitoring results are representative for the level of exposure of the recent period. This is among others of interest for assessing the effects of reduction measurements or the consequences of accidents of bigger amounts of chemicals entering the marine environment. An aspect which is obvious, but often not mentioned, is that the organisms are a living part of the marine environment which is impacted directly and negatively by hazardous substances, so that monitoring them serves as an indicator for the contamination status of the environment as well as of the investigated biota itself.

Typical biota used for monitoring purposes are fish and shellfish. For fish, the liver or muscle tissue is analysed, depending on the substance. For shellfish, the soft body is used for analysis. But analysis of eggs from seabirds which are feeding at sea has also proven to be a reliable way of monitoring certain regions of the marine environment, determined by the feeding area, particularly as seabirds represent the top predator level (Becker, 1989, 1992; Muñoz Cifuentes *et al*., 2003; Becker and Muñoz Cifuentes, 2004; Fängström *et al*., 2005; Lundstedt-Enkel *et al*., 2006; Verreault *et al*., 2007).

9.2 BIOMONITORING IN MARINE MONITORING PROGRAMMES

Many marine areas are covered by programmes for protection of their environment; most of them are situated at or close to the coast. Although the programmes are performed under national responsibility, they are typically part of international programmes within international conventions, organizations or action plans (see Chapters 2 and 3). Regarding pollution, in most cases the focus lies on the avoidance and reduction of hazardous substances entering the marine environment. Information on whether monitoring activities are part of these programmes and, if so, further details about the performance are often not available from open sources like the web. The United Nations launched in 1974 the Regional Seas Programme in the wake of the 1972 United Nations Conference on the Human Environment held in Stockholm. The areas within the Regional Seas programmes can be classified into three categories, as elaborated in Table 9.1: (i) the independent Regional Seas programmes, (ii) the non-UNEP-administered programmes, and (iii) the UNEP-administered Regional Seas programmes (UNEP: United Nations Environment Programme). An extended overview of marine monitoring programmes at the status of 2005 has been compiled by Roose (2005), which is still true for most of the areas.

Table 9.2 compiles those monitoring activities for which chemical monitoring in biota is explicitly described or at least indicated in the information available in the web and used for this review.

In general, chemical marine monitoring in biota is used for spatial or trend monitoring to detect differences and changes of contamination levels to describe the status and development of the marine environment with respect to hazardous substances. The advantages described above for using biota as matrix are the reason that for decades it is typically a core part of marine monitoring programmes, often of mandatory status. Most national programmes include the requirements of international conventions, so that the international programmes reflect well the core demands valid also on a sub-regional level.

The European programmes are examples for continuously performed and developed regional monitoring programmes for which information is easily accessible via the internet in detail (see references). For the European Seas, chemical monitoring in biota is performed within the programmes of the regional conventions for the protection of the marine environment: the HELCOM convention for the Baltic Sea, the OSPAR convention for the North-East Atlantic Ocean, the Barcelona Convention for the Mediterranean Sea and the Bucharest Convention for the Black Sea. The monitoring activities of these conventions will serve also for the EU Marine Strategy Framework Directive (MSFD) and will be co-ordinated with the adjacent coastal waters under the EU Water Framework Directive (WFD) to avoid parallel work. Biological effect monitoring with respect to a more general health status, as well as contaminants, is also part of the HELCOM, OSPAR and MEDPOL programmes.

Monitoring of biological parameters in the open Baltic Sea under HELCOM started in 1979. The guidelines written more than 20 years ago in 1988 demonstrate that the chemical analysis of harmful substances in biota was already at that time a core part of the Baltic monitoring programme (BMP). The BMP covering the open sea and the coastal monitoring programme were united to COMBINE (Cooperative Monitoring in the Baltic Marine Environment) in 1992.

Within the COMBINE programme, biota monitoring is a central part, as stated in the aims of this programme, to measure and compare levels of selected biota in different areas of the Baltic Sea, at specific locations over time and particularly in areas of special concern to detect possible contamination patterns and changing levels in response to changing input or to access possible threat to marine biota respectively. Obligatory species are fish, herring (*Clupea harengus*) and cod (*Gadus morhua*); as second choice and for the coastal zone are the bivalve species Baltic macoma (*Macoma baltica*) and blue mussel (*Mytylus edulis*) and the isopod Saduria (*Mesidotea entomon*). Details of the HELCOM biota monitoring in the Baltic Sea (COMBINE and BSAP) are given in Table 9.3.

The assessments of the status of the Baltic Sea regarding hazardous substances in biota were part of the Periodic Assessment reports published in the series Baltic Sea Environment Proceedings (BSEP; HELCOM, 1986, 1990, 1996, 2001), completed by thematic reports (HELCOM, 2009a,b, 2010a). The most recent holistic assessment (HELCOM, 2010b) also includes requirements of the Baltic Sea Action Plan (HELCOM, 2007) and will also serve as initial assessment for the implementation of the EU Marine Strategy Framework Directive (EU, 2008).

For the North-East Atlantic Ocean, including the North Sea, OSPAR started monitoring activities in the late 1970s. The Joint Monitoring Programme (JMP) included chemical measurements of a suite of hazardous substances, which were known as problematic compounds at that time and analytical methods were in place to determine PCBs, DDTs, HCHs and HCB, Cd, Pb, Hg, Cu and Zn in marine biota. Following the North Sea Quality Status Report 1993 (OSPAR, 1993), the JMP was revised to become the Joint Assessment and Monitoring Programme (JAMP) for the entire North-East Atlantic convention area. That part of the monitoring within the JAMP where the

Table 9.1 Regional Seas programmes.

Region	Convention/programme/action plan
UNEP administered	
North-West Pacific Region	The North-West Pacific Action Plan (NOWPAP)
East Asian Seas	COBSEA
Eastern Africa Region	Nairobi Convention
Mediterranean Region	Barcelona Convention/MEDPOL/MAP
Caribbean Region	Cartagena Convention/Caribbean Action Plan
Western Africa Region	West and Central Africa Region (WACAF) Action Plan
Non-UNEP administered	
Black Sea Region	Black Sea (Bucharest) Convention http://www.unep.org/regionalseas/programmes/nonunep/blacksea/default.asp
North-East Pacific Region	Antigua Convention http://www.unep.org/regionalseas/programmes/nonunep/nepacific/default.asp
Red Sea and Gulf of Aden	PERSGA http://www.unep.org/regionalseas/programmes/nonunep/redsea/default.asp
The Gulf and Gulf of Oman	ROPME www.ropme.com
South Asian Seas	South Asian Seas Action Plan (SASAP) was adopted in March 1995. The South Asia Cooperative Environment Programme (SACEP) http://www.unep.org/regionalseas/programmes/nonunep/southasian/default.asp
South-East Pacific Region	http://www.unep.org/regionalseas/programmes/nonunep/sepacific/default.asp
Pacific Region	Convention on the Conservation of Nature in the South Pacific (Apia Convention); Waigani Convention; Convention for the Protection of the Natural Resources and Environment of the South Pacific region http://www.unep.org/regionalseas/programmes/nonunep/pacific/default.asp
Independent	
Arctic Region	AMAP and PAME under the Arctic Council http://www.unep.org/regionalseas/programmes/independent/arctic/default.asp
Antarctic Region	CCAMLR/CEMP http://www.unep.org/regionalseas/programmes/independent/antarctic/default.asp
Baltic Sea	HELCOM/COMBINE http://www.unep.org/regionalseas/programmes/independent/baltic/default.asp www.helcom.fi

(*continued overleaf*)

Table 9.1 (*continued*)

Region	Convention/programme/action plan
Caspian Sea	Regional Pollution Monitoring Programme (RPMP) http://www.unep.org/regionalseas/programmes/independent/caspian/default.asp
North-East Atlantic Region	OSPAR/JAMP/CEMP http://www.unep.org/regionalseas/programmes/independent/neatlantic/default.asp www.ospar.org
North-West Atlantic Region and North-East-Pacific Region	See US National Status and Trends Program (NS&T); Canada's Oceans Action Plan

Table 9.2 Programmes including chemical monitoring in biota.

Area or country	Convention	Programme/action plan	Performing organization
Baltic Sea	HELCOM	COMBINE, BSAP	National Institutions of the Contracting Parties
Canada		Canadian Shellfish Sanitation Program (CSSP); DFO marine protected areas programme	Canadian Food Inspection Agency (CFIA); Environment Canada (EC), Fisheries and Oceans Canada (DFO)
China			South China Sea Branch (SCSB) of the State Oceanic Administration (SOA)
Mediteranean Sea	Barcelona Convention	MED POL (actual Phase III)	National Institutions of the Contracting Parties
North Sea and NE-Atlantic Ocean	OSPAR	CEMP	National Institutions of the Contracting Parties
United States of America		National Status and Trends Program (NS&T): National Status & Trends Mussel Watch Project; National Status & Trends Benthic Surveillance Project (inactive)	NOAA NOS (National Ocean Service)
South Asian Seas		Asia Pacific mussel watch	

Table 9.3 Biota monitoring in the Baltic Sea (COMBINE and BSAP).

Compound/group	COMBINE indicator species	BSAP indicator species
DDT and metabolites	*Herring/perch muscle (core/coastal core programme)* Cod liver (main programme) *Macoma baltica* soft tissue (main programme) Blue mussel soft tissue (supporting programme)	
CBs (nos. 28, 52, 101, 118, 138, 153 and 180)	*Herring/perch muscle (core/coastal core programme)* Cod liver (main programme) *Macoma baltica* soft tissue (main programme) Blue mussel soft tissue (supporting programme)	
Dioxins (PCDD), furans (PCDF) and dioxin-like polychlorinated biphenyls		In fish (herring or salmon or perch) muscle
Hexachlorobenzene (HCB)	*Herring/perch muscle (core/coastal core programme)* Cod liver (main programme)	
α- and γ-hexachlorocyclohexane (HCH)	*Herring/perch muscle (core/coastal core programme)* Cod liver (main programme) *Macoma baltica* soft tissue (main programme) Blue mussel soft tissue (supporting programme)	
Chlorinated paraffins (SCCP, MCCP)		*Substance of concern, but no BSAP indicator*
Endosulfan		*Substance of concern, but no BSAP indicator*
TBT	*TBT in Mytilus edulis soft tissue (coastal supporting programme)* Imposex in whelk *Buccinum* (coastal supporting programme)	
Brominated flame retardants PBDEs	In herring – tissue not specified – (supporting programme)	*Substance of concern, but no BSAP indicator*
Brominated flame retardants other	Herring – tissue not specified – (supporting programme)	*Substance of concern, but no BSAP indicator*
Perfluorinated compounds (PFOS, PFOA)		PFOS in sediment or fish (species optional) liver
Organotin compounds (TBT, TPhT)		TBT in sediment or biota (fish or mussel) or imposex
Alkylated phenols		*Substance of concern, but no BSAP indicator*

(*continued overleaf*)

Table 9.3 (*continued*)

Compound/group	COMBINE indicator species	BSAP indicator species
Mercury (Hg)	*Herring/perch/flounder muscle (core/coastal core/coastal main programme)* Cod muscle (main programme) *Macoma baltica* soft tissue (main programme) *Mytilus edulis soft tissue (coastal core programme)* Herring – tissue not specified – (supporting programme) different age classes	In fish (herring or flounder or perch) muscle; in bivalve (*blue mussel* or *Baltic clam*) *soft tissue*; in fish (*herring* or flounder or perch) *muscle*/edible part
Cadmium (Cd)	*Herring/perch/flounder liver (core/coastal core/coastal main programme)* Cod liver (main programme) *Macoma baltica* soft tissue (main programme) *Mytilus edulis soft tissue (coastal core programme)* Herring – tissue not specified – (supporting programme) different age classes	In fish (herring or flounder or perch) liver; in bivalve (blue mussel or *Baltic clam*) *soft tissue*; in fish (herring or flounder or perch) muscle/edible part
Lead (Pb)	herring/perch liver (core/coastal core programme) cod liver (main programme) *Macoma baltica* soft tissue (main programme) *Mytilus edulis* soft tissue (coastal core programme) Herring – tissue not specified – (supporting progr.) different age classes	
Copper (Cu)	herring/perch liver (core/coastal core programme) cod liver (main programme) *Macoma baltica* soft tissue (main programme) *Mytilus edulis soft tissue (coastal core programme)* Herring – tissue not specified – (supporting programme) different age classes	
Zinc (Zn)	*Herring/perch liver (core/coastal core programme)* Cod liver (main programme) *Mytilus edulis soft tissue (coastal core programme)*	
Caesium-137	*Obligatory in fish* (species depending on catch composition and amounts per species)	In herring muscle as indicator for whole Baltic Sea; in plaice and flounder muscle for southern Baltic Sea (southwards from Gotland)

national contributions overlap and are co-ordinated was set up as the Coordinated Environmental Monitoring Programme (CEMP). The aim of the CEMP is to deliver comparable data from across the OSPAR maritime area, which can be used in assessments to address the specific questions raised in the JAMP. The CEMP covers chemical monitoring in biota and sediment, as well as general bioeffects and those related to hazardous chemical substances; see Table 9.4. Seawater monitoring is included for nutrients. Compared with the former JMP, the compounds explicitly mentioned in the JAMP now also include dioxins and furans, but classical compounds like DDTs und HCHs were no longer in the focus of the JAMP and, therefore, not in the CEMP when the programme was started. The working groups under the Environmental Assessment and Monitoring Committee of OSPAR responsible for the chemical and bioeffect monitoring within the CEMP developed procedures to adjust the programme to the needs of the hazardous substances strategy and the analytical possibilities, taking into account the requirements of quality assurance and suitable assessment tools, so that the CEMP became a more dynamic programme without neglecting to continue measuring compounds for long-term time series. Additional to those substances listed in Table 9.4 for the CEMP, the participating laboratories, in spite of JAMP, continued to include most of the compounds of the former JMP into their analytical programmes, so that time series for DDTs and HCHs were not interrupted; therefore, for the OSPAR area, data sets comparable to those produced under HELCOM were also produced, of which α- and γ-HCH are the best examples for decreasing trends of hazardous substances in recent decades, in biota as well as in water.

Also for the Convention on the Protection of the Black Sea Against Pollution (1992, entry into force 1994) monitoring of hazardous substances is one tool to fulfil the basic objectives of the convention. The Black Sea Integrated Monitoring and Assessment Programme (BSIMAP) also includes an extended list of pollutants to be measured in different marine biota, mainly to answer the following policy issues: Does pollution of biota exceed the human consumption safety limits? How does pollution and eutrophication affect the major stocks of marine living resources?

The monitoring programme as it was decided at the Thirteenth Meeting of the Commission on the Protection of the Black Sea Against Pollution (15–17 February 2006, Istanbul, Turkey) includes metals and 'classical' organochlorine compounds comparable to other European programmes, in bivalve and turbot tissue; see Table 9.5.

For the Mediterranean Sea, the Convention for the Protection of The Mediterranean Sea against Pollution (Barcelona Convention) was signed on 16 February 1976 under the umbrella of the United Nations Environment Programme (UNEP). The Programme for the Assessment and Control of Pollution in the Mediterranean region (MED POL) was initiated in 1975 'as the environmental assessment component of the Mediterranean Action Plan (MAP)'. Its task is to assist Mediterranean countries in the implementation of pollution-assessment programmes (marine pollution trend monitoring, compliance monitoring and biological effects monitoring). MEDPOL assists Mediterranean countries in the formulation and implementation of pollution monitoring programmes, including pollution control measures and the drafting of action plans aiming to eliminate pollution from land-based sources. The most recent plan includes an agreement on monitoring criteria for MEDPOL Phase III; see Table 9.6. The following contracting parties are not participating: Spain, France, Italy, Egypt and Libya. These countries will continue monitoring within their National Action Plans (NAPs).

9.3 ANALYTICAL TRENDS IN CHEMICAL MONITORING OF MARINE BIOTA

For a long time a selection of organochlorine compounds (PCBs, DDTs, HCHs) and some heavy metals (Cd, Pb, Hg, Cu, Zn) has been typical for chemical marine monitoring; see Table 9.7.

Table 9.4 Biomonitoring in the OSPAR-CEMP.

Status[a]	App. no.[b]	Parameter			Remarks
m (v)	2	Metals	Cd, Cu, Pb (Cu, Zn)	Fish, mussel	
m	3	PCB	CB 28, CB 52, CB 101, CB 118, CB 138, CB 153 and CB 180	Fish, mussel	
v	9		Co-planar CBs: CBs 77, 126 and 169	Fish, mussel	To be monitored only in identified problem areas
v	13	Dioxins	Polychlorinated dibenzo-dioxines and -furanes		Recommendation on congeners to be measured by ICES-MCWG 2009
m	4	PAHs	Anthracene, benz[*a*]anthracene, benzo[*ghi*]perylene, benzo[*a*]pyrene, chrysene, fluoranthene, ideno[1,2,3-*cd*]pyrene, pyrene und phenanthrene	Mussel	
v	14	PAH-Metabolites	1-Hydroxy-pyrene, 1-hydroxy-phenanthrene, 3- hydroxy-benzo[*a*]pyrene	Fish bile	
v	10		Alkylated PAHs	Fish, mussel	C1-, C2- and C3-naphthalenes, C1-, C2- and C3-phenanthrene, and C1-, C2- and C3-dibenzothiophenes and dibenzothiophene
m	8	Brominated flame retardants:	PBDEs, HBCD		PBDEs 28, 47, 66, 85, 99, 100, 153, 154 und 183
v	12	Perfluorinated compounds (PFCs)	PFOS		Monitoring limited in time; decision in 2012
m	5, 11	Organotin compounds	TBT		Alternatively biota or sediment depending on availability of suitable material
m	7		Direct and indirect eutrophication effects:	Plankton	Phytoplankton chlorophyll; phytoplankton species composition; macrophytes; oxygen; benthos species composition.

[a]m = mandatory v = voluntary.
[b]Number of CEMP appendix.

Table 9.5 Biota monitoring of the Black Sea Commission.

	Mandatory	Not mandatory
Species/matrix	Bivalves, meat Turbot, meat	Anchovies, meat Sprat, meat Horse mackerel, meat
Parameter	Mandatory	Operational
Metals	Cd Cu Hg Pb	Co Zn Fe Ni Cr
Organic compounds	DDT DDD DDE Lindane (γ-HCH) PCBs	Chlorinated phenols PAHs
Radionuclides		Cs Sr Tritium Total radioactivity

Chlorinated dioxins and furans are also of high importance, but normally not part of the regular environmental marine monitoring due to the high analytical effort; however, they are subject of investigations on seafood for human consumption. The monitoring of radionuclides (Cs-137, Sr-90, Pu-238, -239, -240, Am-241) is usually organized in separate programmes.

Particularly for organic substances, only a very small selection of thousands of compounds entering the sea can be measured and is represented in the regular programmes; therefore, current monitoring efforts do not cover the actual situation of the presence of organic contaminants well. A range of additional compounds that are likely to enter the sea due to their production, use and physico-chemical properties has been investigated regarding their presence in the marine environment by a directed analysis. These candidate substances are published in so-called priority substances lists. Those organic substances being of interest to be analysed in marine biota, including so-called 'classical' and some new candidate substances, are listed in Table 9.8. They are the subject of projects and research, one-off-surveys or target-screening programmes being performed at intervals of several years. Data on other compounds can also be obtained by means of undirected analytical methods, the so-called non-target screening. The moment such a compound is discovered to be present in the marine environment and that such information can become relevant for possible further action need not correspond to a recent input into marine waters; in many cases the improvement of analytical methods and instruments has enabled the detection and quantification of substances which are produced and used in high quantities for many years; for example, the brominated flame retardants, chlorinated paraffins and perfluorinated compounds (PFCs).

For most of these compounds the analytical details are described in the Technical Annexes to the OSPAR–JAMP Guidelines for Monitoring Contaminants in Biota (OSPAR, 2009) which have been updated and supplemented in recent years by the ICES Marine Chemistry Group (MCWG).

Although volatile organic compounds have been detected in marine biota in coastal areas, often with higher concentrations of many priority substances (microgram per kilogram level), they have

Table 9.6 Biota Monitoring under MEDPOL. Extract from trend (and state) monitoring criteria for MED POL Phase III (Ref.: http://195.97.36.231/medpol/).

	Hazardous Substances		Biological effects	
Area	Coastal/reference areas and hot spots			
Status	Mandatory	Recommended	Mandatory (pilot)[a]	
Parameters	Total Hg and Cd	Other heavy metals, HH+,[b] PAH	DNAx, EROD[c], MT[d], LMS[e]	
Sampling frequencies	Annually at the pre-spawning period	Semi-annually, seasonally or more often depending on the matrix	Quarterly or semi-annually	
Species (tissue)	Mandatory	**Recommended**[f]	**for** EROD, DNAx	**for** MT, LMS
	MG (*Mytilus galloprovincialis*, whole soft tissue) MB (*Mullus barbatus*, fillet)	ME (*Mytilus edulis*), PP (*Perna perna*), DT (*Donax trunculus*) or MC (*Mactra corrallina*, whole soft tissue), MS (*Mullus surmuletus*) or UM (*Upeneus moluccensis*, fillet)	MB (*Mullus barbatus*), if not available *Mugil* sp., DL (*Dicentrarchus labrax)* for caging (liver)	MB (*Mullus barbatus*), if not available *Mugil* sp., DL (*Dicentrarchus labrax*) for caging (liver) MB (Mullus barbatus, fillet) *Mytilus* sp. if not available *Patella* sp. (digestive gland, hepatopancreas for limpets)
Number of samples/specimen	Min. 5 parallel samples for the selected species. Min. 15 specimens to be pooled in each sample for MG[g,h].		Min. 5 parallel samples for the selected species[g]	

[a]planned to become mandatory after a pilot phase
[b]HH+: halogenated hydrocarbons.
[c]*EROD*: Ethoy-resofurin-O-deethylase
[d]*MT*: Metallothionein
[e]*LMS*: Lysosomal membrane stability
[f]If the mandatory species are not available.
[g]Recommended number during pilot phase.
[h]Number will depend on the statistical design of the trend monitoring programme.

Table 9.7 List of metals being of interest to be analysed in marine biota.

			Tissue	Instrumental technique	Methodological guidance
Heavy metals and metaloids	Cd, Pb, Cu, Zn				OSPAR (2009)
	Hg				OSPAR (2009)
Organometallic compounds	Methylmercury				
	Organotin compounds	Tributyltin (TBT); triphenyltin (TPhT)	Fish muscle	GC–MS, GC–FPD, GC–ICP-MS; HPLC	OSPAR (2009)

not become part of the marine monitoring programmes, obviously because of their low potential to bio-concentrate and -accumulate (Roose, 2005).

9.4 BIOTA MONITORING PROGRAMME

9.4.1 Sampling

General information about the objectives of a monitoring programme and the conditions for temporal trend or spatial monitoring affecting the sampling of marine biota are available from the description of those programmes. Most often this information includes the sampling frequency and the spatial coverage, depending on what should be assessed and which level of detail is needed. Regarding the sampling frequency for biota monitoring for long-term temporal trend assessments, at least one sampling period per year is common. As the condition of animals is changing periodically over the year (biological cycle), sampling typically takes place in a stable period regarding environmental conditions and physiological processes, in any case outside the spawning period. This period will differ for species and area. For example, for fish in northern European seas this period is summer to autumn; for mussels it is late autumn to winter (ICES).

9.4.2 Choice of Species and Tissue

The choice of species for biomonitoring depends on the specific objectives of the monitoring programme and can only be determined on the basis of information on the sampling area, the composition of the stock or population of the fish, shellfish or seabirds, and their migration pattern. The species should be representative and sufficiently abundant, ideally throughout the entire monitoring area, and the tendency to migrate should be low, so that results can be related to the surrounding environment.

As regards the selection of the species to be monitored for chemical contaminants, one should take account of the following requirements: the species should accumulate the contaminant in sufficient amounts without being seriously affected; the specimens collected should be of reasonable size, giving adequate amounts of tissue for chemical, and if part of an integrated programme, also for biochemical and physiological analyses; shellfish should be hardy enough to survive for defecation before being prepared for analysis.

The choice of tissue is predominantly determined by the degree of bioaccumulation and the related concentration of the contaminant, so that for most species the liver is preferred for lipophilic

Table 9.8 List of organic substances being of interest to be analysed in marine biota.

			Tissue	Instrumental technique	Methodological guidance
Alkylated phenols		Nonylphenols (NPs); nonylphenolethoxylates (NPEs); octylphenols (OPs); octylphenol ethoylates (OPEs)	Fish muscle		
Brominated flame retardants	Polybrominated dipenylether (PBDEs)	BDE 47, 85, 99, 100, 153, 154	Fish muscle, liver	GC–ECD, GC–MS	OSPAR (2009)
	Hexabromocyclode decane	HBCDD	Fish muscle		OSPAR (2009)
	Tetrabromobisphenol-A	TBBPA			
Chlorinated organic compounds	Chloroalkanes	Short-chain chlorinated paraffins (SCCPs, C10–C13); medium-chain chlorinated paraffins (MCCPs, C14–C17)	Fish liver[a]	GC–MS	Oehme *et al.* (2006)
	Chlordanes	*cis*-Chlordane, *trans*-chlordane, *trans*-nonachlor	Fish liver[a]	GC–ECD, GC–MS	Oehme *et al.* (2006)
	Dibenzo-dioxines, -furans and co-planar CBs	PCDD/Fs		HRGC/HRMS	MCWG (2009)
		DDE, DDD, DDT	Fish liver[a]	GC–ECD, GC–MS	OSPAR (2009)
	Hexachlorobenzene	HCB	Fish liver[a]	GC–ECD, GC–MS	OSPAR (2009)
	Hexachloro-cyclohexanes	α-HCH, β-HCH, γ-HCH, δ-HCH	Fish liver[a]	GC–ECD, GC–MS	OSPAR (2009)

(*continued overleaf*)

Table 9.8 *(continued)*

					Methodological guidance
	Polychlorinated Biphenyls	PCBs	Fish liver[a]	GC–ECD, GC–MS	MCWG (2009)
		Endosulfan		GC–ECD, GC–MS	Theobald *et al*. (2007)
Perfluorinated compounds (PFCs)		Perfluorooctane sulfonate (PFOS); perfluorooctanoic acid (PFOA)	Fish liver[a]	LC–MS	Theobald *et al*. (2007)
					OSPAR (2009)
Polyaromatic hydrocar-bons (PAHs)			Biota	GC–MS, HPLC-UVF	OSPAR (2009)
		PAH-metabolites	Fish bile	HPLC-UVF	Kammann (2006)
Alkylated PAH			Biota		OSPAR (2009)

[a]Or muscle of higher lipid content; when analysing for human consumption, the edible part is preferred.

compounds. For fatty fish with also relatively high lipid content in the fillet it may be of advantage to use the muscle tissue instead of liver, even if the concentrations are lower, as in most cases the lower signal height in the chromatogram of the muscle tissue sample is compensated by a lower influence of the matrix, thus resulting in a better signal-to-noise-ratio and a better peak separation. The decision to analyse individual or pooled samples also depends on the amount of tissue available from the individual sample of a given age, size and gender. For example, in the case of fish, there could be a conflict between the need to get a sufficient amount of material and, for example, (i) the objective to monitor the recent period and, therefore, preferring younger and smaller fish, (ii) to analyse on an individual basis to be able to calculate the within-batch/-catch variation for temporal trend monitoring and (iii) to analyse chemical and biochemical parameters/biomarkers as much as possible from the same individual sample for integrated chemical and bioeffects monitoring purposes.

9.4.3 Choice of Size, Age, Sex and Number of Individuals

During selecting samples from a catch, it is normally not possible to determine the age. As size and age of fish and shellfish are correlated rather well for most species, the length of the fish or the shell can be used as a selection criterion. The age of the fish should be controlled by analysis of the otoliths at a later time. For freshly caught flatfish the sex can be determined without dissection by screening the shape of the gonads shining through when lighted from the back.

Where it has been successfully applied in the past, length-stratified sampling may be maintained. For this procedure, the length range is divided into five equal groups on a logarithmic scale; for each length interval, at least five individuals have to be selected. For some areas the length range has become smaller, due to changes in the composition of the population, particularly for fish, so that data obtained by this procedure are no longer comparable. As the age-to-length correlation and the contaminant load differ between male and female individuals, it is often more appropriate to sample with a view to minimize natural variability within the sample. This can be done by selecting only individuals of the same sex and a narrow length range reflecting younger fish; for example, only young females. This causes a significant reduction of the within-batch variability, and particularly for time trend assessments, more reliable statistical results can be obtained from shorter time series.

For shellfish, a sample should be collected with the number of individuals large enough to be divided into at least three equal pools with each pool consisting of at least 20 animals and enough soft tissue for all analyses. The length of the individuals collected should be constant from year to year at each station, or should at least fall within a very narrow range (e.g. within 5 mm). To reflect recent levels of contamination, young individuals should be chosen. Examples and more details regarding the selection criteria described are given in the JAMP guidelines (OSPAR, 2009).

The minimum number of individuals for the length-stratified sampling is 25, which results in at least five analytical samples when pooled. For individual analysis, statistical calculations of North Sea fish data (Nicholson *et al.*, 1997) recommend a minimum number of 12 fish. Compared with 25 analyses, the small loss of statistical significance is justified by the lower analytical extent and the related costs, whereas a significant improvement would need a much higher analytical effort by more than three times the number of samples to be analysed.

9.5 ANALYTICAL METHODS

The main differences in analytical methods concern the sample preparation, extraction and clean-up methods. Therefore, this section will mainly apply to the special aspects of biota samples. More details on instrumental determination of micro-pollutants are given in Chapter 6.

9.5.1 Sample Preparation

Fish can be dissected for subsampling immediately after being caught, if appropriate conditions are available on board. One of the main conditions that require particular attention is to avoid contamination of the sample. Shellfish individuals in samples should defecate before being used for analysis, which implies that they will not be processed immediately after sampling. Seabird eggs will be opened under clean conditions in the analytical laboratory.

For trace metal analysis it is necessary to dissect and wrap the tissue sample working under clean conditions to avoid contamination by dust and other particles. On board this can be done in a laminar airflow workstation, a so-called clean bench. If there is no clean bench available or no possibility to install such a device temporarily, the ungutted sample has to be wrapped in bags or foil of appropriate properties (e.g. aluminium foil, polythene bags) and immediately frozen and stored for subsequent treatment later in the land-based laboratory. The ideal solution on board would be to work in a fully equipped clean-room container.

For organic contaminants it is possible to dissect in a laboratory on board, if the equipment of this room and the fresh air supply is designed to avoid contamination by exhaust gases or evaporation of volatile organic substances, e.g. from sealing materials or furniture. Good basic conditions are supplied when walls and laboratory furniture are manufactured from seawater-resistant stainless steel, the floor without softeners and the fresh air is sucked in at a place not impacted by exhaust gases from the engine or the galley and filtered.

For dissection of subsamples for trace metal analyses, only metal-free knife or scalpel blades and tweezers must be used; for organic contaminants, stainless steel or also ceramic blades are suitable, but plastic tweezers must be avoided. For cleaning between samples, high-purity water and for fatty samples organic solvents (acetone, alcohol or mixtures, at least analytical-grade purity) should be used. On board at offshore areas, fresh seawater is also suitable for precleaning the equipment.

Subsamples (e.g. of liver) should be stored in precleaned containers made of glass (e.g. borosilicate glass), stainless steel or aluminium, or should be wrapped in precleaned aluminium foil and shock-frozen quickly in liquid nitrogen or a blast freezer. The individual samples should be clearly labelled and stored together in a suitable container placed in a deep freeze at $<-20\,^{\circ}C$ until analysis. Subsamples for enzyme tests should be stored in vials suitable for storage in liquid nitrogen, labelled clearly and stored in liquid nitrogen until analysis.

Further details are given for example in Chapters 5 and 6 of the JAMP Guidelines for Monitoring Contaminants in Biota (OSPAR, 2009).

9.5.2 Extraction and Clean-up

For metals, the classical procedure is to freeze-dry and homogenize the tissue, followed by digestion with strong acids. For very low concentrations a complexation and re-extraction step to reduce the matrix is recommended when determined by graphite furnace atomic absorption spectroscopy (GFAAS). Although procedures are described in the literature about digestion in open vessels, it is recommended to perform digestion under pressure, using quartz tubes heated in closed steel cartridges on a heating block or in closed Teflon vessels in a microwave oven; the latter is nowadays the preferred method owing to the shorter digestion times and an easier and safer handling. The classical determination is by GFAAS, which is increasingly substituted by multi-element methods based on inductively-coupled plasma (ICP) ionization and spectrometric detection like mass spectrometry (MS) or optical emission spectroscopy (OES). A less often applied method is the total reflection X-ray analysis (TXRF), which is also capable of multi-element determination in low-concentrated environmental samples.

For mercury analysis, instruments for the direct determination from the solid or liquid sample without any preceding digestion have been developed recently, which may substitute the common methods of digestion, reduction to elemental mercury by Sn(II) or hydrid generation, followed by amalgamation where appropriate. The determination is performed by cold vapour atomic absorption spectroscopy (CVAAS) or atomic fluorescence spectroscopy (CVAFS). These methods are applied for measuring so-called total mercury. To distinguish between inorganic mercury and organic methyl- and ethyl-mercury compounds, a chromatographic separation is needed. For gas chromatographic separation, the compounds have to be alkylated to become volatile. The alkylated compounds have to be decomposed in a thermic reactor to generate elemental mercury when using AAS or AFS for the detection. A literature research is recommended to find the best combination of existing modifications for extraction, formation of intermediates and detection methods for the material to be analysed and for the currently available equipment, as those are specifications which cannot be described in detail here.

For organic contaminants, the entire analytical procedure has to avoid any losses of target analytes by vaporization or deterioration. In a first step the tissue has to be dried and homogenized. Classical procedures are also freeze-drying with subsequent cryo-milling or milling with water-free sodium sulfate and quartz sand in a mortar grinder, or cutting the tissue mechanically and mixing it with quartz sand and drying the relatively coarse homogenized sample in a common microwave to remove the moisture (Haarich, unpublished standard operation procedure). Lipophilic analytes will be extracted together with the lipids by solvent extraction, which may be performed by column extraction at normal pressure and room temperature (cold extraction), with Soxhlet-extraction under reflux or automated by accelerated solvent extraction (ASE by Dionex®) under pressure at temperatures above the boiling point of the solvent used. Using ASE saves time and solvent and provides a better reproducibility. The necessary reduction of the solvent volume is performed using rotating evaporators or multisample evaporators (e.g. by Buechi®), which may be adapted regarding the tubes to the preceding extraction device, so that a change of the tube is no longer necessary. Column chromatography is applied to separate the analytes from the lipids, by adsorption chromatography with lipid-retaining fixed phase (e.g. Florisil®) or gel permeation chromatography (GPC) using the principle of size exclusion to extract the lipids in front of the analytes. GPC also has the advantage that it is automated, so the samples can be processed mostly unattended and also over night to save time. The final step before measurement is the clean-up by chromatography (e.g. activated silica), combined with fractionation by gradient solvent elution when necessary. However, even after this procedure, the quality of the chromatograms may be not sufficient owing to matrix components which behave very similar to the target compounds and which were not removed before, making it difficult or impossible to fit the signals, when using gas chromatography with electron capture detection (GC–ECD) detection, due to a disturbed baseline. If one can relinquish on certain sensitive compounds like dieldrin, a further extraction with concentrated sulfuric acid may be successful. Otherwise, cleaning-up needs some experimentation to find a suitable chromatographic phase which is able to remove the disturbing substances and which has sufficient purity for trace analysis purposes. A general recommendation cannot be given. Using a GC–MS system will reduce this problem; nevertheless, a sufficient chromatographic separation should be aspired to here also.

The methods described cannot be applied without major modifications for the more polar PFCs like PFOA and PFOS, which are typically determined by LC–MS(/MS). Although methods are described in the literature for biological tissue without removing the lipid and other matrix components to a major extent, clean solutions are necessary for marine samples to achieve low detection limits and to avoid fast soiling of the LC–MS(/MS) equipment. To avoid soiling, biota samples can be extracted with alcohol, and the lipid is removed by sequential freezing, followed by solid-phase extaction clean-up (see Fig. 9.1; Theobald *et al.*, 2007).

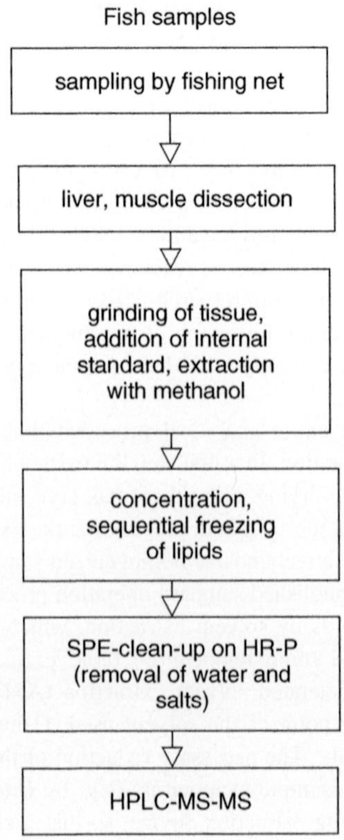

Figure 9.1 Analytical procedure for the determination of PFCs in fish tissue samples (modified from Theobald *et al*. (2007)).

9.5.3 Detection and Quantification

The final instrumental measurement of biological samples, as a part of the overall analytical procedure, does not differ significantly from that applied for water or sediment samples with respect to the instrumental equipment, the use of internal standards, calibration and quantification. The determination can be performed by high-resolution GC–ECD, but more and more laboratories have changed to mass spectrometric detection in different modes and instrumental modifications (GC–MS, GC–MS/MS, in electrothermal, negative chemical or positive chemical ionization mode) or liquid chromatography (high-performance liquid chromatography (HPLC)–MS or – MS/MS). Extensive literature is available for the detailed description of the instrumental parameters; for example, from instrument manufacturers and column suppliers.

9.5.4 Quality Control

Quality control requirements are also described in many places; for example, the guidelines of the regional conventions of HELCOM and OSPAR, and separately in Chapter 11 of this book. One general problem that causes difficulties for laboratories analysing marine biological samples (particularly for organic compounds at very low concentrations) should be mentioned. Only a very limited range of certified reference material appropriate for marine biota monitoring exists regarding the matrix, the concentration level and the selection of certified compounds. Also, the test materials supplied by the organizers of external proficiency testing schemes often do not cover the concentration range detected in the monitoring samples of the individual laboratory, so that conclusions drawn from the results of such tests may not necessarily be transferable to the quality of the monitoring results. As the effort to prepare and certify such materials is high and costly, but the number of customers relatively low, co-operation on a regional level between customer institutes having facilities for sampling suitable materials and institutions or agencies equipped and experienced to process this material may reduce this problem. Proficiency testing schemes for external quality control for chemicals in biota and for bioeffects measurements in the marine area are established on an international level to support the performing institutions; see BEQUALM (www.bequalm.org) and QUASIMEME (www.quasimeme.org).

9.6 INTEGRATION OF CHEMICAL AND BIOLOGICAL EFFECTS MONITORING

When biological effects measurement techniques were introduced in marine monitoring, one of the main goals of their supporters was to be able to substitute chemical investigations. However, more than a decade of investigations has demonstrated that the majority of biological effects methods are not specific enough to meet the requirements that would allow this substitution. So a change of the aim has taken place from a substitution approach to complementarity. In this sense, the chemical measurements aid the interpretation of the biological effects measurements in terms of identifying the chemical causes of the biological effects and establishing concentration responses. Biological effects measurements will serve for early warning, for integrating effects of complex mixtures (synergism and antagonism), for indicating impacts also when concentrations of each chemical compound separately are below their detection limits, for indicating impacts of novel and often unidentified chemicals and for indicating impacts on a cellular or molecular level which may not directly be linked to the measured concentrations of chemicals. Following the strategy of the OSPAR JAMP, biological effects monitoring in relation to hazardous substances is necessary to

> The Contracting Parties shall, in accordance with the provisions of the Convention, take all possible steps to prevent and eliminate pollution and to take the necessary measures to protect the maritime area against adverse effects of human activities so as to safeguard human health and to conserve marine ecosystems and, when practicable, restore marine areas which have been adversely affected.

Within the European marine strategy framework, most of the work to promote the integration of contaminant and bioeffects monitoring, taking into account the existing knowledge and work of expert groups in both fields (e.g. ICES expert groups), has been done by the four ICES/OPAR workshops on Integrated Monitoring of Contaminants and their Effects in Coastal and Open Sea Areas (WKIMON; see Fig. 9.2).

Integration of chemical and bioeffects monitoring implies taking care of several practical aspects when planning and preparing the sampling. The ideal case would be to take subsamples from the same individual fish or mussel for both chemical and bioeffects measurements. However, for this, the available amount of material may not be sufficient (e.g. the liver of a small flatfish) or the pretreatment procedures are incompatible for simply dividing the material, e.g. for enzyme test methods and metal analysis, the latter requiring homogenization as the analytes are not evenly distributed in the tissue. More detailed guideline for contaminant-specific biological effects can be found in OSPAR (2008); see Table 9.9.

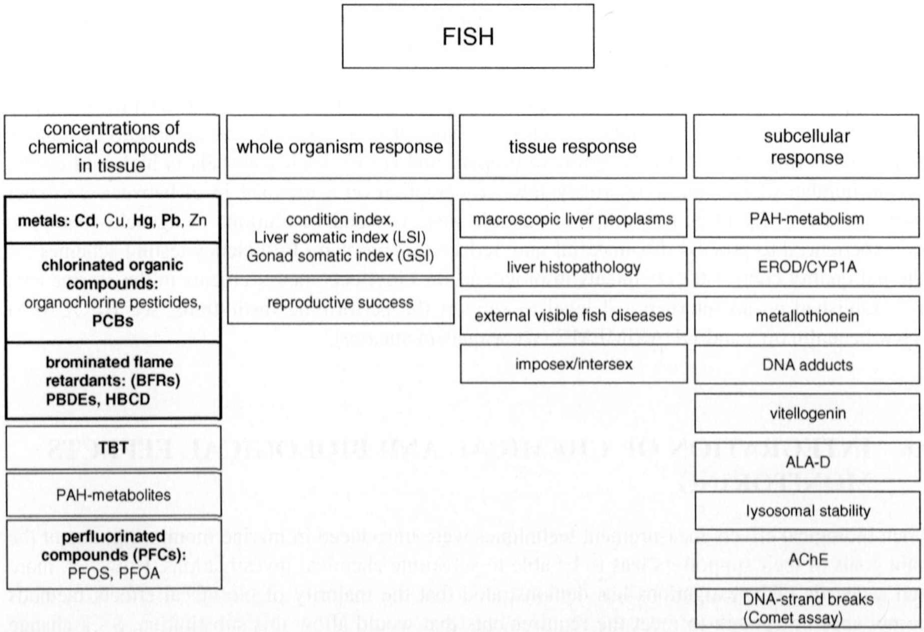

Figure 9.2 Biomonitoring of fish. Modified from WGBEC Report 2006 (WGBEC, 2006). Used with kind permission of the International Council for the Exploration of the Sea.

Table 9.9 Biological effects monitoring in OSPAR-CEMP.

Status[a]	App. no.[b]	Biological effect method	Investigated biota	Compound	Ref.
m		Imposex	Gastropods	TBT	
		Intersex	Bivalves	Metals (Pb, Cd, Cu, Zn), butyltin compounds	Rank (2009)
v	14	PAH metabolites	Fish bile	PAHs	
v	15	Bioassays			
v	15	CYP1a			
v	15	Lysosomal stability			
v	15	Liver histopathology/ macroscopic liver-neoplasmns	Fish		
v	15	External visible fish diseases	Fish		
v	15	Reproductive success in fish	Fish		

[a]m = mandatory v = voluntary
[b]Number of CEMP appendix.

9.7 ASSESSMENT

Assessments for contaminants in biota have been performed and published in the framework of the conventions mentioned on a more or less regular basis. Assessment criteria, against which the monitoring results achieved are compared, to describe the environmental status have been developed in the last decade and used in the recent assessments by HELCOM (2010b) and OSPAR (2010). However, for a number of determinands, assessment criteria are still not formally adopted. At the time of writing, the discussion on the applicability of the existing assessment criteria of the regional conventions in the context of the MSFD is still ongoing.

REFERENCES

Becker, P. H. (1989). Seabirds as monitor organisms of contaminants along the German North Sea coast. *Helgoland Mar. Res.*, **43**(3–4), 395–403.

Becker, P. H. (1992). Egg mercury levels decline with the laying sequence in charadriiformes. *Bull. Environ. Contam. Toxicol.*, **48**(5), 762–767.

Becker P. H. and Muñoz Cifuentes J. (2004). Contaminants in bird eggs: recent spatial and temporal trends, Wadden Sea Ecosystem No. 18, Common Wadden Sea Secretariat, Trilateral Monitoring and Assessment Group, Wilhelmshaven, Germany, pp. 5–25, http://www.waddensea-secretariat.org/news/documents/bird-eggs2004/WSE-18-cont-eggs-%2801.03.05%29.pdf.

EU (2008). Directive 2008/56/EC of the European Parliament and of the Council of 17 June 2008 establishing a framework for community action in the field of marine environmental policy (Marine Strategy Framework Directive), http://eur-lex.europa.eu/LexUriServ/LexUriServ.do?uri=OJ:L:2008:164:0019:0040:EN:PDF.

Fängström B., Athanasiadou M., Athanassiadis I. *et al.* (2005). Polybrominated diphenyl ethers and traditional organochlorine pollutants in fulmars (*Fulmarus glacialis*) from the Faroe Islands. *Chemosphere*, **60**(7), 836–843.

HELCOM (1986). First Periodic Assessment of the State of the Marine Environment of the Baltic Sea Area 1980–1985; General conclusions (1986) Baltic Sea Environmental Proceedings No. 17.

HELCOM (1990). Second Periodic Assessment of the State of the Marine Environment of the Baltic Sea, 1984–1988; Background document (1990) Baltic Sea Environmental Proceedings No. 35.

HELCOM (1996). Third Periodic Assessment of the State of the Marine Environment of the Baltic Sea, 1989–1993; Executive Summary (1996), Baltic Sea Environmental Proceedings No. 64.

HELCOM (2001). Fourth Periodic Assessment of the State of the Marine Environment of the Baltic Sea, 1994–1998; (2001/02), Baltic Sea Environmental Proceedings No. 82.

HELCOM (2007). *The Baltic Sea Action Plan* (BSAP), http://www.helcom.fi/BSAP/en_GB/intro/.

HELCOM (2009a). Hazardous substances of specific concern to the Baltic Sea – Final report of the HAZARDOUS project (2009), Baltic Sea Environmental Proceedings No. 119, 95 pp.

HELCOM (2009b). Radioactivity in the Baltic Sea, 1999–2006 – HELCOM thematic assessment (2009), Baltic Sea Environmental Proceedings No. 117.

HELCOM (2010a). Hazardous substances in the Baltic Sea – an integrated thematic assessment of hazardous substances in the Baltic Sea, Baltic Sea Environmental Proceedings No. 120B.

HELCOM (2010b). Ecosystem health of the Baltic Sea. A holistic assessment of environmental status in the Baltic Sea 2003–2007, Baltic Sea Environmental Proceedings No. 122.

Kammann U. (2006). PAH metabolites in bile fluids of dab (*Limanda limanda*) and flounder (*Platichthys flesus*): spatial distribution and seasonal changes. *Environ. Sci. Pollut. Res.*, **14**(2), 102–108. DOI: http://dx.doi.org/10.1065/espr2006.05.308.

Lundstedt-Enkel K., Asplund L., Nylund K. *et al.* (2006). Multivariate data analysis of organochlorines and brominated flame retardants in Baltic Sea guillemot (*Uria aalge*) egg and muscle. *Chemosphere*, **65**(9), 1591–1599.

MCWG (2009). Report of the Marine Chemistry Working Group (MCWG). ICES SCIENCE COMMITTEE, ICES CM 2009/MHC:03 REF. SCICOM, ACOM. Copenhagen, http://www.ices.dk/workinggroups/.

Muñoz Cifuentes J., Becker P. H., Sommer U., Pacheco P. and Schlatter R., 2003. Seabird eggs as bioindicators of chemical contamination in Chile. *Environ. Pollut.*, **126**(1), 123–137

Nicholson, M. D., Fryer, R. S. and Ross, C. A. (1997). Designing monitoring programmes for detecting temporal trends in fish and shellfish. *Mar. Pollut. Bull.*, **34**(10), 821–826.

Oehme M., Theobald N., Baaß A.-C. *et al.* (2006). Identification of organic compounds in the North and Baltic Seas. Final Report UBA R&D-Project. Project number: FKZ 200 25 224/01, 248 pp. http://www.umweltdaten.de/publikationen/fpdf-l/3509.pdf.

OSPAR Convention (1992). Convention for the Protection of the Marine Environment of the North-east Atlantic. Article 2 (General Obligations), 1(a), p. 8. http://www.ospar.org/html_documents/ospar/html/OSPAR_Convention_e_updated_text_2007.pdf

OSPAR (1993). *North Sea Quality Status Report*, Publication 1993 No. 14, OSPAR Commission, London (ISBN 1 872349 06 4).

OSPAR (2008). JAMP Guidelines for Contaminant-Specific Biological Effects (OSPAR Agreement 2008–09). OSPAR Commission, Monitoring guidelines Ref. No. 2008–9.

OSPAR (2009). *JAMP Guidelines for Monitoring Contaminants in Biota*, Ref.-No. 1992-2, OSPAR Commission, London.

OSPAR, (2010). *Quality Status Report 2010*, OSPAR Commission, London. 176 pp., http://qsr2010.ospar.org/en/index.html.

Rank, J. (2009): Intersex in *Littorina littorea* and DNA damage in *Mytilus edulis* as indicators of harbour pollution. *Ecotoxicol. Environ. Saf.*, **72**(4), 1271–1277.

Roose P. (2005). Monitoring of organic micro-contaminants in the marine environment: principles, programmes and progress made. In *Volatile Organic Compounds and Related Microcontaminants in the Scheldt Estuary and the Southern North Sea* (ed. P. Roose), Academic Proefschrift, Vrije Universiteit, Amsterdam, Chapter 1.2, pp. 14–31.

Theobald N., Gerwinski W., Caliebe C. and Haarich M. (2007). Development and validation of a method for the determination of polyfluorinated organic substances in sea water, sediments and biota – occurrence of these compounds in the North and Baltic seas, Project 202 22 213 of the Federal Environmental Agency, Germany, http://www. umweltbundesamt.de/uba-info-medien/dateien/3509.htm. 133 p. (in German, executive summary in English).

Verreault J., Berger U. and Gabrielsen G. W. (2007). Trends of perfluorinated alkyl substances in herring gull eggs from two coastal colonies in northern Norway: 1983-2003. *Environ. Sci. Technol.*, **41**(19), 6671–6677.

WGBEC (2006). Report of the Working Group on Biological Effects of Contaminants (WGBEC) 27–31 March 2006, ICES Copenhagen.

FURTHER READING

General

Aprea C., Colosio C., Mammone T. *et al.* (2002). Biological monitoring of pesticide exposure: a review of analytical methods. *Chromatogr. B Anal. Technol. Biomed. Life Sci.*, **769**(2), 191–219.

Bartram J. and Balance R. (eds) (1996). *Water Quality Monitoring – A Practical Guide to the Design and Implementation of Freshwater Quality Studies and Monitoring Programmes*. Published on behalf of United Nations Environment Programme and the World Health Organization UNEP/WHO, ISBN 0 419 22320 7 (Hbk) 0 419 21730 4 (Pbk): Chapter 11 – Biological Monitoring.

Conti M. E. (2008). *Biological Monitoring: Theory and Applications – Bioindicators and Biomarkers for Environmental Quality and Human Exposure Assessment*, Boston Wit Press, 288 pp. (ISBN 978-1-84564-002-6).

Loeb S. L. and Spacie A. (1994). *Biological Monitoring of Aquatic Systems*, CRC Press, Boca Raton, FL. 400 pp. (ISBN: 9780873719100).

Vernberg J., Calabrese A., Thurberg F. P. and Vernberg W. B. (eds) (1986). *Biological Monitoring of Marine Pollutants*, Academic Press, http://adsabs.harvard.edu/abs/1986ECSS...23..278P.

Conventions, Monitoring and Assessment Programmes

AMAP

Arctic Monitoring and Assessment Programme: http://www.amap.no

European Union

EU (2010). Report on the recommendations for the Good Environmental Status of European Seas regarding contaminants and pollution effects under the Marine Strategy Framework Directive. MSFD GES Task Group 8

HELCOM

HELCOM Commission, Helsinki: http://www.helcom.fi
HELCOM (1988). Guidelines for the Baltic Monitoring Programme for the Third Stage; Part C. Harmful Substances in Biota and Sediments (1988), BSEP 27C.

ICES

http://www.ices.dk
WKIMON. ICES/OSPAR Workshop on Integrated Monitoring of Contaminants and their Effects in Coastal and Open Sea Areas (WKIMON). http://www.ices.dk/workinggroups/

OSPAR

http://www.ospar.org
OSPAR (1997). JAMP Guidelines for General Biological Effects Monitoring (OSPAR Agreement 1997-7.) OSPAR Commission, Monitoring guidelines Ref. No. 1997-7
OSPAR (2002). JAMP guidelines on quality assurance for biological monitoring in the OSPAR area. OSPAR Commission, Monitoring guidelines Ref. No. 2002-15

UNEP

The Regional Seas Programme: Black Sea, Wider Caribbean, East Asian Seas, Eastern Africa, South Asian Seas, ROPME Sea Area, Mediterranean, North-East Pacific, North-West Pacific, Red Sea and Gulf of Aden, South-East Pacific, MEDPOL = UNEP Mediterranean Action Plan

Canada

CSSP Canadian Shellfish Sanitation Program (CSSP) is led by the Canadian Food Inspection Agency (CFIA) in collaboration with Environment Canada (EC) and Fisheries and Oceans Canada (DFO). http://www.inspection.gc.ca/english/fssa/fispoi/csspccsme.shtml

China

South China Sea Branch (SCSB) of the State Oceanic Administration (SOA) http://www.coi.gov.cn/english/eoverview/enhfj/dire.htm

United States

NOAA (National Oceanic and Atmospheric Administration): Mussel Watch Programme. http://ccma.nos.noaa.gov/about/coast/nsandt/welcome.html

10

Use of In-Situ Methods

Richard Greenwood[a], Graham A. Mills[b], Gary R. Fones[c], and Kees J.M. Kramer[d]

[a] *University of Portsmouth, School of Biological Sciences, Portsmouth, UK*
[b] *University of Portsmouth, School of Pharmacy and Biomedical Sciences, Portsmouth, UK*
[c] *University of Portsmouth, School of Earth and Environmental Sciences, Portsmouth, UK*
[d] *Mermayde, Bergen, The Netherlands*

10.1 INTRODUCTION

Marine environments have been affected by growing population and industrial pressures in the coastal zone. Activities such as dumping of waste (both solid and liquid), construction on the shore line and in harbours, dredging, and extraction of minerals, oil and gas all contribute to changes (often deleterious) in environmental quality. The coastal fringe contains some particularly sensitive ecosystems (e.g. coral reefs and salt marshes). Some fisheries, both inshore and offshore,

---

Chemical Marine Monitoring: Policy Framework and Analytical Trends, First Edition.
Edited by Philippe Quevauviller, Patrick Roose, and Gert Verreet.

can be affected. In order to be able to assess the impacts of the various anthropogenic activities it is necessary to monitor a wide range of properties (chemical, physical and biological) of the marine environment. Increasingly, monitoring is being harmonized within regulatory frameworks and international agreements. In recent years there has been a move towards establishing baseline characteristics and the current quality status of aquatic environments. This information is important for the detection of natural perturbations of the environment, the assessment of environmental risk associated with various human activities, and the identification of trends where remedial actions have been taken. There is a worldwide acknowledgement of the need to monitor concentrations of key nutrients in order to understand and predict effects of changes on global climate change, and a realization of the need for long-term acquisition and archiving of accurate data that reflect the state of water bodies. Monitoring is also used in a forensic mode for the detection of sources of pollution; to provide a sentinel to protect sensitive, vulnerable sites; and to check compliance with legally defined environmental quality standards. Much of the work on developing monitoring methods has been focused on freshwater applications, but now there is an increasing need to adapt these for use in the marine sector. Whilst methods for measuring physical properties are well established in coastal waters, monitoring the chemical quality lags behind. Since pollutants move across national boundaries, this is an international problem, and it is particularly relevant for marine waters. Whilst global monitoring networks have been established for following the movement of persistent organic pollutants in the atmosphere, similar networks are not yet available for oceanic systems.

Monitoring in the marine environment presents a number of challenges that are not encountered in the freshwater environment. The oceans cover a large area, and there are complex systems of currents that shift with time. Large dilutions of contaminants occur, and so many pollutants appear at very low, but often biologically significant, concentrations. In coastal waters, tidal effects are marked and can cause large shifts in, for instance, salinity over a tidal cycle, and can transport large volumes of sediment. In these zones, significant redistribution of pollutants and nutrients between sediment and the water column can occur. Many inshore sites are difficult to reach because of tidal fluxes. Boats are needed to reach sampling sites below the low water mark, and large distances are involved for offshore work. Samples of water, sediment and biota can be transported back to land-based laboratories for analysis or analysed in ship-based laboratories. Thus, the use of sampling coupled with classical chemical or biological analysis is expensive, and time consuming, and often fails to provide robust, representative information on changes over time (short term over a tidal cycle and longer term over seasons, years and decades) and sufficiently detailed spatial information over the large areas involved. A further problem is that prolonged delays between sampling and analysis can compromise sample integrity, especially for labile analytes. This can be reduced where ship-board analysis is available. However, samples taken at significant depths can be modified during transport to the surface, since this can take several hours, and be associated with large changes in temperature and pressure. This is a particular problem for unstable samples, such as those taken from areas such as hydrothermal mixing zones that are characterized by nonequilibrium, pressure- and temperature-dependent chemical systems (Le Bris *et al*., 2000). In regions where changes in composition of the seawater occur over short distances, and at one location over short time periods (e.g. in hydrothermal regions, and in inshore tidal regions), then it is difficult to achieve the necessary resolution using spot samples. These difficulties have driven the development of in-situ methods that offer advantages over the classical methods (Vuillemin *et al*., 2009).

The ideal solution to the problems outlined above is the use of remote in-situ technologies that can provide continuous or semi-continuous observations, and transmit the information telemetrically to shore-based data-gathering establishments. In contrast with methods that depend on collection of samples for analysis in laboratories (land based or ship-board), in-situ methods measure variables directly in the environmental medium. This approach has the advantage that it eliminates the need for costly (>€18 000/ day) sampling from ships, and transport processes, and there can be real-time acquisition of information that decision makers can respond to problems in a short time scale. No matter what methods are applied, it is essential that appropriate quality control (QC) and quality assurance (QA) procedures are in place in order for the information to be useful for end users, including regulators.

The first true in-situ analyser was that for determining salinity on the basis of measurements of conductivity and temperature. This was first reported in 1964, endorsed by the Scientific Committee on Ocean research (SCOR) and the International Association for the Physical Sciences of the Oceans (IAPSO) in 1967, and is considered to represent a turning point in marine chemistry (Wallace, 1974; Wangersky, 2005). Many in-situ techniques are now available and are currently used for monitoring a wide range of hydrological and physicochemical variables in the marine environment. The technologies for monitoring hydrological variables are outside the scope of this chapter. A system of smart buoys, which can be deployed at key sites can house a range of measuring instruments and transmit data from them telemetrically, is available and is in operation in European waters. This forms the basis of network of observation points that can provide long-term data that can be used for detecting trends. Variables routinely monitored include pH, pO_2, pCO_2, turbidity, conductivity and concentrations of nutrients such as phosphate and nitrate. Sensors, like any technology deployed in nutrient-rich environments (such as most coastal waters), suffer from the problem of biofouling, and this can be a limiting factor. There are fewer sensors routinely in use for monitoring concentrations of heavy metals, persistent organic pollutants and algal toxins. In vulnerable areas, such as aquacultural installations, where protection against contamination by pollutants or algal toxins is necessary, it is possible to deploy living organisms to provide warnings in real time.

This chapter aims to provide an outline of current developments in the area of in-situ monitoring of the chemical quality of marine waters. The topics covered include spectroscopic and electrochemical methods, sensors (both physical/chemical sensors and biosensors) and alert systems, and a view on future developments and applications of these technologies.

10.2 IN-SITU AUTOMATIC ANALYSERS

Several approaches have been taken to achieving in-situ measurements. One approach has been to miniaturize existing analytical methods, such as colorimetric and spectroscopic methods. In the former, some quantitative reaction generates a coloured product that can be measured spectroscopically; the latter measures analytes directly in the water. There is considerable experience in using these methods in the laboratory, and more recently on-line, and in the field. Developments in optical-fibre technology, solid-state electronics, long-life, stable power supplies, pumps and valves have enabled the production of robust, miniaturized instruments that are portable and that can be used at sites such as a river bank or in a boat. A good example is provided by the system developed to be used on board research vessels to measure concentrations of hydrogen sulfide in seawater samples (Sakamoto *et al*., 1986). The work to improve the reliability of these miniaturized systems that are based on flow-injection analysis (FIA) methods was accompanied by important developments in the design of housings for instruments to be deployed in situ in robust environments such as fast-flowing water or deep water.

Analysers based on FIA methods for use in situ comprise a pump, detector and narrow-bore tube manifold (Vuillemin *et al*., 2009). Sample is pumped through a tube manifold where it is mixed with appropriate reagents to form a product that can be detected by a spectroscopic method. In some cases the reaction product is coloured and can be detected by visible-wavelength spectroscopy; in others it is detected by fluorimetry. A system of valves allows sample to be replaced by standard solutions from time to time to achieve an in-situ calibration of the instrument and provide QC information. Several problems were encountered during the development of FIA analysers for use in situ in the marine environment. These were discussed by Daniel *et al*. (1995), and include refractive index gradients along the reaction tube due to mixing reagents with sample, and potential impacts of pressure, temperature and salinity. The latter can cause problems due to differences in refractive index between samples associated with changing salinity over a tidal cycle in inshore waters. Refractive index changes can contribute to the signal obtained from the absorption of a coloured reaction product and produce significant bias. This problem was overcome in the development of an in-situ analyser for nitrate and nitrite (Daniel *et al*., 1995) by using dual wavelength monitoring that uses a reference wavelength where there is no impact of the coloured reaction product. No effect of pressure over a range of 0.2–2 MPa was observed on the reaction (reduction of nitrate to nitrite in a cadmium column and conversion of nitrite to an azo dye) in this system. However, the reaction kinetics were affected, with faster reaction rates at higher salinities. However, the effect was minimal over the range of 17 to 34 practical salinity units (a useful range in coastal waters). Temperature can vary over time, with depth, and over short horizontal distances in shallow waters of the coastal zone or in regions of geothermal activity, and can affect the analysis by changing reaction rates and viscosity of sample and reagents (and hence flow rates and mixing times). The impact of these effects on the analysis can be minimized by adjusting the manifold design and stopping the flow after addition of reagents to enable the reaction to develop. Controlling the temperature of the reaction coil is another alternative approach. However, longer stop flow times will decrease the sampling frequency, and use of temperature control will increase power consumption, and decrease maximum deployment time. Adaptation of existing colorimetric methods for use in situ in seawater involves a large amount of development work to optimize the design of the manifold and optical systems. In addition, work is needed to provide a housing that will protect the analyser during deployment.

Several miniaturized analysers have been developed for in-situ measurement of variables, including silicic acid (silicate), nitrate and nitrite, iron(II), total iron and total sulfide, and have been used successfully in the marine environment. One of the driving forces for this has been the need for information on the chemical quality of water in relation to the distribution of biota in the deep ocean, particularly in the region of hydrothermal vents. Some of these systems have incorporated limited temperature regulation. Amongst these is a system based on a colorimetric method (using the production of β-silicomolybdic acid that is reduced (by tin(II)) into an intensely coloured molybdenum blue) for the measurement of silicic acid (silicate) in seawater (Floch *et al*., 1998). This is strongly temperature sensitive, and in order to achieve a short measurement time and a stable response the reaction coil was maintained at 30 °C. A dual-beam system was used to reduce the impact of salinity on the method. The use of temperature regulation allowed a high sample processing rate ($60\,h^{-1}$) and the limit of detection was $29\,\mu g\,L^{-1}$, but this increased power consumption. An FIA analyser for measuring ammonium in fresh and saline waters (Aminot *et al*., 2001) was based on a fluorometric method (using an orthophthaldialdehyde-sulfite reagent), since this was robust to variation in refractive index of the sample over a wide range. The method provided a limit of detection in the range $0.4–0.5\,\mu g\,L^{-1}$ and a linear response up to ammonium concentrations of $0.9\,mg\,L^{-1}$ in seawater. The range could be extended to $1.8\,mg\,L^{-1}$ by using a quadratic calibration curve. Interferences from hydrogen carbonate, dissolved oxygen and turbidity in the ranges normally found in marine environments were low. However, the reaction is slow at ambient temperatures, and strongly temperature dependent. This system used a compromise between extending the time available for the reaction to take place (using stop-flow, which decreases the sampling frequency) and moderate heating of the reaction coil to 30 °C. This achieved an acceptable sampling rate ($9\,h^{-1}$), with only a relatively modest increase in power consumption.

A system with three analysers was developed by Thouron *et al*. (2003) to measure three key nutrients: nitrate, phosphate and silicate. The phosphate assay was based on the reaction of inorganic phosphate with molybdic acid. These workers used assay chemistries (using a cadmium surface to reduce of nitrate to nitrite, and conversion of nitrite to an azo dye; and the formation of yellow β-silicomolybdic acid that is reduced using ascorbic acid in molybdemum blue) similar to those employed by Daniel *et al*. (1995) and Floch *et al*. (1998) for nitrate and silicate respectively. However, they took a different approach to the design of the analyser, using reverse-flow-injection analysis (rFIA), where reagents are injected into the sample stream (rather than the other way round in FIA). This system allows the use of micro flow rates that require only small volumes of standards and reagents compared with FIA. Solenoid-driven sample pumps were used instead of peristaltic pumps used in the FIA systems described above. These modifications eliminate the need to replace the tubing used in the peristaltic pumps and reduce the power required for each sample, as well as the volumes of reagent used, and so prolong the potential deployment time. The system had a mechanism for maintaining equal pressure between sample and analyser, and was unaffected by pressure up to 10 MPa. However, because of the effect of temperature on the analyses, as in the designs described above, it was necessary to control temperature. A temperature of 30 °C was selected, and this was maintained by built-in resistance plates. These were located under the whole manifold for phosphate and silicate analysers, and under the cadmium sheet in the nitrate analyser. This system achieved a sampling rate of 2 h^{-1} and was capable of 400 autonomous measurements per deployment. The nitrate and silicate analysers were found to be reliable and had limits of detection of 6 $\mu g\,L^{-1}$ and 10 $\mu g\,L^{-1}$ respectively and linear response up to 2.4 $mg\,L^{-1}$ and 14 $mg\,L^{-1}$ respectively for the two nutrients. The phosphate analyser had a limit of detection of 10 $\mu g\,L^{-1}$ and linear response up to 0.5 $mg\,L^{-1}$, but was less reliable than the other two systems.

Okamura *et al.* (2001) developed an in-situ analyser for measuring manganese in hydrothermal plumes. This was a through-flow system using chemiluminescence (CL) and was based on manganese(II)-catalysed luminol CL after removal of potentially interfering metals (mainly copper, iron and zinc) using an in-line 7-dodecenyl-8-quinolol resin column. This system was affected by both temperature and pressure, and data were corrected (on the basis of laboratory calibration experiments) for these variables. The analyser had a limit of detection in the region of 13 $ng\,L^{-1}$ and an upper limit of 0.1 $mg\,L^{-1}$ manganese in situ. Since it was based on a flow-through technology, it was possible to provide high-definition profiles of concentrations of manganese in hydrothermal plumes over distances of 1000 m.

An FIA analyser (ALCHIMIST) was developed by IFREMER (France) to measure nitrate plus nitrite and total sulfide in deep marine locations near hydrothermal vents (Le Bris *et al.*, 2000). For the analysis of nitrate and nitrite, this system used a method based on reduction of nitrate to nitrite using a cadmium–copper column and reaction of the nitrite to form a coloured azo-dye, similar to that described above. Total sulfide ($H_2S + HS^- + S^{2-}$) analysis was based on the reaction of sulfides with dimethyl-*p*-phenylene diamine to form methylene blue. A dual-wavelength system was used for both analytes. This system was affected by temperature and pressure, and it was thought that the effects of the latter were due to mechanical perturbation of components of the system. Although in-situ calibration reduces the maximum sampling rate to 22 h^{-1} for this system, it was used to account for the effects of temperature and pressure. This system had in-situ limits of detection of 26 $\mu g\,L^{-1}$ and 27 $\mu g\,L^{-1}$ for total sulfide and nitrite plus nitrate respectively. When mounted on a deep-sea remotely operated vehicle (ROV) it provided sufficient spatial resolution to be able to measure the concentrations of the two analytes in different parts of a 0.25 m^2 quadrat at a depth of 1650 m. This provided information on the chemical quality of the water in the region of different biological populations that could not be obtained using grab sampling.

Recently (Vuillemin *et al.*, 2009), an in-situ analyser (the CHEMINI system) has shown promise for measuring analytes at even greater depths and in difficult environments and demonstrates some of the latest developments in this type of technology. This has built on the experience gained with the systems described above. This system uses a double-layered block into which the manifold is etched, thus reducing the use of potentially vulnerable tubing to a minimum. The detection module has a photodiode detector and low power consumption circuit board in a tank at atmospheric pressure. The quartz flow cell is held at ambient pressure and light is provided by light-emitting diodes and is transferred to the flow cell and returned to the detector through optic fibres. The signal is obtained using synchronous detection to measure very low currents (in the picoamp range) in the presence of high noise levels. The power consumption of the system is low: 20 mA h per measurement at 12 V. In this system, the reagents (up to 2 L) and standards (250 mL) are contained in flexible bags and enable 400 analyses per deployment.

Prototypes of this CHEMINI system have been deployed in deep-sea hydrothermal environments for the measurement of concentrations of dissolved iron and total sulfide. Total sulfide is measured using a reaction with dimethyl-*p*-phenylenediamine in the presence of iron(III) to produce methylene blue (maximum absorbance 667 nm). The limit of detection of this method is 3 $\mu g\,L^{-1}$ and the upper limit of the range is 13 mg L^{-1}, though this can be extended to 40 mg L^{-1} by using a reduced injection volume. Iron(II) is measured using a reaction with ferrozine [3-(2-pyridyl)-5,6-bis(4-phenyl sulfonic acid)-1,2,4-triazine] to form a magenta-coloured complex (maximum absorbance 562 nm). Total iron can be measured in this way by reducing iron(II) to iron(III) using ascorbic acid. The limit of detection of total iron is 17 $\mu g\,L^{-1}$ and the upper end of the range 5.6 mg L^{-1} (extendable to 112 mg L^{-1} using reduced injection volumes). Temperature has a marked effect on the performance of this rig, and a 30% decrease in sensitivity and a significant increase in noise levels were observed for a fall from 20 °C to 5 °C. In part, this may be explained by effects on the properties of the peristaltic pumps (probably the tubes) that lead to a reduced flow rate, and by the increased viscosity of water at the lower temperature.

The CHEMINI system was deployed on an ROV to investigate the chemical environment in mussel beds (*Bathymodiolus azoricus*) in the mixing zone adjacent to hydrothermal vents, in this case at a depth of 1650–1750 m. Here, the temperature varied between 4.4 °C and 14.2 °C, pH between 5.6 and 7.8, and dissolved oxygen between oxygen rich and anoxic. In this difficult environment the analyser provided useful spatial data, and indicated a 50% loss of dissolved iron from the thermal plume in the mixing zone (probably as iron sulfide). Where monitoring in time is required to provide information on temporal changes in the chemical environment, this system can potentially provide eight measurements per day over a 6 month period.

The combination of miniaturization, the availability of protective deployment devices, availability of stable, long-life power supplies and advances in communications capabilities has enabled these analyser systems to be buoy based, located on robotic vehicles or towed by ships. However, the use of reagents sets limits on the time that instruments can be deployed and adds significantly to the bulk and weight of the analytical system to be transported and deployed, and there has been a move to develop reagentless methods where possible (Lacombe *et al*., 2008).

10.3 SPECTROSCOPIC METHODS

A number of spectroscopic- and spectrometric-based in-situ devices have been developed for use in marine applications for measuring, for example, dissolved gases, nutrients and trace metals. These are usually miniaturized versions of conventional laboratory analytical equipment but that have been fitted with robust protective housing to be able to enable the systems to be deployed in harsh environments and to withstand high pressures. A range of optical detectors can used, and these have included absorbance, reflectance, luminescence, fluorescence, refractive index and light scattering. Underwater mass spectrometric detectors can also be employed, usually with a membrane inlet sample introduction system. Many of these in-situ sensors have been reviewed in two text books (Buffle and Horvai, 2000; Varney, 2000) and in papers by Prien (2007) and Moore *et al*. (2009). The following provides a brief discussion of some of different optical- and mass-based technologies available and their specific application to the marine environment.

One of the simplest optical sensing systems measures changes in refractive index. Such detectors can be used in the marine setting to measure bulk particulate loads (e.g. amounts of organic and inorganic particles in the water column). These devices work on the back scattering of light principle and can be useful in estimating phytoplankton loads and degree of sediment resuspension. Probably the most widely used method is optical absorption spectroscopy. This method is often used analytically in conjunction with flow-injection analysers or auto-analysers, as described in the previous section. Absorption spectra (both visible and ultraviolet regions) in seawater are more complex than in pure water owing to the influence of salts. The absorption peaks are broader, and this can cause overlap of spectra between different compounds, making identification harder. An early example of a high-resolution in-situ ultraviolet spectrophotometer (ISUS) that can take rapid (of the order of 1 s) direct measurements was first described by Johnson and Coletti (2002; Johnson *et al.*, 2006). This ISUS was developed at the Monterey Bay Aquarium Research Institute, California, USA, and was used to measure the concentration of nitrate, bisulfide and bromine, as these species have distinctive ultraviolet absorption spectra. The system records high-resolution spectra (<1 nm) using a diode-array system permitting the deconvolution of overlapping spectral peaks. Data for nitrate concentrations obtained from the device fitted well with theoretical models for this compound. The detection limit for nitrate was 0.11 mg L^{-1} for measurements made at 1 Hz, although this could be lowered by taking longer (30 s) measurements (Johnson *et al.*, 2006). The device has the potential to operate at depths of 2 km and can be deployed for over 3 months without any degradation of performance due to biofouling or instrumental drift. The spectral analysis routine was able to compensate for the effect of temperature on performance. This instrument can also be used to measure sulfides that are produced in deep-sea hydrothermal vents, and a future development of the instrument could permit the measurement of carbonate ions so as to help better understand the reasons for the increasing acidity of oceans. It should be possible to decrease the size the instrument and reduce the power consumption by using light-emitting diodes that can reach wavelengths of 210 nm. In-situ ultraviolet/visible spectrometers are frequently used to monitor surface, water, and waste and drinking water-treatment processes. Light attenuation measurements are made over the whole spectrum (e.g. 200–735 nm), and over this region it is possible to measure nitrate, chemical oxygen demand (COD) and total suspended solids (turbidity) directly in the water column (van den Broeke *et al.*, 2006). Another deep-sea (up to 6 km) in-situ ultraviolet spectroscopic device operating over wavelengths in the range 190–360 nm has been developed by TriOS Optical Sensors, Germany (http://www.trios.de/). The instrument was originally designed for nitrate measurements, but it can also be used for estimating biological oxygen demand (BOD), COD, total organic carbon (TOC), humic acids and some phenols. Two versions are available, with a path length of 1–10 mm or 10–60 mm depending on the concentrations being measured. The device also has a specially developed 'nano-coated' optical window that has been shown to resist biofouling on its surface.

In-situ fluorescence spectroscopy can also be used to measure compounds that naturally exhibit this characteristic. Fluorescence spectroscopy is very sensitive; however, the drawback is the intrinsic lack of selectivity in the choice of excitation and emission wavelengths to be used. The broad character of fluorescence peaks usually prohibits finding an explicit excitation/emission wavelength combination that is unique to the analyte being measured. Owing to their chemical structure, polynuclear aromatic hydrocarbons (PAHs) are inherently fluorescent (most can be excited in the ultraviolet range 240–300 nm and emit light in the range 310–400 nm range). A submersible ultraviolet fluorimeter has been designed for the measurement of PAHs in water and is commercially available. The EnviroFlu-HC (TriOS Optical Sensors, Germany) is designed to excite compounds (using a xenon lamp) at $\sim 254 \pm 12.5$ nm and detect the emitted light at $\sim 360 \pm 25$ nm. The instrument weighs about 2 kg and is rated to 500 m depth. Recently, Tedetti *et al.* (2010) evaluated the device for monitoring PAHs in coastal waters off Marseilles (north-western Mediterranean Sea), where there was likely to be inputs of these anthropogenic chemicals. Deployments were made over a period of 1 year and spot samples of seawater were collected in parallel so that the results of the two monitoring methods could be compared. Results showed that the EnviroFlu-HC sensor was not specific to the measurement of PAHs in the water column. Owing to the broadness of the light signal used, tryptophan-like material and, to a lesser extent, humic substances also exhibited a response. The poor sensitivity of the instrument did not permit the measurement of PAHs (dissolved and bound to particles) in unpolluted coastal waters; however, these substances could be detected in polluted environments (total PAHs >1 μg L^{-1}). Teddetti *et al.* (2010) suggested that the sensor was useful in the general assessment of water quality and in locating sources of anthropogenic inputs such as wastewater plumes that contain elevated levels of tryptophan- and humic-like materials.

Optodes, sometimes referred to as optrodes, can also work using fluorescence spectroscopy principles. Here, indicator dyes (e.g. ruthenium or platinum complexes) are employed and these are embedded or immobilized into sol–gel matrices of thin polymer layers. The presence of a test analyte in the light path modifies the properties of the indicator (e.g. emission of fluorescent light) and this forms the basis of operation these sensors. These devices are frequently used to measure oxygen (Prien, 2007), but they can also be used for pH and pCO_2 (Zhu *et al.*, 2005; Schroeder *et al.*, 2007; Sotomayor *et al.*, 2010).

Infrared devices can also be used, and this area was reviewed by Mizaikoff (2003). The application of these devices is mainly for the measurement of volatile chlorinated and aromatic hydrocarbons, and this relies on mid-infrared-range optical systems. This wavelength region is particularly attractive for optical sensing, since molecule-specific information is provided by stimulation of the ground vibrational modes of organic substances. Use of Fourier transform methods enables simultaneous measurements of individual chemicals in multicomponent mixtures at concentrations in the microgram per litre range. Interferences to the signal from changes in salinity and water turbidity also need to be taken into account. An in situ, miniaturized modular Fourier transform infrared (FT-IR) device with a fibre-optic sensor head was developed as part of the EU framework project SOFIE (Kraft and Mizaikoff, 2000). The device can operate at depths of 300 m deployed either on a tow frame or on an ROV. The instrument was used to measure various volatile chlorinated and aromatic hydrocarbons in seawater. The effects of other interfering pollutants (e.g. aliphatic hydrocarbons and phenols) did not significantly affect the operation of the sensor. In addition, the long-term stability and resistance to fouling of the device were good (Kraft *et al.*, 2003).

In-situ monitoring devices based on Raman spectroscopy have been developed. Here, a laser light in the visible region is shone on a sample, the light is scattered and a small fraction (about 1 in 10^7 molecules) of this is shifted in frequency (the Raman shift) as the atoms in the material vibrate. Analysis of the frequency shifts (spectrum) of light reveals the characteristic vibrational frequencies of the atoms and, thus, the chemical composition and structure of the sample being analysed. As a consequence, Raman spectroscopy is highly specific, and spectra can be recorded rapidly (within a few seconds). Raman-active peaks are strongest for compounds that are symmetrical in structure and that do not have a dipole moment. In the marine environment, Raman spectroscopy has many potential applications, but there are difficulties to be overcome. The Raman signal can often be weak, and many chemical species present in seawater are not Raman active or are present at only trace concentrations. The water itself has a subdued Raman signature, unlike its overwhelming infrared signature. Therefore, the ubiquitous signal of seawater does not produce significant interferences in the Raman optical path. Brewer *et al.* (2004), based at the Monterey Bay Aquarium Research Institute, was one of the first groups to develop a deep-ocean Raman in-situ spectrometer (DORISS) for geochemical studies. The instrument uses a Nd:YAG green laser operating at 532 nm to provide the Raman excitation, with a holographic grating as a filter and a charge-coupled device (CCD) array to detect the weak Raman signals that are produced. The DORISS system is capable of being deployed from an ROV and can work to a depth of 3.6 km and at water temperatures of 2 °C. In-situ Raman spectra have been obtained from seawater at various depths and from various gases, liquids and solid mineral samples. An interesting application of the device is in the study of hydrothermal vents and their associated bacterial mats, and this is possible because two of the vent gases (methane and hydrogen sulfide) exhibit strong Raman scatters. However, one drawback to the deployment of DORISS is its high power consumption, which necessitates the use of a large power pack or tether for prolonged operation. A miniaturized version of a Raman system has been developed and tested using simulated hydrothermal vent fluids by Dable *et al*. (2006).

Another Raman-based device has also been developed by European research groups in the EU-funded SOFIE and MISPEC research framework projects. In this instrument, a modified technique, surface-enhanced Raman spectroscopy, is used. Here, the target molecule is adsorbed onto a metal substrate (e.g. gold or silver) and the excitation light excites electrons in the conduction band of the metal. Using this approach, it is possible to increase the sensitivity (six to eight orders of magnitude) of the method and permit the in-situ analysis of organic compounds such as PAHs over a concentration range of micrograms per litre to nanograms per litre (Schmidt *et al*., 2004).

Laser-induced breakdown spectroscopy is an emerging technology with potential in-situ monitoring applications. This uses a laser pulse to create a micro-plasma in the sample; then, a spectrophotometer captures the transient light to identify and quantify the analytes. It is a relatively low-cost method and can be easily adapted for use with automatic or robotic systems. Measurements can be taken rapidly (less than 1 s) and sensitivity (milligrams per litre or better) is similar for most elements. Michel *et al*. (2007) demonstrated the potential of this approach in the laboratory for the detection of calcium, potassium, lithium, manganese and sodium in seawater. Changes in pressures, temperature and concentration of NaCl had little effect on the signal. Preliminary results suggest that this technique could be suitable for the in-situ chemical analysis of the marine environment.

Membrane inlet mass spectrometers have been developed for the in-situ measurements of dissolved gases and volatile organic compounds over a wide range of ocean depths. Devices can be deployed on a wide variety of platforms (tethered, moored, on-board ship or attached to ROVs) in different environments (freshwater, coastal and deep oceans). As they use no reagents for their operation, the spectrometers can be deployed for extended periods. Using mass spectrometers underwater is very challenging, since the devices operate under high vacuum. Analytes must be transported from the pressurized aqueous environment into a vacuum system, and this can place severe demands on pumps, particularly for long-term deployments of the apparatus. It has also been necessary to give consideration to the power requirements of the equipment. Two designs of mass spectrometer are used (linear quadrupole or ion trap), both with electron-impact ionization (Short *et al*., 2001). Both have similar detection limits (about 1 mg L^{-1}); however, ion-trap systems do have a higher mass limit (650 Da) compared with quadrupole detectors (100 Da). Ion traps also have the potential for tandem mass spectrometry experiments. The most common sample introduction method employs a supported, high-pressure polydimethylsiloxane membrane, and such systems are theoretically capable of operating to ocean depths of 2 km. These membranes allow the selective permeation of gases and nonpolar, volatile organic compounds. The thickness of the membrane affects the response time and sensitivity of the instrument. Although thin membranes are often preferred, these can be mechanically weak and not able to withstand harsh field conditions. The effects of hydrostatic pressure and temperature on the permeability of the membrane must be taken into account when making quantitative measurements of dissolved analytes over a range of vertical depth profiles (Bell *et al*., 2007). One solution to overcome the effects of temperature is to use an in-built heater block so as to maintain isothermal conditions around the membrane. The situation with changes in hydrostatic pressure is more complex, however, and depends in part on the molecular size of the analyte. Several models have been developed to try to account for this effect.

Other types of sample inlet system have also been proposed by researchers based at the University of South Florida (St Petersburg, Florida, USA). These include a sediment probe, a syringe pump and various sorbent-trapping procedures to preconcentrate chemicals. Extracted compounds can also be separated on a short gas chromatographic column prior to mass spectrometric analysis. Such sample introduction systems have the potential to extend the range of compounds (e.g. low molecular weight polar pollutants) that can be detected in situ. Another compact in-situ mass spectrometric system (NEREUS) has been developed (Hemond and Camilli, 2002; Camilli and

Hemond, 2004) that enables an efficient, rapid evaluation of underwater spectra. This instrument has double focusing of the ion beam and, hence, can measure mass spectra at relatively high resolution. The power consumption is generally small.

Mass spectrometers can be used for the simultaneous detection of multiple species. Most applications of underwater mass spectrometry systems have been the in-situ detection and quantification of dissolved gases (e.g. CO_2, O_2, methane) and small volatile organic compounds (e.g. dimethylsulfide, toluene) at a range of depths in the water column (Camilli and Hemond, 2004; Short *et al.*, 2006; Schlüter and Gentz, 2008). Much of this work is still at a preliminary stage. Larger molecular weight (i.e. above 200 Da) compounds are not usually detected with these systems, as the permeability of the inlet membrane to such molecules is very low, and this reduces sensitivity. By the use of global positioning technology that is time matched to the concurrent mass spectral data, it is theoretically possible to create high-resolution three-dimensional maps of the spatial distributions of chemicals and for the identification of pollutant plumes. Such systems offer much potential for the future in monitoring the distribution of natural and anthropogenic chemicals. However, in addition to improvements in the mass spectrometer itself, further developments, such as self-calibration of the analytical system, lower limits of detection for other traces gases (e.g. H_2 and N_2O), improved inlet system materials (e.g. potential use of carbon nanotube membranes) and miniaturization of gas or liquid chromatographic columns that can be easily interfaced to the spectrometer system, are needed. The reduction in size of the spectrometer brings added benefits of a reduced vacuum system and associated power demands. The ability to measure low-abundance isotopes (e.g. ^{13}C, ^{14}N, ^{18}O, ^{34}S) of natural elements using in-situ mass spectrometric techniques would be beneficial in improving the understanding of the global oceanic cycling of many compounds. Higher resolution and more-sensitive detectors are needed for this purpose, but several projects in these areas are currently ongoing (Kibelka *et al.*, 2004).

10.4 ELECTROCHEMICAL TECHNIQUES

Electrochemical-based instruments are among the most widely used devices for in-situ chemical analysis (Denuault, 2009; Moore *et al.* 2009), and include conductometric, potentiometric and amperometric/voltammetric electrode systems (Taillefert *et al.*, 2000). Conductometric electrodes are used to measure salinity and are incorporated into a variety of commercially available designs. Examples of potentiometric devices include those to measure pH (Cai and Reimers, 1993), sulfide (e.g. the Ag_2S electrode; Revsbech *et al.*, 1983) and pCO_2 (Cai and Reimers, 1993). Examples of amperometric devices include those for measuring concentrations of O_2 and N_2O (Revsbech *et al.*, 1980, 1988). Electrodes to detect (free) trace metal concentrations so far lack the sensitivity required for use in the marine environment. Furthermore, the analyses are usually affected by changes in salinity.

In potentiometric devices, potential is measured with near zero current passing through an electrochemical cell consisting of a working electrode and a reference electrode. The potential is proportional to the concentration of the analyte. In amperometric devices (based on voltammetry), the current is measured at a fixed potential between the working and reference electrodes. The current is proportional to the concentration of the analyte. The technique of voltammetry has long been the method of choice for trace metal analysis and works by measuring the current while scanning the entire voltage range of the solid-state electrode (I versus E curves). This then allows the measurement of more than one species at a given time in the same region of space (Brendel and Luther, 1995). The technique of voltammetry can be viewed as being analogous to absorbance versus wavelength curves in spectroscopy, because the electrode currents measured for peaks during voltage scans are proportional to the concentration(s) of the analyte(s). Voltammetry is thus a nonselective method that can measure many chemical species, which over the last two decades has included the application to a suite of trace metals, including cadmium, copper, lead and zinc.

An understanding of the distribution of trace metals in the marine environment is of importance for ecological assessments as well as for understanding their roles in different biogeochemical processes (Nriagu, 1990). Usually, the monitoring of trace metals has relied on spot sampling and subsequent laboratory analysis using techniques such as inductively coupled plasma mass spectroscopy, graphite furnace atomic absorption spectroscopy and neutron activation analysis. The major drawback to these techniques is that they measure only total concentrations of the metal present. Voltammetry methods, however, such as anodic stripping voltammetry and competitive ligand equilibration–adsorptive cathodic stripping voltammetry, have been used for over 25 years for determining both total dissolved metals and for understanding metal speciation (Florence and Batley, 1977; Sipos *et al.*, 1977; van den Berg, 1984). In voltammetry, a reference electrode (e.g. Ag/AgCl) is used to apply a voltage against a working electrode (e.g. Au/Hg) while a counter platinum electrode is used to measure current so that no current passes through the reference electrode. Generally, measurements of total metal concentrations are made after sample acidification and ultraviolet irradiation to destroy (complexing) organic matter. Speciation measurements include the determination of complexation properties, such as stability constants and total ligand concentrations, of model or naturally occurring ligands. These are usually obtained by recording the voltammetric peak current intensity of the test metal during the titration of the complexing solution with the metal. The major limitations of conventional approaches to trace metal analyses, even for in-field measurements, are the sample perturbations due to sampling, and possible sample storage and handling effects, and the long analysis times required for metal speciation procedures. The species distribution and properties must be preserved in order to interpret the role of metals in biogeochemical processes, and in particular to assess their ecotoxicological effect. The conventional laboratory-based techniques were adapted (Buffle and Tercier-Waeber, 2005) to work in the field (e.g. on river, lake or sea shore) (Huang and Dasgupta, 1999) or on board a ship (Achterberg and Van den Berg, 1994; Whitworth *et al.*, 1998) to minimize problems encountered with sample storage.

It is now well accepted (Buffle and Tercier-Waeber, 2005; Tercier-Waeber *et al.*, 2008) that the development of rugged, submersible non- or little-perturbing devices and probes, for automatic, in-situ measurements (i.e. measurements performed inside the water and sediment/soil matrix) is an important need. Voltammetry is very well suited to this purpose, since it can be used in low-cost, automated and miniaturized equipment with low energy requirements. However, significant technical developments are required to improve conventional voltammetric devices before they can be deployed truly in situ, rather than being limited to use as on-board analysers. A review of these developments is given in Buffle and Tercier-Waeber (2005), while an initial overview of the concept of in-situ voltammetry is given in Buffle and Tercier-Waeber (2000).

Robust hemispherical mercury-plated microelectrodes are used with voltammetric techniques such as cyclic and square-wave anodic stripping voltammetry. The design of these electrodes for in-situ measurements has progressed in two directions since they were introduced to the aquatic sciences by de Vitre *et al.* (1991), Tercier *et al.* (1995) and Brendel and Luther (1995). These are (1) fabrication on an electro-etched iridium substrate (polished (micro-)electrode 5–16 μm Ø) in combination with an agarose gel antifouling membrane (Tercier and Buffle, 1996) and (2) fabrication on a gold wire substrate (usually 100 μm Ø) without a membrane. The intended field applications of the two designs impose radically different operational and performance requirements. The first technique enables highly sensitive speciation measurements of dissolved trace metals, such as cadmium and lead, without interferences from colloidal and/or macromolecular materials, and was developed primarily for water column studies. The gel layer is usually 0.3–1 mm thick and ensures purely diffusive transport of ions to the microelectrode surface. Sensitivity is dependent on deposition times (typically 5–15 min) that are required before applying square-wave anodic stripping voltage (SWASV) conditions (Tercier *et al.*, 1995). There have been several reviews of this technique published during the past 5 years (Buffle and Tercier-Waeber, 2005; Tercier-Waeber *et al.*, 2008; Tercier-Waeber and Taillefert, 2008).

The simplest integrated micro-analytical system reported is the gel-integrated microelectrode (GIME). It consists of a single or an array of mercury-plated iridium microelectrodes covered with an agarose gel layer that is a few micrometres thick (Tercier and Buffle, 1996; Belmont-Hebert *et al.*, 1998; Tercier-Waeber *et al.*, 2000). These sensors allow the simultaneous measurements of the dynamic fraction (i.e. the fraction potentially bioavailable) of copper(II), lead(II), cadmium(II) and zinc(II) at the nanogram per litre level using SWASV and of manganese(II) and iron(II) at the microgram per litre level using square-wave cathodic sweep voltammetry (SWCSV). Measurements with the GIME are performed in two successive steps (Tercier and Buffle, 1996; Belmont-Hebert *et al.*, 1998): equilibration of the agarose gel with the test solution (typically 5 min) and voltammetric analysis inside the gel. Important features of the GIME are their sensitivity and reliability, which are due to the advantages of micro-sized electrodes and to tight control of their fabrication. Detection limits for interconnected GIME SWASV measurements of cadmium(II), copper(II), lead(II) and zinc(II), using a preconcentration time of 15 min, are 6 ng L^{-1}, 13 ng L^{-1}, 6 ng L^{-1} and 19 ng L^{-1} respectively. The GIME has been used as the basis for the development of another type of gel-integrated micro-sensor, the complexing gel-integrated microelectrode (CGIME). The CGIME is a novel voltammetric device which allows the simultaneous measurements of the true free-ion concentrations of copper(II), lead(II) and cadmium(II) at sub-nanomolar levels in natural waters (Tercier-Waeber *et al.*, 2005; Noel *et al.*, 2006). The determination of the free metal ion concentration is crucial for the assessment of the (eco)toxicological impact of trace metals, since this fraction is related to bioavailability (Campbell *et al.*, 2002). A detailed description of the CGIME preparation has been given by Noel *et al.* (2006).

The only fully commercially available in-situ voltammetry system is the voltammetric in-situ profiling (VIP) system, which incorporates all of the requirements for a reliable voltammetric-based environmental trace metal sensor as outlined by Tercier-Waeber *et al.* (2008). The VIP system has the capacity for monitoring and profiling concentrations of trace metals, down to 500 m, with sub-nanomolar sensitivity, and is based on the interconnected GIME array described above. This system consists of several other units: an online oxygen removal module, a multiparameter probe (Ocean Seven 316, Idronaut-Milan) and a calibration deck unit. A detailed description of the device is given elsewhere (Tercier *et al.*, 1998).

A second type of voltammetric device is the Au/Hg microelectrode. This was originally developed for the concurrent measurement of oxygen, manganese(II), iron(II), iodide and/or forms of reduced sulfur at micromolar concentrations in the pore waters of marine and freshwater sediments (Luther *et al.*, 1998). Such microelectrodes usually consist of a 13, 25 or 100 μm diameter gold wire soldered directly to a copper conductor using a silver-containing solder (Brendel and Luther, 1995). The gold wire is then enclosed in one of a range of casings, depending on the application of interest. For the depth profiling of chemical species in biofilms (Xu *et al.*, 1997), the capillary tubing is pulled to a tip of 100–500 μm diameter. Polyethyletherketone (PEEK) tubing (3 mm diameter and a few centimetres long) filled with epoxy resin is the configuration for the measurement of chemical species in deep-sea hydrothermal vents (Luther *et al.*, 2001) and in the water column of marine and freshwater environments (Luther *et al.*, 2003; Glazer *et al.*, 2006). For the depth profiling of chemical species in freshwater and marine sediments (Sundby *et al.*, 2005) the glass tubing is pulled to a tip of 500–750 mm diameter and approximately 5 cm long and filled with epoxy resin. Preparation and manufacture of the microelectrodes was first published by Brendel and Luther (1995) with updated reviews published in Luther *et al.* (2008) and Tercier-Waeber *et al.* (2008).

Four research systems are under development for making in-situ measurements with Au/Hg electrodes; all are based on commercially available potentiostats in pressure-resistant housings. The first instrument is designed for water depths of 30–50 m (Luther *et al.*, 1999) and consists of a transmitter enclosed in a pressure housing, which is connected to the working, reference and counter electrodes though a 1 m long shielded cable. The second instrument, the in-situ Electrochemical Analyzer I™ (AIS ISEA I™), is designed for real-time measurements in both shallow and deep waters and consists of a potentiostat and a multiplexer to monitor up to four electrodes successively. This has been used on ROVs (Nuzzio *et al*., 2002) and on CTD-rosette deployments (Lewis *et al*., 2007). The third instrument, the AIS ISEA II™, is similar in design but adapted for unattended measurements. In this generation of instrument, the internal computer is programmed to run a sequence of experiments and store data on a memory card until the analyser is recovered. This instrument has been deployed on a free benthic lander for sediment depth profiles and for benthic chamber measurements (Luther *et al*., 2008). The latest generation of ISEA is capable of functioning in both real time and in the programmable mode (Moore *et al*., 2007). All ISEA instruments are controlled by software that controls voltammetric experiments in real time or in a programmable sequence. The software allows the user to run linear sweep, cyclic, normal pulse, differential pulse and square-wave voltammetry, including stripping techniques, between +2 and −2 V at scan rates ranging between 1 and 10 000 mV s^{-1} and sensitivities in the picoamp to milliamp range. These Au/Hg microelectrodes have been used in a number of applications in the marine environment (Luther *et al*., 2008; Tercier-Waeber and Taillefert, 2008).

The Metalyser HM1000 (http://www.wagtech.co.uk/) is another commercially available voltammetric system. This is a hand-held portable heavy-metal analyser which uses the analytical principles of anodic and cathodic stripping voltammetry, and incorporates vibrating microwire electrodes. It is used to measure concentrations of copper, cadmium, lead, mercury and arsenic, with detection limits in the micrograms per litre range. This is not sensitive enough for pristine marine waters, where concentrations of these metals are in the nanogram per litre range.

Difficulties in calibration and validation are limitations to the use of in-situ electrochemical devices. As the systems are measuring in real time it is difficult to calibrate the instrument in situ without undertaking a large number of spot samples for verification. One such intercomparison was used to validate the VIP system. Reliability and performance were assessed in the laboratory using five replicate VIP devices (Braungardt *et al*., 2009). Concentrations of the dynamic metal fractions (cadmium and copper) measured by the VIP system in certified reference materials were close to the stated values. There was good agreement between the concentrations measured by five VIPs simultaneously deployed for periods of several hours in coastal waters. However, there is still a need for more intercalibration and validation studies in order to increase confidence in the reliability and robustness of these arrays of in-situ devices. New developments, such as environmentally friendly disposable probes with micro-fabricated bismuth working electrodes for trace metal analysis (Zou *et al*., 2008), cobalt-based microelectrodes for phosphate (Lee *et al*., 2009) and competitive screen-printed electrode sensors (Zaouak *et al*., 2009, 2010) are still in the laboratory development phase; it will be a challenge to adapt them for in-situ operation.

10.5 SENSORS

A sensor is a device that produces a measurable response to a change in a physical condition, such as temperature or thermal conductivity, or to a change in chemical concentration. Sensors are ideally suited for making in-situ measurements in the marine environment and may help overcome some of the current problems of the undersampling (in both space and time) of coastal waters and the deep ocean.

Many platforms are available on which these sensors can be deployed, and include Apex floats (3260 floats in service on 22 October 2009 – https://www.bodc.ac.uk/data/online_delivery/argo/), autonomous underwater vehicles, gliders, benthic landers, moorings (http://www.eurosites.info/about.php). All of these have the capability to address the current data shortage; however, this will be achieved only when biogeochemical sensor systems have undergone a multiple order-of-magnitude change in size, power consumption, reliability and robustness (Daly *et al.*, 2004). Currently, there is a large amount of research being undertaken to push forward the use of sensors in situ. The Networks of Sensors – Demonstration High Resolution Networks programme (http://www.nerc.ac.uk/research/programmes//background.asp) is a 5-year programme (2010–2014) of research funded by the Natural Environment Research Council in the UK. This aims to deliver up to five collaborative demonstration projects exploiting the potential of distributed high-density networks of sensors, with the objective of discovering new insights into the environment. One of the large research groups within Europe that is taking up the challenge for the development of miniaturized multiparametric biogeochemical sensors for deployment in the oceans is the Centre for Marine Microsystems at the University of Southampton (http://www.soton.ac.uk/cmm/). Here, they are developing a suite of chemical sensors for nitrate/nitrite, phosphate, ammonia, oxygen, methane, pH, alkalinity and iron/manganese using lab-on-a-chip technologies that employ a chelating resin preconcentration step in order to attain the required detection limits (Sieben *et al.*, 2010).

10.5.1 Physical-, Chemical-based Sensors

In-situ sensors have been used for a number of years to measure physical-based parameters such as conductivity, oxygen, pH and carbon dioxide in seawater. The majority of these are available commercially, particularly in sensor packages such as conductivity–temperature–depth (CTD) profiling instruments, where the data can be used to calculate salinity, density, sound velocity and other parameters of interest. The term CTD is often used today to describe a package that includes the actual CTD as well as auxiliary sensors to measure other parameters (such as dissolved oxygen, pH, turbidity, chlorophyll *a*, rhodamine, blue–green algae, ammonia and nitrate) along with a device to collect water samples for subsequent analysis in the laboratory. Often, arrays of sensors are used so that multiple parameters can be measured simultaneously, and these systems can be fully automated. These can either be deployed from a ship in the profiling mode or can be deployed on a mooring for long-term monitoring as part of an observation system.

Commercially available conductivity sensors use an AC voltage applied to nickel electrodes or graphite electrodes. The challenges are to maintain the calibration during long-term deployments and to improve the method to estimate the time-varying drift of measured conductivity from autonomous CTDs on moorings or on Argo floats (Owens and Wong, 2009). Oxygen sensors are also available commercially for a number of applications, such as profiling sediment–water interfaces and water columns. These can be either electrochemically or optically based systems (Glud *et al.*, 2000). Most oxygen sensors are based on the original Clark polarographic electrode (Reimers, 2007); however, there are now optical dissolved-oxygen sensors that utilize luminescent technology (Glud *et al.*, 2000; Reimers, 2007).

There are numerous sensors available (both commercially and as research devices) for the measurement of pH and pCO_2. These parameters are important in climate change (Sabine *et al.*, 2004; The Royal Society, 2005) and, therefore, need to be measured accurately in monitoring programmes, ideally by in-situ measurements. The sensors include electrochemical, potentiometric, photometric and fibre-optic-based instruments (Cai and Reimers, 2000). There are two primary techniques for measuring in-situ pH. The first is electrochemical (potentiometric) using a hydrogen electrode and reference electrode and is the standard method. However, there is a major problem with these devices when ionic strength and composition changes, such as in estuarine environments. The second approach is the photometric/fibre-optic measurement of pH (Cai and Reimers, 2000). This uses an indicator dye that is added directly to the test solution or is confined within a sensing cell or matrix. These techniques have excellent precision (0.001 pH units) and accuracy (0.004 pH units), but ideally are used in flow-through systems and, therefore, are best suited for operation on-board research ships (Bellerby *et al*., 1995, 2002). Instruments for measuring pCO_2 have been used mainly as flow-through systems on-board ships (Cooper *et al*., 1998; Hardman-Mountford *et al*., 2008); however, a pCO_2 system for deployment on ocean moorings has been developed (DeGrandpre *et al*., 1997). This consists of a fibre-optic pH sensor in a flow-through cell connected to a CO_2-permeable membrane equilibrator.

10.5.2 Biosensors

Biosensors are devices that integrate a biological recognition element with a transducer of some form that converts changes associated with interaction between the receptor and the target compound into an output signal that can be measured quantitatively. Several reviews (Ron, 2007; Farré and Barceló, 2009; Farré *et al*., 2009; Roig *et al*., 2009) have provided useful definitions and descriptions of biosensors, and have differentiated them from biological systems (bioassays, biological early warning), where such an integral transduction mechanism is absent. Authors have classified biosensors in a number of ways, usually on the basis of the type of recognition unit or the nature of the transducer. For instance, the biological component can be provided by DNA, enzymes, immunological systems, receptor proteins and whole cells. The transduction component can be provided by acoustic, chemical, electrochemical (potentiometric, amperometric, conductimetric), microbalance, optical (absorbance, bioluminescence, chemiluminescence, total internal reflection, surface plasmon resonance) and piezoelectric mechanisms.

Biosensors were originally developed for biomedical purposes, and glucose biosensors still represent close to 90% of this mature global market (Chiappini *et al*., 2010). The advantages of biosensors, such as high specificity, sensitivity, in some cases biological relevance and ability to work in a wide range of matrices, make them potentially very useful in the area of environmental monitoring. The use of such devices that could be deployed and operate remotely to make repeated in-situ measurements that could be transmitted to a laboratory is seen as the ideal way of meeting the requirements for monitoring the quality of environmental water. A lot of research has been directed towards this target. However, there are few commercially available biosensors for use in the environment so far, and even fewer that can be used in situ. There is currently a large gap between the research output and the commercial availability of appropriate instruments for environmental applications.

The development of biosensors based on enzyme systems is exemplified by those based on enzyme recognition units for the detection and measurement of organophosphate and carbamate insecticides in food and water samples. These groups of compounds inhibit acetylcholine esterase (AChE), and the recognition unit is based on this mechanism. Some systems have used acylhydrolase (pseudocholinesterase) or butyrylcholinesterase instead of AChE, and all have similar properties. Other systems have used organophosphate hydrolase (OPH) that hydrolyses organophosphates instead of being inhibited by them, and has the advantage that, unlike cholinesterases, it does not require a regeneration step. Most systems measure pH changes associated with modification of the production of acetic acid produced by the uninhibited enzyme, and detect this either colorimetrically or amperometrically. Since the development of a flow-injection system for the analysis of paraoxon in surface water and soil extracts (Shi and Stein, 1996), work has continued to improve on sensitivity, stability and reliability. Over the last 15 years the stability and reproducibility have improved, but the sample throughput rate (around $5–6\,h^{-1}$) and the sensitivity (around 10^{-8} M) have not changed markedly. Most of the systems developed are at the proof of concept or prototype stage, and have been tested in the laboratory but are not available for field use. Similar progress has been made with biosensors for heavy metals and some of the key organic pollutants of concern (e.g. herbicides, fungicides, organotins, PAHs, PCBs, and benzene, toluene, ethylbenzene and xylene isomers (BTEX) components). A flow-injection, microbial whole-cell-based biosensor was developed (Han *et al*., 2002) for the detection of trichloroethylene in groundwater and drinking water could not be used in situ, but provided an alternative analytical method to that already in use. One group of researchers (Golden *et al*., 2005) worked to overcome many of the problems associated with the production of commercial systems that could be used in the laboratory or in field. They miniaturized a multiple array system based on immunosensors linked to a fluorescence detection system, and an automated robust multichannel microfluidic system. They produced a flexible, automated, portable multianalyte sensor that has potential for adaptation in a range of areas from biomedical to environmental. Major public investment over many years in Europe has resulted in the development of many potentially useful biosensors (Farré *et al*., 2009), and two optically linked multichannel immunosensor devices (RIANA and AWACSS) that have been used with success in environmental applications. The former has been used in river water and in drinking water treatment plants for the detection of pesticides. The latter has been used mainly in wastewater treatment plants (Rodriguez-Mozaz *et al*., 2006) and can measure up to 32 organic pollutants at low nanogram per litre concentrations simultaneously, at a rate of $4\,h^{-1}$ (including regeneration time), in real environmental samples, without sample pretreatment. Despite the demonstrated utility, these instruments have not yet become widely used.

Although a requirement for specific sensors for measuring individual pollutants has been stated by many workers, there remains a gap between the production of promising prototypes that will achieve this and the delivery of commercially available robust instruments. In part, this is due to the potentially small market for such devices and the high costs of the development to manufacturing stages, and of validation for in-situ environmental use. This contrasts with the successful development of commercial instruments based on biosensors for use in medical applications. An important environmental role for biosensors might be in the in-situ measurement of general water quality (measured as levels of nutrients that can lead to eutrophication, BOD), sub-lethal effects (e.g. endocrine disruption) or general toxicity of the complex mixtures of toxicants that tend to be present in environmental waters (Rodriguez-Mozaz *et al*., 2005). Another function where development of biosensors might be economically favourable is detection of toxins produced by dinoflagellates and algae that can cause commercially relevant damage that affects producers of some sea foods. Here, they could provide alarm systems that would be widely used, and would justify the high development costs.

There is a widespread requirement for measuring nutrients in the marine environment. Currently, most in-situ measurements of nitrate and nitrite in seawater are made by miniaturized wet chemical analysers and spectroscopic systems, as described above, but some biosensors have been developed, mainly for monitoring nutrients in wastewater systems, that may be useful in marine monitoring.

One mature biosensor technology (Damgaard *et al*., 1995; Larsen *et al*., 2000) is based on the use of immobilized bacteria that convert NO_2^- or NO_3^- to N_2O. The bacteria and a reserve of an electron donor are separated from the sample by an ion-permeable membrane that allows the substrates to diffuse into the sensor, and the bacteria are separated from the electrochemical sensor for N_2O by a gas-permeable membrane. This reduces potential interference to a very low level. A commercial version of this sensor is available (www.unisense.dk), and can be modified by using different strains of denitrifying bacteria to measure either NO_2^- and NO_3^-, or NO_2^-. This technology was originally developed for use in wastewater treatment plants, but has been demonstrated to work in seawater. Other types of biosensor have been developed for measuring NO_2^-. One based on immobilized nitrite reductase coupled to an amperometric detector and using a redox-active complex as transducer (Chen *et al*., 2007) was demonstrated to have a low level of detection (4 nM) and acceptable stability. Another (Zhang *et al*., 2009) based on cytochrome c nitrite reductase immobilized on a conductometric electrode surface with a stabilized electron mediator (methylviologen) was stable and had a limit of detection of $0.05\,\mu M$. Although both of these novel devices show great potential for use as in-situ sensors, much development work would be needed to develop validated commercial products that can be used for routine environmental monitoring.

The problems associated with the wide range of marine toxins, and the biosensors that have been developed to measure these, have been comprehensively reviewed by Campàs *et al.* (2007). Available methods for monitoring marine toxins include the mouse bioassay, cytotoxicity assays, receptor binding studies, immunoassays and enzyme inhibition studies. There would be significant advantages to be gained by the use of biosensors for primary screening rather than the currently used mouse bioassay. The latter is time consuming, nonspecific, variable between laboratories and expensive. Further, there are ethical concerns associated with the use of vertebrates in toxicity screening. Campàs *et al.* (2007) consider that, currently, biosensors should not be viewed as replacements for the available analytical methods but should be considered for use in routine screening. The detection limit for saxitoxin of a system based on sodium channels in cultured mammalian neurones was 30 000 times lower than that of the mouse bioassay. This system had a level of detection 300 times below the regulatory limit for brevetoxins. Immunosensors have been developed using a range of transduction methods (including piezoelectric, fibre optic, surface plasmon resonance and electrochemical). The electrochemical device has low limits of detection for okadaic acid, brevetoxin, domoic acid and tetrodotoxin (1.5–2 ng L^{-1} for the first three and 0.02 ng L^{-1} for the last), and has the potential for miniaturization (Campàs *et al.*, 2007). Enzyme-inhibition-based sensors have been incorporated into through-flow systems, and one has a limit of detection of 0.1 ng L^{-1} for okadaic acid.

Although highly desirable, and very sensitive for a range of marine toxins, none of the biosensors is available as a commercial instrument for use in food or environmental laboratories. However, devices that can be used to detect algal species that produce the toxins of concern have shown considerable promise and may be closer to commercial availability. Identification of harmful species using microscopical methods is difficult, and methods that can differentiate between harmful and nonharmful species without the need for long taxonomic training and experience would be very helpful. The biosensor methods for this purpose involve manual stages (isolation of RNA and manipulation of the hybridization steps) in the preparation of the sample for analysis. One multi-probe chip technology is based on the immobilization of DNA probes on a gold sensor (Diercks *et al.*, 2008). These systems have high sensitivity, a reasonable shelf life and the potential to be mass produced. However, there are still technical problems to be overcome in the development of a robust commercial device that can be used in the laboratory to monitor the composition of algal blooms in spot samples of seawater. LaGier *et al.* (2007) used commercially available hand-held instruments as the basis of their biosensor. The group developed electrochemical assays for *Karenia brevis* (a red tide dinoflagellate), a viral pathogen and a range of pathogenic bacterial species that are indicators of faecal pollution. The sensor for the dinoflagellate was demonstrated to be highly sensitive and selective, and to have a linear response over five orders of magnitude of concentration of DNA. The assay was streamlined by substituting a filter polymerase chain reaction (PCR) and crude lysate as a PCR template for the DNA extraction step, and still retained sufficient sensitivity, though there was a loss of linearity at higher cell concentrations. Trials using coastal water samples demonstrated the potential usefulness of the dinoflagellate sensor. However, as for the other classes of sensors, much development work is needed to develop a system that could be used in situ.

A fluorescence-based fibre-optic biosensor has been developed for determining in-situ real-time concentrations of trace metals (copper and zinc) in seawater (Zheng *et al.*, 2003, 2004). The recognition unit is a protein molecule that is site-specifically labelled with a fluorophore that is attached to the distal end of an optical fibre. This binds free copper(II) with a high affinity and selectivity. The transducer reports the metal's concentration as a change in fluorescence intensity or lifetime, using a frequency-domain approach. The sensor has a detection limit of 6 pg L^{-1} free copper(II) in seawater, while the accuracy and precision of the sensor are comparable to those of cathodic ligand exchange/adsorptive cathodic stripping voltammetry. However, these techniques have been mainly utilized in medical applications and are not available commercially.

There has been much disquiet concerning the presence of a wide range of compounds in the environment that can disrupt normal endocrine function, and work to develop biosensors for measuring endocrine disruptors (of sex and thyroid hormone systems) has been ongoing in a number of laboratories. Some biosensors have been targeted on steroids generally (testosterone, oestradiol, cholesterol) and sensitive biosensors have been developed (Xiong *et al*., 2009; Maser and Xiong, 2010). Other work has been directed to the measurement of particular xenoestrogens (in particular bisphenol A, which is commonly found in plastics), and there have been developments that could support the development of biosensors such as an immunosensor (Marchesini *et al*., 2005), oestrogen receptor-α linked to a field-effect transistor (Sánchez-Acevedo *et al*., 2009) and an amperometric biosensor based on a tyrosinase recognition unit (Yin *et al*., 2010). Some mechanisms, such as that based on the insertion of the human oestrogen receptor (Skretas and Wood, 2005) in a bacterial system, can be used to detect general oestrogenic activity and could be developed into a biosensor application. In this system, the oestrogen receptor can be replaced with the ligand-binding domain of the human thyroid receptor to enable ligands related to the thyroid hormones to be detected. A biosensor combining a quartz crystal microbalance the hormone-binding domain of the α-oestrogen receptor (Carmon *et al*., 2005) and one based on the combination of immobilization of that receptor with cyclic voltammetry and impedance spectroscopy (Xia *et al*., 2010) show promise for the direct detection of oestrogenic compounds. A biosensor based on a commercially available antibody was developed to detect sub-nanogram per litre concentrations of testosterone in environmental water samples, and was incorporated in to the RIANA array of biosensors (Tschmelak *et al*., 2006). With the exception of the latter development, most of the other systems are a long way from in-field or in-situ deployment for monitoring endocrine disruptors in environmental water.

Recent developments have produced potentially useful materials, aptamers (Song *et al*., 2008), and molecularly imprinted polymers (MIPs) for use in sensors. The former are truly biological and could be used in biosensors. The latter are not biological, but they could be used to produce chemical sensors with many of the desirable properties of biosensors (Rodriguez-Mozaz *et al*., 2005). Aptamers are nucleic acid sequences that can bind with high specificity to a wide range of target molecules (proteins, peptides, biochemicals and small molecules – including pharmaceuticals and pesticides). These have been immobilized to act as recognition units and linked to a range of transducers (including electrochemical, optical and mass-sensitive systems). Once isolated, aptamers can be produced reliably on a large scale. This, combined with their relative stability and specificity, makes them potentially useful for use in biosensors for both medical and environmental measurements. However, a large amount of development work is needed to produce a working biosensor that could be deployed in situ (Song *et al*., 2008). MIPs mimic the biological activity of antibodies and can be used as recognition units in sensors. They combine some of the desirable properties (specificity) of biological systems but are much more stable chemically, thermally and mechanically (Rodriguez-Mozaz *et al*., 2005). MIPs have been incorporated in biomimetic sensors for a range of analytes (Pichon and Chapuis-Hugon, 2008), including herbicides: haloacetic acids in drinking water (Suedee *et al*., 2004); paraoxon in spiked vegetable and water samples (Alizadeh, 2010); monoamine naphthalenes in drinking water (Valero-Navarro *et al*., 2009); explosives in natural waters and soil (Alizadeh *et al*., 2010); copper(II) ions in waste water (Yang and Zhang, 2009). These developments are exciting; however, the technology is still underdeveloped (Ávila *et al*., 2008) and it will be some time before these sensors become commercially available.

Currently, an array of laboratory-based bioassays is used to assess the overall toxicity of water samples. Biosensors offer the potential to replace the use of bioassays and to provide rapid, automated in-situ monitoring of water quality. Several strategies have been used to develop biosensor systems that can provide rapid, sensitive measurements of the toxicological quality of environmental waters. One approach is to use an array of whole-cell sensors that respond to a range of toxic insults and signal damage to different cellular systems. One such system was developed by Lee *et al.* (2005) and comprised a set of 20 recombinant bioluminescent bacteria with different promoters fused with the bacterial *lux* genes. The bacteria were immobilized in a 384-well microplate, and on chips, and luminescence detected using a cooled CCD camera. The analysis time was 2 h. Different toxicants produced different patterns of response, depending on mode of action, amongst the strains of bacteria, and this has the potential for identifying the presence of particular types of pollutant in environmental samples. A similar but simpler approach using an array of five different whole-cell bacterial sensors (three metabolic and two catabolic) was used in the field to survey the quality of groundwater in an area with a long-term exposure to chlorinated solvents, and to evaluate remediation strategies (Bhattacharyya *et al.*, 2005). The different strains of bacteria gave different responses that could be interpreted in terms of the accompanying chemical analysis. A sensitive flow-through fibre-optic sensor based on two bacterial strains, one responding to DNA damage and the other to the production of heat-shock protein, has been demonstrated to work in flowing tap water, with a response time of 1 h, over a 24 h period (Eltzov *et al.*, 2009). However, the response became attenuated after 20 h in river water, and this problem will need to be addressed before this prototype can be developed into a commercial device. A novel approach that incorporates sulfur-oxidizing bacteria into a biosensor has shown some promise (Oh *et al.*, 2010). These bacteria oxidize sulfur to produce sulfuric acid, and this causes changes in both electrical conductivity and pH that can be measured. The system was tested using artificial stream water and was stable over a period of 30 days; when the water was spiked with a range of concentrations of chromium(VI), the pH increased and electrical conductivity decreased in a dose-dependent manner. The sensitivity was far greater than other bacteria-based sensors, and $5\,\mu g\,L^{-1}$ of chromium(VI) was detected in 1 h, and a median inhibitory concentration (IC_{50}) of $5\,\mu g\,L^{-1}$ was measured after 10 h of exposure. There is still much work necessary to develop this proof of concept into a commercial device. An automated system developed for use in estuarine environments is based on fusion of a cyanobacterium to an amperometric gas-diffusion electrode that measures changes in the production of photosynthetic oxygen (Campanella *et al.*, 2000). This responded in a dose-related manner to a range of pollutants, including mercury, copper, atrazine and carbaryl. The working life of the sensor was around 2 days. A multivariate chemometric approach to analysing the responses of an array of eight screen-printed platinum electrodes with four modified by the immobilization of four different enzymes (tyrosinase, horse radish peroxidase, acetyl choline esterase and butyryl cholinestesterase) demonstrated the potential of this approach (Tønning *et al.*, 2005). The data from the sensor array were analysed alongside standard toxicological assays (Microtox®, *Pseudokirchneriella subcapita* (a freshwater alga) growth inhibition and *Daphnia magna* 48 h immobilization) and measures of water quality (BOD, COD, TOC, inhibition of nitrification and inhibition of respiration). The authors demonstrated that the array could differentiate between wastewater samples from four different sites, though it was necessary to use reference sample solutions to account for drift in response of the system over time.

The use of biosensors for the measurement of general toxicity is a desirable goal, and would enhance significantly the ability of regulators to monitor water quality in a way that would provide early warnings of problems, and timely information on the results of remediation actions. It would also allow a better regulation of discharges than is possible with infrequent spot sampling and chemical analysis for a limited number of compounds of interest that provides no information on the interactions that can occur between toxicants. However, there are currently no biosensor technologies available that could be applied in situ in the marine environment.

10.6 BIOLOGICAL EARLY-WARNING SYSTEMS

In contrast with most of the devices discussed in this chapter, biological early-warning systems (BEWSs) are used to detect events like (accidental) spills, not to detect the average condition of the aquatic environment. They make use of living organisms to detect incidental pollution. Like the canaries used in coal mines to warn (by dying) miners of the presence of toxic and potentially explosive methane gas, living organisms have been used since the early 1970s as sensors to detect pollution in water bodies. Organisms can 'smell' the rapid increase in the occurrence of toxic chemicals. They react swiftly, so that an alarm can be generated to warn of, for example, an accidental spill.

The rationale behind their use to protect drinking water intakes, for example, was that routine monitoring programmes in river/sea water chemically analyse only a small part of the large number of chemicals that can potentially enter the aquatic environment, either by discharge or accidental spill. Organisms may detect, sense, many more compounds than those routinely analysed chemically. Although they will certainly not detect all potentially dangerous compounds, aquatic organisms can be used as wide-spectrum sensors. In addition, where most chemical monitoring is based on spot samples collected at weekly intervals (or even lower frequencies), organisms work around the clock, seven days per week. When an alarm is generated by the BEWS, it may be used to trigger automated sampling so that chemical analyses can be used to find evidence of the causative compound(s), though it is not always possible to define the cause of even a clear biological effect. With the exception of some highly specific biosensors (Kröger *et al*., 2002) described above, organisms in a BEWS are not able to define the nature of the chemical(s). A BEWS is a semi-quantitative sensor, not a qualitative detector.

In order to obtain the necessary early warning, it is necessary to use organisms that can give a rapid biological response (alarm) to a relatively steep increase in the concentration of toxic pollutants within a period of minutes to hours. Unlike bioaccumulation studies such as the 'Mussel watch', originally proposed by Goldberg (1975), where it may take several weeks to reach equilibrium between water and tissue concentration, the response needs to be fast and the biological functions monitored need to change rapidly following changes in the environment such that there is a functional and preferably quantifiable reaction to toxic chemicals in the water column. In practice, this means the detection of changes in either behaviour or in physiological parameters. Behavioural responses that can be used in BEWSs include: activity, locomotory behaviour, avoidance, positive rheotaxis and escape behaviour (Diamond *et al*., 1988; Kramer and Botterweg, 1991; Gerhardt *et al*., 2006). Physiological responses potentially include respiration rate, ventilation frequency (fish), heart rate, pumping rate, bio-electric potential, photosynthesis and growth rate, bio-luminescence, ion flux, blood chemistry and haematology, and death (Kramer and Botterweg, 1991; Handy and Depledge, 1999). Irrespective of which biological function is used, BEWSs are based on the detection of an 'abnormal' situation. The current behaviour/physiological status will always be compared with a past status, either by using a calibration curve or threshold, or (more commonly) from historic recording of the organism's own functioning (e.g. of 1 h before). Comparison with a control system is useful in the laboratory, but cumbersome in a field situation. Once an abnormal situation is detected with sufficient confidence, an alarm is generated by the system.

Requirements of a successful BEWS are a combination and integration of a suitable (preferably endemic) test organism, a method of detection of changes in physiology or behaviour and a means of signal processing, data evaluation and alarm generation. It will be obvious that a biological system can be used only when the conditions of the ambient water do not prevent normal life of the organism. Therefore, the temperature, oxygen content, pH, salt content and the background toxicity should be appropriate for the organisms used. Further, the successful BEWS should be easy to operate, the organisms easily obtained and the system reliable and capable of working for weeks to months with a minimum of false alarms (Baldwin and Kramer, 1994). The system must be robust and fully validated for field application. Too often a monitoring system works well in the laboratory, but fails in environmental application (Gruber *et al*., 1994; von Danwitz *et al*., 1998). Finally, costs, notably of maintenance (e.g. frequency of replacement of organisms), will also determine its success.

Most commercial BEWSs that have passed the laboratory phase and are implemented in the field have been developed for monitoring freshwater/drinking water (Kramer and Botterweg, 1991; Gerhardt *et al*., 2006; Kramer, 2009). Typical organisms and associated alarm responses used in practical applications include several types of fish (ventilation frequency, locomotory behaviour and positive rheotaxis), bivalves (valve movement), small invertebrates (e.g. *Corophium* sp. (locomotory behaviour)), algae (fluorescence) and bacteria (various physiological parameters, including luminescence). Typical applications have been discussed by Kramer (2009) for several case studies.

Application in the marine environment is hampered by the limitation that many of the organisms most commonly used in BEWSs cannot survive in seawater. Without aiming to provide a complete account, several systems have been demonstrated to work successfully in the marine environment; others may be easily adapted for use with marine species.

In the Musselmonitor® (Aquadect, Brouwershaven, NL), the valve movement of eight bivalves is used (Kramer and Foekema, 2001) as an indicator of water quality. For each individual, equipped with a noninvasive electronic sensor on either shell, the present valve movement is compared with its past behaviour. Prolonged closure of the shells is a clear indication of adverse environmental conditions (escape behaviour). Freshwater applications often use the zebra mussel (*Dreissena polymorpha*) for the monitoring of surface or drinking water, but marine bivalve species, like the common mussel (*Mytilus edulis*) and oysters (*Crassostrea* sp.), have been successfully used in field applications (Kramer, 2009). The monitoring system, also available for in-situ operation, can be used with mussels, clams and oysters.

The measurement system of the MFB Multispecies Freshwater Biomonitor (LimCo International, Ibbenbüren, DE) is based on a magnetic induction method. Test chambers are flow-through cells with the two electrode pairs placed on the chamber walls. Movements of the animals will change the conductivity and the electrical field between a second pair of electrodes and, thus, generate specific electrical signals indicating typical behaviour (e.g. swimming/locomotion, ventilation, inactivity). A cell with electrode pairs but without organisms acts as control. Despite its name, the MFB system equally operates in freshwater and marine environments (Gerhardt and Clostermann, 1998). Each cell may contain one or more organisms; even multispecies operation is an option. Applications in the marine environment include the use of the crustacean mud shrimp (*Corophium volutator*) as test organism (Kirkpatrick *et al*., 2006).

Although no practical applications are known to us, there is no reason why BEWSs operating with fish may not work in the marine environment. Selected marine fish can replace the normally used freshwater species in monitoring ventilatory behaviour (ASTM, 1995); for example, in systems like the BioSensor® (Biological Monitoring Inc., Blacksburg VA, US), an early warning system with eight fish in flow-through tanks.

In the Algae Toximeter® (bbe Moldaenke, Kiel-Kronshagen, DE) system the concentration and activity of the naturally present algae are monitored. If a toxic compound is present, its interaction with the photosynthesis centre leads to an inhibition of algal activity. The freshwater species (e.g. *Chlorella* sp.) may be replaced by suitable marine species.

Luminescent bacteria have been used in ecotoxicological studies (Microtox®; Bulich, 1979), initially using *Photobacterium phosphoreum*, a marine species but almost entirely used in freshwater/effluent studies (where addition of salt is required). The TOXcontrol® system (microLAN, Waalwijk, NL) meets the requirements of automated and standalone operation, but marine applications have not been reported so far. The test uses the reduction of luminescent light emitted by the bacteria (*Vibrio fischeri*) as the indicator of toxic effects. This luminescent bacteria monitor is an automated version of the procedure as defined by the ISO 11 348-1 standard (ISO, 2007). There seem no limitations to convert its operation for use in the marine environment.

Since on-line BEWSs may be considered as 'sensors', one might expect that the ISO 15 839 standard on on-line sensors/analysing equipment for water (ISO, 2003) is valid for BEWSs. Unfortunately, this is true only to a limited extent. Unlike most sensors that are the target of ISO 15 839, BEWSs cannot identify individual compounds. Of course, they show a response to one compound in a toxicity test, but the objective of a BEWS is operation in the natural environment and, consequently, they must respond to a mixture of compounds. This makes validation less easy. It is argued that there should be validation protocols for BEWSs that differ from those for the more chemical-analysis-oriented sensors.

For some BEWSs a few guidelines have been developed and validation protocols developed (Kramer, 2009). These guidance documents detail the necessary experiments and provide a good basis for the validation of BEWSs in monitoring practice (e.g. Wagenvoort *et al.*, 2008). For spot sampling and subsequent laboratory analysis, compounds can both be identified and quantified. In contrast, BEWSs can in principle only be semi-quantitative and offer no identification. Kramer (2009) argued that validation methods that are tuned to the test system principles and take into account the level of identification and quantification should be developed.

10.7 FUTURE

As the importance of a healthy marine environment has become more widely recognized, and legislation is being extended to provide protection, the need for reliable and widespread monitoring becomes more urgent. The logistic problems associated with representative sampling over wide areas preclude the use of strategies based on spot sampling that were developed for monitoring freshwater. New strategies need to be developed, and although the task is daunting, significant progress has been made in devising methods that can be used to collect information remotely. However, there are still barriers to be overcome before long-term remote deployment can become routine. Whilst many successful deployment devices (including Argo floats, landers, auto-submarine vehicles and smart buoys) and associated communication systems are available, the range of routine measuring devices to use with them is relatively limited.

Biofouling of instruments and optical and sensor surfaces is amongst the problems to be solved. This is a challenge in all aquatic environments, but is particularly marked in nutrient-rich coastal waters. Approaches that have been used include mechanical scrapers, nontoxic (e.g. silicone greases) and toxic (e.g. tributyltin) barriers, and the use of copper coatings, meshes and shutters. Other solutions are being sought in the use of coatings that peel away over a deployment period and biomimetic surfaces (e.g. peptide–peptoid conjugates and phosphorylcholine-based polymers) that were initially developed for medical applications.

Operational field lifetimes of remotely deployed instruments are currently limited by the power consumption of instruments and the available power supplies. Cabled observatories can provide the power to operate sensor networks for extended periods; however, the establishment of the required infrastructure is expensive and, by necessity, limited in scope. Many in-situ sensors still rely on commercially available batteries. The use of renewable energy systems that can utilize naturally occurring chemicals and conditions in the water body are potential options that hold some promise for the future. These include use of methane hydrate fuel cells, microbial fuel cells, sea-surface photovoltaic cells and motion-to-electricity conversion techniques.

Recent developments in microfabrication, microfluidics and integrated optics in many areas of analytical chemistry are being applied in the development of in-situ monitoring devices for the marine environment. Lab-on-a-chip technologies have significant advantages in terms of a small size and limited reagent and power requirements. The challenge is to ensure that these systems can attain the sensitivity needed for many marine applications where analytes are present at only trace concentrations. However, as the overall cost of the sensing system is reduced, this would potentially enable the deployment of larger numbers of devices and thereby improve the spatial and temporal resolution and extent of off shore monitoring activities.

Currently, there is a lack of recognized validation protocols for many of the methods described in this chapter, and a lack of support networks for QA and QC. The commercial development and wide acceptance of in-situ technologies will depend on the availability of systems such as those available for classical, laboratory-based instrumental analytical procedures. There is a need for innovations in the development of suitable reference materials and interlaboratory trials.

Most of the devices with the potential for use in the in-situ monitoring of marine waters were developed for use in the laboratory, where they could serve as less expensive, more rapid and more sensitive replacements for conventional analytical methods. They could enable frequent screening of water quality to avoid the problems associated with infrequent regulatory sampling. The more expensive regulatory measurements could then be used only when required. However, much development and validation work is necessary to reach this goal. The long-term aim of using these technologies for in-situ monitoring in the marine environment is even further away. Although many devices have been developed to proof of concept or prototype stage, the development work to convert these into robust, reliable, validated instruments that can be deployed in the marine environment for prolonged periods (weeks to months) has not been undertaken. This may be a consequence of the short-term nature of research funding and the failure to provide associated support for the development work necessary to take advantage of exciting research innovations. There have been many lost opportunities by failure to capitalize on significant research developments.

REFERENCES

Achterberg E. P. and Van den Berg C. M. G. (1994). *Anal. Chim. Acta*, **284**, 463–471.

Alizadeh T. (2010). *Thin Solid Films*, **518**, 6099–6106.

Alizadeh T., Zare M., Ganjali M. R. *et al*. (2010). *Biosens. Bioelectron*., **25**, 1166–1172.

Aminot A., Kérouel R. and Birot D. (2001). *Water Res*., **35**, 1777–1785.

ASTM (1995). *ASTM Standard E1768-95. Guide for Ventilatory Behavioral Toxicology Testing of Freshwater Fish*, American Association for Testing of Materials, West Conshohocken, PA, 14 pp. (re-approved 2003).

Ávila M., Zougagh M. and Ríos Á. (2008). *TrAC, Trends Anal. Chem*., **27**, 54–65.

Baldwin I. G. and Kramer K. J. M. (1994). Biological early warning systems (BEWS), in *Biomonitoring of Coastal Waters and Estuaries* (ed. K. J. M. Kramer), CRC Press, Boca Raton, FL, pp. 1–28.

Bell R. J., Short R. T., van Ameron F. H. and Byrne R. H. (2007). *Environ. Sci. Technol*., **41**, 8123–8128.

Bellerby R. G. J., Turner D. R., Millward G. E. and Worsfold P. J. (1995). *Anal. Chim. Acta*, **309**, 259–270.

Bellerby R. G. J., Olsen A., Johannessen T. and Croot P. (2002). *Talanta*, **56**, 61–69.

Belmont-Hebert C., Tercier M. L., Buffle J. *et al*. (1998). *Anal. Chem*., **70**, 2949–2956.

Bhattacharyya J., Read D., Amos S. *et al*. (2005). *Environ. Pollut*., **134**, 485–492.

Braungardt C. B., Achterberg E. P., Axelsson B. *et al*. (2009). *Mar. Chem*., **114**, 47–55.

Brendel P. J. and Luther G. W. (1995). *Environ. Sci. Technol*., **29**, 751–761.

Brewer P. G., Malby G., Pasteris J. D. *et al*. (2004). *Deep-Sea Res. I*, **51**, 739–753.

Buffle J. and Horvai G. (eds), (2000). *In Situ Monitoring of Aquatic Systems: Chemical Analysis and Speciation*, John Wiley and Sons, Ltd., Chichester

Buffle J. and Tercier-Waeber M. L. (2000). In situ monitoring of aquatic systems: Chemical analysis and speciation, in *In Situ Monitoring of Aquatic Systems: Chemical Analysis and Speciation* (eds J. Buffle and G. Horvai), John Wiley and Sons, Ltd., Chichester, pp. 279–405.

Buffle J. and Tercier-Waeber M. L. (2005). *TrAC, Trends Anal. Chem*., **24**, 172–191.

Bulich A. A. (1979). Use of luminescent bacteria for determining toxicity in aquatic environments, in *Aquatic Toxicology* (eds L. L. Marking and R. A. Kimerle), ASTM STP 667, pp. 98–106.

Cai W. J. and Reimers C. E. (1993). *Limnol Oceanogr*., **38**, 1762–1773.
Cai W. J. and Reimers C. E. (2000). In Situ monitoring of aquatic systems: chemical analysis and speciation, in *In Situ Monitoring of Aquatic Systems: Chemical Analysis and Speciation* (eds J. Buffle and G. Horvai), John Wiley and Sons, Ltd., Chichester, pp. 75–119.
Camilli R. and Hemond H. F. (2004). *TrAC, Trends Anal. Chem*., **23**, 307–313.
Campanella L., Cubadda F., Sammartino M. P. and Saoncella A. (2000). *Water Res*., **35**, 69–76.
Campàs M., Prieto-Simón B. and Marty J-L. (2007). *Talanta*, **72**, 884–895.
Campbell P. G. C., Errecalde O., Fortin C. *et al*. (2002). *Compar. Biochem. Physiol. C – Toxicol. Pharmacol*., **133**, 189–206.
Carmon K. S., Baltus R. E. and Luck L. A. (2005). *Anal. Biochem*., **345**, 277–283.
Chen H., Mousty C., Cosnier S. *et al*. (2007). *Electrochem. Commun*., **9**, 2240–2245.
Chiappini S. A., Kormes D. J., Bonetto M. C. *et al*. (2010). *Sens. Actuators B: Chem*., **148**, 103–109. DOI: 10.1016/j.snb.2010.04.039.
Cooper, D. J., Watson, A. J., and Ling, R. D. (1998). *Mar. Chem*., **60**, 147–164.
Dable B. K., Love B. A., Battaglia T. N. *et al*. (2006). *Appl. Spectros*., **60**, 773–780.
Daly K. L., Byne R. H., Dickson A. G. *et al*. (2004). *Mar. Technol. Soc. J*., **38**, 121–143.
Damgaard L. R., Larsen L. H. and Revsbech N. P. (1995). *TrAC, Trends Anal. Chem*., **14**, 300–303.
Daniel A., Birot D., Lehaitre M. and Poncin J. (1995). *Anal. Chim. Acta*, **308**, 413–424.
De Vitre R. R., Tercier M. L. and Buffle J. (1991). *Anal. Proc*., **28**, 74–75.
DeGrandpre M. D., Hammar T. R., Wallace D. W. R. and Wirick C. D. (1997). *Limnol. Oceanogr*., **42**, 21–28.
Denuault G. (2009). *Ocean Sci*., **5**, 697–710.
Diamond J., Collins M. and Gruber D. (1988). An overview of automated biomonitoring – past developments and future needs, in *Automated Biomonitoring: Living Sensors as Environmental Monitors* (eds D. S. Gruber and J. M. Diamond), Ellis Horwood, Chichester, pp. 23–39.
Diercks S., Metfies K. and Medlin L. K. (2008). *Biosens. Bioelectron*., **23**, 1527–1533.
Eltzov E., Marks R. S., Voosr S. *et al*. (2009). *Sens. Actuators B: Chem*., **142**, 11–18.
Farré M. and Barceló D. (2009). Bioassays and biosensors, in *Rapid Chemical and Biological Techniques for Water Monitoring* (eds C. Gonzalez, R. Greenwood and Ph. Quevauviller), John Wiley and Sons, Ltd., Chichester, pp. 125–156.
Farré M., Kantiani L., Pérez S. and Barceló D. (2009). *TrAC, Trends Anal. Chem*., **28**, 170–185.
Floch J., Blain, Birot D. and Treguer P. (1998). *Anal. Chim. Acta*, **377**, 157–166.
Florence T. M. and Batley G. E. (1977). *J. Electroanal. Chem*., **75**, 791–798.
Gerhardt A. and Clostermann M. A. (1998). *Environ. Int*., **24**, 699–701.
Gerhardt A., Ingram M. K., Kang I. J. and Ulitzur S. (2006). *Environ. Toxicol. Chem*., **25**, 2263–2271.
Glazer B. T., Luther G. W., Konovalov S. K. *et al*. (2006). *Deep-Sea Res. Pt II – Top. Stud. Oceanogr*., **53**, 1740–1755.
Glud R. N., Gundersen J. K. and Ramsing N. B. (2000). Electrochemical and optical oxygen microsensors for in situ measurements, in *In Situ Monitoring of Aquatic Systems: Chemical Analysis and Speciation* (eds J. Buffle and G. Horvai), John Wiley and Sons, Ltd., Chichester, pp. 19–73.
Goldberg E. D. (1975). *Mar. Pollut. Bull*., **6**, 111.
Golden J. P., Taitt C. R., Shriver-Lake L. C. *et al*. (2005). *Talanta*, **65**, 1078–1085.
Gruber D., Frago C. H. and Rasnake W. J. (1994). *J. Aquat. Ecosyst. Health*, **3**, 87–92.
Han T.-S., Sasaki S., Yano K. *et al*. (2002). *Talanta*, **57**, 271–276.
Handy R. D. and Depledge M. H. (1999). *Ecotoxicol*., **8**, 329–349.
Hardman-Mountford N. J., Moore G., Bakker D. C. E. *et al*. (2008). *J. Mar. Sci*., **65**, 1498–1503.
Hemond H. and Camilli R. (2002). *TrAC, Trends Anal. Chem*., **21**, 526–533.
Huang H. L. and Dasgupta, P. K. (1999). *Anal.Chim. Acta*, **380**, 27–37.
ISO (2003). *ISO 15839:2003. Water quality – On-line sensors/analysing equipment for water – Specifications and performance tests*, International Organisation for Standardisation, Geneva, Switzerland, p. 30.
ISO (2007). *ISO 11348 1–3:2007. Water quality – Determination of the inhibitory effect of water samples on the light emission of Vibrio fischeri (Luminescent bacteria test) – Part 1: Method using freshly prepared bacteria. Part 2: Method using liquid-dried bacteria. Part 3: Method using freeze-dried bacteria*., International Organisation for Standardisation, Geneva, Switzerland, pp. 21–23.
Johnson K. S. and Coletti L. J. (2002). *Deep-Sea Res. I*, **49**, 1291–1305.

Johnson K. S., Coletti L. J. and Chavez F. P. (2006). *Deep-Sea Res. I*, **53**, 561–573.
Kibelka G. P. G., Short R. T., Toler S. K. *et al*. (2004). *Talanta*, **64**, 961–969.
Kirkpatrick A. J., Gerhardt A., Dick J. T. A. *et al*. (2006). *Ecotoxicol. Environ. Saf*. **64**, 298–303.
Kraft M. and Mizaikoff B. (2000). *Int. J. Environ. Anal. Chem*., **78**, 367–383.
Kraft M., Jakusch M., Karlowatz M. *et al*. (2003). *Appl. Spectrosc*., **57**, 591–599.
Kramer, K. J. M. (2009). Continuous monitoring of waters by biological early warning systems, in *Rapid Chemical and Biological Techniques for Water Monitoring* (eds C. Gonzalez, R. Greenwood and Ph. Quevauviller), John Wiley and Sons, Ltd., Chichester, pp. 197–219.
Kramer K. J. M. and Botterweg J. (1991). Aquatic biological early warning systems: an overview, in *Bioindicators and Environmental Management* (eds D. W. Jeffrey and B. Madden), Academic Press, London, pp. 95–126.
Kramer K. J. M. and Foekema E. M. (2001). The 'Musselmonitor®' as biological early warning system: the first decade, in *Biomonitors and Biomarkers as Indicators of Environmental Change: A Handbook*, Vol II (eds F. M. Butterworth, M. E. Gonsebatt-Bonaparte and A. Gunatilaka), Kluwer/Plenum, New York, pp. 59–87.
Kröger S., Piletsky S. and Turner A. P. (2002). *Mar. Pollut. Bull*., **45**, 24–34.
Lacombe M., Garçon V., Thouron D. *et al*. (2008). *Talanta*, **77**, 744–750.
LaGier M. J., Fell J. W. and Goodwin K. D. (2007). *Mar. Pollut. Bull*., **54**, 757–770.
Larsen L. H., Damgaard L. R., Kjær T. *et al*. (2000). *Water Res*., **34**, 2463–2468.
Le Bris N., Sarradin P.-M., Birot D. and Alayse-Danet A.-M. (2000). *Mar. Chem*., **72**, 1–15.
Lee J. H., Mitchell R. J., Kim B. C. *et al*. (2005). *Biosens. Bioelectron*., **21**, 500–507.
Lee W. H., Seo Y. and Bishop P. L. (2009). *Actuators B: Chem*., **137**, 121–128.
Lewis B. L., Glazer B. T., Montbriand P. J. *et al*. (2007). *Mar. Chem*., **105**, 296–308.
Luther G. W., Brendel P. J., Lewis B. L. *et al*. (1998). *Limnol. Oceanogr*., **43**, 325–333.
Luther G. W., Reimers C. E., Nuzzio D. B. and Lovalvo D. (1999). *Environ. Sci. Technol*., **33**, 4352–4356.
Luther G. W., Rozan T. F., Taillefert M. *et al*. (2001). *Nature*, **410**, 813–816.
Luther G. W., Glazer B. T., Ma S. F. *et al*. (2003). *Aquat. Geochem*., **9**, 87–110.
Luther G. W., Glazer B. T., Ma S. F. *et al*. (2008). *Mar. Chem*., **108**, 221–235.
Marchesini G. R., Meulenberg E., Haasnoot W. and Irth H. (2005). *Anal. Chim. Acta*, **528**, 37–45.
Maser E. and Xiong G. (2010). *J. Steroid Biochem. Mol. Biol*., **121**, 633–640.
Michel A. P., Lawrence-Snyder M., Angel S. M. and Chave A. D. (2007). *Appl. Opt*., **46**, 2507–2515.
Mizaikoff, B. (2003). *Anal. Chem*., **75**, 258A–267A.
Moore T. S., Nuzzio D. B., Deering T. W. *et al*. (2007). *Electroanalysis*, **19**, 2110–2116.
Moore T. S., Mullaugh K. M., Holyoke R. R. *et al*. (2009). *Annu. Rev. Mar. Sci*., **1**, 91–115.
Noel S., Tercier-Waeber M. L., Lin L. *et al*. (2006). *Electroanalysis*, **18**, 2061–2069.
Nriagu, J. O. (1990). *Environment*, **32**, 6–9.
Nuzzio D. B., Taillefert M., Cary S. C. *et al*. (2002). In Situ voltammetry at deep-sea hydrothermal vents, in *Environmental Electrochemistry: Analyses of Trace Element Biogeochemistry* (eds M. Taillefert and T. F. Rozan), American Chemical Society, Washington, DC, pp. 40–53.
Oh S. E., Hassan S. H. A. and Van Ginkel S. W. (2010). *Sens. Actuators B: Chem*., **154**, 17–21.
Okamura K., Kimoto H., Saeki K. *et al*. (2001). *Mar. Chem*. **76**, 17–26.
Owens W. B. and Wong A. P. S. (2009). *Deep-Sea Res*., **56**, 450–457.
Pichon V. and Chapuis-Hugon F. (2008). *Anal. Chim Acta*, **622**, 48–61.
Prien R. D. (2007). *Mar. Chem*., **107**, 422–432.
Reimers C. E. (2007). *Chem. Rev*., **107**, 590–600.
Revsbech N. P., Sorensen J., Blackburn T. H. and Lomholt J. P. (1980). *Limnol. Oceanogr*., **25**, 403–411.
Revsbech N. P., Jorgensen B. B., Blackburn T. H. and Cohen Y. (1983). *Limnol. Oceanogr*., **28**, 1062–1074.
Revsbech N. P., Nielsen L. P., Christensen P. B. and Sorensen J. (1988). *Appl. Environ. Microbiol*., **54**, 2245–2249.
Rodriguez-Mozaz S., López de Alda M. and Barceló D. (2006). *Talanta*, **69**, 377–384
Rodriguez-Mozaz S., López de Alda M., Marco M.-P. and Barceló D. (2005). *Talanta*, **65**, 291–297.
Roig B., Bazin I., Bayle S. *et al*. (2009). Biomolecular recognition systems for water monitoring, in *Rapid Chemical and Biological Techniques for Water Monitoring* (eds C. Gonzalez, R. Greenwood and Ph. Quevauviller), John Wiley and Sons, Ltd., Chichester, pp. 175–195.

Ron E. Z. (2007). *Curr. Opin. Biotechnol.*, **18**, 252–256.
Sabine C. L., Feely R. A., Gruber N. *et al.* (2004). *Science*, **305**, 367–371.
Sakamoto C. M., Johnson K. S. and Beecher, C. L. (1986). *Limnol. Oceanogr.*, **31**, 894–900.
Sánchez-Acevedo Z., Riu J. and Rius F. X. (2009). *Biosens. Bioelectron.*, **24**, 2842–2846.
Schimdt H., Ha N. B., Pfannkuche J. *et al.* (2004). *Mar. Pollut. Bull.*, **29**, 229–234.
Schlüter M. and Gentz T. (2008). *J. Am. Soc. Mass Spectrom.*, **19**, 1395–1402.
Schroeder C. R., Neurauter G. and Klimant I. (2007). *Microchim. Acta*, **158**, 205–218.
Shi R. and Stein K. (1996). *Anal. Chim Acta.*, **324**, 21–27.
Short R. T., Fries D. P., Kerr M. L. *et al.* (2001). *J. Am. Soc. Mass Spectrom.*, **12**, 676–682.
Short R. T., Toler S. K., Kibelka G. P. G. *et al.* (2006). *Trends Anal. Chem.*, **25**, 637–646.
Sieben V. J., Floquet C. F. A., Ogilvie I. R. G. *et al.* (2010). *Anal. Meth.*, **2.**, 484–491.
Sipos L., Valenta P., Nurnberg H. W. and Branica M. (1977). *J. Electroanal. Chem.*, **77**, 263–266.
Skretas G. and Wood D. W. (2005). *J. Mol. Biol.*, **349**, 464–474.
Song S., Wang L., Li J. *et al.* (2008). *Trends Anal. Chem.*, **27**, 108–117.
Sotomayor M. P. D. T., Raimundo Jr I. M., Rohwedder J. J. R. and Oliveira Neto G. (2010). *Ecl. Quím., São Paulo*, **35**, 33–43.
Suedee R., Srichana T., Sanpagai C. *et al.* (2004). *Anal. Chim Acta.*, **504**, 89–100.
Sundby B., Caetano M., Vale C. *et al.* (2005). *Environ. Sci. Technol.*, **39**, 2080–2086.
Taillefert M., Luther G. W. and Nuzzio D. B. (2000). *Electroanalysis*, **12**, 401–412.
Tedetti M., Guigue C. and Goutx M. (2010). *Mar. Pollut. Bull.*, **60**, 350–362.
Tercier, M. L. and Buffle J. (1996). *Anal. Chem.*, **68**, 3670–3678.
Tercier M. L., Parthasarathy N. and Buffle J. (1995). *Electroanalysis*, **7**, 55–63.
Tercier M. L., Buffle J. and Graziottin F. (1998). *Electroanalysis*, **10**, 355–363.
Tercier-Waeber M. L. and Taillefert M. (2008). *J Environ. Monit.*, **10**, 30–54.
Tercier-Waeber M. L., Pei J., Buffle J. *et al.* (2000). *Electroanalysis*, **12**, 27–34.
Tercier-Waeber M. L., Confalonieri F., Riccardi G. *et al.* (2005). *Mar. Chem.*, **97**, 216–235.
Tercier-Waeber M. L., Confalonieri F., Koudelka-Hep M. *et al.* (2008). *Electroanalysis*, **20**, 240–258.
The Royal Society (2005). *Ocean Acidification Due to Increasing Atmospheric Carbon Dioxide*. Policy Document 12/05, June, 60 pp.
Thouron D., Vuillemin R., Philippon X. *et al.* (2003). *Anal. Chem.*, **75**, 2601–2609.
Tønning E., Sapelnikova S., Christensen J. *et al.* (2005). *Biosens. Bioelectron.*, **21**, 608–617.
Tschmelak J., Kumpf M., Käppel N. *et al.* (2006). *Talanta*, **69**, 343–350.
Valero-Navarro A., Salinas-Castillo A., Fernández-Sánchez J. F. *et al.* (2009). *Biosens. Bioelectron.*, **24**, 2305–2311.
Van den Berg C. M. G. (1984). *Mar. Chem.*, **15**, 1–18.
Van den Broeke J., Langergraber G. and Weingarter A. (2006). *Spectrosc. Europe*, **18**, 1–4.
Varney M. S. (ed.) (2000). *Chemical Sensors in Oceanography*, Francis & Taylor, Abingdon, pp. 1–348.
Von Danwitz B., Blübaum-Gronau E., Diehl P. *et al.* 1998. Recommendation on the deployment of continuous biomonitors for the monitoring of surface waters, Working Group of the Federal States on Water Problems (LAWA), Berlin, Germany, 46 pp.
Vuillemin R., Le Roux D., Dorval P. *et al.* (2009). *Deep-Sea Res. I*, **56**, 1391–1399.
Wagenvoort A., de Hoogh C., Kramer K. *et al.* (2008). Standardisation, quality assurance and data evaluation of on-line biological alarm systems. 3, Mosselmonitor® (Delta Consult, Kapelle, NL). KWR report BTO 2008.040, Kiwa Water Research, Nieuwegein, Netherlands, 51 pp.
Wallace W. J. (1974). *The Development of the Chlorinity/Salinity Concept in Oceanography*, Elsevier, Amsterdam, 227 pp.
Wangersky P. J. (2005). *Sci. Mar.*, **69**, 75–84.
Whitworth D. J., Achterberg E. P., Nimmo M. and Worsfold P. J. (1998). *Anal. Chim. Acta*, **371**, 235–246.
Xia W., Li Y., Wan Y. *et al.* (2010). *Biosens. Bioelectron.*, **25**, 2253–2258.
Xiong G., Luo Y., Jin S. and Maser E. (2009). *Chemico-Biol. Interact.*, **178**, 215–220.
Xu K., Dexter S. C. and Luther G. W. (1997). *Corrosion*, **300**, 1–18.
Yang Z.-P. and Zhang C.-J. (2009). *Sens. Actuators B: Chem.*, **142**, 210–215.
Yin H., Zhou Y., Xu J. *et al.* (2010). *Anal. Chim Acta.*, **659**, 144–150.
Zaouak O., Authier L., Cugnet C. *et al.* (2009). *Electroanalysis*, **21**, 689–695.
Zaouak O., Authier L., Cugnet C. *et al.* (2010). *Electroanalysis*, **22**, 1159–1165.

Zhang Z., Xia S., Leonard D. *et al*. 2009. *Biosens. Bioelectron*., **24**, 1574–1579.
Zheng H. H., Thompson R. B., Maliwal B. P. *et al*. (2003). *Anal. Chem*., **75**, 6807–6812.
Zheng H. H., Thompson R. B. and Fierke C. A. (2004). *Biophys. J*., **86**, 606A–607A.
Zhu Q., Aller R. C. and Fan Y. (2005). *Environ. Sci. Technol*., **39**, 8906–8911.
Zou Z. W., Jang A., MacKnight E. *et al*. (2008). *Sens. Actuators B: Chem*., **134**, 18–24.

Part 5
Quality Assurance

11

Certified Reference Materials for Marine Monitoring

Francesca Pellizzato[a], Evin McGovern[b], and Philippe Quevauviller[c]

[a]*Marine Scotland, Marine Laboratory, Aberdeen, UK*

[b]*Marine Institute, Galway, Ireland*

[c]*Vrije Universiteit Brussel (VUB), IUPWARE, Department of Hydrology and Hydrological Engineering, Brussels, Belgium*

Chemical Marine Monitoring: Policy Framework and Analytical Trends, First Edition.
Edited by Philippe Quevauviller, Patrick Roose, and Gert Verreet.

11.1 INTRODUCTION

The awareness about the need to improve and control the quality of analytical results from marine monitoring programmes has increased since the 1970s with interlaboratory studies being initiated by various organizations and the setting up of proficiency testing schemes (Wells *et al.*, 1993). This is discussed in Chapter 1 of this book and will not be repeated in the present chapter. In addition, quality assurance/control principles are widely discussed in a separate book of the Wiley's *Water Quality Measurements* series (Quevauviller, 2002a). The focus in the present chapter is on certified reference materials (CRMs), which are widely recognized tools for the verification of the accuracy of analytical techniques (Quevauviller and Maier, 1999), forming an integral part of quality control systems used, for example, in accreditation schemes. The availability of CRMs for the quality control of marine chemical measurements in environmental matrices has drastically improved over the last three decades, and the materials cover a large range of matrices and parameters (Ulberth, 2006; Wise *et al.*, 2006). This chapter recalls general principles about the production and use of reference materials (RMs) and provides a (nonexhaustive) description of matrix CRMs used in marine monitoring analyses, highlighting improvements, recent addition and future needs.[1]

11.2 PRODUCTION AND USE OF REFERENCE MATERIALS

11.2.1 Definitions

CRMs are also referred to as 'standard reference materials' (SRMs) by the National Institute of Standards and Technology (NIST), USA. They allow analytical laboratories to link their results

[1] The tables included in this chapter have been compiled for information only. Despite the authors making every effort to ensure the accuracy in the information provided, the users of CRMs should consult the related certificate provided by the vendor for full and accurate information. The tables are not intended to be complete, and some of the CRMs indicated may no longer be commercially available.

The following abbreviations will be used in the tables. NIST: USA National Institute of Standards and Technology; IAEA: International Atomic Energy Agency; BCR: EC Bureau of Community Reference, now EC–Joint Research Center–Institute for Reference Materials and Measurements (EC-JRC-IRMM); NRC: Canada National Research Council; NWRI: National Water Research Institute, Environment Canada; CIL: Cambridge Isotope Laboratories, USA; NIES: National Institute for Environmental Standards, Environment Agency, Japan; LGC: Laboratory of the Government Chemist (UK); RTC RT Corporation (USA); NRCCRM National Research Center for CRMs (China).

with those of internationally recognized standards. In addition, CRMs enable the user to verify their performance at any desired moment in terms of accuracy. CRMs can be:

1. Pure substances or solutions to be used for calibration and/or identification. Pure substances are usually certified by establishing the maximum amount, in mass fractions, of the impurities which remain in the purified substance. For metals, it is possible to estimate the mass fraction of other elements within the limits of detection of the available analytical methods, typically some fractions of a per cent. For inorganic salts or oxides, this estimate may already be more difficult. The stoichiometry also has to be assessed (e.g. water of crystallization).
2. Materials of a known matrix composition for the calibration of certain types of comparative measuring instrument. Certified calibration solutions have to be prepared on a mass basis within specially trained and skilled laboratories. Certification will be based on a metrologically valid weighing procedure after a proper purity and stoichiometry assessment.
3. Matrix RMs which, as far as possible, represent the matrix being analysed by the user and which have a certified content. Such materials are mainly intended for the verification of a measurement process; matrix CRMs are composed of a natural unknown or only partially known matrix in which the amounts of a certain number of substances are certified. Usually, analysts prefer fully natural materials which are as similar as possible to real samples and not artificially enriched materials. In addition, spiking of solids with solutions containing the contaminants leads to unknown losses in the preparation. Therefore, gravimetric results can only rarely be used to certify the material. Certification can only be based on an analytical approach using complex procedures with limited precision compared with mass determinations.
4. Methodologically defined RMs. In such materials, the certified value is defined by the applied method following a very strict analytical protocol; for example, the RM may be certified following a standardized method (e.g. single or sequential extraction procedure).

CRMs are products of very high added value. Their production and certification are very costly. Therefore, except for calibration materials for comparative methods, they should be reserved for selected verification of analytical procedures, not for daily checks such as, for example, intralaboratory statistical control, nor for interlaboratory studies (round-robins).

11.2.2 Requirements for the Preparation of Reference Materials

11.2.2.1 General Requirements

This section describes the general requirements which are applicable to all kinds of CRMs, including those used in marine monitoring. A series of requirements have to be respected for the preparation of RMs and CRMs. They include three basic properties: (i) the representativity of the sample; (ii) the homogeneity; and (iii) the long-term stability. The preparation procedure has to be adapted to the type of material to be prepared and to the objective it has to fulfil.

In order to arrive at sound conclusions on the analytical performance of a method or of a laboratory, the RM should be as similar as possible to samples currently analysed by the laboratory in its daily practice. This means that the RM should pose similar difficulties; that is, induce the same sources of error that can be encountered in analysing the real samples. Materials which would pose more difficulties than the daily samples may also fulfil the requested task of evaluating the method's or laboratory's performance as long as they do not require additional handling, which may induce large additional uncertainties.

The requirement of representativity of the RM means in most cases similarity of:

- the matrix composition;
- the chemical form and concentrations of the analytes;
- the way of binding of these analytes;
- the fingerprint pattern of possible interferences;
- the physical status of the material.

In preparing an RM, these items should be taken into full consideration. In many cases, and for practical reasons, the similarity cannot always be entirely achieved. The material has to be homogeneous and stable in order to ensure that samples delivered to the laboratories are the same and compromises often have to be made at the stage of the preparation. Certain parameters of interest which characterize the difficulty of some real natural samples may disappear; for example, inhomogeneity of real samples may need a special technique or strategy of subsampling. Some other problems may arise or may be enhanced due to the treatment necessary to homogenize or stabilize the material. It is up to the producer and to the user to define the degree of acceptability of compromises. The user should be informed of the real status of the sample, on the treatment performed and perhaps on the treatment to be applied to bring the sample back to a more representative form prior to the analysis.

11.2.2.2 Material Collection

The amount of material to be collected has to be adapted to the purpose of the analysis and is a function of the analytical sample size, stability, shelf life, frequency of use and potential market. It is sometimes better to prepare a limited batch of samples so that the stock lasts a sufficient time and to prepare a new batch of material when new requirements for modern analytical techniques appear or when regulations have changed. The amount collected may vary from a few kilograms of material (e.g. sediments) or even grams (e.g. biota), from some litres (e.g. seawater) or millilitres (e.g. liver oil extracts) for the preparation of an RM for the statistical control of a method, up to hundreds of kilograms of solid material or several cubic metres of water in case of an RM for large interlaboratory studies or CRM production. The producer needs to be equipped to treat the necessary amount of material without substantially changing the representativity. The treatment of 3–5 kg raw material is already the limit for usual laboratory equipment and manual processing; for larger batches, and especially larger volumes of material, it is necessary to scale up to half or even industrial-size machines.

Typical operations for the preparation of a reference material are: stabilization, crushing, grinding, sieving (for solid materials), filtering (for liquid materials) and mixing or homogenization of the material. They can only be performed in specialized laboratories or industries.

11.2.2.3 Material Stabilization and Stability Control

A very sensitive and difficult step of the preparation is the stabilization of the raw material which may affect its representativity. The stabilization is necessary to guarantee that the materials remain unchanged in time. It has to be adapted to each particular case with regard to the matrix and substance to be determined and should be studied in detail before processing the batch of the RM. Pure and synthetic mixtures of solid materials are usually stable and do not need a particular treatment. Natural products or synthetic liquid or gaseous mixtures are often highly unstable. Solid materials are often dried to avoid chemical or microbiological activity. This may be achieved by heat-drying (e.g. for sediments to be analysed for trace elements) or by freeze-drying (e.g. plant or animal tissues, sediments samples to be analysed for trace organics, etc.), or by fixing the water of the material with chemical additives.

Some materials can be sterilized by gamma-irradiation (^{60}Co source). In reality, this treatment is mainly possible for materials to be used for element determinations, as many compounds decay upon gamma-irradiation (e.g. tin compounds; Quevauviller *et al.*, 1994). Sterilization by pasteurization or similar methods or addition of chemical preservatives are other alternatives. Stabilization by simple deep-freezing is also possible, but induces difficulties in transport and storage. In addition, the material can only be used once, as defreezing–refreezing may not lead to a homogeneous material.

The composition of the material and the parameters investigated should remain unchanged over the entire period of use of the material. The study of the material stability in time will mainly depend on its role. If the material is foreseen to be used in a short-term interlaboratory study, the stability may be monitored only over the real duration of the exercise and additionally mimic situations which may be encountered during its short lifetime; for example, transport under severe climatic conditions. This may vary from some hours (e.g. microbiological samples) to several years for a CRM. The (in)stability should be studied or known before the RM is produced and should be monitored on the batch of RM for its whole lifetime. The stability can be estimated by evaluating the characteristics of the material under accelerated ageing conditions; for example, elevated temperatures over long periods of time.

11.2.2.4 Material Homogenization and Homogeneity Control

A material can only serve as a reference when at each occasion of its analysis an identical portion is available. Therefore, when stabilized, the material must be homogenized to ensure a sufficient within- and between-vial homogeneity for the property value to be certified. The inhomogeneity of the material should not affect significantly the total uncertainty of the measurement. Homogenization is not the most difficult problem for gases and liquids, despite the fact that some samples (e.g. estuarine water) may be prone to flocculation during the homogenization process (Kramer *et al.*, 1994). Solid materials or any material composed of various phases (e.g. aerosols, suspensions, etc.) are obviously more difficult to homogenize or to maintain homogeneous. Experience shows that for small particle sizes (i.e. less than 125 μm), narrow particle size distribution and homogeneous density of the particles, a good homogeneity is achieved and can be maintained or re-obtained even after long storage periods with simple shaking or mixing with laboratory tools. Unfortunately, the low particle size presents some drawbacks, as it leads to materials which are usually easier to analyse than real samples (better extractability of analytes or easier matrix digestion because of the large contact surface).

The quantity of material in the bottle or ampoule should be sufficient to possibly perform several determinations. The more subsamples that are taken, the higher the risk that the remaining bulk is no longer identical to the initial material. In addition, the producer must guarantee that, from the first to the last vial filled, the material is fully identical. Therefore, a verification of the homogeneity within and between vials has to be performed. During the filling procedure the producer should set aside vials at regular intervals. The homogeneity of the material should be tested within a single vial to assure that successive test portions from the vial lead to similar results (within-vial homogeneity). The same study should be performed to verify that there is no difference from test portions taken from various vials (between-vial homogeneity). Besides substances of interest, the study should include some matrix constituents or major and minor constituents (in the case of a matrix RM). It is current practice for producers of CRMs (e.g. BCR) to test this homogeneity for different sample intakes. The (in)homogeneity may be estimated by examining the coefficient of variation (CV) of replicate measurements obtained on samples from different vials and comparing those obtained from test portions from one vial. These CVs are also compared with the CV due to the analytical method. This CV is usually obtained by measuring the same extract or digest of one sample several times.

11.2.2.5 Material Storage and Transport

The above-mentioned homogeneity and stability considerations were implicitly related to the storage vial. Vials may be sealed glass ampoules (e.g. for materials containing organic contaminants, solutions, etc.), bottles for solids (e.g. sediments, aquatic plants, animal tissues, etc.), stainless steel or aluminium bottles for gases, or any kind of protecting vial. It is advisable to avoid light or radiation interaction by using, as often as possible, amber glass or high-density polymers. The temperature of storage should be adequate to ensure sufficient stability. Low temperatures are often desirable, but are not always necessary (e.g. large, cool rooms for stocks of CRM). Cooling sometimes can even be harmful, leading, for example, to precipitation of dissolved compounds. The storage conditions together with the delivery system should be deduced from a properly conducted stability study of the material and possibly a preliminary study of the material under various conditions and in different storage vials (especially for CRM). The transport should be done in the shortest period possible. Rapid delivery systems are unfortunately expensive and are solely used for certain particular cases. Accompanying each test sample, the user should receive a reporting form to be sent back to the organizer on which is indicated the status of the samples as received. Temperature indicators may be added to the samples so that too high temperatures occurring in transport are detected upon arrival of the sample.

In some circumstances, storage in deep-frozen conditions seems to be the only way to stabilize unstable compounds over a long-term period (e.g. phenyltin compounds; see Quevauviller and Ariese, 2001); the availability of RMs certified for these compounds will, hence, result in expensive storage needs and transport conditions which will have a direct consequence on the price of the material. In other cases, deep-freezing storage has been shown to create adverse effects on sediment materials; for example, results obtained on sequential extraction for trace metal determinations in sediments showed that the extractability may change at very low temperature and, hence, the 'speciation' may be altered when materials are stored at −20 °C or below (Quevauviller, 2002b); in such a case, the recommendation is to store materials in cool conditions (e.g. +4 °C) but not to freeze them.

11.2.3 Procedures to Certify and Assign Values

The certification of RMs follows strict rules which are described in a special ISO Guide 35: 'the certified value should be an accurate estimate of the true value with a reliable estimate of the uncertainty compatible with the end use requirements'. Depending on the type of property value to be certified and the type of CRM, there may be differences in the approach applied. The certification of primary calibrants, such as pure compounds, or calibration solutions relies on the identification, the purity and stoichiometry assessment and on gravimetric methods. Matrix CRMs cannot be certified on the basis of direct gravimetric methods, since samples have to be analysed after a total transformation or removal of the matrix; in this case, the certification will rely, for example, on analyses within one laboratory using two or more independent methods by two or more independent analysts or analyses by interlaboratory studies using one or several different methods possibly including definitive methods (e.g. isotope dilution mass spectrometry). In all cases, only laboratories of the highest and proven quality should be involved. The different certification approaches are described in another volume (Quevauviller, 2002a) and will not be repeated here. A certification of an RM leads to a description sheet (certificate) which should cover, in particular:

- administrative information on the producer and the materials;
- brief description of the material with main properties and its preparation;
- intended use of the materials;
- information for correct use and storage of the CRM;
- certified values and confidence limits;
- other noncertified values (optional);
- analytical method(s) used for certification and technical evaluation of the results;
- identification of certifying institute(s);
- legal notices and signature of certifying body.

11.2.4 Correct Use of Reference Materials

Reference materials might be used for different purposes in measurement quality, such as for method validation, calibration, estimation of measurements uncertainty, training, internal quality control and proficiency testing. Different types of RM are required for different functions. Method validation and measurement uncertainty make substantial use of RMs for the estimation of bias (the difference between the measured value and the true value). Traceable certified values would be desirable for this aim, but lower quality CRMs may also be used provided matrix type and analyte concentration are within the scope of the method to be validated. RMs can be used for troubleshooting when some unexpected results are obtained, for training purposes or for checking the correct use of a method.

Pure-substance RMs are generally used for calibration of the measurement stage of a method. The uncertainty associated with the RM purity will contribute to the total uncertainty of the measurement. RMs also give a major contribution in quality control and quality assurance. For this purpose, homogeneity and stability are essential, while the certified property value is desirable, but often not possible. Certifying the property values of proficiency testing samples often is, in fact, too expensive; therefore, consensus mean values are used instead, carrying some undisclosed elements of uncertainty.

As the certification of RMs is a complex and time-consuming process, the number of materials and their availability is limited and generally smaller than their demand. Therefore, the user must carefully consider and choose the most suitable material available which might not completely match the original requirements. In assessing the appropriateness and fitness for purpose of any RM based on the customer and analytical requirements, the factors to be considered are diverse: measurand including analyte, measurement range, matrix match and potential interferences, sample size, homogeneity, stability, measurement uncertainty and value assignment procedure.

Once the choice of the material has been made, supporting information on the use of the RM are reported in the certificate (possibly ISO guide 31 compliant) and in the report (possibly ISO guide 35 compliant) on characterization, certification and statistical analysis procedures. In the certificate, the user can find instructions on how to correctly handle the material (if special precautions are necessary), how to mix the sample when opening it (to assure homogeneity), how to dry the material (for the determination of the dry mass correction), how to correctly store the material (to ensure the stability within the shelf life indicated in the certificate), what is the minimum sample intake of the material, as well as the correct use of the certified value and uncertainty.

11.3 CRMS FOR TRACE ELEMENTS AND NUTRIENTS

11.3.1 CRMs for Trace Elements in Seawater

Analysis of trace elements in seawater is a challenge owing to the low concentrations of the analytes and the strong interference of major constituent ions in seawater. Seawater CRMs, therefore, are essential to underpin high-quality measurements. Nonetheless, there are few seawater CRMs for trace metals available; see Table 11.1. Estuarine water CRMs include LGC6016 (LGC UK), BCR-505 (JRC-IRMM, EC) and SLEW-3 (NRC Canada). LGC6016 was collected from the Severn Estuary, UK (offshore from a heavily industrialized area near Avonmouth). The material is certified for cadmium, copper, manganese, nickel and lead and has higher concentrations than the BCR and NRC estuarine water CRMs. BCR-505 (salinity $S = 12.1$) was collected in 1992 and is certified for cadmium, copper, nickel and zinc (Quevauviller *et al*., 1996a). SLEW-3 ($S = 15$) was sampled in San Francisco Bay at a depth of 5 m and is certified for 11 trace elements, including elements listed above and arsenic.

Coastal water CRMs are available from JRC-IRMM and NRC, Canada. BCR-579 is only certified for total mercury and, indeed, is the only RM certified for mercury (Kramer *et al*., 1998). CASS-5 (NRC Canada, $S = 33.5$) is a nearshore seawater sampled from Halifax Harbour, Nova Scotia, and is certified for 10 trace elements.

There is a lack of CRMs for offshore and ocean waters. NRC Canada has produced NASS-6, which was collected in Sandy Cove, Nova Scotia. The concentrations of trace elements in available CRMs are higher than generally found in surface waters of the open seas (Grasshoff *et al*., 1999; Millero, 2006), so these materials are not optimal for supporting trace element analysis in open ocean studies.

The Water Framework Directive (WFD) (2000/60/EC; European Commission, 2000) requires monitoring of priority substances and other relevant pollutants in coastal and transitional waters and environmental quality standards (EQSs) have been set at community level for cadmium, lead, mercury and nickel (Directive 2008/105/EC; European Commission, 2008). The three NRC RMs have certified concentrations for cadmium, lead and nickel that are below the EQSs, while BCR-505 estuarine material is certified for cadmium and nickel at levels above the EQSs. Concentrations in LGC6016 well exceed the EQSs for these elements. The certified concentration of mercury in the only available material, CASS-5, is also lower than the WFD EQS.

Analytically, estuarine waters may behave differently to offshore waters owing to, for instance, higher concentrations of dissolved organic matter (DOM) and complexation (Kramer *et al*., 1994). Consequently, in selecting the most appropriate CRM for seawater analysis, it is important to match matrix and concentrations with samples in so far as practical. Seawater CRMs are generally filtered and acidified during preparation. It is critical to follow the producer's guidance when subsampling CRMs, and this may require use of clean facilities to avoid contamination. Furthermore, accurate CRM determinations do not necessarily infer good-quality analysis, as errors that may arise in the sampling and/or sample-handling process may not be accounted for. This can be critical for trace element analysis in seawater.

11.3.2 CRMs for Nutrients and Inorganic Carbon in Seawater

MOOS-1 (NRC, Canada) is a seawater certified for orthophosphate, silicate, nitrite and total oxidized nitrogen (TOxN). The material was collected at a depth of 200 m off Cape Breton Island, Nova Scotia. VKI RMs (Eurofins Miljø A/S, Denmark) also produces RMs for estuarine and coastal waters. The estuarine water ($S = 11.1$) is certified for ammonium, nitrite, TOxN and total nitrogen (QC SW3.1) and orthophosphate, total phosphate and silicate (QC SW3.2). The seawater RM (QC SW 4.1 and QC SW 4.2, $S = 35$) is certified for the same parameters with lower concentrations.

Table 11.1 Reference materials available for trace elements in seawater. Values marked with an asterisk are non-certified; all other values are certified.

Code	SLEW-3	NASS-6	CASS-5	BCR-505	BCR-579	LGC6016
Organization	NRC	NRC	NRC	EU–JRC–IRMM	EU–JRC–IRMM	LGC
Country of origin	Canada	Canada	Canada	EU	EU	UK
Matrix	Estuarine Water	Seawater	Nearshore Seawater	Estuarine water	Coastal seawater	Estuarine water
Units	$\mu g\,L^{-1}$	$\mu g\,L^{-1}$	$\mu g\,L^{-1}$	$nmol\,kg^{-1}$	$ng\,kg^{-1}$	$mg\,kg^{-1}$
[±] expressed as	±95% CI	±95% CI	±95% CI	Half width of 95% CI	±95% CI	±95% CI
Units of issue	500 mL	500 mL	500 mL	1 L	1 L	50 mL
Salinity	15	33.5	33.5	~12	~28	
Arsenic	1.36 ± 0.09	1.43 ± 0.12	1.24 ± 0.09	41.0*		
Cadmium	0.048 ± 0.004	0.0311± 0.0019	0.0215± 0.0018	0.8 ± 0.04		101 ± 2
Chromium	0.183 ± 0.019	0.118 ± 0.008	0.106± 0.013	3.97*		

(continued overleaf)

Table 11.1 *(continued)*

Code	SLEW-3	NASS-6	CASS-5	BCR-505	BCR-579	LGC6016
Cobalt	0.042 ± 0.01	0.015*	0.095	0.99 ± 0.26		
Copper	1.55 ± 0.12	0.248 ± 0.025	0.38 ± 0.028	29.4 ± 1.5		190 ± 4
Iron	0.568 ± 0.059	0.495 ± 0.046	1.44 ± 0.11	19 ± 4		
Lead	0.009± 0.0014	0.006± 0.002	0.011 ± 0.002	0.24 ± 0.14		196 ± 3
Manganese	1.61 ± 0.22	0.530 ± 0.050	2.62 ± 0.2	13.2*		976 ± 31
Mercury					1.9 ± 0.5	
Molybdenum	5.1*	9.9*	9.8*			
Nickel	1.23 ± 0.07	0.301 ± 0.025	0.33 ± 0.023	24.1 ± 2		186 ± 3
Silver	0.003					
Uranium	1.8*	3.0*	3*			
Vanadium	2.57 ± 0.32	1.46 ± 0.17	1.32 ± 0.14	63.8*		
Zinc	0.201 ± 0.037	0.257 ± 0.020	0.719 ± 0.068	172 ± 11		55*

There has been increased interest in measuring dissolved inorganic carbon in recent years. Andrew Dickson in Scripps Institution of Oceanography, California, produces RM batches to support this, stating values and uncertainties for total dissolved inorganic carbon (DIC) and total alkalinity (TA). These parameters, alongside others, may be used to calculate pH and pCO_2. Salinity and dissolved nutrient data are also provided in the certificates of analyses.

11.3.3 CRMs for Trace Elements in Aquatic Biota

There are a number of (certified) RMs commercially available to support analytical quality control in laboratories engaged in monitoring trace elements in marine biota. Concentrations of selected elements in molluscs, crustacea and marine algae are presented in Table 11.2 and concentrations in fish reference materials are presented in Table 11.3. Mussel tissue is a widely used monitoring matrix for trace elements in the marine environment. There are various mussel reference materials available, although the frozen SRM 1974b (NIST) is not available in Europe. The oyster tissue SRM 1566b (NIST) certificate was revised and reissued in 2010 and is certified for 23 elements and methylmercury. NRC produces two lobster hepatopancreas materials, one of which is defatted. Other reference materials include one marine alga (BCR-279) (EU–JRC–IRMM) and a fish otolith material (NIES-22) with reference values for cadmium, copper, lead and zinc (Yoshinaga *et al.* 2000). Consequently, there are very limited CRM options for laboratories that determine metal concentrations in marine macroalgae.

There are a number of available fish materials, including some freshwater fish reference materials, that can be used in support of monitoring programmes. These include three BCR tuna reference materials. Fish liver is often a target matrix for monitoring programmes, and OSPAR's JAMP Guidelines for Monitoring Contaminants in Marine Biota (OSPAR, 1999) stipulates liver as the preferred matrix for determination of target trace elements other than mercury. Nonetheless, DOLT-4 produced by NRC in Canada is the only liver tissue currently available. Laboratories should be aware that CRMs may differ from the matrix routinely analysed due to processing of the RM. Most materials are (freeze-)dried and some CRMs have also undergone acetone extraction to remove fat.

Most certified/reference values are for total metal concentrations, and there are many materials available for the trace elements that are most commonly included in monitoring, such as total mercury, cadmium, lead, arsenic, copper, zinc and nickel. Furthermore, with techniques such as inductively couple plasma mass spectrometry now widely used, a much broader suite of elements can be measured at little extra cost. Many materials have certified and/or reference concentrations for a broad range of elements. Nonetheless, NIST SRMs are the only shellfish RMs with certified/reference values for all of nine target elements required by the European Shellfish Waters Directive 2006/113/EC (European Commission, 2006).

Many elements (e.g. chromium, arsenic and mercury) can exist in various oxidation states and/or in compounds that can exhibit different environmental behaviour and toxicological properties. These are therefore of interest to monitoring authorities. Cost and analytical issues often preclude their incorporation into routine monitoring, and the limited availability of appropriate RMs to underpin quality control may be a barrier. Most RMs are certified for total mercury and many are also certified for methylmercury (Quevauviller *et al.*, 1996b). With some notable exceptions, organic species of arsenic, such as arsenobetaine in fish and arseno-sugars in marine macroalgae, typically predominate in marine biota, and these forms are generally considered of low toxicological risk to consumers. Consequently, monitoring the more toxic chemical species, and in particular inorganic arsenic, is of interest to laboratories engaged in seafood testing. Most available materials are certified or provide reference concentrations for total arsenic, but only BCR-627 (tuna) (EU–JRC–IRMM), provides certified values for other arsenic species, namely arsenobetaine and dimethylarsinic acid.

Table 11.2 Reference materials available for trace elements in shellfish, crustacea and marine algae. Values marked with an asterisk are non-certified; all other values are certified.

Code	BCR-668	ERM-CE278	SRM 2976	SRM 2977	SRM 1974b	SRM 1566b	TORT-2	LUTS-1	BCR-279
Organization	EU–JRC –IRMM	EU–JRC –IRMM	NIST	NIST	NIST	NIST	NRC	NRC	EU–JRC –IRMM
Country of Origin	EU	EU	USA	USA	USA	USA	Canada	Canada	EU
Matrix	Mussel tissue *Mytilus edulis*	Mussel tissue *Mytilus edulis*	Mussel tissue *Mytilus gallopro-vincialis*	Mussel tissue *Perna perna*	Mussel tissue *Mytilus edulis*	Oyster tissue *Cras-sostrea virginica*	Lobster Hepatopan-creas	Non-defatted Lobster Hepatopan-creas	Sea lettuce *Ulva lactuca*
Units	$mg\,kg^{-1}$	$mg\,kg^{-1}$	$mg\,kg^{-1}$	$mg\,kg^{-1}$	$mg\,kg^{-1}$	$mg\,kg^{-1}$	$mg\,kg^{-1}$	$mg\,kg^{-1}$	$mg\,kg^{-1}$
As	Dry wt	Dry wt	Dry wt	Dry wt	Wet wt	Dry wt	Dry wt	Dry wt	Dry wt
[±] expressed as	Standard deviation	Half width 95% CI					±95% CI	±95% CI	±95% CI
Unit of issue	10 g	8 g	25 g	10 g	5 × (8–10) g	25 g	35 g	6 × 10 g	35 g
Form	Freeze-dried	Freeze-dried	Freeze-dried	Freeze-dried	Wet	Freeze-dried	Defatted and dried	Wet	Dried
Al			134 ± 34*	400*		197.2 ± 6.0			
As	7.1 ± 0.5*	6.07 ± 0.13	13.3 ± 1.8	8.83 ± 0.91*	0.796 ± 0.049*	7.65 ± 0.65	21.6 ± 1.8	19.0 ± 0.9	3.09 ± 0.21
Ag			0.011 ± 0.005*	4.58 ± 0.33*	0.028 ± 0.003*	0.666 ± 0.009		3.89 ± 0.33	
Cd	0.275 ± 0.011*	0.348 ± 0.007	0.82 ± 0.16	0.179 ± 0.003	0.155 ± 0.005*	2.48 ± 0.08	26.7 ± 0.6	14.2 ± 1.0	0.274 ± 0.022

Cr	0.37 ± 0.06*	0.78 ± 0.06	0.50 ± 0.16*	3.91 ± 0.47*	0.233 ± 0.010*		0.77 ± 0.15	0.53 ± 0.08	
Cu		9.45 ± 0.13	4.02 ± 0.33	9.42 ± 0.52	0.967 ± 0.016*	71.6 ± 1.6	106 ± 10	107 ± 8	13.1 ± 0.4
Hg		0.196 ± 0.009	0.061 ± 0.0036	0.101 ± 0.004*	0.017 ± 0.0011	0.0371 ± 0.0013	0.27 ± 0.06		
Mn		7.69 ± 0.23	33 ± 2*	23.93 ± 0.29		18.5 ± 0.2	13.6 ± 1.2	8.02 ± 0.86	
Ni			0.93 ± 0.12*	6.06 ± 024	0.109 ± 0.005*	1.04 ± 0.09	2.50 ± 0.19	1.34 ± 0.23	
Pb		2.00 ± 0.04	1.19 ± 0.18	2.27 ± 0.13	0.752 ± 0.026*	0.308 ± 0.009	0.35 ± 0.13	0.069 ± 0.011	13.5 ± 0.4
Se		1.84 ± 0.10	1.80 ± 0.15	1.78 ± 0.16*	0.224 ± 0.015*	2.06 ± 0.15	5.63 ± 0.67	4.30 ± 0.36	0.59 ± 0.04
Sn			0.096 ± 0.039	1.47 ± 0.27*	0.028 ± 0.002*	0.031 ± 0.008*	0.04*		
V				1.1*		0.577 ± 0.023	1.64 ± 0.19		
Zn	70.7 ± 0.4*	83.1 ± 1.7	137 ± 13	135 ± 5*	12.3 ± 0.3*	1424 ± 46	180 ± 6	82.9 ± 5.4	51.3 ± 1.2
CH_3-Hg^+		0.0281 ± 0.0003	0.0366 ± 0.0010	0.0071 ± 0.00044	0.0132 ± 0.0007	0.152 ± 0.013			

BCR-668: certified/reference values also available for Ce, Co, Cs, Dy, Er, Eu, Fe, Gd, Ho, La, Lu, Mo, Nd, Pr, Sc, Sm, Tb, Th, Tm, U, Y, Yb. SRM 2976: certified/reference values also available for Ca, Ce, Co, Cs, Eu, Fe, K, Mg, Na, Rb, Sc, Sr, Th, Tl. SRM 2977: certified/reference values also available for Au, Ba, Ca, Ce, Co, Cs, Fe, K, La, Mg, Na, P, Rb, Sb, Sc, Sm, Sr, Th, Tl, U. SRM 1974b: reference values also available for Fe. SRM 1566b: certified/reference values also available for B, Ba, Ca, Co, Fe, K, Mg, Na, Rb, S, Sb,Sr, Th, U. TORT-2: certified values also available for Co, Fe, Mg, Mo, Sr. LUTS-1: certified values also available for Ca, Co, Fe, K, Sr. Note SRM 1974b not available in Europe. Methylmercury concentrations are as inorganic mercury except SRM 1974b.

Table 11.3 Reference materials available for trace elements in finfish. Values marked with an asterisk are non-certified; all other values are certified.

Code	BCR-463	BCR-627	ERM-CE464	SRM 1946	SRM 1947	DOLT-4	DORM-3	NIES 22[a]
Organization Country of origin	EU–JRC –IRMM EU	EU–JRC –IRMM EU	EU–JRC –IRMM EU	NIST USA	NIST USA	NRC Canada	NRC Canada	NIES Japan
Matrix	Tuna fish	Tuna fish	Tuna fish	Lake Superior fish tissue	Lake Michigan fish tissue	Dogfish liver *Squalus acanthias*	Fish protein	Fish otolith
Units	$mg\ kg^{-1}$	$mg\ kg^{-1}$	$mg\ kg^{-1}$	$mg\ kg^{-1}$	$mg\ kg^{-1}$	$mg\ kg^{-1}$	$mg\ kg^{-1}$	$mg\ kg^{-1}$
As	Dry wt	Dry wt	Dry wt	Wet wt	Wet wt	Dry wt	Dry wt	Dry wt
[±] expressed as	Half width 95% CI	Half width 95% CI	Half width 95% CI	±95% CI	±95% CI	Expanded uncertainty	Expanded uncertainty	
Units of issue	15 g	10 g	15 g	5× (7–9) g	5 × 8 g	25 g	20 g	3 g
Form	Freeze-dried	Dried	Freeze-dried	Wet frozen	Wet frozen	Cooked and sterilized	Dried	Dried
Al						200*	1700*	
As		4.8 ± 0.3		0.277 ± 0.010	0.732 ± 0.039	9.66 ± 0.62	6.88 ± 0.30	
Ag						0.93 ± 0.07	0.04*	
Cd				0.00208 ± 0.00026*		24.3 ± 0.8	0.290 ± 0.020	0.0028*

Cr						1.4*	1.89 ± 0.17	
Cu				0.476 ± 0.060*	0.411 ± 0.029	31.2 ± 1.1	15.5 ± 0.63	0.74*
Hg	2.85 ± 0.16		5.24 ± 0.10	0.433 ± 0.009	0.254 ± 0.005	2.58 ± 0.22	0.382 ± 0.060	
Mn					0.076 ± 0.004	4.6*		
Ni						0.97 ± 0.11	1.28 ± 0.24	
Pb						0.16 ± 0.04	0.395 ± 0.050	0.023*
Se				0.491 ± 0.043*	0.475 ± 0.084	8.3 ± 1.3	3.3*	
Sn						0.17*	0.066 ± 0.012	
V						0.6*		
Zn				3.10 ± 0.18*	2.66 ± 0.08	116 ± 6	51.3 ± 3.1	0.47*
Arsenobetaine (μmol kg^{-1})		52 ± 3						
Dimethylarsinic acid (μmol kg^{-1})		2.0 ± 0.4						
CH_3-Hg^+	3.04 ± 0.16		5.50 ± 0.17	0.394 ± 0.016	0.233 ± 0.010	1.33 ± 0.12	0.355 ± 0.056	

SRM 1946: certified/reference values also available for Ca, Fe, K, Mg, Na, P. SRM 1947: certified values also available for Fe, Rb. DOLT-4: certified/reference values also available for Ca, Co, Fe, K, Mg, Mo, Na, Sr. DORM-3: certified value also available for Fe. NIES 22: certified values also available for Ba, Ca, K, Mg, Na, Sr. Methylmercury concentrations are as inorganic mercury except BCR 463 and ERM-CE-464.

[a]NIES 22 not available overseas.

11.3.4 CRMs for Trace Elements in Sediment

There are many aquatic sediment materials with certified and/or reference values for trace elements and too many to list here. A nonexhaustive overview of the available (certified) RMs for inorganic constituents in estuarine or harbour sediments and in marine sediments is given in Tables 11.4 and 11.5 respectively. These materials are produced by IRMM, NIST, IAEA, NRC, NIES, LGC, NWRI and NRCCRM (China, available through LGC), although the material produced by NIES is, however, not available overseas. Sediment testing laboratories should be aware of the many reference materials available (including those not listed here) and choose materials suitable for their own needs.

Except for ERM-CC580 (EC–JRC–IRMM), which is certified only for mercury and methylmercury and represents high mercury contamination levels (Quevauviller *et al.*, 1998), all other materials are provided with certified, reference and indicative values for many inorganic constituents. Certified, reference, information or indicative values are listed in Tables 11.4 and 11.5 only for the main elements of interest for marine environmental monitoring (heavy metals and some other transition metals). For other constituents, such as other transitional metal, post-transition metals, lanthanoids and actinoids, certified, reference or information values are listed in the footnote of the table. Heavy metals are well characterized, being certified in most of the materials. Lanthanoids are certified in BCR-667 (EU–JRC–IRMM). In this material, heavy metal concentrations are indicative. NIST produce a marine sediment reference material (SRM 2703) which is primarily intended for use in evaluating analytical methods for the direct determination of selected elements in solid samples of aquatic sediment and dissolution techniques using small sample sizes. For larger sample sizes, the equivalent SRM 2702 should be used and the latter material generally has much tighter uncertainties quoted. LGC6137 is a sediment from the Severn Estuary certified for extractable metals (metal soluble in hot aqua regia using methods based on ISO 11466 (1995)). IRMM has also certified sediment reference materials for their extractable trace metal contents using a three-step sequential extraction (Quevauviller *et al.*, 1997).

Materials produced by NRC Canada (HISS-1, MESS-3 and PACS-2) provide good coverage for trace elements, including certified values for heavy metals, as well as other transitional, alkali and alkaline earth metals.

The highest concentrations for the most routinely monitored metals are found in PAC-2 (NRC) and SRM 1944 (NIST), and the lowest in SRM 1646a (NIST) and HISS-1 (NRC).

11.4 CRMS FOR ORGANIC NON-HALOGENATED COMPOUNDS

11.4.1 CRMs for PAHs in Biota and Sediment

Available RMs to support monitoring of polycyclic aromatic hydrocarbons (PAHs) in biota (fish, shellfish and seaweed) and sediment (marine and freshwater) are presented in Tables 11.6 and 11.7 respectively. A comparison of currently available CRMs for PAHs in sediment and biota with those available in 2000 (De Boer and McGovern, 2001) shows that there are some additional materials now available, but also that a number of certified materials are no longer available. This is especially the case for fully certified materials. The available RMs for biota and sediment typically provide certified/reference values for parent PAH with between two and six rings. All the biota materials, with the exception of the EDF-2524 (CIL) for biota and LGC-6188 for sediment, have some values for alkylated PAHs. Primarily, these values are for methyl-naphthalenes and methyl-phenanthrenes. IAEA materials also provide recommended and information values for a number of other petroleum hydrocarbon parameters.

The most comprehensive coverage of PAHs in a biota CRM is NIST SRM 1974b. This mussel tissue is certified for 22 PAHs and reference concentrations are provided for another 16 PAHs.

Table 11.4 Reference materials available for trace elements in estuarine and harbour sediments. Values marked with an asterisk are non-certified; all other values are certified.

Code	BCR-277R	BCR-667	ERM-CC580	SRM 1646a	PACS-2	TH-1	TH-2	LGC6137	BCR-320R	SRM 1944
Organization	EU–JRC–	EU–JRC–	EU–JRC–	NIST	NRC	NWRI	NWRI	LGC	EU–JRC–	NIST
Country of origin	IRMM EC	IRMM EC	IRMM EC	USA	Canada	Canada	Canada	UK	IRMM EC	USA
Matrix	Estuarine sediment	Estuarine sediment	Estuarine sediment	Estuarine sediment	Esquimalt harbour sediment	Harbour sediment	Harbour sediment	Extractable metals Estuarine sediment	Channel sediment	New York/ New Jersey waterway sediment
Units	$mg\,kg^{-1}$	$mg\,kg^{-1}$	$mg\,kg^{-1}$	$mg\,kg^{-1}$	$mg\,kg^{-1}$	$mg\,kg^{-1}$	$mg\,kg^{-1}$	$mg\,kg^{-1}$	$mg\,kg^{-1}$	$mg\,kg^{-1}$
As	Dry wt	Dry wt	Dry wt	Dry wt	Dry wt				Dry wt	Dry wt
[±] expressed as	±95% CI				±95% CI				±95% CI	
Units of issue	40 g	40 g	50 g	70 g	65 g	100 g	100 g		40 g	50 g
Form	Dried	Dried	Air dried	Freeze-dried	Freeze-dried				Dried	Freeze-dried
Al				2.297 ± 0.018	6.62 ± 0.32	55\,597	64\,799			5.33% ±0.49%
Ag				<0.3*	1.22 ± 0.14		5.47*			6.4 ± 1.7*

(continued overleaf)

Table 11.4 *(continued)*

Code	BCR-277R	BCR-667	ERM-CC580	SRM 1646a	PACS-2	TH-1	TH-2	LGC6137	BCR-320R	SRM 1944
As	18.3 ± 1.8			6.23 ± 0.21	26.2 ± 1.5	9.59*	8.70	12.4	21.7 ± 2.0	18.9 ± 2.8
Cd	0.61 ± 0.07	0.67 ± 0.11*		0.148 ± 0.007	2.11 ± 0.15	5.41*	5.22		2.64 ± 0.18	8.8 ± 1.4
Cr	188 ± 14	178 ± 16*		40.9 ± 1.9	90.7 ± 4.6	125	123	47	59 ± 4	266 ± 24
Cu	63 ± 7	60 ± 9*		10.01 ± 0.34	310 ± 12	103	124		46.3 ± 2.9	380 ± 40*
Hg	0.128 ± 0.017		132 ± 3	0.04*	3.04 ± 0.20	0.44*	0.62		0.85 ± 0.09	3.4 ± 0.5*
Mn		920 ± 40*		234.5 ± 2.8	440 ± 19	583	584	665	910 ± 50	505 ± 25
Ni	130 ± 8	128 ± 9*		23*	39.5 ± 2.3	45.5	43		27.1 ± 2.2	76.1 ± 5.6
Pb		31.9 ± 1.1*		11.7 ± 1.2	183 ± 8	257	194	73.0	85 ± 5	330 ± 48
Se				0.193 ± 0.028	0.92 ± 0.22	0.911*	0.825		0.96 ± 0.18*	1.4 ± 0.2*
V				44.84 ± 0.76	133 ± 5	89.2	89.2	47.0	46.5 ± 2.8	100 ± 9(%)*
Zn	178 ± 20	175 ± 13*		48.9 ± 1.6	364 ± 23	1527	908	231	319 ± 20	656 ± 75
Co	22.5 ± 1.4	23.0 ± 1.3*		5*	11.5 ± 0.3	15.6*	15.7		9.7 ± 0.6	14 ± 2(%)*
Fe		44 800 ± 1000*		2.008 ± 0.039(%)	4.09 ± 0.06	35\,032	35\,361	30 700	25 700 ± 1300	3.53 ± 0.16
K				0.864 ± 0.016(%)	1.24 ± 0.05		22\,313*	5010		1.6 ± 0.2(%)*
Mo				1.8*	5.43 ± 0.28		1.18*			
Sc				5*			13.5*		5.2 ± 0.4	10.2 ± 0.2(%)*
Sr				68*	276 ± 30		273*			
Sn	6.5 ± 1.8*			1*	19.8 ± 2.5		10.77*		9.4 ± 1.7*	42 ± 6*

BCR 667 is also certified for Ce, Dy, Er, Eu, Gd, Ho, La, Lu, Nd, Pr, Sc, Sm, Tb, Tm, Yb, Th, U. Indicative values are also reported for Br, Cs, Sb, Se, Ta, Zn. Additional information: As, Au, Sr, Y; PACS-2 has also certified values for Sb, Be, Li, Ag, Al, Ca, Mg, Na, P, S, Ti, and information values for Tl, U, Cl, Si, C. LGC6137 also certified values for Ba, Be, Ca, Li, Mg, Na, and indicative values for Cd, Mo, Se, Sn, Ti. BCR-320R is also certified for Th, Tl, U. SRM 1944 reference values are reported for Si, Be, Se, Th, Ca, Cl, Na, Br, Ce, Rb, Sc, Ti, information values for Sb, Ce, Eu, Au, La, Th, U.

Table 11.5 Reference materials available for trace elements in marine sediments. Values marked with an asterisk are non-certified; all other values are certified.

Code	SRM 2702	SRM 2703a	IAEA-405	IAEA-433	HISS-1	MESS-3	NIES 12[a]	NCS DC74301	NCS DC75301	NCS DC75304	NCS DC75305
Organization	NIST	NIST	IAEA	IAEA	NRC	NRC	NIES	NRCCRM	NRCCRM	NRCCRM	NRCCRM
Country of origin	USA	USA	Monaco	Monaco	Canada	Canada	Japan	China	China	China	China
Matrix	Marine sediment	Marine sediment	Marine sediment	Marine sediment	Hibernia Shelf marine sediment	Beaufort Sea sediment	Marine sediment	Marine sediment	Offshore marine sediment	Offshore marine sediment	Offshore marine sediment
Units	$mg\,kg^{-1}$	$mg\,kg^{-1}$	$mg\,kg^{-1}$	$mg\,kg^{-1}$	$mg\,kg^{-1}$	$mg\,kg^{-1}$	$mg\,kg^{-1}$	$\mu g\,g^{-1}$	$\mu g\,g^{-1}$	$\mu g\,g^{-1}$	$\mu g\,g^{-1}$
As	Dry wt	Dry wt	Dry wt	Dry wt	Dry wt	Dry wt	Dry wt				
[±] expressed as	±95% CI		±95% CI	±95% CI	±95% CI	±95% CI					
Units of issue	50 g	5 g		10 g	100 g	50 g		50 g	75 g	50 g	50 g
Form	Freeze-dried	Freeze-dried		Freeze-dried	Freeze-dried	Freeze-dried	Dried				
Al	8.41% ±0.22%	8.33% ±0.22%	77 900 ± 5200*	0.133 ± 0.006*	0.73 ± 0.05	8.59 ± 0.23	7.22%*				
Ag	0.622 ± 0.078*	0.59 ± 0.12*		78.2 ± 1.4*	0.016 ± 0.002	0.18 ± 0.02					
As	45.3 ± 1.8	45.5 ± 1.7	23.6 ± 0.7*	18.9 ± 0.5*	0.801 ± 0.099	21.2 ± 1.1		5.8 ± 0.8	10.3	07.1	4.6
Cd	0.817 ± 0.011	0.811 ± 0.0076	0.73 ± 0.05*	0.153 ± 0.008*	0.024 ± 0.009	0.24 ± 0.01	3.0*		0.20		
Cr	325 ± 22		84 ± 4*	136 ± 2*	30.0 ± 6.8	105 ± 4	201*	58.4 ± 1.3	86	59	38

(continued overleaf)

Table 11.5 *(continued)*

Code	SRM 2702	SRM 2703a	IAEA-405	IAEA-433	HISS-1	MESS-3	NIES 12[a]	NCS DC74301	NCS DC75301	NCS DC75304	NCS DC75305
Cu	117.7 ± 5.6*	120 ± 15*	47.7 ± 1.2*	30.8 ± 0.6*	2.29 ± 0.37	33.9 ± 1.6	104*	424 ± 8	31	357	231
Hg	0.4474 ± 0.0069	0.474 ± 0.066	0.81 ± 0.4*	0.168 ± 0.04*	0.01*	0.091 ± 0.009	1.16*		0.048	0.95	0.13
Mn	1757 ± 58	1734 ± 48	495 ± 11*	316 ± 4*	66.1 ± 4.2	324 ± 12	837*				
Ni	75.4 ± 1.5		32.5 ± 1.4*	39.4 ± 0.7*	2.16 ± 0.29	46.9 ± 2.2	57.6*	150 ± 4	34.3	167	108
Pb	132.8 ± 1.1	130 ± 11	74.8 ± 2.2*	26.0 ± 0.6*	3.13 ± 0.40	21.1 ± 0.7	101*	29.3 ± 1.1	25	37	22
Se	4.95 ± 0.46*		0.44 ± 0.12*	0.78 ± 0.08*	0.050 ± 0.007	0.72 ± 0.05			0.16		
V	357.6 ± 9.2	360 ± 13	95 ± 5*	160 ± 4*	6.80 ± 0.78	243 ± 10	134*	112 ± 5		101	69
Zn	485.3 ± 4.2	480 ± 22	279 ± 7*	101 ± 2*	4.94 ± 0.79	159 ± 8	738*	160 ± 3	87	137	142
Co	27.76 ± 0.58	27.70 ± 0.50	13.7 ± 0.7*	12.9 ± 0.3*	0.65*	14.4 ± 2.0	16.6*	76.7 ± 1.2	14.2	81	53
Fe		7.38(%) ±0.32(%)	37 400 ± 700*	40.8 ± 0.5*	0.246 ± 0.009	4.34 ± 0.11	4.31 (%)*				
K		2.08(%) ±0.24(%)	24 900 ± 7200*	16.6 ± 3.0*	0.332 ± 0.013	2.6*	1.56 (%)*				
Mo	10.8 ± 1.6*				0.13*	2.78 ± 0.07		7.2 ± 0.5		14	5.7
Sc	25.9 ± 1.1	25.95 ± 0.68	13.52 ± 2.01*	14.6 ± 1.2*				25.6 ± 2.9		23	15
Sr	119.7 ± 3.0	118 ± 18	118 ± 14*	302 ± 6*	96.9 ± 11.2	129 ± 11	126*	267 ± 15	150	298	667
Sn	31.6 ± 2.4*		7.6 ± 1.3*	2.32 ± 0.15*	0.11*	2.50 ± 0.52					

SRM 2702 reference values are also reported for Ba, Ce, La, Na, P, Rb, Sb, Sr, Th, Ti, Tl, Ca, Ga. Information values are reported for Mg, Mo, Se, Be, C, Cs, Hf, Li, Nb, Nd, S, Sm, U, W. SRM-2703 certified values also for Ba, Ce, La, Na, Rb, Sb, Th, Ti, U, reference values for Ca, Cs, Hf, Sm, W. IAEA-405 has also recommended values for Li, Mg, Sb and information values for Br, Cs, Eu, Hf, La, Lu, Sm, Tb, Th, U, Yb. IAEA-433 has also recommended values for Ba, Br, Cs, La, Li, Mg, Ru, Sb, Th, U, and information values for Ce, Eu, Hf, Lu, Na, Nd, Sm, Ta, Tb, Yb, Zr. HISS-1 has also certified values for Be, Li, Ag, Al, Ca, Mg, Na, Ti, and information values for Sb, Tl, U, Cl, Si. MESS-3 has also certified values for Sb, Be, Li, Ag, Al, Ca, Ti, Tl, and information values for U, Cl, Si, Mg, Na, P, S, Si, C. NCS DC74301 also certified for several oxides, B, Ce, Dy, Er, Eu, Ga, Gd, Ho, La, Li, Lu, Nd, Pr, Rb, Sb, Sm, Tb, Th, Tm, U, W, Y, Yb. NCS DC75301 certified for Ba and several oxides and indicative values for Ga, Mo, B, Zr, La, Ce, Nd, Sm, Eu, Tb, Ho, Yb, Lu, Y, Pr, Gd, Dy, Er, Tm, Rb, Sc, Th, Nb, Ta, Hf, Cs, W, Sb, U. NCS DC75304 and NCS DC75305 certified also for F, B, Be, Bi, Br, Ce, Cs, Dy, Er, Eu, Ga, Gd, Hf, Ho, La, Li, Lu, Nb, Nd, Pr, Rb, Sb, Sm, Ta, Tb, Th, Tm, U, W, Y, Yb, Zr.

[a]NIES 21 not available overseas.

Table 11.6 Reference materials available for PAHs in biota. Values marked with an asterisk are non-certified; all other values are certified.

Code	SRM 1974b	SRM 2977	IAEA 432	IAEA-435	IAEA-406	EDF-2524	IAEA-140/OC
Organization	NIST	NIST	IAEA	IAEA	IAEA	CIL	IAEA
Country of origin	USA	USA	Monaco	Monaco	Monaco	Monaco	Monaco
Matrix	Mussel tissue	Mussel tissue	Mussel tissue	Tuna homogenate	Fish homogenate	Clean fish reference material	*Fucus* (sea plant homogenate)
Units	μg kg^{-1}	μg kg^{-1}	μg kg^{-1}	μg kg^{-1}	μg kg^{-1}	ng kg^{-1}	μg kg^{-1}
As	Dry wt	Dry wt	Dry wt	Dry wt	Dry wt		Dry wt
[±] expressed as	±95% CI	±95% CI	±95% CI	±95% CI	±95% CI	±2 std dev	±95% CI
Units of issue	5× (8–10) g	10 g	40 g	30 g	35 g	10 g	35 g
Form	Frozen	Freeze-dried	Freeze-dried	Freeze-dried	Freeze-dried	Slurry	Freeze-dried
Acenaphthene	2.7 ± 0.53*	4.9 ± 1.2*		3.4 (3.5)*	1100 (0.29–2300)*	967 ± 604	3.3–7*
Acenaphthylene	4.7 ± 1.2*			8 (12)*	8 (1.1–26)*	516 ± 290	
Anthracene	5.2 ± 0.71	6.2 ± 1.4*	1.5 (1.1)*	7.9 (10)*	19 (15–43)*	592 ± 284	4–93*
Benz[*a*]anthracene	46.8 ± 5.2	20.19 ± 0.87	3.8 (3.1)*	3.4 (3.9)*	0.81 (0.24–0.91)*		14–32*
Benzo[*a*]fluoranthene	6.26 ± 0.73						
Benzo[*a*]pyrene	27.6 ± 3.8	5.3 ± 0.61	0.9 (0.5)*	26 (42)*	2.9 (0.62–3)*		16–22*

(*continued overleaf*)

Table 11.6 *(continued)*

Code	SRM 1974b	SRM 2977	IAEA 432	IAEA-435	IAEA-406	EDF-2524	IAEA-140/OC
Benzo[*b*]chrysene	5 ± 0.30*						
Benzo[*b*]fluoranthene	63.8 ± 5.8	11.1 ± 0.50	4.8 (1.7)*	1.8 (2.1)*	2.3 (0.78–9.5)*	794 ± 157	33–37*
Benzo[*c*]chrysene	3.14 ± 0.42*						
Benzo[*c*]phenanthrene	18.3 ± 2.1*						
Benzo[*e*]pyrene	102 ± 11	13.29 ± 0.43	4.5 (1.7)*	23 (28)*			19–33*
Benzo[*ghi*]perylene	30.8 ± 3.3	9.45 ± 0.37		0.89 (0.65)*	0.78 (0.28–1.3)*		17–35*
Benzo[*j*]fluoranthene	29.5 ± 2.9	4.48 ± 0.15					
Benzo[*k*]fluoranthene	31.2 ± 1.8	4.02 ± 0.75*	1.9 (1.1)*		2.8 (0.38–130)*	222 ± 7.60	15–27*
Biphenyl	6 ± 1.4*	6 ± 1.3*		6.6 (1.4)*			
Chrysene	62.2 ± 9.9	42.2 ± 5.5*	5.5 (3.8)*	2.8 (2.6)*	2.3 (2.2–4.4)*	720 ± 314	25–49*
Cyclopenta[*cd*]pyrene	2.24 ± 0.10*						
Dibenz[*a,c*]anthracene	2.09 ± 0.13*						
Dibenz[*a,h*]anthracene	3.23 ± 0.31	1.47 ± 0.33*		8.7 (11)*			2.6–160*
Dibenz[*a, j*]anthracene	4.61 ± 0.47*						
Fluoranthene	169 ± 7	38.9 ± 0.63	12 (6.5)*	3.9 (4.2)*	4.9 (4–9.6)*	4,930 ± 1,310	57–110*
Fluorene	4.88 ± 0.36	10.3 ± 0.13	4.1 (2.2)*	1.6 (1.3)*		4,400 ± 3,530	4.6–1600*

Indeno[1,2,3-*cd*]pyrene	21.1 ± 1.1	4.76 ± 0.15		2.2 (2.6)*			20–53*
Naphthalene	24 ± 1.2	21.1 ± 1.4*	15 (18)*	7.8 (5.9)	24 (16–440)*	15,600 ± 15,800	9–43*
Perylene	9.8 ± 1.4	3.69 ± 0.38	5 (2.8)*				2.7–9.8*
Phenanthrene	25.5 ± 1.1	36.2 ± 2.5	27 (21)	8.4 (4.9)*	22 (8–220)*	12,000 ± 14,000	40–110*
Picene	7.4 ± 1.6*						
Pyrene	178 ± 6	77.4 ± 2.1	13 (6)*	11 (15)*	4.5 (2.1–9.1)*	6,300 ± 1,630	46–79*
Triphenylene	42.7 ± 7.1	36.1 ± 2.4*					
1-Methylnaphthalene	6.06 ± 0.49*	15.6 ± 1.5*	8.8 (9.5)*	17 (1)*			6.5–15*
1-Methylphenanthrene	9.66 ± 1.3	39 ± 1.9	4.2 (2.8)*	4.8 (3.6)*	12 (3.6–92)*		9–14*
2,3,5-Trimethylnaphthalene	3.95 ± 0.32*						
2,6-Dimethylnaphthalene	3.3 ± 1.6*			24 (2.9)*			
2-Methylanthracene	2.29 ± 0.04*						
2-Methylnaphthalene	12.3 ± 0.9*	17.3 ± 1.7*	14 (13)*	26 (2.9)*			9–23*
2-Methylphenathrene	24 ± 1.2		9.4 (4.9)*	24 (28)*	4.7 (3.3–10)*		15–40*
3-Methylphenanthrene	12.5 ± 0.4						
4-Methylphenanthrene and 9-methylphenanthrene	15.8 ± 1.8*						

Table 11.7 Reference materials available for PAHs in sediments. Values marked with an asterisk are non-certified; all other values are certified.

Code	BCR-535	SRM 1944	SRM 1941b	IAEA-159	IAEA-417	IAEA-408	CRM104	LGC6188
Organization	EU–JRC–IRMM EU	NIST	NIST	IAEA	IAEA	IAEA	RTC	LGC
Country of origin		USA	USA	Monaco	Monaco	Monaco	USA	UK
Matrix	Freshwater harbour sediment	New York, New Jersey waterway sediment	Marine sediment Baltimore Harbor	Estuarine sediment – River Clyde	Sediment – Venice Lagoon	Estuarine sediment – mudflats from the Tagus estuary	Sediment	River sediment
Units	mg kg^{-1}	mg kg^{-1}	μg kg^{-1}	μg kg^{-1}	μg kg^{-1}	μg kg^{-1}	mg kg^{-1}	mg kg^{-1}
As	Dry wt	Dry wt	Dry wt	Dry wt	Dry wt	Dry wt	Dry wt	Dry wt
[±] expressed as	Half width of 95% CI	±95% CI	±95% CI	±95% CI	±95% CI	±95% CI		
Units of issue	40 g	50 g	50 g	45 g	40 g	40 g		30 g
Form	Air dried	Freeze-dried	Freeze-dried	Dried	Deep-frozen dried	Freeze-dried		
Acenaphthene		0.57 ± 0.03*	38.4 ± 5.2*	6 (4)*	180 (62)*	3.3 (2.0–17)*	0.544 (0.501–0.587)	0.07 (0.02)
Acenaphthylene			53.3 ± 6.4*	6.4 (5.4)*	42 (35)*	3.6 (2.1–4.7)*		0.1*
Acephenanthrene			30.5 ± 1.9*					
Anthanthrene		0.9 ± 0.1*						
Anthracene		1.77 ± 0.33	184 ± 18	11 (5.1)*	630 (240)*	9.8 (8–13)*	0.365 (0.326–0.404)	0.36 (0.11)
Benz[*a*]anthracene	1.54 ± 0.10	4.72 ± 0.11	335 ± 25	54 (20)*	3200 (1200)*	53 (35–60)*	5.41 (5.01–5.82)	0.83 (0.18)

Benzo[*a*]fluoranthene		0.78 ± 0.12	73 ± 18*	15 (7.1)*				
Benzo[*a*]pyrene	1.16 ± 0.10	4.3 ± 0.13	358 ± 17	58 (26)*	2800 (1200)*	48 (30–63)*	0.396 (0.34–0.452)	0.65 (0.14)
Benzo[*b*]chrysene		0.63 ± 0.1	53 ± 12					
Benzo[*b*]fluoranthene	2.29 ± 0.15	3.87 ± 0.42	453 ± 21	100 (42)*	4100 (2000)*	46 (32–69)*	5.22 (4.86–5.58)	0.82 (0.19)
Benzo[*c*]phenathrene		0.76 ± 0.1	58 ± 15*					
Benzo[*e*]pyrene	1.86 ± 0.13	3.28 ± 0.11	325 ± 25	82 (8.2)*	3000 (830)*	55 (51–81)*		
Benzo[*ghi*]perylene		2.84 ± 0.1	307 ± 45	95 (45)*	2300 (1300)*	38 (20–52)*	0.378 (0.341–0.416)	0.36 (0.13)
Benzo[*j*]fluoranthene		2.09 ± 0.44	217 ± 5*					
Benzo[*k*]fluoranthene	1.09 ± 0.15	2.3 ± 0.2	225 ± 18	49 (14)*	2000 (300)*	46 (26–61)*	3.49 (3.22–3.76)	0.5 (0.08)
Biphenyl		0.32 ± 0.07*	74 ± 8*	11 (3.6)*	42 (4)*			
Chrysene		4.86 ± 0.1	291 ± 31	58 (26)*	3600 (1700)*	35 (25–56)*	6.59 (6.14–7.05)	0.83 (0.16)
Dibenz[*a,c*]anthracene		0.335 ± 0.013	36.7 ± 5.2					
Dibenz[*a,h*]anthracene		0.424 ± 0.069	53 ± 10	25 (14)*	1100 (1100)*	11 (7.8–14)*	1.08 (0.99–1.17)	0.13 (0.05)
Dibenz[*a, j*]anthracene		0.5 ± 0.044	48.9 ± 4.6					
Dibenzothiophene		0.62 ± 0.01*		6.5 (2.8)*	280 (16)*			
Fluoranthene		8.92 ± 0.32	651 ± 50	110 (32)*	7700 (3000)*	84 (53–110)*	9.2 (8.58–9.82)	1.79 (0.35)
Fluorene		0.85 ± 0.03*	85 ± 15	13 (7.7)*	230 (110)*	6.7 (4.6–24)*	0.626 (0.58–0.673)	0.12 (0.04)
Indeno[1,2,3-*cd*] fluoranthene			9.63 ± 0.34*					

(*continued overleaf*)

Table 11.7 *(continued)*

Code	BCR-535	SRM 1944	SRM 1941b	IAEA-159	IAEA-417	IAEA-408	CRM104	LGC6188
Indeno[1,2,3-*cd*]pyrene	1.56 ± 0.14	2.78 ± 0.1	341 ± 57	120 (34)*	2700 (370)*	51 (45–53)*		0.37 (0.14)
Naphthalene		1.65 ± 0.31	848 ± 95	23 (13)*	150 (110)*	27 (16–47)*	0.565 (0.513–0.617)	0.22 (0.11)
Pentaphene		0.288 ± 0.026	25.3 ± 1*					
Perylene		1.17 ± 0.24	397 ± 45	27 (11)*	1200 (460)*	320 (140–420)*		
Phenanthrene		5.27 ± 0.22	406 ± 44	59 (29)*	3900 (1500)*	35 (21–43)*	4.66 (4.36–4.96)	1.04 (0.3)
Picene		0.518 ± 0.093	46.6 ± 4.7					
Pyrene	2.52 ± 0.18	9.7 ± 0.42	581 ± 39	100 (38)*	6000 (2200)*	77 (57–93)*	7.43 (6.88–7.97)	1.48 (0.5)
Triphenylene		1.04 ± 0.27	108 ± 5	110 (84)*				
1,2-Dimethylphenanthrene		0.28 ± 0.01*						
1,6-, 2,9- and 2,5-dimethylphenanthrene		1.67 ± 0.03*						
1,7-Dimethylphenanthrene		0.62 ± 0.02*						
1,8-Dimethylphenanthrene		0.24 ± 0.01*						
1,9- and 4,9-dimethylphenanthrene		1.2 ± 0.03*						
1-Methylfluoranthene		0.66 ± 0.02*	42.4 ± 2.1*					
1-Methylnaphthalene		0.52 ± 0.03*	127 ± 14*	20 (10)*	53 (21)*	7.5 (4.7–12)*		
1-Methylphenanthrene		1.7 ± 0.1*	73.2 ± 5.9	20 (9.6)*	320 (320)*	10 (9.2–11)*		
1-Methylpyrene		1.29 ± 0.03*	52.5 ± 2.3*	17 (4.8)*				

2,3,5-Trimethylnaphthalene		25.5 ± 5.1*				
2,6-Dimethylnaphthalene		75.9 ± 4.5*	21 (4.2)*			
2,6-Dimethylphenanthrene	0.79 ± 0.02*					
2,7-Dimethylphenanthrene	0.67 ± 0.02*					
2-Methylanthracene	0.58 ± 0.04*	36 ± 15*				
2-Methylnaphthalene	0.95 ± 0.05*	276 ± 53*	30 (19)	81 (44)*	14 (13–33)*	0.28 (0.25–0.31)
2-Methylphenanthrene	1.9 ± 0.06*	128 ± 14	26 (6.5)*	580 (230)*	12 (12–4)*	
2-Methylpyrene	1.81 ± 0.04*	78.7 ± 4*				
3,5-Dimethylphenanthrene	1.31 ± 0.04*					
3,9-Dimethylphenanthrene	2.42 ± 0.05*					
3-Methylfluoranthene	2.46 ± 0.07*	28.8 ± 1.3*				
3-Methylphenanthrene	2.1 ± 0.1*	105 ± 13				
4-Methylphenanthrene and 9-methylphenanthrene	1.6 ± 0.2*	63.5 ± 2.5*				
4-Methylpyrene	1.44 ± 0.03*	66.4 ± 2.6*				
7-Methylfluoranthene	0.69 ± 0.02*	45.4 ± 1.5*				
8-Methylfluoranthene	0.86 ± 0.02*	49.5 ± 2.7*				

However, this material is frozen and not available in Europe. NIST freeze-dried mussel tissue CRMs are not available at present, including SRM 2977, which has lower concentrations than SRM 1974b. The IAEA produce four biota materials with recommended values for certain PAHs and information values for other PAHs, although reported 95% confidence intervals are often wide. These materials are a mussel tissue, two fish tissues and a seaweed (*Fucus* sp.). Cambridge Isotope Laboratories (CIL)/Cerilliant Corporation have produced a 'clean' fish material with reference concentrations for 11 parent PAH compounds. It is perhaps surprising to see the lack of mussel tissue available given the widespread use of mussels for PAH monitoring programmes globally. Furthermore, there are three materials for fish tissue, although PAH determination in fish tissue is not widely incorporated in monitoring programmes due to the capacity of fish to rapidly metabolize PAHs.

There are a limited number of CRMs for sediment shown in Table 11.7, and materials formerly available from NRC Canada (De Boer and McGovern, 2001) are no longer available. Currently available sediment materials for PAHs are from harbours and freshwater or estuarine environments, with only SRM 1941b, taken from the mouth of Baltimore Harbor, listed as a marine sediment. Thus, the materials are not ideal for use in wider marine monitoring programmes where lower concentrations may be routinely measured. Again, the NIST materials provide greatest coverage of PAHs, with 52 values listed for SRM 1944 of which 24 are certified. BCR-535 is a freshwater sediment certified for seven parent PAH (Wegener *et al*., 1999). In addition to the above-mentioned materials, there are eight reference sediments, the EC-series, available from NWRI (not tabulated), although these are not certified.

11.4.2 CRMs for Organotin Compounds

Monitoring of the antifoulant tributyltin (TBT) by chemical means in biota and sediment normally includes determination of its degradation products dibutyltin (DBT) and monobutyltin (MBT). The pesticide triphenyltin (TPhT) is also often incorporated in organotin analysis; it should be noted, however, that this compound is prone to instability (Morabito *et al*., 1999; Quevauviller and Ariese, 2001). CRMs for organotins in biota and sediment are produced by JRC–IRMM (EC), NRC and NIES (Japan), although NIES materials are not available outside Japan; see Table 11.8. There are three sediment materials available from NRC. High concentrations of TBT and DBT are evident in PACS-2 (NRC), a harbour sediment from British Columbia. SOPH-1 and HIPA-1 (NRC) were produced by diluting PACS-2 with HISS-1, a sediment collected off the coast of Newfoundland that contains negligible concentrations of organotins. CRM 462, a mussel tissue produced by EU–JRC–IRMM and certified for TBT, DBT and MBT, is currently the only CRM available for organotins in marine biota (Morabito *et al*., 1999). BCR-462 and BCR-646 are sediments certified for TBT and DBT. The former is a coastal sediment and the latter a freshwater sediment from the North Sea Canal with higher TBT levels. BCR-646 is also certified for MBT and for TPhT and its breakdown products.

11.5 CRMS FOR ORGANIC HALOGENATED COMPOUNDS

11.5.1 CRMs for PCBs

The (certified) RMs available for polychlorinated biphenyls (PCBs) in biota are reported in Tables 11.9 and 11.10. There are in total 20 materials: 5 mussel tissues, 1 whale blubber, 3 fish oils, 10 fish tissues and 1 sea plant (*Fucus*). The materials are produced by NIST, IAEA, NRC, EU–JRC–IRMM, CIL and Wellington Laboratories. The whale blubber SRM 1945 (NIST) provides the highest number of certified PCBs concentrations (i.e. 36 PCB congeners), although

Table 11.8 Reference materials available for organotin compounds in biota and in sediments. Values marked with an asterisk are non-certified; all other values are certified.

Code	ERM-CE477	NIES11	CRM-462	NIES 12	HIPA-1	SOPH-1	PACS-2	BCR-646
Organization Country of origin	EU–JRC–IRMM EU	NIES Japan	EU–JRC–IRMM EU	NIES Japan	NRC Canada	NRC Canada	NRC Canada	EU–JRC–IRMM EU
Matrix	Mussel tissue	Fish tissue	Coastal sediment	Marine sediment	Sediment	Sediment	Sediment	Freshwater sediment
UNITS	$mg\,kg^{-1}$	$\mu g\,g^{-1}$	$\mu g\,kg^{-1}$	$mg\,kg^{-1}$	$ng\,g^{-1}$ (as Sn)	$ng\,g^{-1}$ (as Sn)	$ng\,g^{-1}$ (as Sn)	$\mu g\,kg^{-1}$
As	Dry wt	Dry wt	Dry wt	Dry wt				Dry wt
[±] expressed as	Half width of 95% CI		±95% CI		±95% CI	±95% CI	±95% CI	±95% CI
Units of issue	14 g	20 g	25 g	30 g	25 g	12 g	65 g	40 g
Form	Freeze-dried	Freeze-dried	Dried	Air dried	Freeze-dried	Freeze-dried	Freeze-dried	Dried
Triphenyltin (TPhT)		*6.3		*0.008				29 ± 11
Tributyltin (TBT)	2.2 ± 0.19	1.3 ± 0.1	54 ± 15	0.19 ± 0.03	78 ± 9	125 ± 7	890 ± 105	480 ± 80
Dibutyltin (DBT)	1.54 ± 0.12		68 ± 12			174 ± 9	1047 ± 64	770 ± 90
Monobutyltin (MBT)	1.50 ± 0.28						600*	610 ± 120

Reference value not certified: NIES 11 mg TPhT (as chloride) kg^{-1}; PACS 2 MBT $ng\,g^{-1}$ (as Sn);
NIES 11 and NIES 12 not available overseas from Japan. BCR-646 also certified for TPT, DPT and MPT.

Table 11.9 Reference materials available for PCBs in shellfish and marine algae. Values marked with an asterisk are non-certified; all other values are certified.

Code	SRM 1974b	SRM 2974a	SRM 2977	IAEA-432	BCR 682	IAEA 140/OC
Organization	NIST	NIST	NIST	IAEA	EU–JRC–IRMM	IAEA
Country of origin	USA	USA	USA	Monaco	EU	Monaco
Matrix	Mussel tissue	Mussel tissue	Mussel tissue	Mussel homogenate	Mussels	*Fucus* (sea plant homogenate)
Unit	μg kg^{-1}	μg kg^{-1}	μg kg^{-1}	μg kg^{-1}	μg kg^{-1}	μg kg^{-1}
As	Dry wt	Dry wt	Dry wt	Dry wt	Wet wt	Dry wt
[±] expressed as	±95% CI	±95% CI	±95% CI	±95% CI	±95% CI	±95% CI
Units of issue	5× (8–10) g	8 g	10 g	40 g	Tin, 70 g	3 5g
Form	Frozen	Freeze-dried	Freeze-dried	Freeze-dried	Wet, sterilized	Freeze-dried
PCB 8	3.7 ± 1.1*	2.01 ± 0.08	1.99 ± 0.14	0.57 (0.32)*		
PCB 18	8.30 ± 1.3	4.03 ± 0.22	2.24 ± 0.74*	1.6 (1.7)*		1.5 (1.2–1.9)*
PCB 28	33.9 ± 2.5	12.4 ± 0.2*	5.17 ± 0.36	0.32 (0.26)*	0.30 ± 0.07	1.7 (1.3–2.5)*
PCB 31	28.4 ± 2.3	9.24 ± 0.13*	3.86 ± 0.29	1 (1.04)*		1.8 (1.7–2.1)*
PCB 44	38.0 ± 2.0	16.24 ± 0.71	3.22 ± 0.21	0.7 (0.68)*		2 (1.9–2.1)*
PCB 45	4.9 ± 1.8*	1.02 ± 0.06*				

PCB 49	55.9 ± 2.3	17.1 ± 1.2	2.44 ± 0.27	0.29 (0.08) ref		1.6 (1.1–1.9)*
PCB 52	61.8 ± 3.7	22.42 ± 0.92	8.02 ± 0.56	1.2 (1.2)*	0.78 ± 0.09	3.8 (2.6–4.9)*
PCB 66	62.9 ± 3.7	20.6 ± 1.1	3.55 ± 0.18	0.38 (0.16)*		0.89 (0.65–1.10)*
PCB 70	59.3 ± 2.2	15.45 ± 0.64		0.64 (0.35)*		1.8 (1.1–4.0)*
PCB 74	35.0 ± 2.3	9.02 ± 0.37		0.13 (0.03)*		
PCB 82	11.5 ± 1.4	4.56 ± 0.21*		0.11 (0.08)*		
PCB 87	42.7 ± 3.6	14.36 ± 0.56		0.27 (0.11)*		1.6 (1.2–2.0)*
PCB 92	27.2 ± 5.7*					
PCB 95	59.6 ± 3.6	23.72 ± 0.49	5.17 ± 0.53	1.1 (0.79)*		
PCB 97				0.31 (0.16)*		
PCB 99	58.4 ± 2.7	24.51 ± 0.54	3.0 ± 1.2*	2.1 (2.4)*		0.9 (2.7–9.8)*
PCB 101	106 ± 11	39.84 ± 0.96	10.6 ± 0.9	1.2 (0.49)*		2.4 (1.9–3.2)*
PCB 105	39.5 ± 1.8	16.47 ± 0.43	2.93 ± 0.46*	0.5 (0.39)*		0.49 (0.4–0.7)*
PCB 107	10.2 ± 1.2					
PCB110	99.1 ± 7.1	35.88 ± 0.87		1.12 (0.4)*		
PCB118	102 ± 4	42.9 ± 2.1	10.0 ± 0.41	1.09 (0.42)*	2.6 ± 0.3	1 (0.97–1.1)*
PCB123						
PCB128	17.7 ± 1.2	8.24 ± 0.33	2.38 ± 0.28	0.25 (0.08)*		
PCB 130		3.05 ± 0.14*				
PCB 132	24.0 ± 2.5					
PCB 138	91 ± 14	61.5 ± 2.3	7.94 ± 0.63	2.2 (0.84)*	4.6 ± 0.8	1.7 (1.2–2.6)*
PCB 138/163/164	133.5 ± 9.5					
PCB 141				0.19 (0.22)*		
PCB 146	19.0 ± 1.6	8.07 ± 0.40		0.59 (0.11) info		
PCB 149	69.2 ± 2.8	31.77 ± 0.95	8.95 ± 0.67	1.4 (0.41)*	5.7 ± 0.9	1.2 (1.1–1.6)ref
PCB 151	18.4 ± 1.6	5.99 ± 0.20		0.4 (0.04)*		
PCB 153	121 ± 8	78.8 ± 2.5	14.1 ± 1.3	2.8 (0.99)*	9.2 ± 0.8	1.7 (1.3–3.1)*

(continued overleaf)

Table 11.9 *(continued)*

Code	SRM 1974b	SRM 2974a	SRM 2977	IAEA-432	BCR 682	IAEA 140/OC
PCB 156	7.09 ± 0.79	5.80 ± 0.25	0.950 ± 0.036	0.35 (0.31)*		0.17 (0.08–0.36)*
PCB 157	2.33 ± 0.24*	0.952 ± 0.059*				
PCB 158	9.86 ± 0.95			0.17 (0.07)*		
PCB 163	19.9 ± 0.5*					
PCB 167		2.12 ± 0.14*		0.32 (0.1)*		
PCB 170	2.66 ± 0.34	2.04 ± 0.08	2.74 ± 0.25	0.57 (0.51)*	0.17 ± 0.05	0.21 (0.16–0.76)*
PCB 174				0.26 (0.27)*		
PCB 177		5.48 ± 0.20		0.26 (0.14)*		
PCB 178		2.40 ± 0.13*				
PCB 180	11.5 ± 1.0		6.32 ± 0.72	0.2 (0.11)*	0.77 ± 0.07	0.43 (0.38–0.52)*
PCB 183	12.3 ± 0.3	7.06 ± 0.26		0.24 (0.05)*		
PCB 187	29.0 ± 1.5	15.52 ± 0.48	4.47 ± 0.32	0.87 (0.13)*		0.38 (0.30–1.2)*
PCB189				0.12 (0.15)*		
PCB 194		0.485 ± 0.040	0.881 ± 0.032	0.27 (0.28)*		
PCB 195				0.22 (0.18)*		

SRM 1947b: certified also for PCB 187/159/182 and reference values for PCB 56 and 67; SRM 2974a certified values also for 180/193 and reference values also for PCB 56/60, 92/84/89, 202.

Table 11.10 Reference materials available for PCBs in fish and blubber. Values marked with an asterisk are non-certified; all other values are certified.

Code	SRM 1588b	SRM 1945	BCR-349	ERM-BB350	IAEA-406	IAEA-435	CARP-2	EDF-2524	EDF-2525	EDF-2526	WMF-01	SRM 1946	SRM 1947	BCR-718	BCR-719
Organization	NIST	NIST	EU–JRC–IRMM	EU–JRC–IRMM	IAEA	IAEA	NRC	CIL	CIL	CIL	Wellington Laboratories	NIST	NIST	EU–JRC–IRMM	EU–JRC–IRMM
Country of origin	USA	USA	EU	EU	Monaco	Monaco	Canada	USA	USA	USA	Canada	USA	USA	EC	EC
Matrix	Cod liver oil	Whale blubber	Cod liver oil	Fish oil	Fish homogenate	Tuna homogenate	Ground whole carp	Clean fish	Contaminated fish	Fortified fish	Fish tissue	Lake Superior fish tissue	Lake Michigan fish tissue	Canned fresh herring	Chub
Unit	μg kg^{-1}	μg kg^{-1}	μg kg^{-1}	μg kg^{-1}	μg kg^{-1}	μg kg^{-1}	μg kg^{-1}	ng kg^{-1}	ng kg^{-1}	ng kg^{-1}	pg g^{-1}	μg kg^{-1}	μg kg^{-1}	μg kg^{-1}	ng kg^{-1}
As		Wet wt			Dry wt	Dry wt	Wet wt	Wet wt	Wet wt	Wet wt	Dry wt	Wet wt	Wet wt	Wet wt	
[±] expressed as	±95% CI	±95% CI	±95% CI	±95% CI	±95% CI	±95% CI	±95% CI	±95% CI	±95% CI	±95% CI		±95% CI	±95% CI	±95% CI	±95% CI
Units of issue	5 × 1.2 mL	2 × 15 g	2 g	2 g	35 g	30 g	6 × 9 g	1 × 10 g	1 × 10 g	1 × 10 g	10 g	5 × 7–9 g	5 × 8 g	70 g	70 g
Form		Frozen			Freeze-dried	Freeze-dried	Slurry	Slurry	Slurry	Slurry	Freeze-dried	Frozen	Frozen		
PCB 8							4.8 ± 1.8*								
PCB18	7.96 ± 0.45	4.35 ± 0.56			0.38 (0.11–0.93)*	4.3 (4.6)*	27.3 ± 4.0	86.0 ± 60.4	1390 ± 970	100 ± 49.0		0.84 ± 0.11*	2.72 ± 0.95*		
PCB28	27.8 ± 1.4	13.1 ± 1.1	68 ± 8	21.3 ± 1.1	0.57 (0.43–1.3)*	1.4 (0.55)*	86.6 ± 25.9	253 ± 244	7100 ± 1260	245 ± 268		2.00 ± 0.24*	14.1 ± 1.0	0.41 ± 0.04	

(*continued overleaf*)

Table 11.10 *(continued)*

Code	SRM 1588b	SRM 1945	BCR-349	ERM-BB350	IAEA-406	IAEA-435	CARP-2	EDF-2524	EDF-2525	EDF-2526	WMF-01	SRM 1946	SRM 1947	BCR-718	BCR-719
PCB 31	8.52 ± 0.50	3.56 ± 0.36			0.29 (0.23–0.47)*	1.5 (0.66)*			4000 ± 71.6			1.46 ± 0.20*	10.4 ± 1.4		
PCB 44	32.4 ± 2.7	12.1 ± 0.5			0.46 (0.46–0.52)*	3.0 (2.8)*		274 ± 192	14 200 ± 9660			4.66 ± 0.86	20.4 ± 1.7		
PCB 45	2.17 ± 0.29*				0.18 (0.03–0.33)*								1.76 ± 0.76*		
PCB 49	29.8 ± 0.6	18.3 ± 0.1			0.41 (0.36–0.44)*	3.1 (1.7) info		176 ± 41.6	13 600 ± 9100			3.80 ± 0.39	27.3 ± 3.8		
PCB 52	82.4 ± 1.7	40.7 ± 1.3	149 ± 21	37.4 ± 2.2	1.3 (1.0–2.2)*	4.4 (2.5)*	138 ± 43	653 ± 200	27 100 ± 12 100	369 ± 124		8.1 ± 1.0	36.4 ± 4.3	1.00 ± 0.04	
PCB 66	51.7 ± 7.5	22.4 ± 0.5			0.67 (0.49–0.96)*	5.0 (1.2)*		218 ± 212	56 500 ± 20 800			10.8 ± 1.9	69.4 ± 5.3		
PCB 70	15.9 ± 0.7*	11.0 ± 1.1				4.6 (2.1)*			44 400 ± 3860			14.9 ± 0.6	50 ± 12*		
PCB 74	40 ± 11*	21.6 ± 7.8*		23.0 ± 1.9				348 ± 398	23 100 ± 8440			4.83 ± 0.51	33.7 ± 3.1		
PCB 81								1.27 ± 2.52	161 ± 74.0	3.00 ± 5.60					13.6 ± 0.4
PCB 82	8.2 ± 1.6*	4.5 ± 1.1											3.87 ± 0.67*		
PCB 87	54.6 ± 2.1	20.6 ± 2.6			0.82 (0.69–1.4)*	6.9 (3.9)*			38 400 ± 24 000			9.4 ± 1.4	27.9 ± 1.5		
PCB 92	36.2 ± 8.2*	23.4 ± 3.9											32.6 ± 5.2*		

PCB 95	37.7 ± 8.2	33.9 ± 0.5			1.1 (1.0–1.2)*	8.3 (6.3)*						11.4 ± 1.3	33.6 ± 5.1*	
PCB97									29 800 ± 14 700					
PCB 99	101 ± 12	58.5 ± 5.2		62 ± 6	0.91 (0.89–1.2)*	16 (4.1)*		588 ± 120	94 300 ± 25 200	215 ± 204		25.6 ± 2.3	78.0 ± 6.0	
PCB 101	127 ± 9	78 ± 12	372±18	111 ± 5	3.1 (2.2–3.4)*	23 (10)*		1130 ± 274	82 700 ± 21 400	579 ± 362		34.6 ± 2.6	90.8 ± 0.3	2.12 ± 0.06
PCB 105	59.2 ± 1.2	28.6 ± 1.2		25.8 ± 2.1	0.71 (0.48–0.88)*	6.7 (3.6)*	53.2 ± 15.6	280 ± 80.4	50 100 ± 15 700	108 ± 73.0	49050 ± 14200	19.9 ± 0.9	50.3 ± 3.7	0.63 ± 0.06
PCB 107	16.2 ± 3.9*	5.04 ± 0.94										8.86 ± 0.20*	17.1 ± 1.2	
PCB110	69.3 ± 4.8	33.8 ± 2.3		54.1 ± 2.8	1.4 (1.3–2.0)*	15 (8.3)*		789 ± 170	84 900 ± 19 100	288 ± 112			94.6 ± 4.3	
PCB118	172 ± 7	76.5 ± 2.9	460 ± 40	84 ± 4	2.5 (1.9–3.7)*	24 (9.3)*	148 ± 33	692 ± 104			130100 ±32500	52.1 ± 1.0	112 ± 6	1.78 ± 0.07
PCB123								11.4 ± 9.24	3280 ± 2020	7.38 ± 9.58				
PCB128	39.9 ± 6.0	23.0 ± 1.1			0.8 (0.51–1.4)*	9.5 (4.7)*	20.4 ± 4.4	127 ± 62.2	28 200 ± 9460			22.8 ± 1.9	31.6 ± 2.1	0.62 ± 0.10
PCB 130	7.35 ± 0.27*	10.2 ± 0.6*												
PCB 132		21.1 ± 4.8*										5.83 ± 0.76*	20.8 ± 2.1	
PCB 137	11.8 ± 0.8*	7.7 ± 1.4*						31.8 ± 26.8	7250 ± 2440					

(*continued overleaf*)

Table 11.10 *(continued)*

Code	SRM 1588b	SRM 1945	BCR-349	ERM-BB350	IAEA-406	IAEA-435	CARP-2	EDF-2524	EDF-2525	EDF-2526	WMF-01	SRM 1946	SRM 1947	BCR-718	BCR-719
PCB 138	212 ± 29	146 ± 13		137 ± 10	4 (2.5–6.3)*	70 (32)*		1110 ± 400	178 000 ±27800	395 ± 184		115 ± 13	162.0 ± 6.9	2.97 ± 0.11	
PCB 138/163/164	263.5 ± 9.1	131.5 ± 7.4					103 ± 30*								
PCB 141					0.41 (0.40–0.43)*			152 ± 119	22 040 ± 3500						
PCB 146	47 ± 14*	45.1 ± 4.0						261 ± 69.2	39 500 ± 17 000			30.1 ± 3.5	40.5 ± 2.0		
PCB 149	90 ± 11	89.0 ± 6.9		88 ± 9	2 (1.7–2.4)*	29 (7.1)*		608 ± 788	69 800 ± 24 600			26.3 ± 1.3	67.1 ± 3.7	2.58 ± 0.11	
PCB 151	48.1 ± 5.4	28.6 ± 1.3			0.67 (0.62–0.75)*	12 (2.7)*		279 ± 212	24 900 ± 11 100	99.8 ± 17.1			23.3 ± 5.3*		
PCB 153	275 ± 4	228 ± 10	940 ± 40	220 ± 11	3.7 (2.9–6.0)*	81 (37)*	105 ± 22	1360 ± 516	226 000 ±71200	490 ± 334		170 ± 9	201 ± 3	4.62 ± 0.10	
PCB 154	3.48 ± 0.29*	3.97 ± 0.22											3.51 ± 0.46*		
PCB 156	18.0 ± 2.1	11.4 ± 0.9		20.1 ± 1.3	0.27 (0.21–0.59)	3.4 (1.3)*		64.7 ± 18.4	13 100 ± 2620	23.3 ± 23.8	14890 ± 5020	9.52 ± 0.51	13.3 ± 0.9	0.19 ± 0.09	
PCB 157	4.3 ± 1.2*	3.27 ± 0.67*						19.2 ± 8.68	3380 ± 1010	9.30 ± 9.16			4.08 ± 0.77*		
PCB 158	11.3 ± 1.8*	9.1 ± 1.0						73.2 ± 52.6	11 600 ± 1870			7.66 ± 0.88*	11.3 ± 0.9		
PCB 163	42.2 ± 3.2*	46.7 ± 8.1*										31.8 ± 0.8*	40.0 ± 5.2*		

PCB 167	12.2 ± 0.8*							24.7 ± 17.1	7060 ± 3020	12.0 ± 9.54				
PCB 170	41.9 ± 3.8	42.6 ± 2.2			0.54 (0.38–0.81)*	11 (5.9)*		119 ± 42.0	35 100 ± 12 700			25.2 ± 2.2	29.2 ± 2.4	0.350 ± 0.026
PCB 170/190		40.6 ± 2.6					20.6 ± 2.9*							
PCB 172	8.39 ± 0.33*	12.4 ± 3.4*						38.6 ± 12.0	8450 ± 1600					
PCB 174	15.0 ± 2.8*	25.2 ± 0.1			0.3 (0.29–0.37)*	9.1 (3.1)*						9.3 ± 1.3*	18.6 ± 1.7	
PCB 177	5.18 ± 0.36*	28.6 ± 1.2*		25.8 ± 2.0	0.18 (0.16–0.19)*	7.9 (4.2)*		126 ± 47.8	18 800 ± 4140					
PCB 178		20.7 ± 1.7*						68.2 ± 5.80	12 100 ± 1840					
PCB 180	98.5 ± 6.3	138 ± 10	282 ± 23	67 ± 4	1.2 (1.0–1.2)*	32 (13)*	53.3 ± 13.0	412 ± 182	108 000 ±23600	116 ± 64.4		74.4 ± 4.0	80.8 ± 5.0	0.795 ± 0.027
PCB 183	28.9 ± 1.8	38.0 ± 1.8		22.5 ± 1.8	0.32 (0.28–0.35)*	11 (2.9)*		125 ± 59.4	28 300 ± 6740			21.9 ± 2.5	23.3 ± 1.9	
PCB 187	43.6 ± 7.2	121 ± 11		67 ± 5	1.1 (0.96–1.2)*	31 (10)*		357 ± 222	62 900 ± 21 600			55.2 ± 2.1	54.8 ± 2.6	
PCB189								6.16 ± 2.72	1440 ± 498	3.51 ± 2.76				

(continued overleaf)

Table 11.10 *(continued)*

Code	SRM 1588b	SRM 1945	BCR-349	ERM-BB350	IAEA-406	IAEA-435	CARP-2	EDF-2524	EDF-2525	EDF-2526	WMF-01	SRM 1946	SRM 1947	BCR-718	BCR-719
PCB 194	13.5 ± 1.2	53.5 ± 5.2		23.4 ± 1.5	0.13 (0.10–0.25)*	3.9 (1.1)*	10.9 ± 3.1	48.1 ± 25.6	12 700 ± 3200			13.0 ± 1.3	13.2 ± 0.9		
PCB 195	4.6 ± 0.6*	14.3 ± 2.2			0.04 (0.03–1.5)*	2.2 (1.3)*			4620 ± 1450			5.30 ± 0.45	4.95 ± 0.77		
PCB 196		74.5 ± 8.4*		41 ± 7				44.8 ± 56.6	7720 ± 3240						
PCB 199	13.3 ± 0.2*	84.2 ± 8.0*			0.16 (0.013–0.17)*			81.7 ± 67.4	16 700 ± 2400						
PCB 200		4.30 ± 0.32*			0.1 (0.009–1.1)*										
PCB 201	12.18 ± 0.46	16.8 ± 1.3			0.12 (0.07–0.23)*	4.7 (1.6)*						2.83 ± 0.13*	3.59 ± 0.43*		
PCB 203									13 800 ± 2360						
PCB 206	3.75 ± 0.07	44.9 ± 4.2			0.05 (0.03–3.0)*	2.8 (1.4)*	4.4 ± 1.1	10.4 ± 2.06	4960 ± 768			5.40 ± 0.43	6.24 ± 0.88		
PCB 208		23.9 ± 1.7*							2370 ± 350						
PCB 209	3.21 ± 0.26	17.2 ± 1.9			0.07 (0.07–1.1)*	2.0 (0.84)*	4.6 ± 2.0*	14.4 ± 1.08	3510 ± 982			1.30 ± 0.21	2.45 ± 0.68*		

SRM 1588b: reference values reported also for PCB 16/32, 17, 22, 33, 41/71, 42, 46, 55/60, 63, 64, 83, 85, 91, 97, 114, 136, 141, 178/129, 179, 185, 187/159/182, 189, 191, 193, 196/203, 207. SRM 1945: reference values also for PCB 56, 119, 175, 176, 185, 191, 193, 197, 202, 205, 207. IAEA-406: noncertified value also for PCB 198. CARP-2 reference values also for PCB 66/95, 101/90, 187/182. EDF-2524 certified values also for PCB 37 and PCB 114. EDF-2525 certified values also for PCB 32, 37 and 47. WMF-01 certified values also for PCB 81, 114, 123, 157, 167. SRM-1946 reference values reported also for PCB 56, 63, 193. SRM-1947 certified values also for PCB 63 and PCB 193, and reference value for PCB 56.

there are difficulties in distributing this material overseas due to trade restrictions on marine mammal products. The other NIST materials also offer a wide range of certified PCBs, as do the materials produced by CIL, although the latter quote a much higher uncertainty. European materials certified include fish oils (BCR-349 and ERM-BB350), mussel (BCR-682) and herring (BCR-718) (van Leeuwen *et al*. 2006).

There are 10 (certified) RMs available for PCBs in sediments – see Table 11.11; three materials are produced by NIST, three by IAEA, one by EC–JRC–IRMM, CIL, Wellington Laboratories and RTC. Of the 10 materials available, only SRM 1941b (NIST) is a marine sediment, all the others being freshwater, estuarine or lagoon sediments. The most comprehensive CRMs are those from NIST, with SRM 1944, SRM 1941b and SRM 1939a, certified for 25, 28 and 64 PCBs congeners respectively. The heavily contaminated sediment EDF-5184 by CIL offers also certified values for many PCB congeners, but wide uncertainties are quoted. The materials provided by IAEA provide recommended or informative values.

11.5.2 CRMs for PCDDs, PCDFs and Dioxin-like PCBs

The (certified) RMs for dioxins and furans are reported in Tables 11.12 and 11.13 for biota and sediment respectively. With regard to biota, a comparison with the materials available in 2000 (De Boer and McGovern, 2001) indicates that, with the exception of CARP-2 replacing CARP-1 and the new fish tissue WMF-01 by Wellington Laboratories, the situation is almost unchanged. This confirms the difficult to certify these analytes.

The materials available are the cod liver oil SRM 1588a (NIST), carp tissue CARP-2 (NRC) and three fish tissue materials with different degrees of contamination (CIL), and the above-mentioned fish tissue WMF-01 (Wellington Laboratories). SRM 1588b by NIST has reference values for several dioxin congeners and octachlorodibenzofuran (OCDF). The carp tissue has reference values only for dioxins. The three CIL materials have certified values for a range of furans and dioxin congeners, although the reported uncertainties are wide. WMF-01 has certified values only for three dibenzo-*p*-dioxins and two dibenzofurans; all others are provisional values for information only, which might be certified at a later stage.

The dioxin-like PCBs are certified in the CIL and Wellington Laboratories materials and in BCR-719. Reference values are reported in SRM 1974b (NIST), SRM 1945 (NIST) and in the sea plant homogeneate IAEA-140 OC for PCB 77, in SRM 1945 (NIST) for PCB 126 and PCB 169.

The availability of (certified) RMs for dioxins and furans in sediments has decreased considerably since 2000. A number of materials are no longer available, in particular those produced by NWRI. Therefore, there are currently only three materials with values for dioxin and furans. Those are the SRM 1944 (NIST), EDF-5184 (CIL) and WMS-01 (Wellington Laboratories). The values are reference for the first and certified for the second and third. Dioxin-like PCBs are certified in EDF-5184 (CIL) and WMS-01 (Wellington Laboratories), and SRM-1941b (NIST) provides a reference value for PCB 77.

11.5.3 CRMs for BFRs

The (certified) RMs for brominated flame retardants commercially available are reported in Table 11.14 for biota and Table 11.15 for sediment. These materials were not available in 2000 (De Boer and McGovern, 2001). For biota, eight materials are listed. NIST has provided certified or reference values in four materials (cod liver oil, whale blubber, fish tissue and mussel tissue),

Table 11.11 Reference materials available for PCBs in sediments. Values marked with an asterisk are non-certified; all other values are certified.

Code	SRM 1944	SRM 1941b	SRM 1939a	BCR-536	EDF-5184	IAEA 417	IAEA 408	IAEA 159	CNS329-050	WMS-01
Organization	NIST	NIST	NIST	EU–JRC–IRMM	CIL	IAEA	IAEA	IAEA	RTC	Wellington Laboratories
Country of origin	USA	USA	USA	EU	USA	Monaco	Monaco	Monaco	USA	Canada
Matrix	New York New Jersey waterway sediment	Marine sediment	River sediment A	Freshwater harbour sediment	Heavily contaminated sediment	Venice Lagoon sediment	Tagus estuary sediment	Sediment	Fresh water sediment	Lake sediment
Units	μg kg^{-1}	μg kg^{-1}	μg kg^{-1}	μg kg^{-1}	μg kg^{-1}	ng g^{-1}	ng g^{-1}	ng g^{-1}	μg kg^{-1}	Pg g^{-1}
As	Dry wt	Dry wt		Dry wt		Dry wt	Dry wt	Dry wt		Dry wt
[±] expressed as	95% CI	95% CI	95% CI	Half-width of 95% CI of mean	2 st dev	95% CI	95% CI	95% CI		
Units of issue	50 g	50 g	50 g	40 g	10 g	40 g	40 g	45 g	50 g	25 g
Form	Freeze-dried	Freeze-dried	Air dried	Dried		Dried	Freeze-dried	Dried		Freeze-dried
PCB 8	22.3 ± 2.3	1.65 ± 0.19	5049 ± 294			1.8 (1.2)*	0.24 (0.22–4.6)*	0.37 (0.21)*		
PCB 18	51.0 ± 2.6	2.39 ± 0.29	2126 ± 264		27.6 ± 11.2	3.7 (2.4)*	0.74 (0.13–1.9)*	0.74 (0.53)*		
PCB 28	80.0 ± 2.7	4.52 ± 0.57	1676 ± 91	44 ± 5	54.2 ± 15.5	5.7 (2.8)*	0.79 (0.35–0.98)*	0.57 (0.28)*	54.1*	

PCB 31	78.7 ± 1.6	3.18 ± 0.41	5698 ± 135			4.1 (1.5)*	0.43 (0.09–1.5)*	0.28 (0.05)*		
PCB 44	60.2 ± 2.0	3.85 ± 0.20	766 ± 53		657 ± 159	9.7 (4.6)*	0.47 (0.23–3.0)*	0.35 (0.14)*		
PCB 49	53.0 ± 1.7	3.85 ± 0.20	3314 ± 317		476 ± 155	7.8 (2.9)*	0.35 (0.22–0.38)*	0.32 (0.09)*		
PCB 52	79.4 ± 2.0	5.24 ± 0.28	3691 ± 68	38 ± 5	1340 ± 260	17 (7.9)*	0.6 (0.38–0.93)*	0.67 (0.25)*	230*	
PCB 66	71.9 ± 4.3	4.96 ± 0.53	556 ± 25		403 ± 40.8	23 (16)*		0.41 (0.07)*		
PCB 87	29.9 ± 4.3	1.14 ± 0.16	166 ± 21		1810 ± 1110	19 (6.0)*	0.87 (0.26–1.1)*	0.44 (0.25)*		
PCB 95	65.0 ± 8.9	3.93 ± 0.62	859 ± 29			20 (1.5)*		1.26 (1.17)*		
PCB 99	37.5 ± 2.4	2.90 ± 0.36	386 ± 81*		1160 ± 496	19 (7.8)*		0.40 (0.20)*		
PCB 101		5.11 ± 0.34	476 ± 42	44 ± 4	3140 ± 552	42 (15)*	1.2 (0.81–1.7)*	0.52 (0.16)*	390*	
PCB 105	24.5 ± 1.1	1.43 ± 0.10	173 ± 47*	3.5 ± 0.6	1050 ± 314	22 (4.5)*	1.2 (0.81–1.7)*	0.41 (0.31)*		3998 ± 951
PCB 110	63.5 ± 4.7	4.62 ± 0.36	1008 ± 118		3340 ± 768	42 (15)*	0.83 (0.50–0.90)*	0.54 (0.16)*		
PCB 118	58.0 ± 4.3	4.23 ± 0.19	385 ± 103*	27.5 ± 2.2	2520 ± 904	43 (18)*	1.2 (0.9–1.6)*	0.52 (0.21)*	75*	8115 ± 1663
PCB 128	8.47 ± 0.28	0.696 ± 0.044	76 ± 24*	5.4 ± 1.2	694 ± 181	12 (4.2)*	0.33 (0.27–0.53)*	0.27 (0.27)*		
PCB 137			21.7 ± 4.1*		164 ± 106					

(continued overleaf)

Table 11.11 (*continued*)

Code	SRM 1944	SRM 1941b	SRM 1939a	BCR-536	EDF-5184	IAEA 417	IAEA 408	IAEA 159	CNS329-050	WMS-01
PCB 138		3.60 ± 0.28	190 ± 52*	27 ± 5	3970 ± 2820	45 (22)*	1.6 (1.1–2.1)*	0.60 (0.31)*	226*	
PCB 141					1010 ± 346	5.2 (2.1)*				
PCB 149	49.7 ± 1.2	4.35 ± 0.26	402 ± 105*	49 ± 4	3390 ± 838	25 (11)*	1.4 (1.3–1.6)*	0.52 (0.14)*		
PCB 151	16.93 ± 0.36		181 ± 49*		1410 ± 788	3.9 (0.18)*				
PCB 153	74.0 ± 2.9	5.47 ± 0.32		50 ± 4	3880 ± 902	39 (20)*	1.9 (0.98–2.1)*	0.56 (0.09)*	133*	
PCB 156	6.52 ± 0.66	0.507 ± 0.090	40.7 ± 4.4	3.0 ± 0.4	457 ± 189	5.9 (1.2)*	0.36 (0.31–0.39)*	0.32 (0.43)*		715 ± 248
PCB 170		1.35 ± 0.09	97 ± 27*	13.4 ± 1.4	1250 ± 334	8.1 (4.3)*	0.47 (0.34–0.59)*	0.15 (0.04)*		
PCB 174						3.1 (0.97)*	0.34 (0.20–0.44)*			
PCB 177			73.3 ± 6.8		743 ± 238	1.8 (0.61)*	0.35 (0.23–0.43)*	0.20 (0.10)*		
PCB 180	44.3 ± 1.2	3.24 ± 0.51		22.4 ± 2.1	2940 ± 774	16 (7.4)*	1.1 (0.85–1.2)*	0.26 (0.10)*	104*	
PCB 183	12.19 ± 0.57	0.979 ± 0.087	38 ± 11*		810 ± 394	3.4 (1.4)*	0.32 (0.26–0.33)*	0.27 (0.32)*		

PCB 187		2.17 ± 0.22	41.1 ± 5.7*		1520 ± 232	8.1 (3.3)*	0.68 (0.49–0.94)*	0.39 (0.05)*		
PCB 189			3.40 ± 0.47		50.2 ± 18.2	0.41 (0.23)*				85.2 ± 17.8
PCB 194	11.2 ± 1.4	1.04 ± 0.06	41.1 ± 5.7		622 ± 146	2.7 (1.3)*	0.2 (0.2–0.23)*	0.09 (0.04)*		
PCB 195	3.75 ± 0.39	0.645 ± 0.060	17.8 ± 3.0*		268 ± 73.8	1.2 (0.53)*	0.12 (0.06–0.16)*	0.08 (0.03)*		
PCB 199			7.61 ± 0.92		691 ± 226	1.2 (0.79)*				
PCB 206	9.21 ± 0.51	2.42 ± 0.19	33.4 ± 2.4		152 ± 35.4	1.8 (0.96)*		0.34 (0.47)*		
PCB 209	6.81 ± 0.33	4.86 ± 0.45	8.32 ± 0.60		6.03 ± 3.1	1.2 (0.7)		0.14 (0.21)*		

SRM 1944 also certified for PCB 101/90, 138/163/164, 170/190, 187/159/182. SRM.1941b has reference values also for PCB 45, 56, 63, 70, 74, 77, 107, 132, 146, 158, 163, 174, 193, 201. SRM 1939a also certified for PCB 45, 56, 63, 70, 74, 82, 92, 107, 112, 114, 119, 130, 146, 152/132, 154, 157, 158, 163, 167, 172, 174, 175, 176, 178, 180/193, 185, 191, 201, 202, 207, 208. EDF-5184: certified values also for PCB 37, 74, 81, 97, 114, 123, 146, 157, 158, 167, 172, 178, 203, 208. BCR 536: certified also for PCB 163. IAEA-417: indicative values also for PCB 22, 40, 45, 70, 74, 91, 97, 114, 158, 185, 200, 201. IAEA-159: indicative values also for PCB 70, 74, 201. IAEA-408: indicative values also for PCB 196. WMS-01 is also certified for PCB 114, 123, 157, 167, reference value for PCB 81.

The total sum of PCBs is also reported as non-certified values in the series of sediments EC-1 to EC-8 by NWRI with values varying from 0.021 to 2.00 $\mu g\, g^{-1}$. In EC-1 to EC-5 reference values are also reported for 22 individual congeners.

Table 11.12 Reference materials for chlorinated dioxins, furans and non-*ortho* PCBs in biota. Values marked with an asterisk are non-certified; all other values are certified.

Code	SRM 1588b	SRM 1946	CARP-2	EDF-2524	EDF-2525	EDF-2526	WMF-01	BCR-719
Organization	NIST	NIST	NRC	CIL	CIL	CIL	Wellington Laboratories	EU–JRC –IRMM
Country of origin	USA	USA	Canada	USA	USA	USA	Canada	EU
Matrix	Cod liver oil	Lake Superior fish tissue	Common carp	Clean fish	Contami-nated fish	Fortified fish	Fish tissue	Chub
Units	µg kg^{-1}	µg kg^{-1}	ng kg^{-1}	ng kg^{-1}	ng kg^{-1}	ng kg^{-1}	pg g^{-1}	ng kg^{-1}
As		Wet wt	Wet wt	Wet wt	Wet wt	Wet wt	Dry wt	
[±] expressed as	±95% CI	±95% CI	±95% CI	±2 st dev	±2 st dev	±2 st dev		±95% CI
Units of issue	4 × 1.2 mL	5× (7–9) g	6 × 9 g	10 g	10 g	10 g	10 g	
Form		Frozen	Slurry	Slurry	Slurry	Slurry	Freeze-dried	
2,3,7,8-TeCDF			18.2 ± 1.6*	2.42 ± 0.74	24.3 ± 4.74	18.7 ± 5.58	13.1 ± 4.9	
1,2,3,7,8-PCDF			5.6 ± 0.3*	0.09 ± 0.06	4.58 ± 1.42	39.0 ± 7.36	1.53 ± 1.4*	
2,3,4,7,8-PCDF				0.21 ± 0.14	14.5 ± 4.04	37.8 ± 10.2	7.15 ± 2.2	
1,2,3,4,7,8-HxCDF				0.09 ± 0.14	5.95 ± 1.52	83.3 ± 23.0	0.86 ± 1.0*	
1,2,3,6,7,8-HxCDF				0.08 ± 0.12	1.73 ± 0.54	62.8 ± 19.6	0.51 ± 0.7*	
1,2,3,7,8,9-HxCDF					0.10 ± 0.20	57.3 ± 10.9	0.25 ± 0.4*	

2,3,4,6,7,8-HxCDF				0.08 ± 0.05	1.04 ± 0.30	58.6 ± 14.2	0.68 ± 1.2*	
1,2,3,4,6,7,8-HpCDF				0.17 ± 0.28	0.59 ± 0.44	81.6 ± 13.7	1.01 ± 1.9*	
1,2,3,4,7,8,9-HpCDF					0.16 ± 0.32	76.7 ± 26.6	0.30 ± 0.5*	
OCDF	1.00*			0.24 ± 0.58	0.38 ± 0.50	185 ± 57.4	1.38 ± 2.1*	
1,2,7-TCDD	0.32*							
1,2,3,4-TeCDD	0.38*							
2,3,7,8-TeCDD	0.21*		7.4 ± 0.7*	0.07 ± 0.06	17.0 ± 3.90	19.8 ± 4.18	13.1 ± 4.4	
1,2,3,7,8-PCDD			5.3 ± 1.3*	0.15 ± 0.03	3.71 ± 0.90	39.9 ± 10.6	2.72 ± 1.3	
1,2,3,4,7,8-HxCDD			1.6 ± 0.3*	0.06 ± 0.03	0.33 ± 0.18	54.9 ± 7.80	0.22 ± 0.3*	
1,2,3,6,7,8-HxCDD	0.39*		5.8 ± 0.8*	0.24 ± 0.14	2.03 ± 0.60	51.1 ± 19.3	0.88 ± 0.4	
1,2,3,7,8,9-HxCDD	0.22*		0.78 ± 0.12*	0.07 ± 0.05	0.30 ± 0.14	52.9 ± 18.1	0.27 ± 0.4*	
1,2,3,4,6,7,8-HpCDD			6.4 ± 0.9*	0.29 ± 0.54	0.48 ± 0.36	70.7 ± 23.2	0.59 ± 0.7*	
OCDD	1.01*		9.4 ± 1.7*	0.59 ± 0.82	1.71 ± 1.38	181 ± 53.4	3.91 ± 6.2*	
PCB 77	1.31 ± 0.02*	0.327 ± 0.025		8.82 ± 4.16	1,850 ± 834	451 ± 179	2233 ± 720	196 ± 6
PCB 126	1.31 ± 0.02*	0.380 ± 0.017		2.14 ± 1.24	628 ± 242	431 ± 17.9	739 ± 260	20.0 ± 0.8
PCB 169	0.22 ± 0.01*	0.106 ± 0.014		0.65 ± 0.46	52.1 ± 14.0	512 ± 160	76 ± 30	1.80 ± 0.15

Table 11.13 Reference materials available for chlorinated dioxins and furans and non-*ortho* PCBs in sediments. Values marked with an asterisk are non-certified; all other values are certified.

Code	SRM 1944	SRM 1941b	EDF-5184	WMS-01
Organization	NIST	NIST	CIL	Wellington Laboratories
Country of origin	USA	USA	USA	Canada
Matrix	New York New Jersey waterway sediment	Marine sediment	Heavily contaminated sediment	Lake sediment
Units	$\mu g\,kg^{-1}$	$\mu g\,kg^{-1}$	$Ng\,kg^{-1}$	$Pg\,g^{-1}$
As	Dry wt	Dry wt	Dry wt	Dry wt
[±] expressed as	±95% CI	±95% CI	2 st dev	
Units of issue	50 g	50 g	10 g	25 g
Form	Freeze-dried	Freeze-dried	Sterilized	Freeze-dried
2,3,7,8-TCDF	0.039 ± 0.015*		219 ± 47.8	52.5 ± 16
1,2,3,7,8-PCDF	0.045 ± 0.007*		122 ± 24.0	12.6 ± 5.0
2,3,4,7,8-PCDF	0.045 ± 0.004*		164 ± 50.4	18.5 ± 6.1
1,2,3,4,7,8-HxCDF	0.22 ± 0.03*		277 ± 42.8	67.3 ± 24
1,2,3,6,7,8-HxCDF	0.09 ± 0.01*		159 ± 23.6	20.3 ± 8.7

1,2,3,7,8,9-HxCDF	0.019 ± 0.018*		7.44 ± 7.38	2.68 ± 4.0*
2,3,4,6,7,8-HxCDF	0.054 ± 0.006*		48.4 ± 18.7	16 ± 8.0
1,2,3,4,6,7,8-HpCDF	1.0 ± 0.1*		346 ± 45.6	299 ± 73
1,2,3,4,7,8,9-HpCDF	0.040 ± 0.006*		80.2 ± 30.4	15.1 ± 4.6
OCDF	1.0 ± 0.1*		301 ± 50.6	509 ± 157
2,3,7,8-TCDD	0.133 ± 0.009*		1.96 ± 1.10	17.7 ± 5.6
1,2,3,7,8-PCDD	0.019 ± 0.002*		5.79 ± 2.12	7.96 ± 2.8
1,2,3,4,7,8-HxCDD	0.026 ± 0.003*		5.61 ± 2.72	8.66 ± 2.7
1,2,3,6,7,8-HxCDD	0.056 ± 0.006*		10.9 ± 3.50	20.8 ± 4.8
1,2,3,7,8,9-HxCDD	0.053 ± 0.007*		6.88 ± 1.94	17.3 ± 8.0
1,2,3,4,6,7,8-HpCDD	0.80 ± 0.07*		5.61 ± 2.72	293 ± 63
OCDD	5.8 ± 0.7*		2050 ± 580	1899 ± 456
Total toxic equivalent (TEQ)	0.25 ± 0.01*			
PCB 77		0.31 ± 0.03*	11 700 ± 2600	1717 ± 520
PCB 126			2540 ± 1080	84.9 ± 35
PCB 169			139 ± 92.4	7.97 ± 5.3

WMS-01 also has certified values for the sum of the tetra-, penta-, hexa-, hepta-, dibenzofurans and dibenzo-*p*-dioxins.

Table 11.14 Reference materials for polybrominated flame retardants (BFRs) in biota. Values marked with an asterisk are non-certified; all other values are certified.

Code	SRM 1588b	SRM 1945	SRM 1947	SRM 2977	EDF-2524	EDF-2525	EDF-2526	WMF-01
Organization	NIST	NIST	NIST	NIST	CIL	CIL	CIL	Wellington Laboratories
Country of origin	USA	USA	USA	USA	USA	USA	USA	Canada
Matrix	Cod liver oil	Whale blubber	Lake Michigan fish tissue	Mussel tissue	Clean fish	Contami-nated fish	Fortified fish	Fish tissue
Units	$\mu g\,kg^{-1}$	$\mu g\,kg^{-1}$	$\mu g\,kg^{-1}$	$\mu g\,kg^{-1}$	$\mu g\,kg^{-1}$	$\mu g\,kg^{-1}$	$\mu g\,kg^{-1}$	$pg\,g^{-1}$
As		Wet wt	Wet wt	Dry wt	Wet wt	Wet wt	Wet wt	Dry wt
[±] expressed as	±95% CI	±95% CI	±95% CI	±95% CI	±2 st dev	±2 st dev	±2 st dev	
Units of issue	4 × 1.2 mL	2 × 15 g	5 × 8 g	10 g	10 g	10 g	10 g	10 g
Form		Frozen	Frozen	Freeze-dried	Slurry	Slurry	Slurry	Freeze-dried
BDE 17				1.04 ± 0.19*				
BDE 28		2.66 ± 0.37*			26.6 ± 45.8	312 ± 202		3124 ± 290

BDE 28/33	1.08 ± 0.23*		2.26 ± 0.46*	2.54 ± 0.40				
BDE 47	17.8 ± 2.0*	39.6 ± 0.2	73.3 ± 2.9	36.5 ± 4.0	712 ± 818	9080 ± 2620		123 200 ± 24 800
BDE 49	2.25 ± 0.24*		4.01 ± 0.10	1.20 ± 0.19		524 ± 274		
BDE 66			1.85 ± 0.13	0.453 ± 0.046	23.4 ± 21.6	262 ± 81.0		
BDE 99	0.56 ± 0.20*	18.9 ± 2.3	19.2 ± 0.8	4.68 ± 0.92*	184 ± 86.2	2280 ± 472		37 500 ± 4220
BDE 100	1.89 ± 0.45*	10.3 ± 1.1	17.1 ± 0.6	1.82 ± 0.64*	113 ± 93.6	1720 ± 566		35 870 ± 14 500
BDE 153		8.34 ± 0.55	3.83 ± 0.04	0.16 ± 0.04*	21.6 ± 14.7	2030 ± 506	7.48 ± 14.7	17 040 ± 8000
BDE 154	0.495 ± 0.069*		6.88 ± 0.52	0.20 ± 0.09*	30.9 ± 39.8	2550 ± 1000		19 790 ± 2880
BDE155		4.75 ± 0.93*	0.45 ± 0.10*					
BDE 183						137 ± 47.8		532 ± 400*
BDE209						545 ± 1999		
2′-MeOPDE 68		53.5 ± 3.3*						
6′-MeO-PBDE 47		66.5 ± 4.7*						

Table 11.15 Reference materials available for BFR in sediments. Values marked with an asterisk are non-certified; all other values are certified.

Code	CNS329-050	EDF-5184
Organization	RTC	CIL
Country of origin	USA	USA
Matrix	Freshwater sediment	Heavily contaminated sediment
Units	$\mu g\,kg^{-1}$	$ng\,g^{-1}$
As		
[±] expressed as		±2 st dev
Units of issue		10 g
Form		Sterilized
BDE 28		25.8 ± 31.2
BDE 47	149*	94.7 ± 218
BDE 49		14.5 ± 34.8
BDE 66		32.0 ± 112
BDE 77		106 ± 66.6
BDE 85		14.4 ± 45.0
BDE 99	192*	95.1 ± 206
BDE 100	108*	17.6 ± 38.0
BDE 138		12.2 ± 40.6
BDE 153	160*	22.4 ± 59.4
BDE 154	108*	25.3 ± 73.8
BDE 183	52.6*	43.3 ± 82.8
BDE 209	(81.7)*	9990 ± 14 300

providing certified or reference values for some 15 bromodiphenyl ether (BDE) congeners. CIL have certified certain BDE congener concentrations in their three fish tissue materials, although the reported uncertainties are wide. Wellington Laboratories have recently made commercially available the new CRM WMF-01.

There are only two (certified) RMs for brominated flame retardants available for sediments, and only EDF-5184 (CIL) provides certified values.

11.5.4 CRMs for OCPs

The (certified) RMs for organochlorine pesticides (OCPs) in biota are reported in Table 11.16. There are currently 15 materials available, which represents an improvement in comparison with the only eight materials available in 2000 (De Boer and McGovern, 2001). Of those 15 materials, seven are produced by NIST (three mussel tissue, one cod liver oil, one whale blubber and two lake fish tissues), four by IAEA (fish, mussel, tuna and *Fucus* homogenates), one BCR (cod liver oil), one by NRC (carp) and two (fish tissues) by CIL. SRM 1588b and SRM 1945, 1946 and 1947 by NIST offer the greatest coverage of certified values for OCPs. EDF-2524 and EDF-2525 also offer a very extensive coverage, but with uncertainties that are much higher than for the NIST materials. Indicative or recommended values are provided for the IAEA materials and for CARP-2 (NRC).

Table 11.16 Reference materials available for OCPs in biota. Values marked with an asterisk are non-certified; all other values are certified.

Code	SRM 1974b	SRM 1588b	SRM 1945	SRM 1946	SRM 947	SRM 2974b	SRM 2977	IAEA-406	IAEA-432	IAEA-435	IAEA-140/OC	BCR-598	CARP-2	EDF-2524	EDF-2525
Organization	NIST	NIST	NIST	NIST	NIST	NIST	NIST	IAEA	IAEA	IAEA	IAEA	EU–JRC	NRC	CIL	CIL
Country of origin	USA	USA	USA	USA	USA	USA	USA	Monaco	Monaco	Monaco	Monaco	–IRMM EU	Canada	USA	USA
Matrix	Mussel tissue	Cod liver oil	Whale blubber	Lake Superior fish tissue	Lake Michigan fish tissue	Mussel tissue	Mussel tissue	Fish homogenate	Mussel homogenate	Tuna homogenate	*Fucus* (sea plant homogenate)	Cod liver oil	Ground whole carp	Clean fish	Contaminated fish
Units As	μg kg^{-1} Dry weight	μg kg^{-1}	μg kg^{-1} Wet weight	μg kg^{-1} Wet weight	μg kg^{-1} Wet weight	μg kg^{-1} Dry weight	μg kg^{-1} Dry weight	ng g^{-1} Dry Weight	ng g^{-1} Dry Weight	ng g^{-1} Dry Weight	ng g^{-1} Dry Weight	μg kg^{-1}	μg kg^{-1} Wet weight	ng kg^{-1} Wet weight	Ng kg^{-1} Wet weight
[±] expressed as	±95% CI	±95% CI	±95% CI	±95% CI	±95% CI	±95% CI	±95% CI	±95% CI	±95% CI	±95% CI	±95% CI	±95% CI	±2 standard deviation	±2 standard deviation	±2 standard deviation
Units of issue	3 × 15 g	5 × 1.2 mL	2 × 15 g	5× (7–9) g	5 × 8 g	5× (8–10) g	10 g	35 g	40 g	30 g	35 g	5 g	6 × 9 g	10 g	10 g
Form	Frozen		Frozen	Frozen	Frozen	Freeze-dried	Freeze-dried	Freeze-dried	Freeze-dried	Freeze-dried	Freeze-dried		Slurry	Slurry	Slurry
Hexachlorobenzene		163 ± 16	30.6 ± 1.5	7.25 ± 0.83	7.48 ± 0.66	0.113 ± 0.007		1.5 (0.95–2.0)*	0.2 (0.1)*	2.6 (1.0)*	*1.3 (0.35–3.3)*	55.7 ± 2.0		783 ± 360	18 100 ± 15 300

(*continued overleaf*)

Table 11.16 *(continued)*

Code	SRM 1974b	SRM 1588b	SRM 1945	SRM 1946	SRM 947	SRM 2974b	SRM 2977	IAEA-406	IAEA-432	IAEA-435	IAEA-140/OC	BCR-598	CARP-2	EDF-2524	EDF-2525
α-HCH		99 ± 15	16.9 ± 1.4	5.72 ± 0.65	1.06 ± 0.12			0.79 (0.23–1.7)*	0.23 (0.25)*	0.76 (0.73)*	1.4 (1.3–1.5)*	42 ± 3		267 ± 226	1400 ± 1140
β-HCH								0.75 (0.20–2.3)*	0.22 (0.11)*	1.3 (1.4)*	4.6 (2.4–9.5)*	16 ± 3			834 ± 436
γ-HCH		23.3 ± 1.7	3.18 ± 0.01	1.14 ± 0.18	0.355 ± 0.095*			0.27 (0.11–0.80)*	0.58 (0.54)*	1.1 (0.91)*	*11 (5.4–16)*	23 ± 4		390 ± 136	492 ± 216
Aldrin								0.75 (0.61–1.2)*	1.5 (1.4)*	1.1 (0.64)*	0.76 (0.5–4.5)*				
trans-Chlordane	11.3 ± 1.7		11.8 ± 0.5	8.36 ± 0.91	12.8 ± 1.2	7.12 ± 0.15	2.01 ± 0.39*	0.7 (0.63–1.0)*	0.25 (0.1)*	0.48 (0.34)*		6.9 ± 1.6	4.5 ± 0.7*		11 500 ± 7240
cis-Chlordane	13.4 ± 1.0	186 ± 22	48.1 ± 1.6	32.5 ± 1.8	49.0 ± 5.5*	8.54 ± 0.17	1.14 ± 0.39*	2.8 (2.0–4.1)*	0.5 (0.1)*	8.0 (1.4)*	1.4 (0.36–2.8)*	24.4 ± 1.8			30 100 ± 19 000
Heptachlor	2.09 ± 0.83*							0.32 (0.23–0.46)*	1 (1.1)*	1.1 (2.1)*	*3 (0.99–4.4)*				1970 ± 1110
Heptachlor epoxide		30.0 ± 1.9	10.7 ± 0.09	5.50 ± 0.23	13.4 ± 0.8			0.99 (0.37–1.6)*	11 (18)*	2.8 (2.5)*	0.79 (0.32–1.4)*				8210 ± 1560
trans-Nonachlor	12.8 ± 1.4	222 ± 10	198 ± 16	99.6 ± 7.6	127 ± 6	5.60 ± 0.39	1.25 ± 0.17	4.1 (3.9–4.1)*				39 ± 4	11.0 ± 0.9*	1130 ± 542	57 700 ± 51 000
Dieldrin	6.1 ± 1.3*	156 ± 4	50.1 ± 4.1	32.5 ± 3.5	80.8 ± 3.8		5.55 ± 0.61	3.5 (1.4–7.0)*	1.3 (0.94)*	3.8 (2.1)*	1.7 (0.72–2.8)*	59 ± 4	8.3 ± 0.8*	488 ± 74.6	54 500 ± 17 300
cis-Nonachlor	6.3 ± 1.6*	92.4 ± 3.0	45.8 ± 3.3	59.1 ± 3.6	54.1 ± 7.3	1.91 ± 0.10		0.86 (0.77–1.4)*	0.2 (0.17)*	13 (3.5)*				211 ± 126	27 700 ± 6400
Oxychl-ordane	3.57 ± 0.71*	37.5 ± 4.5	21.2 ± 1.1	18.9 ± 1.5	23.6 ± 1.5							11.0 ± 1.8			18 100 ± 11 200
2,4′-DDE	3.32 ± 0.43	15.6 ± 5.5*	14.2 ± 1.4	1.04 ± 0.29*	3.39 ± 0.28			0.76 (0.48–1.3)*	0.62 (0.64)*	2.9 (2.3)*			2.9 ± 0.5*		

4,4′-DDE	41.0 ± 3.8	676 ± 36	497 ± 19	373 ± 48	720 ± 43	17.37 ± 0.82	11.8 ± 1.2	9.2 (6.2–11)*	2.1 (1)*	91 (58)*	1.2 (0.86–1.6)*	610 ± 40	158 ± 14*	10 100 ± 2440	587 000 ± 140 000
2,4′-DDD	10.8 ± 1.6	34.8 ± 1.6	19.5 ± 1.2	2.20 ± 0.25	3.31 ± 0.16		3.15 ± 0.25	0.88 (0.43–3.0)*	1.8 (2.7)*	2.5 (0.52)*		30 ± 4	21.8 ± 0.7*		
4,4′-DDD	33.0 ± 2.2	285 ± 37	120 ± 5	17.7 ± 2.8	45.9 ± 3.6	13.56 ± 0.58	3.92 ± 0.56	2.8 (2.0–3.7)*	0.88 (0.49)*	12 (7.4)*	0.72 (0.61–0.90)*	400 ± 30	90.9 ± 8.5*	1640 ± 756	97 600 ± 33 200
2,4′-DDT	8.83 ± 0.56*	162 ± 8	91 ± 14	22.3 ± 3.2*	15.7 ± 0.89			2.9 (0.90–4.4)*	1.8 (2)*	7.8 (4.8)*					
4,4′-DDT	3.91 ± 0.94*	570 ± 27	233 ± 8	37.2 ± 3.5	59.5 ± 6.7	6.78 ± 0.32	1.32 ± 0.16	3 (1.8–5.6)*	0.67 (0.46)*	18 (10)*	2.2 (1.4–3.6)*	179 ± 18		976 ± 1390	9100 ± 2700
Mirex		11.8 ± 0.6	31.0 ± 3.4	6.47 ± 0.77	5.09 ± 0.73										93 700 ± 23 200
Endrin								1.9 (0.86–5.1)*	5.4 (6.9)*	12 (12)*	0.71 (0.43–1.6)*				2420 ± 434
α-Endosulfan								3.5 (0.94–4.7)*	4 (3.6)*	7.7 (6.1)*	0.9 (0.58–4.4)*			534 ± 378	1310 ± 722
β-Endosulfan								1.4 (1.0–1.6)*	1.8 (2)*	17 (17)*					10 100 ± 1620
Endosulfan sulfate								3.6 (3.1–5.6)*	1.8 (1.3)*	13 (10)*					

The availability of OCP reference materials for sediment matrices has also increased in comparison with year 2000. There are currently six materials with certified or indicative values for OCPs; see Table 11.17. These include three NIST materials, such as the marine sediment SRM 1941b, and three sediments by IAEA. The best coverage of certified values is offered by SRM-1941b (NIST). All others materials provide only noncertified values.

11.5.5 CRMs for Chlorinated Benzenes

Chlorinated benzenes are ubiquitous hydrophobic chlorinated organic pollutants which can enter the aquatic environment as solvents, by-products of phenol and pesticide manufacturing. Among these compounds, hexachlorobenzene (HCB), owing to its persistency, bioaccumulation potential and toxicity, has been placed on various lists as a substance of concern. These include listing in Annex X of the WFD as a priority hazardous substance and as one of the initial 12 POPs under the Stockholm Convention. The analysis of this class of compounds is still quite problematic. As a consequence, the availability of CRMs for chlorinated benzenes in biota and sediments is very scarce. Only HCB is certified in some organochlorine pesticide biota materials. IAEA provides some biota materials with recommended or information values for HCB; see Table 11.16. With regard to sediments, HCB is certified in SRM 1944; information or recommended values for HCB are reported for three sediments produced by IAEA; see Table 11.17. NWRI materials (EC-1 to EC-8) provided reference concentrations for many di- and tri-chlorobenzenes, pentachlorobenzene, hexachlorbenzene, hexachlorbutadiene, hexachloroethylene and octachlorostyrene, with EC-2 and EC-3 providing certification of a number of these analytes (De Boer and McGovern, 2001). Although many of these materials can still be acquired from NWRI to support laboratories undertaking these analyses, they are no longer certified owing to insufficient participation in proficiency testing exercises to underpin the certification process.

11.6 FUTURE NEEDS OF CRMS

Although progress has been made in the last decade and some needs have been fulfilled, such as new CRMs for polybrominated diphenyl ethers, there have been regulatory developments and additional emerging pollutants have been included in monitoring frameworks such as the OSPAR JAMP and WFD. For instance, there are very few seawater CRMs available to laboratories required to determine WFD Annex X substances in coastal and transitional waters to assess compliance with EQSs (Bercaru *et al.*, 2003). Therefore, there is the necessity to produce new certified materials or certify already existing materials for those substances. In the following subsections, some of the emerging class of organic compounds are listed.

11.6.1 Polyfluorinated Compounds (PFCs)

This is a class of pollutants comprising per- and poly-fluorinated alkyl substances constituted by an alkyl chain (typically from 4 to 18 carbon atoms) partly or fully fluorinated and a hydrophilic functional group. They are used in the surface treatment of consumer applications (including textiles, carpets, furniture, leather and paper products) and as additives in fluoropolymer production. Some ionic PFCs, such as perfluoroalkane sulfonates and in particular perfluorooctane sulfonate (PFOS), exhibit extreme persistence because of the high stability of the carbon–fluorine bond. Other compounds are bioaccumulative and toxic. These properties, together with the worldwide

Table 11.17 Reference materials available for OCPs in sediments. Values marked with an asterisk are non-certified; all other values are certified.

Code	SRM 1944	SRM 1941b	SRM 1939a	IAEA-417	IAEA-159	IAEA-408
Organization	NIST	NIST	NIST	IAEA	IAEA	IAEA
Country of origin	USA	USA	USA	Monaco	Monaco	Monaco
Matrix	New York New Jersey waterway sediment	Baltimore Harbor marine sediment	River sediment	Venice Lagoon sediment	Clyde estuary sediment	Tagus estuary sediment
Units	$\mu g\,kg^{-1}$	$\mu g\,kg^{-1}$	$\mu g\,kg^{-1}$	$ng\,g^{-1}$	$ng\,g^{-1}$	$ng\,g^{-1}$
As	Dry wt	Dry wt	Dry wt	Dry wt	Dry wt	Dry wt
[±] expressed as	±95% CI	±95% CI	±95% CI	±95% CI	±95% CI	±95% CI
Units of issue	50 g	50 g	50 g	40 g	45 g	40 g
Form	Freeze-dried	Freeze-dried	Air dried	Deep-frozen dried	Dried	Freeze-dried
Hexachloro-benzene	6.03 ± 0.35	5.83 ± 0.38		1.2 (0.8)*	0.17 (0.15)*	0.41 (0.30–0.57)*
cis-Chlordane	16.51 ± 0.83	0.85 ± 0.11	1.70 ± 0.61*	1.8 (1.7)*		0.27 (0.06–0.48)*
trans-Nonachlor	8.20 ± 0.51	0.438 ± 0.073	0.91 ± 0.31*			
γ-HCH (Lindane)				0.54 (0.35)*	0.57 (0.60)*	0.19 (0.11–0.20)*
α-HCH	2.0 ± 0.3*			0.28 (0.19)*	0.10 (0.06)*	0.61 (0.21–1.5)*
β-HCH				2.8 (2.3)*	0.24 (0.17)*	0.55 (0.38–1.7)*

(continued overleaf)

Table 11.17 (*continued*)

Code	SRM 1944	SRM 1941b	SRM 1939a	IAEA-417	IAEA-159	IAEA-408
trans-Chlordane	8 ± 2*	0.566 ± 0.093	1.81 ± 0.33*	1.2 (0.72)*		
cis-Nonachlor	3.7 ± 0.7*	0.378 ± 0.053		1.7 (1.3)*		
2,4′-DDE	19 ± 3*	0.38 ± 0.12*	1.64 ± 0.41*		0.19 (0.09)*	
2,4′-DDD	38 ± 8*			11 (5.0)*	0.39 (0.22)*	0.19 (0.18–0.33)*
4,4′-DDE	86 ± 12*	3.22 ± 0.28	5.46 ± 0.83*	14 (6.9)*	0.89 (0.50)*	1.4 (0.88–2.0)*
4,4′-DDD	108 ± 16*	4.66 ± 0.46		21 (11)*	0.82 (0.58)*	0.87 (0.56–1.7)*
4,4′-DDT	119 ± 11	1.12 ± 0.42*		19 (12)*	0.71 (0.69)*	0.67 (0.48–0.98)*
2,4′-DDT				2.5 (2.6)*	2.32 (4.33)*	0.38 (0.09–3.8)*
Heptachlor					0.31 (0.21)*	0.42 (0.23–0.70)*
Heptachlor epoxide				4.8 (4.3)*	2.84 (3.31)*	0.64 (0.43–1.5)*
Endrin				7.1 (6.4)*	0.77 (0.81)*	0.57 (0.14–1.2)*
Endrin aldheyde						
Endosulfan sulfate				8.4 (7.6)*	3.46 (3.69)*	1.6 (0.5–8.2)*
Dieldrin				2.3 (2.5)*	0.48 (0.37)*	0.3 (0.30–0.48)*
Aldrin				1.6 (1.1)*	0.17 (0.15)*	0.41 (0.2–2.3)*
α-Chlordane					0.17 (0.12)*	0.12 (0.10–0.34)*
β-Chlordane					0.67 (0.80)*	
α-Endosulfan				14 (19)*	1.11 (1.02)*	1.6 (0.3–6.2)*
β-Endosulfan				5.1 (3.7)*	1.49 (1.93)*	

HCB is also certified in NWRI materials EC-2 and EC-3. Non-certified values are available for EC-1, EC-4, EC-5, EC-6, EC-7 and EC-8.

distribution in humans and in the environment, made these compounds one of the most important class of emerging POPs. Calibrant solutions have recently been put on the market, but unfortunately, at present, CRMs for PFCs analysis are not yet available.

11.6.2 Polychlorinated Naphthalenes (PCNs)

This class comprises 75 congeners containing from one up to eight chlorine atoms per naphthalene molecule. They are persistent, bioaccumulative and toxic, and some of them show dioxin-like toxicity. Frequently detected in a wide variety of environmental matrices, they remain a relatively little known group of environmental pollutants because of the difficulty of analysis. This difficulty is also reflected in the lack of matrix CRMs. Calibrant solutions are available.

11.6.3 Chlorinated Paraffins (CPs)

CPs are highly complex technical mixtures of polychlorinated *n*-alkanes with a chlorination degree varying between 30 and 70% and subdivided according to their carbon chain length into short-chain CPs (SCCPs, C10–13), medium-chain CPs (MCCPs, C14–17) and long-chain CPs (LCCPs, >C17). They are used in a wide range of industrial applications, such as flame retardants and plasticizers, and as additives in metal-working fluids, in sealants, paints and coatings. SCCPs are classified as persistent, have a high potential for bioaccumulation, are toxic to aquatic organisms and carcinogenic. In 2001, the European Union (EU) included SCCPs in the list of priority substances in the field of water policy amending the WFD 2000/60/EC, and a global ban on SCCPs is being considered under the Stockholm POPs convention. The analytical determination of CPs is very challenging because of the complexity of their mixtures. There is a lack of calibrant solutions and matrix reference materials for these substances, impeding accuracy and traceability of the analytical measurement. These problems result in poor laboratory performance and comparability.

Tomy *et al.* (1998) have found SCCPs in two SRMs from NIST (SRM 1588 organics in cod liver and SRM 1945 organics in whale blubber), and suggested that these materials could be possible candidates for future certifications, once a method for their determination is validated.

11.6.4 Alkylphenols and Alkylphenol Ethoxylates

Alkylphenol ethoxylates are synthetic surfactants used in some detergents and cleaning products. In anaerobic conditions they easily break down into alkylphenols, which are used as antioxidants in plastics and rubber products. The most common alkylphenol ethoxylates are nonylphenol ethoxylates. The most common alkylphenols are nonylphenols. They are toxic to aquatic organisms, bioaccumulative and they possibly have endocrine-disrupting properties. They were included in the OSPAR List of Chemicals for Priority Action. Nonylphenol is also included on the list of priority hazardous substances under the EC WFD. There are no matrix reference materials certified for these compounds, but calibrant solutions are available.

REFERENCES

Bercaru O., Gawlik, B. M., Ulberth F. and Vandecasteele C. (2003). Reference materials for the monitoring of the aquatic environment-a review with special emphasis on organic priority pollutants. *J. Environ. Monit.*, **5**, 697–705.

De Boer J. and McGovern E. (2001). Certified reference materials for organic contaminants for use in monitoring of the aquatic environment. *Trends Anal. Chem.*, **20**(3), 140–159.

European Commission (2000). European Commission, Directive 2000/60/EC of the European Parliament and of the Council of 22 October 2000 establishing a framework for Community action in the field of water policy. *Off. J. Eur. Commun.*, **L 327** (2000) 1.

European Commission (2006). European Commission, Directive 2006/113/EC of the European Parliament and of the Council of 12 December 2006 on the quality required of shellfish waters *Off. J. Eur. Commun.*, **L 376** (2006).

European Commission (2008). European Commission, Directive 2008/105/EC of the European Parliament and of the Council of 16 December 2008 on Environmental quality standards in the field of water policy, amending and subsequently repealing Council Directives 82/176/EEC, 83/513/EEC, 84/156/EEC, 84/491/EEC, 86/280/EEC and amending Directive 2000/60/EC of the European Parliament and of the Council. *Off. J. Eur. Commun.*, **L 348** (2008).

Grasshoff K., Kremling K. and Ehrhardt M. (1999). *Methods of Seawater Analysis*, third edn, Wiley–VCH.

Kramer K. J. M., Vinhas T. and Quevauviller Ph. (1994). Estuarine reference material for trace elements: effect of flocculation upon homogeneity and stability. *Mar. Chem.*, **46**, 77–87.

Kramer K. J. M., Quevauviller Ph., Dorten W. S. *et al*. (1998). Certification of total Hg in a seawater reference material, CRM 579. *The Analyst*, **123**, 959–963.

Millero, F. J. (2006). *Chemical Oceanography*, 3rd edn, Marine Science Series, 30. CRC Press, Taylor and Francis Group, Boca Raton, FL.

Morabito R., Muntau H., Cofino W. and Quevauviller Ph. (1999). A new mussel certified reference material (CRM 477) for the quality control of butyltin determinations in the marine environment.. *J. Environ. Monitor.*, **1**, 75–82.

OSPAR (1999). *Guidelines for Monitoring Contaminants in Biota* (Agreement 1999-2), OSPAR Commission, London.

Quevauviller Ph. (2002a). *Quality Assurance for Water Analysis*, Water Quality Measurement Series, vol. 3, (series ed. Ph. Quevauviller), John Wiley & Sons, Ltd, Chichester, pp. 252 (ISBN: 0-471-89962-3).

Quevauviller Ph. (ed.) (2002b). *Methodologies for Soil and Sediment Fractionation Studies*, The Royal Society of Chemistry, Cambridge, pp. 180 (ISBN: 085404-453-1).

Quevauviller Ph. and Ariese F. (2001). New sediment reference material for the quality control of butyltin and phenyltin analysis. *Trends Anal. Chem.*, **20**, 207–218.

Quevauviller Ph. and Maier E. A. (1999). *Interlaboratory Studies and Certified Reference Materials for Environmental Analysis*, Elsevier, Amsterdam, pp. 558 (ISBN: 0-444-82389-1).

Quevauviller Ph., Astruc M., Ebdon L. *et al*., 1994. Certified reference material (CRM 462) for the quality control of dibutyl- and tributyltin determinations in coastal sediment. *Appl. Organometal. Chem.*, **8**(5), 629–637.

Quevauviller Ph., Kramer K. J. M. and Vinhas T. (1996a). Certified reference material (CRM 505) for the quality control of Cd, Cu, Ni and Zn in estuarine water. *Fresenius' J. Anal. Chem.*, **354**, 397–404.

Quevauviller Ph., Drabaek I., Bianchi M. *et al*. (1996b). Certified reference materials (CRMs 463 and 464) for the quality control of methyl-mercury in tuna fish. *Trends Anal. Chem.*, **15**, 390–397.

Quevauviller Ph., Rauret G., López-Sánchez J.-F. *et al*. (1997). Certification of trace metal extractable contents in a sediment reference material (CRM 601) following a three-step sequential extraction procedure. *Sci. Total Environ.*, **205**, 223–234.

Quevauviller Ph., Fortunati G., Filippelli M. *et al*. (1998). Certification of total mercury and methyl mercury in an estuarine sediment, CRM 580. *Appl. Organometal. Chem.*, **12**, 531–539.

Tomy G. T., Fisk A. T., Westmore J. B. and Muir D. C. (1998). Environmental chemistry and toxicology of polychlorinated *n*-alkanes. *Rev. Environ. Contam. Toxicol*. **158**, 53–128.

Ulberth F. (2006). Certified reference materials for inorganic and organic contaminants in environmental matrices. *Anal. Bioanal. Chem*. **286**, 1121–1136.

Van Leeuwen S. P. J., Van Cleuvenbergen R., Abalos M. *et al*. (2006). New certified and candidate certified reference materials for the analysis of PCBs, PCDD/Fs, OCPs and BFRs in the environment and food. *Trends Anal. Chem.*, **25**(4), 397–409.

Wegener J. W. M., Cofino W. P., Maier E. A. and Kramer G. N. (1999). The preparation, testing and certification of two freshwater sediment reference materials for polycyclic aromatic hydrocarbons and polychlorinated biphenyls: BCR CRM 535 and CRM 536. *Trends Anal. Chem.*, **18**(1), 14–25.

Wells D. E., Cofino W. P., Quevauviller Ph. and Griepink B. (1993). Quality assurance of information in marine monitoring: a holistic approach. *Mar. Pollut. Bull.*, **26**(7), 368–375.

Wise S., Poster D. L., Kucklick J. R. *et al*. (2006). Standard reference materials (SRMs) for determination of organic contaminants in environmental samples. *Anal Bioanal. Chem*. **386**, 1153–1190.

Yoshinaga J., Nakama A., Morita M. and Edmonds J. S. (2000). Fish otolith reference material for quality assurance of chemical analyses. *Mar. Chem*. **69**, 91–97.

Part 6

Research and Developments

12

The Role of Sediments in Coastal Monitoring

Grażyna Kowalewska[a], Maria J. Belzunce-Segarra[b], Birgit Schubert[c], Peter Heininger[c], and Susanne Heise[d]

[a]*Institute of Oceanology, Marine Pollution Laboratory, Polish Academy of Sciences, Sopot, Poland*

[b]*Marine Research Division, AZTI-Tecnalia, Pasajes (Gipuzkoa), Spain*

[c]*Federal Institute of Hydrology, Koblenz, Germany*

[d]*Hamburg University of Applied Sciences, Hamburg, Germany*

Chemical Marine Monitoring: Policy Framework and Analytical Trends, First Edition.
Edited by Philippe Quevauviller, Patrick Roose, and Gert Verreet.

12.1 INTRODUCTION

Sediment is an inherent part of aquatic systems, both containing freshwater and seawater. Sediment consists of mineral and organic particles originating from the weathering of rocks, the erosion of soils, river banks and coast, the particulate detritus matter production of organisms and from wet and dry atmospheric deposition, introduced directly or via surface runoff. The sediment is not a static component of an aquatic system, but the inorganic and organic material is susceptible to transportation.

In the coastal zone, sediment geochemistry is controlled by different natural factors, like hydrology, morphology, oxic conditions, and so on. After deposition, the sediment undergoes mineralization, disaggregation and various biogeochemical processes. Changes in flow velocity and direction of the water course result in the resuspension of already settled particles. The fine sediment particles are more susceptible to resuspension than the coarse sediment. As a result, at some distance from the particulate input, a fractionation occurs. In areas of more laminar water movement the bed contains more fine sediment grains than those of more turbulent flow (Owens, 2007).

Ions of trace elements and hydrophobic organic compounds, including the most harmful contaminants, usually have greater affinity to fine particles than to the coarse sediment grains. As a result, contaminants concentrate in suspended particulate matter (SPM) and fine-grained sediment fractions (OSPAR, 2001), which are more liable to transfer by water than coarse sediment, though also a fractionation of contaminants occurs, depending on particular contaminant affinity to sediment particles (Kowalewska *et al*., 2003; Tsakovski *et al*., 2009; Lubecki and Kowalewska, 2010). Sediments are thus depleted in the more hydrophilic compounds, which dissolve better in water and are transported in the dissolved form (Owens, 2007; Alargarsamy and Zhang, 2010).

Annually, rivers bring as much as $\sim 13.5 \times 10^9$ t of suspended matter into river mouths, together with $1-2 \times 10^9$ t of bedload (Milliman and Meade, 1983). The slower flow at the coast causes deposition of particulates transported from upper reaches of the river, to sediments. A high discharge of suspended matter by rivers to the sea may result in formation of deltas. At coasts with high tidal ranges, estuaries are formed. Resuspension of already settled particles in estuaries is caused mainly by river mixing driven by tides. Deltas, estuaries and other coastal regions are often places favourable for human settlement and have become economically important. All the above circumstances result in intensive municipal, transport and industrial human activities enhancing the contaminant load trapped in sediments and introduced by the river to the sea (Muxica *et al*., 2007). Human activities include those connected with relocation of sediments (like dredging to maintain or deepen waterways and harbour basins) or hydromorphological changes to the water course (e.g. by river dams, watersides and harbour constructions) (Bortone and Palumbo, 2007), which influence the quality, quantities and fate of contaminants associated with sediments as they are introduced to the sea. Dredged material can be disposed of at sea or ashore (Bolam *et al*., 2006; Spencer *et al*., 2008). Sediments are an important source of different construction materials. The pollution of the marine coastal zone is affected by these factors. The natural dynamics, such as high-energy meteorological phenomena (e.g. storms, intensive rains resulting in floods, high tides) determine further the fate of contaminated sediments in the marine environment (Heise, 2007).

Primary production in estuaries is often extremely high, and many marine animals, like fish, either pass the estuaries during their migration up the stream or spawn there, and sediments can play an important role in their life cycles. In polluted coastal ecosystems, biodiversity may decline, even of the unicellular organisms (Cherchi *et al*., 2009). High nutrient and organic matter levels can intensify phytoplankton blooms, resulting in eutrophication and associated with it high detritus concentration, hypoxia/anoxia, and so on (Ducrotoy and Elliot, 2008). The high concentration of plankton and detritus play an important role in transfer of contaminants to sediments and their accumulation therein (Kowalewska, 1999; Konat and Kowalewska, 2001; Kowalewska *et al*., 2003). Moreover, intensive flocculation occurs in the mixing zone of freshwater that carries a lot of inorganic and organic colloids and seawater. The aggregates formed scavenge both the molecular and ionic contaminants from the water column to sediments. Conversely, sediment bioturbation and microbial activity release contaminants to pore water and the superjacent column of water by resuspension and remineralization of sediments (Donnini *et al*., 2007).

These dynamics explain why estuaries, lagoons and other aquatic basins at the shoreline may serve as important buffers in the transport of contaminants to the sea by acting as natural traps and by providing a variable level of natural biodegradation treatment of waste substances (Edlund *et al*., 2006). Contaminants sorbed by coastal sediments may be released either by desorption or resuspension of the fine sediment fraction, and their subsequent transfer at first to the water column, and distributed further along and off the shore with the local currents. So, the coastal marine sediments can generally be considered as accumulation areas of contaminants, which can pose a real threat to the marine life and to the health of people (e.g. Carafa *et al*., 2006; Hartwell and Hameedi, 2006; Boszke and Astel, 2007; Zonta *et al*., 2007).

From the elements above, it becomes evident that chemical monitoring of coastal sediments is crucial. However, sediment monitoring of the coastal environment was not so readily mainstreamed in regular observation programmes. Before the adoption of the Water Framework Directive (WFD; EU, 2000), different marine conventions established the basis for contaminant monitoring of the marine environment, including the sediments in the coastal zone. This was not done in the same way in the different regions. When the WFD was introduced, 'coastal monitoring' was also first understood as 'surface water monitoring'. However, further work by experts (e.g. the SedNet project; Salomons and Brils, 2004) resulted in further elements for the monitoring of priority hazardous substances addressed in the daughter directive 2008/105/EC (EU, 2008a). This directive requires monitoring of contaminants in sediments for temporal trend assessment. Agreed in the same year, the Marine Strategy Framework Directive (EU, 2008b – see Borja (2006) and Mee *et al*. (2008)), treats the marine waters as encompassing 'waters, the seabed and subsoil', not only as water (Tueros, 2009). Now that the legislative basis is completed, the development has begun to harmonize as much as possible the various monitoring programmes, with the establishment of contaminant lists for monitoring, sampling and analytical procedures (including reference material preparation), as well as the development of new, more effective and less-costly monitoring methods and techniques to be used in the coastal zone (e.g. Borja *et al*., 2004a; Quevauviller, 2006; Barcelo and Petrovic, 2007). It is worth underlining that such developments were not already very easy for seawater alone (Borja, 2005). Despite these complexities, it is not the intention of this chapter to tackle all aspects of sediment monitoring, but rather to highlight the role of sediments in coastal monitoring on the basis of four topics that are of high current operational or scientific interest.

12.2 SEDIMENT MONITORING WITHIN THE EUROPEAN WFD

12.2.1 Introduction

The WFD (EU, 2000) was published in December 2000 with the aim of establishing a new legislative framework for European Community action in order to manage, use, protect and restore continental, transitional (estuaries) and coastal waters in Europe. So far, the WFD represents the most important European water legislation and has profoundly influenced the environmental regulations in EU member states, as it represents a big step from the earlier sector approach to a more integrated approach of water management (Cardoso *et al*., 2006; Quevauviller, 2007).

The objective of the WFD is to ensure the quality of the EU waters. For this purpose, all member states are required to achieve the good ecological status in all water bodies by 2015. In addition, the WFD states that comprehensive monitoring programmes had to be operational by 2006. On the basis of this monitoring, the ecological status must be assessed of all surface waters in the EU.

The concepts of ecological status (ES) and chemical status (CS) are developed by the directive for the assessment of the quality of water masses. Surface water quality is assessed on both its ecological status and chemical status. Ecological status is assessed by several characteristics of aquatic flora and fauna (biological elements), along with the supporting physical and chemical elements. The biological elements are very important and supported by the others. The physicochemical elements terminology is used to describe the general conditions of the water column and includes values of variables such as dissolved oxygen, nutrients, thermal conditions, transparency and contaminants in water. The chemical status for priority dangerous substances is determined by the levels of contaminants in water, compared with environmental quality standards (EQSs) listed under the so-called EQS Directive (EU, 2008a). This directive repeals previous relevant directives and amends the WFD (EU, 2000), representing a step further, since in the definition of chemical status the new directive considers not only waters, but also sediment and biota. It thus gives, together with the WFD, the legal basis for the monitoring of priority substances in sediment and biota.

Prior to the publication of the EQS Directive there were no clear indications about which matrices were to be sampled for the physicochemical elements (Borja *et al.*, 2004a), or, in other words, which matrices are more representative of the chemical quality of a specific water body. These authors pointed out that matrices such as 'sediment' or 'biota' (biomonitors), are mentioned explicitly only seven and four times respectively, while 'water' is mentioned on 373 occasions throughout the WFD. On three occasions the terms sediment and biota are used in connection with the derivation of EQSs, as was outlined by Crane (2003).

Nowadays, there is a general consensus within the scientific community on the important role that sediments play for the quality of the water systems, as they can act as a source or as sink of contaminants. For the purpose of trend monitoring, sediment and also biota are, for many substances, the most suitable matrices because they integrate in time and in space the level of contamination of a specific water body. The changes of these levels in these matrices are not as fast as in the water column. Hence, they represent better the chemical status of marine coastal waters. From this point of view, scientists and group of experts, such as the ICES Working Group on Marine Sediments or the European Network SedNet, have argued the necessity of considering sediments in monitoring programmes: sediment contamination should be included in water quality assessments.

The first step on this way was to develop a methodology for including sediments in the definition of the chemical status concept. As mentioned above, under the WFD, the classification of the chemical status of water includes priority substances in water only. The overall chemical status is derived following the 'one out, all out' principle. This approach is based upon the assumption that the worst status of the elements (or priority substances) used in the assessment determines the final status (Heiskanen *et al.*, 2004; Borja, 2005). This means that any metal concentration in water over the EQS results in the whole station failing to achieve good chemical status; in turn, if all concentrations lie below the EQS, the good chemical status conditions are met. From this initial situation a series of concepts, terminologies and methodologies were developed for a comprehensive definition of the chemical status including sediment information. This involved some controversy and debate within the scientific community (Crane, 2003; Borja *et al.*, 2004a,b; Borja, 2005; Borja and Heinrich, 2005; Simboura, 2005). The final objective was to provide methods to assess the chemical and ecological status of European waters in a pragmatic, holistic, rational and science-based approach, achieving a broad consensus between scientists, policy makers and stakeholders (Borja, 2005).

12.2.2 Methodologies to Integrate Sediments in Water Quality Assessments

Several scientists have proposed different methodologies to integrate sediments in the evaluation of the chemical status of water bodies. For example, Crane (2003) proposed development of sediment quality guidelines that, in a tiered risk assessment approach, are used to screen out nonhazardous sediments and trigger further site-specific studies on potentially hazardous ones. Borja *et al.* (2004a,b, 2009) proposed to integrate chemical elements in water, sediment and biota, followed by a decision-tree approach, as is illustrated in Figure 12.1.

In this approach, quality objective levels for metals and organic compounds published in European and Spanish legislation for waters and biota have been used to assess whether they meet or not the good chemical quality EQS Directive (EU, 2008a) for priority substances for waters and the Spanish Law of Biodiversity (Decree 42/2004), for biota. Furthermore, quality objective levels for metals in sediments have been developed and background levels are used in assessing high chemical status, as is detailed in Figure 12.2 (Rodríguez *et al.*, 2006).

<table>
<tr><th>WATER</th><th>SEDIMENT</th><th>BIOMONITORS</th><th>STATUS</th></tr>
<tr><td rowspan="3">All variables meet</td><td>All variables meet</td><td rowspan="3"></td><td rowspan="2">Achieved</td></tr>
<tr><td>1 variable does not meet</td></tr>
<tr><td>≥2 variables do not meet</td><td>Failed</td></tr>
<tr><td rowspan="4">1 variable does not meet</td><td>All variables meet</td><td></td><td rowspan="2">Achieved</td></tr>
<tr><td>1 variable does not meet</td><td>All variables meet</td></tr>
<tr><td>1 variable does not meet</td><td>≥1 variable does not meet</td><td rowspan="3">Failed</td></tr>
<tr><td>≥2 variables do not meet</td><td rowspan="2"></td></tr>
<tr><td>≥2 variables do not meet</td><td></td></tr>
</table>

Figure 12.1 Proposed scheme to integrate water, sediment and biota for establishing the chemical status of water bodies (adapted from Borja *et al.* (2009)). Reprinted with kind permission from Marine Pollution Bulletin, Using multiple ecosystem components, in assessing ecological status in Spanish (Basque Country) Atlantic marine waters by Angel Borja, Juan Bald, Javier Franco *et al.*, 59, 1–3, 54–64 Copyright (2009) Elsevier Ltd.

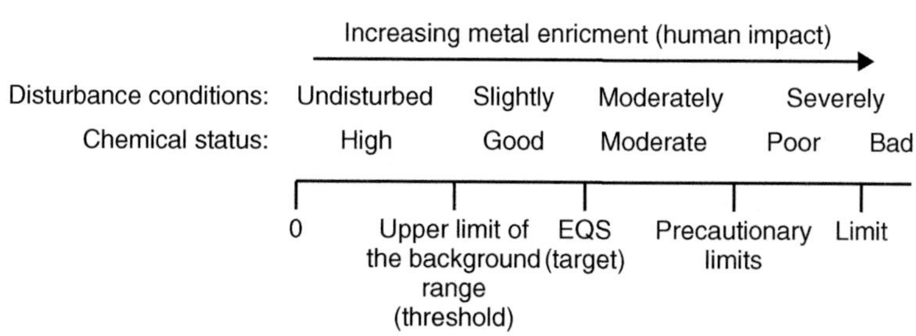

Figure 12.2 Relationships between increasing disturbance, due to metal contamination, and chemical status, according to the WFD and EMS terminology (from Rodríguez *et al.* (2006)). Reprinted with kind permission from Science of the Total Environment, Maximum likelihood mixture estimation to determine metal background values in estuarine and coastal sediments within the European Water Framework Directive by J.G. Rodríguez, I. Tueros, A. Borja *et al.*, 370, 2–3, 278–293 Copyright (2006) Elsevier.

This approach contributes to the assessment of regional undisturbed sediment conditions. The authors propose the incorporation of sediments and biomonitors for determining the quality standards in the context of the WFD, as this integration considers the biotic responses within the water bodies and, therefore, is more realistic and discriminative than the 'one out, all out' principle.

In recent studies of the quality of Basque coastal and estuarine waters in northern Spain (Tueros *et al*., 2009), Spanish researchers have demonstrated that the integrative methodology described above improves the final ecological quality determination. The study focused on the long-term trend (from 1995 to 2007) of water and sediment contamination by metals (As, Cd, Cu, Cr, Hg, Ni, Pb and Zn) and the response of the waters to water treatment programmes. The authors assessed the chemical status of these water bodies using two approaches: (1) following the 'one out, all out' principle under the WFD and (2) combining the chemical quality of both the surface waters and the underlying sediment, using a methodology proposed by Borja *et al*. (2004a,b, 2009). This study has demonstrated that the quality of surface water is best assessed using the chemical status of both the water and underlying sediment. Furthermore, if the sediment assessment is not included, the water bodies' status assessment will be less comprehensive and will reduce the water managers' ability to allocate resources most effectively. In addition, this approach reflects the diminution in contamination of river catchments in recent years, which has improved water quality in many water bodies. However, more research is needed determining EQS and threshold values, both for water and sediments, together with the application of this methodology to other European countries.

12.2.3 Sediment Monitoring within the WFD

Since the publication of the WFD, member states have to ensure the establishment of programmes for the monitoring of waters in order to establish a coherent and comprehensive overview of water status within each river basin district (including freshwater, transitional and coastal waters). Three types of monitoring programme have been defined, as described below (from Borja *et al*. (2008)):

1. Surveillance monitoring, undertaken on a routine basis for periodically evaluating the status of waters and long-term changes.
2. Operational monitoring, in order to establish the status of those water bodies identified as being at risk of failing to meet their environmental objectives and to assess any changes in the status of such bodies resulting from the programmes of measures.
3. Investigative monitoring carried out where the reason of any exceedance for ecological and chemical status is unknown; where surveillance monitoring indicates that the objectives for a water body are not likely to be achieved (and to determine the causes); or to ascertain the magnitude and impacts of 'accidental' pollution.

Surveillance and operational monitoring have some implementation guidelines in the WFD, in terms of the elements to be monitored, parameters, frequency, selection of sites, and so on. However, investigative monitoring should be adapted to each particular case. Although member states have to establish programmes for the monitoring of water bodies and in spite of the requirement that surveillance and operational monitoring networks had to be operative in December 2006, however, monitoring programmes have not been investigated so much within the WFD.

Some researchers have analysed alternatives for a successful implementation of the directive. For example, Allan *et al* (2006) have reviewed the limitations and advantages of current monitoring practices and pointed out the necessity of using emerging low-cost tools, such as biomarkers, biosensors, biological early-warning systems and whole-organism bioassays for biological monitoring, while sampling and analytical tools developed for chemical assessment comprise biosensors, immunoassays, passive samplers, and sensors. The authors discuss the suitability of these techniques for different types of monitoring and argue that these techniques may complement current monitoring by providing additional information with the aim to obtain a more representative picture of the quality of a water body. In other studies, de Jonge *et al*. (2006) consider the development of ecological indices integrating community structure and system functioning for full assessment of the ecosystem status at the EU level.

As an example of investigative monitoring, an integrative sediment quality assessment approach has been applied in a Basque coastal area impacted by slag disposal that produces the risk of not achieving good status due to a sediment quality problem (Borja *et al*., 2008). By using the TRIAD approach (Chapman, 1990) that combines chemical analyses, acute toxicity assessments and biological parameters, an integrated view of the problem and a clear estimation of the potential risk associated with the slag disposal were obtained. This study demonstrated that the slag dumping along the Basque coast, during most of the twentieth century, has not produced toxicological or harmful effects in the biota. The authors concluded that this situation, hence, does not represent a risk for achieving good ecological status, *sensu* WFD.

Since the publication of EQS Directive (EU, 2008a), member states were required in 2008 to include monitoring of sediment and/or biota in their WFD monitoring programme. Some of the main elements of this directive are:

- Member states may develop and adopt EQSs for contaminants in sediment and biota, provided that they are at least as protective as the EQSs for water included in the Annex I to the directive.
- Member states should improve the knowledge and data available on sources of priority substances listed in Annex I to the directive.
- Member states are required to monitor sediment and biota as appropriate to provide data for the long-term trend analysis of concentrations of those priority substances listed in Part A of Annex I to the directive that tend to accumulate in sediment and/or biota. Member states shall take measures aimed at ensuring that such concentrations do not significantly increase in sediment and/or relevant biota.

The main aim is to demonstrate that conditions within a water body are compliant with the EQS adopted for sediment and biota and with the 'no deterioration' requirement.

In 2008, a group of experts was formed, in the context of the Common Implementation Strategy for the WFD, in order to develop a further guidance document on a harmonized implementation of the chemical monitoring strategy in the EU. The guidance document on *Chemical Monitoring of Sediment and Biota under the WFD* was published in 2010 (Guidance Document, 2010) and is in part based on information from the International Council for the Exploration of the Sea (ICES) and the OSPAR Commission. MEDPOL guidelines are also taken into consideration for biota monitoring. This document presents an informal consensus position on best practice agreed by the participant experts. However, the document does not necessarily represent the official position of any of the participants. Thus, the views expressed in the document do not necessarily represent the views of the European Commission.

It gives a set of guidelines on how member states might best carry out such monitoring. However, it must be taken into account that the application of the guidelines shall be adapted to regional and local circumstances. The document reports on the best available techniques and standard procedures for sampling and analysis of sediments and biota; for example, for sediments, it gives guidance on the number of sample stations, number of replicates, frequency of sampling, sediment volume, sediment fraction, sediment samplers, best procedures for transport and conservation, and so on. The document contains descriptions of upcoming techniques, such as passive samplers and sediment bioassays for investigative monitoring; descriptions of processes for normalization of sediment concentrations, for background concentration in sediments, and sediment and biota monitoring for trend analysis. It also describes quality assurance and quality control procedures.

12.3 CHEMICAL MONITORING IN ESTUARIES FOR COASTAL MANAGEMENT

In 2008, the European Union adopted the Marine Strategy Framework Directive (MSFD). Although it does not apply to estuaries directly, its implementation will result in requirements on the quality of sediments in the water bodies. In order to achieve the desired economic potential of European coastal areas in harmony with the marine environment, a sustainable sediment management is indispensable in many respects. As indicated in the introduction of this chapter, the sediment quality is a result of anthropogenic activities typically taking place in the surrounding areas and, often even more important, of socio-economic activities and the biophysical conditions in the whole catchment (Salomons, 2005). A chemical monitoring of sediments has to be designed providing the stakeholders with reliable data both on the status and functioning of the ecosystem and in support of such important uses like management of navigation, tourism or fishing and agriculture. In estuaries with large tides, distinct contamination gradients are usually observed from the inner to the outer parts. These gradients are missing in areas with limited tidal range and small offshore currents, such as the Mediterranean or Baltic seas. Contaminant concentrations in coastal sediments are usually low, and particularly for organic contaminants the analytical uncertainty of measured concentrations is high. Temporal trends in the sediment quality may be detected only over a long time period.

Important common features of estuaries also determining their sediment dynamics are the tidal currents, the transition from freshwater to saline water and a zone with high turbidity. However, they vary widely in terms of size, tidal range, freshwater inflow and fluvial sediment input. The estuarine particulate matter is subject to complex hydrodynamic processes. Once deposited, its reworking can take place almost immediately, or can last years, decades or even still much longer. Any monitoring programme and the assessment of contaminant monitoring data have to take into consideration these specific features (Jay *et al.*, 1997; Ackermann, 1998; ICES, 2005). And vice versa, monitoring data can be valuable to better understand the complexity of transport, mixture, deposition and erosion of fine particulate matter and the associated contaminants in estuaries. An improved understanding of sediment transport in estuaries may support the optimization of sediment management, and it could render the assessment of the input of particle-bound contaminants to the marine environment more reliable (e.g. Heininger *et al.*, 2002).

12.3.1 Sediment Transport in Estuaries

The transport characteristics of estuarine sediments were the subject of several studies (e.g. Kappenberg *et al.*, 1996; Dyer *et al.*, 2001; Huntley *et al.*, 2001). Owing to tidal currents, in the main body of estuaries (transportation zone), particulate matter is transported bidirectionally, and it represents a mixture of varying percentages of marine and fluvial sediments. The extent and the range of upstream transport may differ greatly, depending on the characteristics of the different estuaries. Generally, in the outer estuaries, marine influence prevails, and the inner part is dominated by fluvial influence. In the middle of estuaries, marine and fluvial water and solids are mixed intensively, and there is a steep gradient in the ratio of marine to fluvial sediments. It depends on the discharge characteristics of the river and can change seasonally as a result. In the inner and middle reaches of many estuaries, a turbidity maximum is located, accumulating large amounts of SPM (e.g. Müller and Förstner, 1975; Grabemann *et al.*, 1995; Kappenberg *et al.*, 1996). Turbidity zones may extend up to several kilometres, and mainly freshwater inflow governs their extent and location. Compared with the transport of water masses, the transport of particulate matter to the sea is delayed mainly by the retention in the turbidity zone (e.g. Schubel, 1984; Zwolsman, 1994; Millward and Glegg, 1997; Ridgway and Shimmield, 2002). For the Humber (Millward and Glegg, 1997) and for the Gironde (OSPAR, 2000) estuaries retention times of 18 years and 2 years respectively were estimated. Thus, these estuaries represent long-term sources of contamination to the adjacent marine environment. Major SPM input to the sea occurs mainly in case of extreme floods (Ridgway and Shimmield, 2002; Ackermann, 2004). In addition, dredging and disposal of dredged sediments within estuaries may accelerate the transport of sediments with elevated contaminant concentrations from the estuary towards the sea.

In particular, the fine-grained particulate matter may be accumulated in net deposition areas of low hydrodynamic energy (e.g. in intertidal mudflats, salt marshes or in branches of estuaries), and thus contribute to the retention of particulate matter in estuaries in the long term (Zwolsman *et al.*, 1993; Ackermann, 1998; Scrimshaw and Lester, 2001). These sediments are most likely to be resuspended only due to extreme events or human intervention.

12.3.2 Specifics of Particle-Bound Contaminant Distribution in Estuaries

Provided that local contaminant sources in estuaries are low, and that the fluvial contamination level exceeds the marine level, spatial and temporal variations of contaminant concentrations according to the mixing ratio of marine and fluvial sediments are typical for many estuaries (Förstner *et al.*, 1990; Grant and Middleton, 1993; Prange, 1997). In many tidal rivers, contaminant concentrations decrease from the inner part of estuaries towards their mouth to the sea, owing to the increasing percentage of marine sediments in the longitudinal profile. In addition, contaminant concentrations in recent sediments at fixed monitoring sites in the mixing zone of marine and fluvial particulate matter may show considerable within-year variations. The reason stems from changes in the freshwater inflow, as demonstrated for the Elbe estuary in Figure 12.3. The episodic downstream transport at high freshwater discharge of highly contaminated solids from locations in the inner part of the estuary upstream of Hamburg is a well-known phenomenon (Salomons *et al.*, 1988; Knauth *et al.*, 1993). In contrast, during summer the increased upstream transport of uncontaminated marine sediments periodically results in a decrease of contaminant concentrations in the inner part of estuaries at low freshwater inflow. In such situations the percentage of fine marine particulate matter may reach a portion of about 80% at monitoring sites which are located $\sim$100 km upstream of the mouth (Ackermann and Schubert, 2007).

However, for some rivers the contamination distribution in the estuary is quite independent of the longitudinal position and the freshwater discharge; for example, for the rivers Ems, Humber and Severn (Grant and Middleton, 1993; Ackermann, 2004). There, the contaminant concentrations in the estuary are almost at the marine level. A qualitative model for three large German estuaries gives an indication of the extent of upstream transport of fine-grained sediments; see Figure 12.4. In the Ems, for example, marine sediments dominate from the river mouth up to the tidal weir.

In estuaries, net deposition areas of low hydrodynamic energy (e.g. mudflats) may both remove fine-grained particulate matter from further transport to the sea temporarily or over the long term and act as secondary sources of contamination under extreme floodwater conditions or due to alterations in the hydraulic regime and direct interventions, like dredging. This should be taken into consideration when estimating the sediment transport and particle-bound contaminant input to the sea. The comparison of results from surface samples in deposition areas with those from

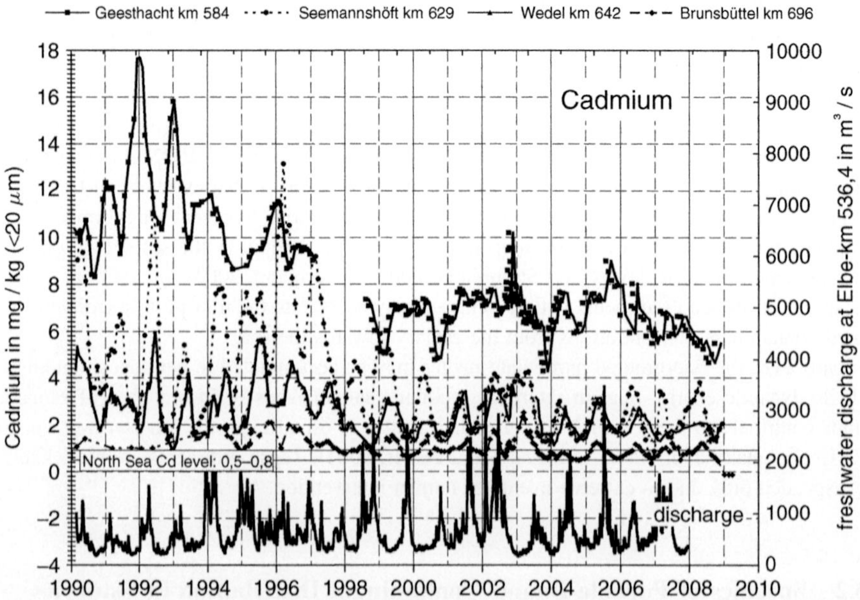

Figure 12.3 Cadmium concentrations (in fraction <20 μm) at monitoring stations of the Elbe estuary and freshwater discharge at Neu-Darchau, Elbe-km 536.4 (Ackermann and Schubert, 2007; Schubert *et al.*, 2009).

recent sediments of the main body of the estuary may provide information on current deposition or erosion trends.

12.3.3 Monitoring Design for Particle-Bound Contaminants in Estuaries

As particle-bound contaminants strongly tend to accumulate in fine-grained particles and organic matter, a correction for differences in grain-size distribution (normalization) should be performed prior to data assessment, unless samples have a comparable composition (Ackermann *et al.*, 1983; Kersten and Smedes, 2002; OSPAR, 2002). The monitoring guidelines of the OSPAR Commission recommend additional normalization with co-factors to compensate for further differences in sample composition (e.g. organic matter and clay content, Fe/Mn oxide coatings).

The design of a programme for monitoring contaminants in mobile sediments of the main body of estuaries should take into consideration (1) the hydrodynamic conditions of the estuary, (2) the expected variability of normalized contaminant concentrations, (3) potential local sources of contamination and (4) sedimentation rates.

For trend monitoring, recently deposited surface sediments, representing the time period under consideration, should be sampled; for example, from areas with low energy and high sedimentation rates (e.g. O'Reilly Wiese *et al.*, 1997a,b). In the main body of estuaries, the upper layers of sediments are likely to be well mixed to depths of up to 10–20 cm by continuous deposition and erosion processes, and thus to represent the current contamination status at the sampling site. Where areas of low energy are lacking, an appropriate alternative to sediment sampling may be the sampling of SPM. For the latter, preferably passive sampling devices, like sedimentation basins, are used. SPM samples have the advantage over sediment samples to represent a defined time period

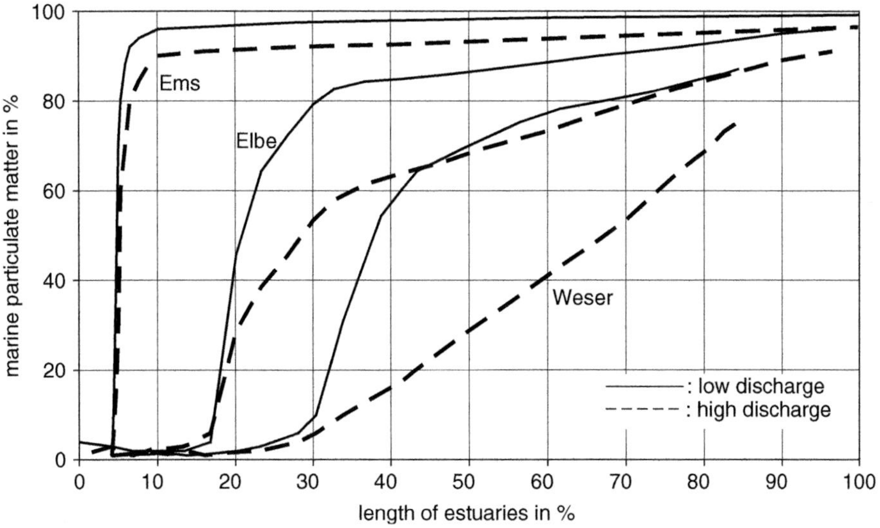

Figure 12.4 Percentage of marine particulate matter in German North Sea estuaries (Schubert *et al*., 2009).

of usually 1–4 weeks. Furthermore, in areas with low deposition rates, sampling of sediments might include contaminated deposits from the past or natural sediments with regional background concentrations.

At fluvially dominated sampling sites near the entrance of an estuary or at mainly marine-influenced stations near the mouth of estuaries, contaminant concentrations in particulate matter usually are not affected significantly by changing freshwater discharge, and a sampling frequency of two to four times per year should be sufficient for detecting a temporal trend due to a change of contaminant input. However, at sampling sites near the river mouth, concentrations often are near marine levels and difficult to detect. At sampling sites in the mixing zone of marine and fluvial particulate matter, where contaminant concentrations usually are subjected to significant variations depending on freshwater discharge, a higher frequency of 12 samples per year is strongly recommended. For the assessment of temporal trends due to changes in contamination input, either samples at similar freshwater inflow or annual mean values should be used. Sampling sites in the mixing zone of estuaries are of special interest, as monitoring data may be used to derive information on the dynamics of fine-grained particulate matter.

Sediment cores from depositional areas may be used for retrospective temporal trend monitoring by analysing separate layers. However, the assessment should take into consideration diagenetic processes in deeper layers which tend to be anoxic (Farmer, 1991; Gobeil *et al*., 1997; Kersten and Smedes, 2002). Furthermore, contaminant loads retained in depositional areas may be estimated from analyses of sediment cores. Analyses of contaminant concentrations in surface layers in depositional areas may give indications of current deposition or erosion tendencies (e.g. following construction work in estuaries).

12.3.4 Summary

First, monitoring provides a survey of temporal trends and spatial distribution of contaminants in fine-grained particulate matter in estuaries. These may be used, for example, to assess the effectiveness of measures to reduce contaminant input or to check compliance with the nondeterioration principle of the EQS Directive (EU, 2008a). Furthermore, potential local sources might be detected.

Second, contaminants may be used as tracers for the transport of fine-grained particulate matter and for validating the results of numerical modelling of sediment transport. An improved understanding of sediment transport processes and controlling factors in estuaries is of utmost importance for planning and optimizing a sustainable sediment management; for example, of construction work or dredging and disposal activities. In estuaries, sediment management for navigation is of special interest, as tidal currents result in large amounts of sediments to be dredged for maintaining the navigation channels. Monitoring results in deposition areas as potential secondary sources of contamination should be considered, when measures for improving water bodies in the course of the implementation of the WFD are planned, in order to avoid a potential release of the contaminant burden of the past.

Third, information on the impact on sediment dynamics of management measures which had been carried out may be derived from changes in results from chemical monitoring. Since 2000, for example, a distinct change in the concentration pattern of contaminants in the Elbe estuary was observed, which was attributed to an increase in upstream transport of fine, lowly polluted marine sediments, probably due to the deepening of the Elbe estuary and other construction work (Ackermann and Schubert, 2007).

Fourth, monitoring data may contribute to a better knowledge of the retention of contaminated particulate matter in estuaries, and to provide a more reliable assessment of particle-bound contaminant input to the sea. A reliable input assessment is required for the implementation of the MSFD.

12.4 INTEGRATION OF BIOAVAILABILITY IN MONITORING CONCEPTS

12.4.1 The Matter of Bioavailability

O'Connor and Paul (2000) stated that, currently, the prediction capability of biological effects from chemical data is poor. We cannot predict the effect of contaminated sediments from describing the physicochemical environment alone. On the other hand, the authors challenge the derivation of sediment quality guidelines on the basis of toxicological systems, arguing that every sediment could be evaluated as toxic by choosing the right (in this case very sensitive) test organism. What are some of the major obstacles in predicting bio-effective or bioavailable concentrations from data on the physico-chemical environment?

1. *Ageing processes.* Hatzinger and Alexander (1995) called the soil–compound interactions by which these compounds become decreasingly bioavailable with time 'ageing'. As being reviewed by a number of authors, ageing is influenced by various factors, such as entrapment within soil micropores (Steinberg *et al.*, 1987) or organic matter (Pignatello and Xing, 1995), pollutant concentration (DiVincenzo and Sparks, 1997) or microbial activity (Carmichael and Pfaender, 1997). A major effect, however, seems to derive from the amount and nature of organic matter to which compounds are sorbed. The amount of a compound that is desorbed with time affects the bioavailable fraction, as compound uptake by organisms is considered to be more rapid from fluid than from sorbed states even though uptake from solid phases may occur either directly or be mediated by microbial activity.

2. *The influence of confounding factors on the bioavailability of contaminants.* In the estuarine and marine environment, salinity, pH, temperature and dissolved organic carbon (DOC) are variable confounding factors with a potential impact on the bioavailability of contaminants which often derive from freshwater systems. An extensive study to assess the combined effects of varying temperature, salinity and dissolved humic substances (DHSs) concentration on the sorption of polycyclic aromatic hydrocarbons (PAHs) to estuarine SPM was done by Tremblay *et al*. (2005). They calculated an increase of the fraction of contaminants sorbed to particles after a decrease in water temperature and an increase in salinity (as encountered during a transition from river water to seawater) by 40 to 64%. However, the differences in sorption behaviour that were caused by salinity, temperature and DHSs were dwarfed by variations caused by different qualities of the SPM (Tremblay *et al*., 2005).

While pH has been identified, in principle, as a governing factor of metal partitioning, well-buffered systems with stable pH values, such as the marine environment, may be impacted most by other environmental factors when it comes to metal solubility and potential bioavailability. Warren and Zimmerman (1994) found that decreasing water temperature and increasing NaCl concentrations increased trace metal (Cd, Zn, Cu) partitioning to the dissolved and potentially more bioavailable phase. In addition to the determination of bioavailable chemical concentrations, the effects that are caused by processes in organisms are difficult to predict.

3. *Mixture toxicity.* Human activities mostly lead to exposure of biota to mixtures. This can be due to chemicals that are originally applied as mixtures (e.g. in pesticide mixtures) or because of transport from different sources. Mixture toxicity is particularly an issue for marine sediments, as they are the sink for a large number of substances, transported over decades by rivers, for example. Even though high dilution rates often result in low concentrations, contaminants with similar modes of action may add up to a significant effect (concept of concentration addition – CA) (Kortenkamp *et al*., 2009). During the workshop 'Risk Assessment in European River Basins – State of the Art and Future Challenges' held in Leipzig, Germany (12–14 November 2007), participants achieved consensus about extrapolation techniques that enable a reasonable predictability of combined effects – assuming that the contaminants are known, that there are no recovery periods between pulsed exposures and that there are no nonchemical (or additional unknown chemical) stressor interferences (Altenburger and Greco, 2009).
4. *Inter-species variability.* Organisms are exposed to sediment-bound contaminants by different exposure pathways depending on their physiology and way of living. There may be the direct exposure via membrane surfaces, such as gills in fish and some invertebrates, or the outer exposure via cuticula in the case of, for example, sediment-dwelling worms, and/or the uptake from the gastrointestinal tract after ingestion of sediment. Resulting toxicities depend on the amount that has actually been taken up, where it has been taken up and what happens to it after having been taken up. These processes are summarized under 'toxicokinetics'. Accordingly, a contaminated sediment may affect species within the biological community very differently. This complicates the use of the term 'bioavailability'. Accordingly, Mayer *et al*. (1996) showed that, even with species using the same primary mode of uptake (in that case via the gastrointestinal tract), digestive fluids from two different animal species solubilized different amounts of metals, leading to different impacts on these organisms.

12.4.2 Organisms Can Render Contaminants Bioavailable

Complicating the issue of bioavailability even further, microorganisms can render previously unavailable contaminants bioavailable. This process is of specific importance for risk management considerations, as sites where contaminants have been evaluated as being unavailable may not get a high priority, while organisms actively increase the bioavailable fraction. Biodegradation studies have shown that bacterial features may enhance the transfer of poorly bioavailable substrates (Wick *et al*., 2001):

- by reducing the mean distances between pollutants and bacteria – for example, by adhesion to sorbents;
- by causing active transfer of the pollutant into the aqueous phase;
- by using uptake systems with high specific affinity.

Hence, concluding on bioeffects of contaminants especially in marine sediments from chemical concentrations is complicated and at present cannot be done conclusively. Ecotoxicity tests, in which test organisms are exposed to water and/or sediment samples, have thus become an additional line of evidence, adding biological responses to the information on chemical composition of sediment. From the variety of available marine biotest assays, which has been compiled by Nendza (2002), a biotest battery should be applied. A suggestion has been made by Peters *et al*. (2002). The specific sensitivities of the biotests should be well known and the results interpreted in an integrated way rather than focusing on the most sensitive organisms, as has been practised in the 'Dredged material management guidelines for coastal sediments' in Germany (BfG, 1999).

How to carry out an integrated assessment of biotest battery responses has been a subject of discussion. We recommend considering the different biotest responses as different lines of evidence for an ecotoxicological response (Heise *et al*., 2000; Hollert *et al*., 2002; Leipe *et al*., 2005; Keiter *et al*., 2009).

It had already been suggested by Long and Chapman (1985) to regard chemical and ecotoxicological data, and information on alterations of the residence community as three lines of evidence within a 'sediment triad'. There have also been attempts to widen the scope of this triad approach and add additional lines of evidence, such as bioaccumulation, toxicity identification evaluation (TIE) or assessment of sediment mobility, as has been reviewed by Chapman and Hollert (2006). Many attempts have been made to improve the aspect of 'alteration of the residence community', as this turns out to be the most ambiguous aspect of the three. While the community of the marine benthos may be relatively stable compared with freshwater benthos, the estuarine environment with its tides and changing salinity already requires such tolerance and variability of its community that alterations are difficult to detect outside of extreme events.

In the long run, the most promising next steps to address bioavailability and bioeffectivity of contaminants in marine sediments are probably improving measurement of the dissolved fraction in pore water, which is the easiest pathway to the organism, and the evaluation of ecotoxicologial responses.

A number of sampling methods have been applied to determine pore water contamination, all having their advantages and disadvantages (Heringa and Hermens, 2003). Currently, a lot of work has been done using solid-phase micro extraction (SPME), which was developed by Arthur and Pawliszyn (1990) as a simple extraction method, using a polymer coating of an optical silica fibre. The method is precise, inexpensive and can be easily automated. It has frequently been applied to marine sediments (Yang *et al*., 2007; Trimble *et al*., 2008; Witt *et al*., 2009) and provides a matrix- and chemical-independent estimation of bioaccumulation for hydrophobic contaminants without extensive biotransformation (You *et al*., 2006). For heavy metals, modelling approaches are of increasing interest when predicting potential risk due to metal concentration in the environment, the most accepted being the biotic ligand model. It has improved a lot for freshwater, but it still needs to be validated or adapted to marine systems and sediments (Ahlf *et al*., 2009).

These tools can improve the chemical assessment part of the sediment triad extensively; they do not, however, replace the necessity to measure effects on organisms. In ecotoxicity studies, toxicokinetic processes and chronic exposures with impacts on different stages of the life cycle can be observed, as well as chemicals and chemical mixtures detected, that have not been expected.

Integrating biotest batteries into marine monitoring, therefore, still seems to be the most reasonable step. Interpreting the biotest responses in an integrated way, and thus using all the information that can be derived from a biotest battery, should be preferred when compared with only focusing on the most sensitive organisms. In the latter case, apart from decisions that may be overprotective, no comparative data would be received if evaluations were carried out on the basis of different organism-tests.

12.5 RESEARCH-DRIVEN SEDIMENT MONITORING

Routine chemical monitoring, which needs prior agreements, standardization, certified laboratories and repeatable analyses, is a costly permanent obligation. Decision makers have a tendency to minimize costs as much as possible, and thus sometimes limit official monitoring often only to seawater and some biota. However, monitoring is still carried on for research purposes. This enables both to solve the questions posed by policy makers concerning levels (Hong *et al*., 2006; Hu *et al*., 2009), sources, fate and toxicity of contaminants in the environment (e.g. Liu *et al*., 2006; McCready, 2006; Pandit *et al*., 2006) and their concentration trends (Moralles-Casselles *et al*., 2008), but also to detect new serious problems to be solved (e.g. Di Leonardo, 2007). Multiple research papers indicated that monitoring of contaminants in sediments provide data that are supplementing and sometimes even substituting water analysis (e.g. Colombo *et al*., 2006; Gevao *et al*., 2006).

First, the majority of compounds most harmful to the marine environment are hydrophobic and because of that their concentration in water is very low. Not only organic compounds, but also toxic trace element ions concentrate in the finest fraction of sediments in the sea. And often only there do they occur at such concentrations that they can be detected. Often the seawater analyses indicate that an environment is free of pollutants and after analysing sediments one can see that there is a real threat (Barcelo and Petrovic, 2007; HELCOM, 2009).

Second, monitoring of sediments collected at carefully selected sites and times gives information on averaged contamination of the whole basin (Colombo *et al*., 2006). It can be a cheaper, quicker and more representative method for characterizing an area than water monitoring that is limited in space and time.

From the analysis of the depth profiles of contaminants and simultaneously of sediment characteristics (such as sedimentation rate, structure and way of formation of sediments and hydrodynamics in the area) one can get information about the trends in the past (Gong *et al*., 2007; Thomas and Brooks, 2010). Analysis of correlation with different environmental parameters can give a lot of information on processes going on in the environment: the sources, ways of transport and fate of contaminants in the environment, information on interactions between particular contaminants, their toxicity and synergic effects (e.g. Boonyatumanond *et al*., 2006; Johnson-Restrepo *et al*., 2008).

There are a great variety of possible contaminants, especially halogen derivatives, which include thousands of possible compounds, and it is impossible to analyse all of them in each sample. Some of those substances belong to the most harmful for life in the sea and human health. Therefore, other ways of solving this problem are practised. The first one, often used in the past and still applied, is based on the total chromatogram treated like a 'fingerprint' of a particular mixture, such as petroleum (Salas *et al*., 2006), or a trade product containing halogenated organic compounds (Sundquist *et al*., 2009). However, at present, since chromatographic techniques have developed so intensively in the last two decades, the analysis of particular compounds is preferred (e.g. Castells *et al*., 2008). These markers, representing groups of similar compounds of organic contaminants, are studied by researchers and recommended for routine monitoring (Cardellichio *et al*., 2006) and are useful for comparison of pollution at different locations (Hong *et al*., 2009). Ratios of marker compounds applied for aromatic hydrocarbons (Cardellichio *et al*., 2007) and halogenated compounds (Liu *et al*., 2006; Moon *et al*., 2007) indicate sources of pollution. It is worth noting that no sediment quality EU guidelines exist, so it is difficult to evaluate and compare toxicity at different sites based on common reference values. Usually, those proposed by US Environmental Protection Agency are applied. The last, but not least, role of research monitoring is to determine the compounds which were not studied before to evaluate their role at particular sites and as a general marker of pollution (e.g. Jara-Marini and Garcia-Rico, 2006; Di Leonardo *et al*., 2009).

The fate of contaminants in the coastal zone depends on so many factors (e.g. Taillefert *et al*., 2007), and the marine sediments form such a complicated, rich organic matrix (Mille *et al*., 2006; Mazur-Marzec *et al*., 2007; Szymczak-Zyła and Kowalewska, 2007; Stephens *et al*., 2007; Unlu and Alpar, 2006) that neither laboratory experiments nor mathematical models can give as full and reliable information as the results of case studies on contaminants in the natural environment. Such study should not include solely the chemical analyses (Falandysz *et al*., 2006), but also investigate the relations of the contaminants studied with the environmental conditions, including biological (Nakata *et al*., 2006; Carafa *et al*., 2006), geological (Lee *et al*., 2006), hydrodynamic (Bone *et al*., 2006) and other factors (e.g. Romano *et al*., 2009). And only the comparison of different case studies will facilitate drawing general conclusions on the key factors governing sources, fate, toxicity and past and future trends of contaminant abundance in the coastal marine environment (David *et al*., 2009; Hou *et al*., 2009; Nakata *et al*., 2009).

From the above, it can be concluded that chemical research monitoring of sediments is still essential for understanding of the relevant environmental processes. Ultimately, all relevant information should be combined to obtain as full as possible a picture of the situation for the parameters studied.

REFERENCES

Ackermann F. (1998). Dynamik der Schwermetallbelastung in feinkörnigen Sedimenten und Schwebstoffen im Tidebereich von Ems, Weser und Elbe, Report BfG-1188, Bundesanstalt für Gewässerkunde, Koblenz.

Ackermann F. (2004). Schwermetalle als Tracer für den Feinpartikeltransport in den Nordsee-Ästuaren, in: Auswirkungen des Ausbaus deutscher Nordsee-Ästuare auf die Gewässergüte und das Baggergutmanagement, Report BfG 2/2004, Bundesanstalt für Gewässerkunde, Koblenz.

Ackermann F. and Schubert B. (2007). Trace metals as indicators for the dynamics of (suspended) particulate matter in the tidal reach of the River Elbe, in *Sediment Dynamics and Pollutant Mobility in Rivers* (eds U. Förstner and B. Westrich), Springer-Verlag, Berlin.

Ackermann F., Bergmann H. and Schleichert U. (1983). *Environ. Technol. Lett*., **4**, 317.

Ahlf W., Drost W. and Heise S. (2009). *J. Soils Sedim*. **9**, 411.

Alargarsamy R. and Zhang J. (2010). *Environ. Monit. Assess*., **161**, 161.

Allan I. J., Vrana B., Greenwood R. *et al*. (2006). *Talanta*, **69**, 302.

Altenburger R. and Greco W. R. (2009). *Integr. Environ. Assess. Manag*., **5**, 62.

Arthur C. L. and Pawliszyn J. (1990). *Anal. Chem*., **62**, 2145.

Barcelo D. and Petrovic M. (eds) (2007). *Sustainable Management of Sediment Resources. Sediments Quality and Impact Assessment of Pollutants*, Elsevier.

BfG (1999). Handlungsanweisung für den Umgang mit Baggergut im Küstenbereich (HABAK-WSV). BfG-Nr. 1100. Bundesanstalt für Gewässerkunde.

Bolam S. G., Rees H. L., Somerfield P. *et al*. (2006). *Mar. Pollut. Bull*., **52**, 415.

Bone S. E., Gonneea M. E. and Charette M. A. (2006). *Environ. Sci. Technol*., **40**, 3273.

Boonyatumanond R., Wattayakorn G., Togo A. and Takada H. (2006). *Mar. Pollut. Bull*., **52**, 942.

Borja A. (2005). *Continent. Shelf Res*., **25**, 1768.

Borja A. (2006). *Mar. Pollut. Bull*., **52**, 239.

Borja A. and Heinrich H. (2005). *Mar. Pollut. Bull*., **50**, 486.

Borja A., Franco J., Valencia V. *et al*. (2004a). *Mar. Pollut. Bull*., **48**, 209.

Borja A., Valencia V., Franco J. *et al*. (2004b). *Mar. Pollut. Bull*. **49**, 8.
Borja A., Tueros I., Belzunce M. J. *et al*. (2008). *J. Environ. Monitor*., **10**, 453.
Borja A., Bald J., Franco J. *et al*. (2009). *Mar. Pollut. Bull*. **59**, 54.
Bortone G. and Palumbo L. (eds) (2007). *Sustainable Management of Sediment Resources. Sediment and Dredged Material Treatment*, Elsevier.
Boszke L. and Astel A. (2007). *J. Environ. Sci. Health Pt. A-Toxic/Hazard. Subst. Environ. Eng*., **42**, 847.
Carafa R., Marinov D., Dueri S. *et al*. (2006). *Mar. Pollut. Bull*., **52**, 1231.
Cardellichio N., Buccolieri A., Giandomenio S. *et al*. (2006). *Ann. Chim*. **96**, 51.
Cardellichio N., Buccolieri A., Giandomenico S. *et al*. (2007). *Mar. Pollut. Bull*., **55**, 451.
Cardoso A. C., Solimini, A. G. and Premazzi G. (2006). Biological monitoring of rivers and European water legislation, in *Biological Monitoring of Rivers: Applications and Perspectives* (eds G. Ziglio, G. Flaim and M. Siligardi), John Wiley & Sons, Ltd., Chichester, UK, pp. 229–240.
Carmichael, L. M. and Pfaender, F. K. (1997). *Environ. Toxicol. Chem*., **16**, 666.
Castells P., Parera J., Santos F. J. and Galceran M. T. (2008). *Chemosphere*, **70**, 1552.
Chapman P. M. (1990). *Sci. Tot. Environ*., **97–98**, 815.
Chapman P. M. and Hollert H. (2006). *J. Soils Sedim*. **6**, 4.
Cherchi A., da Pelo S., Ibba A. *et al*. (2009). *Mar. Pollut. Bull*., **59**, 181.
Colombo J. C., Cappelletti N., Lasci J. *et al*. (2006). *Environ. Sci. Technol*., **40**, 734.
Crane M. (2003). *Toxicol. Lett*., **142**, 195.
David A., Fenel H. and Gomez E. (2009). *Mar. Pollut. Bull*., **58**, 953.
De Jonge V. N., Elliot M. and Brauer V. S. (2006). *Mar. Pollut. Bull*., **53**, 5.
Di Leonardo R., Bellanca A., Capotondi L. *et al*. (2007). *Sci. Tot. Environ*., **388**, 168.
Di Leonardo R., Bellanca A., Neri R. *et al*. (2009). *Chemosphere*, **77**, 778.
DiVincenzo, J. P., and Sparks, D. L. (1997). *Environ. Sci. Technol*., **31**, 977.
Donnini F., Dinelli E., Sangiorgi F. and Fabri E. (2007). *Environ. Int*. **33**, 919.
Ducrotoy J.-P. and Elliot M. (2008). *Mar. Pollut. Bull*., **57**, 8.
Dyer K. R., Robinson M.-C. and Huntley D. A. (2001). Suspended sediment transport in the Humber estuary, in *Land–Ocean Interaction – Measuring and Modelling Fluxes from River Basins to Coastal Seas* (eds D. A. Huntley, G. J. L. Leeks and D. E. Walling), IWA Publishing, London, pp. 169–183.
Edlund A., Soule T., Sjoling S. and Jansson J. K. (2006). *Environ. Microbiol*., **8**, 223.
EU (2000). EU Directive 2000/60/EC of the European Parliament and of the Council of 23 October 2000, establishing a framework for Community action in the field of water policy (Water Framework Directive).
EU (2008a). EU Directive 2008/105/EC of the European Parliament and of the Council of 16 December 2008 on environmental quality standards in the field of water policy, amending and subsequently repealing Council Directives 82/176/EEC, 83/513/EEC, 84/156/EEC, 84/491/EEC, 86/280/EEC and amending Directive 2000/60/EC.
EU (2008b) EU Directive 2008/56/EC of the European Parliament and of the Council of 17 June 2008 establishing a framework for community action in the field of marine environmental policy (Marine Strategy Framework Directive).
Falandysz J., Albanis T., Bachmann J. *et al*. (2006). *J.Environ. Sci. Health Pt. A – Toxic/Hazard. Subst. Environ. Eng*., **41**, 2127.
Farmer J. G. (1991). *Environ. Geochem. Health*, **13**, 76.
Förstner U., Schoer J. and Knauth H.-D. (1990). *Sci. Tot. Environ*. **97**, 347.
Gevao B., Beg M. U., Al-Omair A., *et al*. (2006). *J. Archiv. Environ. Contam. Toxicol*. **50**, 166.
Gobeil C., Macdonald R. W. and Sundby B. (1997). *Geochim. Cosmochim. Acta*, **61**, 4647.
Gong X., Qi. S., Wang Y. *et al*. (2007). *Mar. Pollut. Bull*., **54**, 1434.
Grabemann I., Kappenberg J. and Krause G. (1995). *Neth. J. Aquat. Ecol*., **29**, 217.
Grant A. and Middleton R. (1993). *Neth. J. Aquat. Ecol*., **27**, 111.
Guidance Document no. 25 (2010). Guidance on Chemical Monitoring of Sediment and Biota under the Water Framework Directive. Technical Report – 2010.3991, http://circa.europa.eu/Public/irc/env/wfd/library?l=/framework_directive/guidance_documents/guidance_monitoring/_EN_1.0_&a=d.
Hartwell S. I. and Hameedi M. J. (2006). *Environ. Monitor. Assess*., **121**, 181.
Hatzinger P. B. and Alexander M. (1995). *Environ. Sci. Technol*., **29**, 537.
Heininger P., Schild R., de Beer K. *et al*. (2002). International pilot study for the determination of riverine inputs of polycyclic aromatic hydrocarbons (PAHs) to the maritime area on the basis of a harmonised methodology. Federal Environmental Agency (Umweltbundesamt), Dessau/Rosslau, http://www.umweltbundesamt.de.

Heise S. (ed.) (2007). *Sustainable Management of Sediment Resources. Sediment Risk Management and Communication*, Elsevier.

Heise S., Maaß V., Gratzer H. and Ahlf W. (2000). ecotoxicological sediment classification –capabilities and potentials – Presented for Elbe River Sediments. BfG- Mitteilungen Nr. 22 – Sediment Assessment in European River Basins, pp. 96–104.

Heiskanen A. S., van de Bund W., Cardoso A. C. and Nöges P. (2004). *Water Sci. Technol*. **49**, 169.

HELCOM (2009). Hazardous substances of specific concern to the Baltic Sea. Final report of the HAZARDOUS project, Baltic Sea Environment Proceedings No.119.

Heringa M. B. and Hermens J. L. M. (2003). *TrAC*, **22**, 575.

Hollert H., Heise S., Pudenz S. *et al*. (2002). *Ecotoxicology*, **11**, 311.

Hong S. H., Yim U. H., Shim W. J. *et al*. (2006). *Chemosphere*, **64**, 1479.

Hong S. H., Munschy C., Kannan N. *et al*. (2009). *Chemosphere*, **77**, 854.

Hou L. J., Chen H., Yang Y. *et al*. (2009). *Chemosphere*, **76**, 1114.

Hu L., Zhang G., Zheng B. *et al*. (2009). *Chemosphere*, **77**, 663.

Huntley D. A., Leeks G. J. L. and Walling D. E. (eds) (2001). *Land–Ocean Interaction – Measuring and Modelling Fluxes from the River Basin to the Coastal Seas*. IWA Publishing, London.

ICES (2005). Report of the Working Group on Working Group on Marine Sediments in Relation to Pollution (WGMS), 28 February–4 March, Nantes, France. International Council for the Exploration of the Sea, ICES CM 2005/E:01. pp. 55–74.

Jara-Marini, M. E. and Garcia-Rico L. (2006). *Bull. Environ. Toxicol*., **76**, 677.

Jay D. A., Geyer W. R., Uncles R. J. *et al*. (1997). *Estuaries*, **20**, 262.

Johnson-Restrepo B., Olivero-Verbel J., Lu S. J. *et al*. (2008). *Environ. Pollut*., **151**, 452.

Kappenberg J., Schymura G., Kühn H. and Fanger H.-U. (1996). *Arch. Hydrobiol. Spec. Issues Advanc. Limnol*. **47**, 323.

Keiter S., Braunbeck T., Heise S. *et al*. (2009). *J. Soils Sedim*. **9**, 168.

Kersten M. and Smedes F. (2002). *J. Environ. Monitor*., **4**, 109.

Knauth H.-D., Gandraß J. and Sturm R. (1993). *Vorkommen und Verhalten organischer und anorganischer Mikroverunreinigungen in der mittleren und unteren Elbe*, Forschungsbericht 10204363 des Bundesministeriums für Umwelt, Naturschutz und Reaktorsicherheit, Erich Schmidt Verlag.

Konat J. and Kowalewska G. (2001). *Sci. Tot. Environ*., **280**, 1.

Kortenkamp A., Backhaus T. and Faust M. (2009). *State of the Art Report on Mixture Toxicity*, The School of Pharmacy, University of London, London, 391 pp.

Kowalewska G. (1999), *ICES J. Mar. Sci*., **56**(Suppl.), 219.

Kowalewska G., Konat-Stepowicz J., Wawrzyniak-Wydrowska B. and Szymczak-Zyla M. (2003). *Mar. Pollut. Bull*., **46**, 703.

Lee S. J., Kim J. H., Chang Y. S. and Moon M. H. (2006). *Environ. Pollut*., **144**, 554.

Leipe T., Kersten M., Pohl C. *et al*. (2005). *Mar. Pollut. Bull*. **50**, 446.

Liu W. X., Chen, J. L., Lin X. M. and Tao S. (2006). *Environ. Pollut*., **140**, 4.

Lubecki L. and Kowalewska G. (2010). *Oceanologia*, **52**, 1.

Long E. R. and Chapman P. M. (1985). *Mar. Pollut. Bull*., **16**, 405.

Mayer L. M., Chen Z., Findlay R. H. *et al*. (1996). *Environ. Sci. Technol*., **30**, 2641.

Mazur-Marzec H., Tyminska A., Szafranek J. and Plinski M. (2007). *Environ. Toxicol*., **22**, 101.

McCready S., Birch G. F. and Long E. R. (2006). *Environ. Int*., **32**, 455.

Mee L. D., Jefferson R. L., Laffoley dan d'A. and Elliott M. (2008). *Mar. Pollut. Bull*., **56**, 187.

Mille G., Guiliamo M., Asia L. *et al*. (2006). *Chemosphere*, **64**, 1062.

Milliman J. D. and Meade, R. H. (1983). *J. Geol*., **91**, 1.

Millward G. E. and Glegg G. A. (1997). *Estuar. Coast. Shelf Sci*., **44**(Suppl. A), 97.

Moon H. B., Kannan K., Choi M. and Choi H. G. (2007). *Mar. Pollut. Bull*., **54**, 1402.

Moralles-Casselles C., Kalman J., Micaelo C. (2008). *Chemosphere*, **71**, 1207.

Müller G. and Förstner U. (1975). *Environ. Geol*., **1**, 33.

Muxica I., Borja A. and Bald J. (2007). *Mar. Pollut. Bull*., **55**, 16.

Nakata H., Kannan K., Nasu T. *et al*. (2006). *Environ. Sci. Technol*., **40**, 4916.

Nakata H., Murata S. and Filatreau J. (2009). *Environ. Sci. Technol*., **43**, 6920.

Nendza M. (2002). *Chemosphere*, **48**, 865.

O'Connor T. P. and Paul J. F. (2000). *Mar. Pollut. Bull*. **40**, 59.

O'Reilly Wiese S. B., Emmerson R. H. C., MacLeod C. L. and Lester J. N. (1997a). *Estuaries*, **20**, 484.

O'Reilly Wiese S. B., MacLeod C. L. and Lester J. N. (1997b). *Estuaries*, **20**, 483.
OSPAR (2000) *Quality Status Report 2000*, OSPAR Commission, London.
OSPAR (2001). OSPAR Guidelines for the Monitoring of Sediments. A 13.1 Guidelines for the Sampling and Analyses of Sediments under the Joint Monitoring Programme.
OSPAR (2002). JAMP Guidelines for monitoring contaminants in sediments, Technical Annex 5 –Normalisation of contaminant concentrations; Ref. No. 2002-16, OSPAR Commission, London.
Owens P. (ed.) (2007). *Sustainable Management of Sediment Resources. Sediment Management at the River Basin Scale*, Elsevier.
Pandit G. G., Sahu S. K., Sharma S. and Puranik V. (2006). *Environ. Int.*, **32**, 240.
Peters C., Becker S., Noack U. *et al.* (2002). *Ecotoxicology*, **11**, 379.
Pignatello J. J. and Xing B. (1995). *Environ. Sci. Technol.*, **30**, 1
Prange A. (1997). Erfassung und Beurteilung der Belastung der Elbe mit Schadstoffen. Teilprojekt 2: Schwermetalle – Schwermetallspezies. Zusammenfassende Aus- und Bewertung der Längsprofiluntersuchungen in der Elbe. BMBF-Forschungsvorhaben: 02-WT 9355/4, GKSS-Forschungszentrum Geesthacht GmbH, Geesthacht.
Quevauviller Ph. (2006). *J. Soils Sedim.*, **6**, 2.
Quevauviller Ph. (2007). *Environ. Sci. Pollut. Res.*, **14**, 297.
Ridgway J. and Shimmield G. (2002). *Estuar. Coast. Shelf Sci.*, **55**, 903.
Rodrfguez J. G., Tueros I., Borja A. *et al.* (2006). *Sci. Tot. Environ.*, **370**, 278.
Romano E., Bergamin L., Ausili A. *et al.* (2009). *Mar. Pollut. Bull.*, **59**, 245.
Salas N., Ortiz L., Gilcoto M. *et al.* (2006). *Mar. Environ. Res.*, **62**, 388.
Salomons W. (2005). *J. Soils Sedim.*, **5**, 2.
Salomons W. and Brils J. (eds) (2004). Contaminated sediments in European River Basins, 2004, SedNet (EVK-CT-2001-2002), www.SedNet.org.
Salomons W., Schwedhelm E., Schoer J. and Knauth J. H. (1988). *Water Sci. Technol.*, **20**, 89.
Schubel J. R. (1984). Estuarine circulation and sedimentation: an overview, in *Oceanography of Arabian Sea and Coastal Pakistan* (eds B. U. Hag and J. D. Milliman), Van Nostrand Reinhold, New York, pp. 114–136.
Schubert B., Pies C. and Heil C. (2009). Schadstoffmonitoring von Schwebstoffen und Sedimenten in Ästuaren, in: Aspekte des Schadstoffmonitorings an Schwebstoffen und Sedimenten in der aquatischen Umwelt, Report BfG 7/2009, Bundesanstalt für Gewässerkunde, Koblenz.
Scrimshaw M. D. and Lester J. N. (2001). *Environ. Sci. Technol.*, **35**, 2676.
Simboura N., Panayotidis P. and Papathanassiou E. (2005). *Ecol. Indicators*, **5**, 253.
Spencer K. L., Cundy A. B., Davies-Hearn S. *et al.* (2008). *Estuar. Coast. Shelf Sci.*, **76**, 608.
Steinberg S. M., Pignatello J. J. and Sawhney B. L. (1987). *Environ. Sci. Technol.*, **21**, 1201.
Stephens J. L., dos Santos J. H. Z., Joao Henrique Z. *et al.* (2007). *J. Environ. Sci. Health Pt. A. – Toxic/Hazardous Subst. Eng.*, **42**, 1553.
Sundquist K. L., Tysklind M., Geladi P. *et al.* (2009). *Chemosphere*, **77**, 612.
Szymczak-Zyła M. and Kowalewska G. (2007). *Mar. Pollut. Bull.*, **55**, 512.
Taillefert M., Neuhuber S. and Bristow G. (2007). *Geochem. Trans.*, **8**, 6.
Thomas K. V. and Brooks S. (2010). *Biofouling*, **26**, 73.
Tremblay L., Kohl S. D., Rice J. A. and Gagné J.-P. (2005). *Mar. Chem.*, **96**, 21.
Trimble T. A., You J. and Lydy M. J. (2008). *Chemosphere*, **71**, 337.
Tsakovski S., Kudlak B., Simeonov V. *et al.* (2009). *Talanta*, **80**, 935.
Tueros I., Larreta J., Borja A. *et al.* (2009). *Mar. Pollut. Bull.*, **58**, 1389.
Unlu S. and Alpar B. (2006). *Chemosphere*, **64**, 764.
Warren L. A. and Zimmerman A. P. (1994). *Water Res.*, **28**, 1921.
Wick L. Y., Springael D. and Harms H. (2001) Bacterial strategies to improve the bioavailability of hydrophobic organic pollutants, in *Treatment of Contaminated Soil* (eds R. Stegmann, G. Brunner, W. Calmano and G. Matz), Springer, Berlin, pp 203–218.
Witt G., Liehr G. A, Borck D. and Mayer P. (2009). *Chemosphere*, **74**, 522.
Yang Z.-Y., Zeng E. Y., Maruya K. A. *et al.* (2007). *Chemosphere*, **66**, 1408.
You J., Landrum P. F. and Lydy M. J. (2006). *Environ. Sci. Technol.*, **40**, 6348.
Zonta R., Guerzoni S., Perez-Ruzafa A. and de Jonge V. (2007). *Mar. Pollut. Bull.*, **55**, 403.
Zwolsman J. J. G. (1994). North Sea estuaries as filters for contaminants, Report No. T1233, Delft Hydraulics, The Netherlands.
Zwolsman J. J. G., Berger G. W. and van Eck G. T. M. (1993). *Mar. Chem.*, **44**, 73.

13

Passive Sampling Technologies

Graham A. Mills[a], **Gary R. Fones**[b], **Kees Booij**[c], **and Richard Greenwood**[d]

[a] *University of Portsmouth, School of Pharmacy and Biomedical Sciences, Portsmouth, UK*

[b] *University of Portsmouth, School of Earth and Environmental Sciences, Portsmouth, UK*

[c] *NIOZ Royal Netherlands Institute for Sea Research, Texel, The Netherlands*

[d] *University of Portsmouth, School of Biological Sciences, Portsmouth, UK*

13.1 INTRODUCTION

Some human activities (e.g. oil and gas extraction, dredging, coastal construction work, discharges of domestic and industrial waste) have had a deleterious effect on the chemical and hence biological quality of the marine environment. Over the last 40 years a range of national and international

Chemical Marine Monitoring: Policy Framework and Analytical Trends, First Edition.
Edited by Philippe Quevauviller, Patrick Roose, and Gert Verreet.

agreements has been reached with the aim of reducing or reversing adverse anthropogenic impacts on the marine environment. Hazardous chemicals of concern have been identified and maximum environmental concentrations defined within legislation. In some cases targets have been set for reductions in discharges. The enforcement of such regulations involves monitoring to demonstrate compliance, and to measure changes resulting from remedial actions. There has been significant progress in implementing monitoring in the freshwater environment, and in groundwater, and the experience gained in these areas will underpin the extension of formal monitoring activities in marine waters.

The rationale and requirements for monitoring concentrations of pollutants in water were described by Wu *et al.* (2008). The primary objectives were identified as detection of spatial and temporal changes to identify sources of contaminants and trends in their concentrations (e.g. following remedial actions); checks for enforcement of regulations and for compliance with maximum concentrations defined in legislation; assessment of possible adverse effects (ecological and human health); provision of data to underpin risk assessments.

Active and passive sampling methods can be used for monitoring concentrations of pollutants in the aquatic environment. The former involve the use of low-volume (50 mL to 2 L) spot (grab or bottle) sampling, and in some cases pumps to obtain a large volume (2–100 L) of water that can be passed through an extraction device to concentrate the analytes of interest. Passive methods include the use of passive samplers and living organisms (usually sedentary species such as bivalve molluscs, or caged mobile animals such as fish). This chapter aims to describe the properties of the currently available passive samplers and their potential for use in monitoring the chemical quality of marine waters.

13.1.1 Marine Monitoring Scope

There is a large (in excess of 100 000) and diverse set of chemicals manufactured as primary products or by-products in various quantities worldwide. In the European Union the REACH legislation that is currently being phased in will eventually consider the risk associated with all compounds produced or used in quantities of 1 t or more in Europe. This is projected to include some 30 000 compounds. Risk combines hazard and exposure, where hazard is assessed as toxicity (usually lethality) to humans or to relevant environmental organisms. Other properties, such as environmental persistence (remaining unmodified in the environment for sufficient time for bioaccumulation to occur) and bioaccumulation (the accumulation in the tissues of humans and environmental organisms) are now taken into account in prioritizing the selection of chemicals for testing. Bioaccumulation can lead to transmission up the food chain so that very high concentrations can occur in top predators (including humans). Some sub-lethal toxicological effects (e.g. endocrine disruption) are also used in the prioritization process.

In the past, the lists (e.g. that of the OSPAR Joint Assessment and Monitoring Programme (JAMP)) of high-priority pollutants contained mainly high-volume, nonpolar industrial chemicals such as polyaromatic hydrocarbons (PAHs), polychlorinated biphenyls (PCBs), nonpolar pesticides such as the cyclodienes and heavy metals such as cadmium, lead and mercury, and in some cases their organo-metallic species (e.g. tributyltin (TBT)). There is an agreement between cooperating states to phase out discharges of some of these compounds to achieve a total cessation by 2020, so that concentrations in the marine environment are reduced to background levels. Now, as screening activity is widened to incorporate a far wider range of compounds, new chemicals are becoming of concern to regulators. These compounds of emerging concern come from a far wider range of chemical classes, and have a far wider range of physicochemical properties. Many polar compounds, such as personal care and household product components, pharmaceuticals and detergents, are discharged to rivers via domestic waste treatment and industrial plants. Polar pesticides (herbicides, fungicides and some insecticides) can enter rivers through field application followed by run-off, spray drift and illegal dumping. These substances then enter the marine environment at estuaries, where there is mixing with seawater. These compounds are of particular concern in the estuaries and inshore waters, and can be potentially damaging to vulnerable ecosystems such as coastal reefs. Further offshore the risk associated with these compounds is likely to be significantly less, since dilution factors are large and bioaccumulation is small compared with that of nonpolar pollutants. Few polar compounds will be of concern in the main marine system, apart from those, such as detergents, that are used in oil and gas production, and can appear in produced water. The passive sampling of radionuclides has recently become of interest, but concentrations are low and discharges are usually point sources associated with the nuclear power and defence industries. Some pollutants of concern are not anthropogenic in origin. Toxins produced by algae can cause large-scale deaths of fish, shellfish and other marine animals. These compounds also pose serious threats to human health and can have negative economic impacts on commercial aquaculture. The incidence of such problems has increased over recent years. Generally, biomonitoring has been used to provide warnings and measurements of the occurrence of toxic algal blooms in the areas around aqua-cultural installations.

13.2 MEASUREMENT CURRENT PRACTICE

Methodologies for the regulatory monitoring of chemical quality of environmental waters have been mostly developed for use in surface waters, and have been based on spot (bottle) sampling, linked with analytical measurements in the laboratory. The methodology for the extraction and analysis of chemicals from various compartments (water, sediment and biota) in the marine environment have been comprehensively reviewed (Roose and Brinkman, 2005; Hyötyläinen and Riekkola, 2007), and standard methods are available for most of the nonpolar contaminants of concern. Currently available analytical instrumentation permits low detection limits, in the region of low nanograms per litre for some analytes. However, there are problems associated with handling very nonpolar compounds that have very low water solubilities, that are present at very low dissolved concentrations and that bind to many surfaces, including glass and filter components. However, the robustness and sensitivity of the chemical analysis is not sufficient on its own to provide useful measures of the chemical quality of an aquatic environment, particularly where there is marked spatial or temporal variation. Sound sampling strategies are also necessary. Recently, Wu *et al*. (2008) made a thorough critical examination of current monitoring practice for the measurement of concentrations of persistent organic pollutants (POPs) in the marine environment and found that the majority of sampling designs were not fit for the purpose for which they were intended. Many of the studies reported concentrations measured in single samples (at one site or one time) or in pooled samples. Generally, there was insufficient replication to provide statistically sound discrimination between different sites or between different sampling events. This is particularly concerning where data are used to underpin risk assessments, since an incorrect decision could be very costly, either because unnecessary actions are taken or because necessary measures are not taken.

In recent times, in the area of freshwater monitoring there has been a realization that the use of infrequent spot sampling does not necessarily provide a representative measure of the quality of a body of water, especially where concentrations fluctuate (for instance, with seasonal use of pesticides or discontinuous discharges of industrial or domestic waste). One possible way of reducing the large uncertainties associated with such estimates of the average chemical quality of a body of water is to increase the number of replicates, as outlined above, and sampling frequency, but this would be a costly solution. Another possibility is the use of automated sampling equipment or continuous monitoring instrumentation. However, these can be used only in secure sites and are equally costly. Similar problems face those with responsibility for monitoring the quality of estuarine, coastal and offshore marine waters. Here, problems can be caused by large tidal flows and fluctuating currents. One sample taken at a fixed point in a tidal cycle would provide little useful information about water quality. The cost of repeated spot sampling would be higher in the marine environment than in inland surface waters because of the difficulty of access to tidal sampling sites, and in many cases the necessity of using a boat. The additional difficulties pose significant challenges to the successful implementation of routine monitoring, and alternative, lower cost, methods are being sought. One solution that has been in place for many years is the use of the accumulation of nonpolar pollutants in biota (typically bivalve molluscs, for instance in the Mussel Watch programme). Bioaccumulation by these test organisms provided biologically relevant data (based on equilibrium over deployment times) on available concentrations of pollutants in the water. Most experience to date has been with POPs where times to equilibrium are long. In contrast, equilibrium times for more polar pollutants are of the order of a few days (Söderström *et al*., 2009), and further work will be necessary to determine the utility of this approach for this class of contaminants. The uptake from the environment can be via two routes: trans-epithelial (mainly across the respiratory surface), driven by the concentration of freely dissolved chemical in the water, and by feeding on suspended particles to which pollutants are bound. The relative importance of the two routes will depend on the feeding mechanism and biology of the species. However, even for the dissolved material, accumulation depends not only on the external concentrations, but also on the ability of the test species to eliminate particular compounds. Where there are marked inter-specific differences in the biochemical and physiological capabilities, then the use of biomonitoring information from one species may not be relevant for other, possibly more vulnerable, species. Further, the current practice of pooling samples from biota in order to reduce costs fails to provide robust estimates that can be used with confidence in risk assessments (Wu *et al*., 2008).

An alternative approach that is under consideration is the use of buoy-based sensors for analytes of interest, and telemetric transmission of analytical data. This approach has been successful and is well established (see Chapter 10) for the real-time measurement of properties such as oxygen concentration, pH, turbidity, chlorophyll content and concentrations of some nutrients, but, while this has been regarded as the ideal solution, currently there are few proven sensors for the wide range of organic pollutants and most metals. Another alternative solution is the use of passive samplers to provide representative measurements of concentrations of the pollutants of interest or concern to regulators. Over the past 20 years several designs of passive samplers have been developed to monitor a range of pollutant types (nonpolar organic, polar organic, organo-metals and metals) present in the aquatic environment. Depending on the sampler design and time of deployment in the field, the mass of a pollutant accumulated by a device can reflect either the equilibrium concentration or the time-weighted average (TWA) concentration of the pollutant in the water column. These samplers can be used to accumulate contaminants over extended deployment periods (days to months). This means that large volumes of water can be cleared by the devices and, hence, the samplers effectively pre-concentrate pollutants, lowering the effective level of detection of environmental chemicals. This feature could be particularly important for compounds that are sparingly soluble in water and also in the marine environment where large dilution factors can reduce concentrations to ultra-trace levels. Uptake by passive samplers is driven by the freely dissolved concentrations of chemicals, and so this technology measures only that fraction. This contrasts with the use of water samples, where, depending on the pre-treatment procedures (e.g. ultrafiltration, filtration (0.45 μm) and centrifugation), different fractions can be measured. Currently, environmental quality standards (EQSs) are framed in terms of total or filtered samples for different classes of pollutants. If passive samplers were to be used in a regulatory context, then the EQSs would have to be redefined to reflect the fraction measured by these devices. This concentration has been often described as the biologically relevant, or available, fraction. Since the 1970s, similar devices have been used extensively for monitoring air quality and for measuring workplace exposure to toxic chemicals. Several reviews (Vrana *et al*., 2005a; Kot-Wasik *et al*., 2007; Esteve-Turrillas *et al*., 2008) and two text books (Huckins *et al.*, 2006; Greenwood *et al.*, 2007a) have been published on the state of the art of different passive sampling methods for monitoring both organic and inorganic pollutants in water and their field applications. The following sections provide an overview of the types of device currently available and describe their potential for monitoring different classes of pollutants in the marine environment.

13.3 PASSIVE SAMPLERS

Passive samplers consist of a receiving phase with a high affinity for the pollutant of interest that is exposed to the aquatic medium for a defined period of time. In some cases the receiving phase is separated from the bulk water phase by a diffusion-limiting membrane. This can help decrease the rate of diffusion, and in some environments limit the degree of biofouling of the surface of the receiving phase. Three broad classes of sampler can be distinguished: those for measuring nonpolar organics, polar organics and metals.

Nonpolar samplers consist of an amorphous polymer for which nonpolar analytes have a high affinity. Semi-permeable membrane devices (SPMDs) were the first nonpolar samplers to be used for environmental monitoring on a worldwide scale (Huckins *et al.*, 1990). These samplers consist of low-density polyethylene (LDPE) lay-flat tubing that is filled with a small quantity of the triglyceride triolein as the receiving phase. SPMDs have a large sampling area and can be used to monitor trace levels of pollutants with a log K_{ow} over the range 3.5–8. The extracts from the devices need a clean-up step (e.g. using gel-permeation chromatography or silica column chromatography) to remove any residual triolein prior to instrumental analysis. More recently, polymer films and sheets have been used as nonpolar samplers; examples include LDPE (Adams *et al.*, 2007), silicone rubber (Leslie *et al.*, 2002) and poly(oxymethylene) (Jonker and Koelmans, 2001). These single-phase partition-type samplers have high uptake rates, similar to those of the SPMD, for many nonpolar priority pollutants. For example, 2 g of LDPE can effectively sample, depending on the physico-chemical properties of the analyte of concern, between 20 and 2000 L of water over a deployment period. Other samplers for hydrophobic compounds include the nonpolar version of the Chemcatcher (consisting of an *n*-octanol-soaked C_{18}3M Empore™ disk overlain by a thin LDPE membrane) (Vrana *et al.*, 2005b) and various configurations of the membrane-enclosed sorptive coating (MESCO) sampler (Paschke *et al.*, 2007).

The second group of samplers targets polar analytes. These samplers consist of an adsorption phase overlain by a thin polar diffusion-limiting membrane. Examples are the polar organic chemical integrative sampler (POCIS) (Alvarez *et al.*, 2004a) and the polar version of the Chemcatcher (Kingston *et al.*, 2000; Greenwood *et al.*, 2007b). Receiving phases that have been used include particulate Oasis HLB solid-phase extraction sorbent and activated carbon for the POCIS, and C_{18}, styrene-divinylbenzene or ion-exchange resins bound into 47 mm 3M Empore™ disks for the polar Chemcatcher. A polyethersulfone (PES) microporous diffusion-limiting membrane is used to cover the receiving phase in both devices. These samplers have smaller sampling areas in comparison with the SPMD and the commonly used polymeric devices. Naked Empore™ disk (with no diffusion membrane) samplers have also been used with varying degrees of success (Gunold *et al.*, 2008; Shaw *et al.*, 2009a).

The third sampler group targets freely dissolved inorganic ions (e.g. trace metals, phosphate and sulfide). In the diffusive gradients in thin films (DGT) sampler, a sorption phase is covered by a polyacrylamide hydrogel (thickness ~1 mm) and a 0.45 µm filter membrane. Sorption phases include cation-exchange resins for trace metals, iron oxides and titanium dioxide for phosphate, and silver iodide for sulfide (Zhang and Davison, 1995; Zhang *et al.*, 1998; Teasdale *et al.*, 1999; Panther *et al.*, 2010). The metals version of the Chemcatcher employs a 47 mm 3M Empore™ chelating disk that is covered by a 0.45 µm cellulose acetate filter (Persson *et al.*, 2001).

All samplers operate in a similar fashion. Target analytes diffuse from the bulk water phase to the receiving phase when the chemical activity (i.e. the effective concentration) of the analytes in the water is higher than their activity in the sampler. Far away from the sampler, all transport is by turbulent mixing. At short distances from the sampler–water interface, a transition zone (the water boundary layer (WBL)) exists where turbulence gradually decreases to zero, and where transport is increasingly governed by molecular diffusion. Molecular diffusion also dominates transport within the membrane and within the receiving phase. In some cases a layer of biofouling (consisting of periphyton or even macrofauna) develops on the surface of the sampler where analyte transport is also by molecular diffusion. Water ventilation by organisms trapped in the biofilm layer may alter the hydrodynamic structure of the WBL. Further details on the various transport steps involved for the different available designs of passive sampler can be found elsewhere (Namiesnik *et al.*, 2005; Huckins *et al.*, 2006; Booij *et al.*, 2007; Warnken *et al.*, 2007).

In the initial stage of the exposure, the flux to the receiving phase is linearly proportional to the analyte's free concentration (i.e. excluding the fraction that is bound to colloids and particles) in the water column. When the concentration of an analyte in the water column is constant, the amount of analyte in the sampler increases linearly with time. When the concentration in the water varies with time, the accumulated amounts reflect the TWA concentration $C_{\mathrm{w,TWA}}$. This initial stage is known as the linear uptake stage or kinetic sampling stage. The amounts N of analyte sequestered are related to their aqueous concentrations by

$$N = C_{\mathrm{w,TWA}} R_{\mathrm{s}} t \tag{13.1}$$

where t is the exposure time (days) and R_{s} is the water sampling rate (L day^{-1}) that is measured in laboratory-based calibration studies. The product $R_{\mathrm{s}}t$ can be interpreted as an extracted water volume, and thereby forms a conceptual link between passive sampling and batch water sampling. R_{s} values need to be measured for each analyte of interest, and these values can vary according to water temperature, water turbulence and degree of biofouling on the diffusional surface of a sampler. For long exposure times, the flux to the sampler may decrease, because the analyte approaches its equilibrium concentration in the sampler. In the case of nonpolar samplers, the equilibrium conditions are well characterized by partition coefficients. This sampling stage is therefore known as equilibrium sampling. The time to reach equilibrium is a function of the compound's structure and can often be predicted based on K_{ow} values.

For polar and trace metal samplers the situation is more complicated, because adsorption dominates over absorption (partition). This sorption mechanism is characterized by nonlinear distribution coefficients and competition between target and nontarget analytes. Although the occurrence of nonlinear and competitive sorption is widely recognized in physical chemistry, the modelling of these processes is still in its infancy, and samplers for polar analytes and metals are (almost) exclusively used in the linear uptake stage, when the receiving phase can be regarded as an infinite sink.

Analytes that have a high affinity for the diffusion-limiting membrane experience a large concentration drop over the WBL. This implies that the sampling rates are limited by transport through the WBL and, therefore, increase with increasing water flow rates, and hence are WBL controlled. This is generally the case for the sampling of highly hydrophobic compounds by nonpolar samplers. By contrast, analytes that have a low affinity for the membrane experience the largest concentration drop over the membrane, and sampling rates are membrane controlled (e.g. sampling of compounds with low hydrophobicities by nonpolar samplers, and sampling by polar and metal samplers). In addition to flow effects, the sampling rates may depend on water temperature and in some cases on pH and salinity. When samplers are used for only short deployment periods, consideration also has to be given to the lag phase. This is the time needed for a compound present in the bulk water phase to be transported to the bulk receiving phase. Lag times are both sampler and compound specific and can be determined in laboratory calibration studies.

13.3.1 Nonpolar Organics

The SPMD has been available since 1990 and much attention has subsequently been paid to the underpinning theory of operation, modelling of uptake kinetics and laboratory calibration of nonpolar passive samplers. The generalized equation governing the absorption of a nonpolar analyte with time is given by (Huckins *et al.*, 2006):

$$N_s = K_{sw} V_s C_w \left[1 - \exp\left(-\frac{R_s t}{K_{sw} V_s} \right) \right] \tag{13.2}$$

where N_s is the amount of analyte absorbed, K_{sw} is the sampler–water partition coefficient (volume/volume units), V_s is the sampler volume, C_w is the aqueous concentration and t is exposure time. For short exposure times, Equation (13.2) reduces to Equation (13.1). Accurate K_{sw} values should be available from calibration studies. In-situ sampling rates should also be known; however, the R_s of many nonpolar compounds depends on the water flow and water temperature at the exposure site. With this design of passive sampler, these sampling rates may be estimated from the dissipation of performance reference compounds (PRCs) that are spiked into the sampler prior to deployment and that do not occur in the environment (e.g. usually isotopically labelled analogues of the pollutants of interest). The PRC method for estimating uptake rates has been shown to account quantitatively for differences in water temperature, flow rates (degree of water turbulence) and biofouling (Huckins *et al.*, 2002; Booij *et al.*, 2006).

Although the operating principles of nonpolar samplers are well understood, some issues are unresolved. First, theory predicts that the R_s of analytes with log $K_{ow} > 6$ should decrease only weakly with increasing molecular size. This weak dependence has been confirmed for silicone samplers (Rusina *et al.*, 2010), but for SPMDs and nonpolar Chemcatchers a much stronger decrease has been observed (Huckins *et al.*, 2006; Vrana *et al.*, 2007). This stronger decrease is likely to be artificial, but experimental evidence for this assumption is lacking. Second, Allan *et al.* (2009) observed that C_w estimates from co-deployed samplers of differing designs varied by up to a factor of 5 around the sampler mean value. This variation partly originates from different relationships between R_s and molecular size that are reported for the various samplers, partly from inaccuracies in the K_{sw} values and partly from analytical variability among the laboratories that participated in this study. Variability of C_w estimates within sampler type was much lower (approximately a factor of 1.1 to 1.4). Third, K_{sw} values of reliable accuracy are not always available for all analytes. The selection of a sampler for the monitoring of particular analytes should, therefore, be based on a careful consideration of the availability of accurate K_{sw} values, and verified models that relate R_s to molecular size.

Since the introduction of the SPMD some 20 years ago there have been in excess of 250 papers describing its use in various field trials (Huckins *et al.*, 2006). Many of the applications have recently been reviewed by Esteve-Turrillas (2008). Most of the work has concerned deployments in surface waters. More recently there have been applications of SPMDs and polymeric samplers in the marine environment. The following give some recent examples of the use of these devices and other samplers in measuring TWA concentrations of nonpolar anthropogenic pollutants in a marine context.

Harman *et al.* (2009) deployed SPMDs to measure concentrations of PAHs in waters surrounding a Norwegian oil platform in the North Sea. SPMDs were spiked with PRCs prior to deployment to enable uptake rates to be corrected for large changes in environmental conditions over the 6 week trial. A range of polyaromatic compounds were detectable at concentrations in the low picograms per litre range, but the profile was mainly dominated by alkylated naphthalenes, phenanthrenes and dibenzothiophenes. As expected, the highest concentrations were found nearest to the point discharges. However, measuring some of the more volatile polyaromatics (e.g. naphthalene and alkylated naphthalenes) at trace levels was difficult because of high background contamination of the blanks. SPMDs were also used by Roach *et al.* (2009), who deployed the samplers along a 25 km transect around Sydney Harbour, Australia. The devices were able to measure ultra-trace concentrations (sub-picograms per litre) of a range of polychlorinated dibenzo-*p*-dioxins, polychlorinated dibenzofurans and PCBs in the water column over a 28 day trial. Even at these trace levels, statistically significant spatial and temporal trends in concentration of the pollutants within the harbour system were identifiable. Care must be taken when deploying SPMDs in field situations (e.g. shallow reefs) where the devices can be exposed to high levels of sunlight over the deployment period. In a trial located off the Great Barrier Reef, Komarova *et al.* (2009) found that some PAHs were highly photo-degraded. Concentrations in SPMDs that were unprotected and exposed to sunlight were up to five times lower than those in control devices that were shielded by a mesh deployment cage.

SPMDs were originally designed as monitoring tools to mimic the uptake on hydrophobic chemicals by biota. Several studies have compared the uptake of PAHs, PCBs and polybrominated diphenylethers (PBDEs) by SPMDs with that by blue (*Mytilus edulis*) or green (*Perna viridis*) mussels (Richardson *et al.*, 2001; Esteve-Turrillas *et al.*, 2008). Most workers found that mussels are more efficient at sequestering PAHs, PCBs and organochlorine pesticides (OCPs) from the water column than SPMDs. Caution must be taken in directly interpreting the results from the two monitoring approaches, as SPMDs measure only the water-soluble fraction of pollutants, whilst mussels can sequester pollutants present in both the water-soluble and particle-bound fractions. This fact was emphasized by Boehm *et al.* (2005) when using both SPMDs and mussels (*Mytilus trossulus*) as intertidal monitors to investigate the environmental effects of PAHs accidentally released at oil spill sites around the Prince William Sound, Alaska. However, both methods can give complementary information on bioavailability of POPs available to higher marine organisms and transfer of these chemicals through the food chain. Uptakes of PAHs and PCBs by *Mytilus edulis* and by large silicone rubber sheet samplers were compared by Smedes (2007). Eight sampling stations in the Wadden Sea, Eastern and Western Scheldt off the coast of the Netherlands were used. Overall, there was a close relationship between concentrations found in mussels and those derived from the passive sampler exposures. Seasonal variations in pollutant profiles and concentrations were reflected in both monitoring methods.

Chemicals concentrated in the receiving phase of SPMDs can also be combined with a variety of short-term bioassay procedures to assess the biological effects of the aqueous contaminants. The identification of the most suitable bioassay technique (e.g. Microtox, Mutatox, *Daphnia pulex* immobilization assay) to use is an important factor for toxicity screening purposes. The Microtox assay appears to be particularly compatible with the SPMD extracts (Esteve-Turrillas *et al.*, 2008). When receiving-phase extracts are to be used with bioassay procedures, samplers must be deployed without any PRCs added in order to avoid the toxic effects of these chemicals, since these would directly influence the results. Solvent extracts of LPDE and silicone rubber samplers may also be appropriate for use in toxicity screening procedures and may provide a simpler matrix to work with.

There is an increasing use of SPMDs, LPDE, PDMS tubing or silicone rubber strip samplers in large-scale monitoring programmes to map the distribution of hydrophobic pollutants. However, owing to their lower costs and comparative ease of analysis, it is expected that single-phase polymer-based samplers will become more attractive to use in large-scale surveys. To date, SPMDs have been used by a number of governmental agencies in the USA (e.g. United States Geological Survey, United States Environmental Protection Agency), the UK (Environment Agency for England and Wales) and the Czech Republic (ITRC, 2005). A passive sampling trial survey with silicone strip samplers (12 participating marine laboratories from the International Council for the Exploration of the Sea (ICES) area) was organized by the ICES Working Group on Marine Sediments and the ICES Marine Chemistry Working Group (http://www.passivesampling.net). This survey showed that the cooperation of monitoring agencies and passive sampling scientists allows for an efficient transfer of knowledge on the capabilities and limitations of passive sampling technology.

Recently, Lohmann and Muir (2010) have proposed the establishment of a global passive sampling network (AQUA-GAPS) to monitor the concentrations of POPs in aquatic environments throughout the world. This would be a natural extension of the successful global atmospheric monitoring scheme (GAPS) currently in operation for this class of pollutants. Low-cost polymeric samplers rather than SPMDs have being targeted for this purpose. Samplers could be deployed on the existing network of marine buoys in areas such as the North, Baltic, Mediterranean, Caspian and Black seas and also coastal areas of the USA and Asia, where mussel watch activities are currently ongoing. A number of challenges would, however, need to be addressed. These include the need for a major interlaboratory exercise with the co-deployment of different samplers and the training of local scientists and other personnel in the use of these devices.

A similar study (International Pellet Watch) is already ongoing, but in this case the sampler consists of small samples of polyethylene pellets that are washed along the shoreline throughout the world (Ogata *et al.*, 2009). Such cylindrical resin pellets (<5 mm diameter) are used globally in various manufacturing processes and can be unintentionally released in run-off to surface water and eventually into the ocean and then distributed by currents worldwide. Samples have been collected from 30 sites in 17 countries and despatched for analysis for organochlorine compounds by a central laboratory in Japan. By this approach the distribution of these pollutants could be mapped globally and the concentrations measured could be related to their anthropogenic use in the various participating countries.

13.3.2 Lipophilic Toxins

Special passive samplers have been developed for the detection and monitoring of algal toxins in the marine environment. This work was in response to a need of commercial producers of shellfish for human consumption. Many countries have legislation in place that requires monitoring of contamination levels in order to protect human health. The method most commonly used is the analysis of the concentrations of algal toxins in shellfish tissues. However, this is a difficult and expensive procedure, and since some toxins are metabolized by shellfish, the analysis is made more complex. An alternative is the analysis of phytoplanktonic samples. Early in the 2000s a passive sampler, the solid-phase absorption toxin tracker (SPATT), was developed to provide information on relative accumulation of toxins by biota before, during and after algal blooms (MacKenzie *et al.*, 2004). These samplers comprise a polyester mesh bag containing activated polystyrene divinylbenzene resin. Other receiving phases (e.g. methacrylic ester copolymer resins for the more hydrophilic toxins) can be used to match the toxins to be monitored. These samplers have been used quantitatively to provide estimates of concentrations of a range of marine lipophilic toxins, including azaspiracids, okadaic acids, pectenotoxins, yessotoxins and spirolides in mussels (Rundberget *et al.*, 2009). The extraction and analysis of the toxins is much simpler than for biota, and provides clean samples with few interfering compounds.

Applications of the SPATT sampler have been reviewed (MacKenzie, 2010) and have been extended to monitoring of cyanotoxins produced by planktonic and benthic freshwater cyanobacteria, and that can potentially contaminate potable waters. Recently, the SPATT sampler has been trialled on the west coast of Ireland (Fux *et al.*, 2008, 2009) and compared with indigenous and transplanted marine mussels (*Mytilus edulis*). The accumulation rates of toxins by the SPATT samplers were well correlated with those in the mussels, and uptake was linear (micrograms per SPATT bag, or per gram receiving phase per day) over 7 days. This work provided valuable information on the behaviour of the lipophilic toxins and identified the presence of toxins that had not been recorded previously in these Irish waters. Similar conclusions were drawn from studies using SPATT samplers and mussels in Norwegian fjords (Rundberget *et al.*, 2009). Currently, little work has been done on monitoring some of the more polar toxins (e.g. saxitoxins and domoic acid) (MacKenzie, 2010) of interest. It may be necessary to take advantage of the flexibility of passive samplers by customizing the receiving phase to optimize uptake and quantification of these compounds. Other designs of passive sampler (e.g. POCIS and Chemcatcher) may be appropriate for this type of application.

13.3.3 Polar Organics

As stated above, nonpolar passive sampling devices have high uptake rates (up to several litres per day) for compounds in the range of log K_{ow} 4.5 to 6. However, below log $K_{ow} = 4$ the sampling rate falls off rapidly, and these samplers are not useful for polar pollutants. As instrumental analytical techniques have improved (e.g. using liquid chromatography–mass spectrometry methods), greater numbers of polar organic contaminants (including polar pesticides, antiparasitic compounds, pharmaceuticals, antibiotics, personal care products and detergents) have been detected in the aquatic environment and have subsequently become of interest to regulatory bodies. Such substances have been termed emerging pollutants, or pollutants of emerging concern. In many cases little is known of their ecotoxicological properties. Many of these compounds have been detected at high nanograms per litre concentrations in surface water and wastewater. This group of compounds includes many that have dissociable groups and are difficult to monitor with a single sampler.

Passive samplers have been developed to monitor the more polar pollutants (log $K_{ow} < 4$), and two main designs (Chemcatcher and POCIS) are available (Section 13.3), though others are being developed, such as one with a cellulose membrane and organic solvent receiving phase for polar herbicides (Hyne and Aistrope, 2008). In comparison with many of the samplers for nonpolar pollutants, these have relatively small sampling surface areas and associated lower sampling rates (less than a litre per day, and typically in the tens to hundreds of millilitres per day) (MacLeod *et al*., 2007; Mazella *et al*., 2007; Harman *et al*., 2008; Wong and MacLeod, 2010). The variants of these samplers with diffusion-limiting membranes tend to remain integrative over periods of up to 30 days (Kingston *et al*., 2000; Alvarez *et al*., 2004a).

One feature of the Chemcatcher design that uses an immobilized receiving phase, rather than a solvent or loose powder, is that it can be used without a diffusion-limiting membrane. This variant with naked receiving phases operates in an integrative mode for much shorter time periods (typically 5–10 days), but can be used for short monitoring periods, and to detect short-term fluctuations, such as pulses of pesticide concentrations following storm events. Chemcatcher samplers without membranes have been calibrated for monitoring pharmaceuticals in a range of flow regimes in artificial channel systems (Vermeirssen *et al*., 2008). Naked disks give TWA concentrations only up to 5 days exposure, and sampling rates were markedly flow dependent. Similar findings have been published by authors working on monitoring short-term events: event monitoring of diuron, atrazine and simazine (Stephens *et al*., 2009), short pulses of thiacloprid (Schäfer *et al*., 2008a) and 10 polar and semi-polar pesticides (Schäfer *et al*., 2008b). Laboratory tank calibration trials of Chemcatcher (no membrane) (Gunold *et al*., 2008) showed that uptake rates of a range of polar and semi-polar pesticides remained linear over 10 days of exposure. Linked field trials comparing Chemcatcher (no membrane) with flow-weighted and suspended particle sampling for a range of polar and semi-polar pesticides indicated that this sampler could be useful for monitoring pesticides commonly found in surface waters, though the impact of biofouling is potentially a more serious problem for samplers without a diffusion-limiting membrane to protect the receiving phase.

The samplers with membranes tend to have lag phases of up to a day or more for some pollutants, and this, combined with low sampling rates, precludes their application for short exposures or the detection of short pollution events. However, they can potentially provide information on TWA concentrations over longer term exposures of 14–28 days. Compared with the samplers for nonpolar organics, relatively little reliable calibration data is available, and many of the reported applications are either qualitative (presence or absence of a pollutant) or semi-quantitative (nanograms accumulated in the receiving phase). The numbers of compounds potentially of concern are far greater than those of the nonpolar pollutants generally considered as priority pollutants. Some calibration data are available from laboratory studies, but the calibration techniques and conditions used by different groups have varied in the conditions applied. Since uptake rates are affected by temperature, turbulence and potentially biofouling, it is difficult to transfer laboratory data to the field. This problem is further compounded by the lack of PRCs for the field deployment of these types of sampler. This is due to the use of receiving phases that accumulate pollutants by adsorption rather than partitioning. Uptake and offloading of pollutants are not isotropic. This was demonstrated by Shaw *et al.* (2009a), who investigated the potential utility of a range of compounds (deuterated diuron, dimethylphthalate and chlorpyriphos) as PRCs in Chemcatcher samplers with and without membranes. This group suggested the co-deployment of nonpolar PDMS disks as surrogate PRC samplers. Mazella *et al.* (2007) suggested, however, that deisopropylatrazine could be used as a potential PRC for the POCIS when monitoring herbicides. They demonstrated a measurable desorption of this compound from the sorbent over the sampling timeframe. Development of PRCs for polar analytes is an urgent and important challenge to those working in this field.

Most applications of polar samplers have been in freshwater and wastewater systems, though some studies have been in marinas, harbours, estuaries and inshore reefs. To date, most information is available for the POCIS device. However, one of the first marine applications of polar samplers was that of Kingston *et al.* (2000) that investigated the concentrations of two herbicides (Irgarol and diuron) used in antifouling paints in two marinas. The concentrations found in spot samples were seen to fluctuate widely over the tidal cycle, and the TWA concentrations measured using the Chemcatcher reflected those obtained from the repeated bottle sampling. More recently, French research groups have used the POCIS to measure concentrations of industrial pollutants, pesticides and pharmaceuticals in estuarine waters (Togola and Budzinski, 2007). This group investigated the effects of salinity on sampling rates and found that it had no effect on acidic pharmaceuticals, but a significant effect on basic compounds. This indicates the importance of including salinity in the factors investigated in calibration studies. An application of POCIS to measure endocrine-disrupting compounds (alkyl phenols and ethoxylates, and steroids) in coastal regions of rivers, marine harbours and offshore (1.5 km) waters found integrative sampling of these compounds for up to 28 days (Arditsoglou and Voutsa, 2008). The sampling rates for the suite of 14 compounds ranged between 6 and 222 mL day^{-1}, and the study indicated the potential utility of this approach in marine waters, but caution was expressed concerning the translation of laboratory-based calibrations to the field situation. A study by Stephens *et al.* (2009) in an Australian estuarine harbour system demonstrated that the Chemcatcher passive sampler (with and without membrane) could be used to monitor fluctuations in the concentrations of diuron and simazine associated with rainfall events. Many aqua-cultural installations are located in estuaries and inshore marine situations, and these are potential sources of sporadic pollution with, for instance, veterinary medicines and food components. The POCIS sampler was calibrated in a batch system in seawater and used to monitor a wide range of veterinary medicines (including antibiotics, pesticides and biocides) over a period of a year in the region of a fish farm in the Mediterranean Sea (Bueno *et al.*, 2009). Average analyte concentrations over the sampling period were in the range 0.01–46 ng L^{-1}; these fell below the EQS values for these compounds within the WFD. This study indicates the potential of this technology for monitoring the impacts of aqua-cultural activities without the need for frequent spot sampling. The impact of offshore discharges of produced water from oil platforms in the North Sea was assessed using the POCIS sampler to monitor polar analytes (alkyl phenols) (Harman *et al.*, 2009), and elevated concentrations of short-chain alkyl phenols were detected up to 2 km from the discharge point. Although the concentrations measured were of little biological significance (being orders of magnitude below those that cause acute and sub-lethal effects), the study clearly demonstrated the potential of passive sampling for monitoring pollutants released as a result of anthropogenic activity in offshore waters. Although the usefulness of both the Chemcatcher and POCIS in marine monitoring applications has been demonstrated, the low uptake rates and the low concentrations that are observed in most offshore marine waters may result in detection problems that hamper the identification of temporal and spatial trends. Their specific value will be in monitoring concentrations of polar pollutants in estuarine waters where concentrations fluctuate over a tidal cycle, making the use of spot sampling unreliable. The devices can also be used in ecologically sensitive environments, such as offshore reefs, to gain a better understanding of the effects of the run-off pollutants on delicate marine species (Shaw *et al*., 2010). However, care must be taken in laboratory-based calibration studies to allow for and understand the influence of changes in salinity on uptake rates. This will be of increasing importance where samplers are deployed along a long estuarine transect. As demonstrated with nonpolar samplers, solvent extracts of the receiving phase can be used in short-term bioassay tests to gain a further understanding of the toxicological relevance of individual fractions of the complex mixture of sequestered pollutants. Although less work has been undertaken in this area for polar analytes, Shaw *et al*. (2009b) recently demonstrated the usefulness of this approach. Naked SDB-RPS Empore™ disks were exposed at three sites (Russell–Mulgrave River mouth and two inshore reefs) on the Great Barrier Reef near Cairns. The area is extensively used for growing sugar cane. Extracts were then analysed for herbicides (including phenylureas and triazines) and also applied to four different bioassays targeting different life stages or functions of coral reef biota. The concentrated extracts obtained from the exposed disks were shown to have potential to cause detrimental effects on coral settlement, sea urchin larval development, viability of bacterial communities and photosynthesis in phytoplankton.

13.3.4 Organo-metallics

Most organo-metallic compounds are more toxic than inorganic complexes or free forms of the parent metal and have a high potential for bioaccumulation. The main organo-metallic species of concern in the marine environment are those of tin. TBT was used for a number of years as a biocidally active ingredient in antifouling paints that were applied to the hulls of boats and ships. In many countries the use of TBT for this application has been banned. Other tin species (e.g. monobutyltin (MBT) and dibutyltin (DBT)) can be formed as environmental degradation products of TBT, but these transformations are generally slow. MBT and DBT are also used as heat stabilizers in PVC. TBT has very low water solubility and, hence, is present at only very low concentrations in the water column. However, owing to its toxicity, the EU WFD EQS values for surface waters are set at very low values (0.2 ng L^{-1} EQS-AA and 1.5 ng L^{-1} EQS-MAC) (Lepom *et al*., 2009) for this substance. Most TBT in the aquatic environment is found bound to bottom sediments; it can become redistributed back into the water phase if the benthic layer is disturbed through, for example, dredging activities in a harbour or through large tidal movements.

It is difficult to measure TBT reliably at the EQS concentrations when using low-volume spot samples (1–2 L) combined with routine analytical methods. The use of passive samplers to provide a 'concentrated' extract for subsequent laboratory analysis may help to overcome this problem. Følsvik *et al*. (2000, 2002) reported the use of SPMDs for monitoring the soluble, bioavailable, fraction of organotin compounds. Both DBT and TBT were accumulated by the devices, but not MBT during a several-week exposure trial in a Norwegian fjord. Using this method, it was possible to identify concentration gradients of organotin compounds at the sampling site. However, as the devices had not been fully calibrated for these substances, they could be used only in semi-quantitative or qualitative modes for the estimation of relative concentrations at different points within a sampling area.

A variant of the Chemcatcher sampler has been developed and calibrated for the TWA concentration measurement of organotin compounds (Aguilar-Martínez *et al*., 2008a, b). This sampler combined a C_{18} 3M Empore™ disk-receiving phase with a thin cellulose acetate diffusion-limiting membrane. A linear uptake was obtained for the range of organotin compounds over the 14-day test period. Once accumulated in the receiving phase the compounds were stable over prolonged periods. The effects of pH, salinity and biofouling on the rate of uptake of tin compounds by the sampler were evaluated. Owing to anisotropic exchange kinetics, PRCs could not be used with this passive sampling system to compensate for changes in uptake rate due to variations in water temperature and turbulence and biofouling of the surface of the diffusion-limiting membrane. The performance of the devices was tested in a polluted marine harbour in Spain. Owing to the lower sampling rate for TBT of the Chemcatcher (30–200 mL day^{-1} depending on site conditions) compared with that of the SPMD, application of this device in open-water field sites would be of limited utility.

13.3.5 Metals and Radionuclides

Cadmium (from metallurgic processes, fossil fuels and fertilizers), copper (from mining and antifoulants), lead (from mining and petrol) and mercury (from fossil fuels, incineration and metallurgic processes) are the major heavy metal inputs into the marine environment. Radionuclides (caesium, polonium and technetium – from reprocessing plants and phosphate production) are additional anthropogenically derived inputs considered in this section. As with the organic pollutants, concentrations of these substances in marine waters are mainly determined in spot samples and, thus, it is possible to miss important episodic events where levels can exceed those specified in the WFD and OSPAR regulations. The use of passive sampling monitoring methods may help overcome this problem. In addition, the concentrations of metals and radionuclides in the marine waters are very low, and the use of passive samplers that have the ability to, in effect, sequester analytes from large volumes of water may help to lower the detection limits achieved for these pollutants.

The two main passive samplers that have been extensively used over the last 15 years for the measurement of metals and radionuclides in natural waters, including marine environments, are the DGT and Chemcatcher devices. DGT was developed in 1993 by Davison and Zhang at Lancaster University, UK. Since then, over 200 papers have been published on the application of DGT for a wide range of inorganic analytes in sediments, soils and natural waters. The DGT technique is based on a simple device that accumulates solutes on a binding agent after passage through a hydrogel that acts as a well-defined diffusion layer (Zhang and Davison, 1995). A binding agent such as a resin (e.g. for trace metals Chelex 100) is immobilized in a thin layer of a hydrogel (binding-gel) (Davison and Zhang, 1994). This is separated from the bulk solution by an ion-permeable hydrogel layer (diffusive gel) of known thickness Δg. In-between the diffusive gel and the bulk solution there is a WBL in which transport of ions is solely by molecular diffusion. This is a critical point for both the DGT and Chemcatcher passive samplers, as various assumptions have to be made regarding flow rate and turbulence to nullify the importance of the WBL. When the DGT sampler is deployed, a steady-state gradient is established that can be exploited mathematically to measure in-situ concentrations. The principles of how the DGT operates have been published elsewhere (Davison *et al.*, 2000). Briefly, the flux J of an ion through the gel is given by Fick's first law of diffusion (Equation (13.3)), where D is the diffusion coefficient and dC/dx is the concentration gradient:

$$J = -D\frac{dC}{dx} \tag{13.3}$$

If the diffusion coefficients of ions in the diffusive gel are the same as in water and if there is rapid equilibrium with a binding agent that has a large binding constant and if the thickness of the

WBL is negligible relative to that of the hydrogel, then Equation (13.3) can be rewritten as

$$J = \frac{DC}{\Delta g} \tag{13.4}$$

The DGT device is then deployed for a fixed time t. On retrieval, the binding gel layer is removed, extracted with nitric acid and the mass of ions accumulated in this layer measured using an appropriate analytical technique (Zhang and Davison, 1995; Davison *et al.*, 2000; Murdock *et al.*, 2001) to determine the concentration C_e in the acid extract. The ratio of the eluted/extracted ion is known as the elution factor f_e, which has been reported to be between 0.7 and 0.98 for a suite of trace metals (Garmo *et al.*, 2003). Taking the elution factor into account, the accumulated mass M of ions in the binding layer can be calculated from Equation (13.5), where V_g is the volume of gel in the binding-layer and V_e is the volume of the eluate:

$$M = C_e \frac{V_g + V_e}{f_e} \tag{13.5}$$

M can then be used to calculate the flux through the known area A of the exposed diffusive layer:

$$J = \frac{M}{At} \tag{13.6}$$

Equations (13.4) and (13.6) can be rearranged to give

$$C = \frac{M \Delta g}{DtA} \tag{13.7}$$

This equation enables the concentration C in the bulk solution to be calculated from the known values of Δg and A, the deployment time t and the accumulated mass M. It is crucial to have accurate estimates of D, and these are normally obtained from published diffusion coefficients or from laboratory experiments (see below).

The other main passive sampler that has been used for metals is the Chemcatcher (Greenwood *et al.*, 2007b). Although this device has been used mainly for nonpolar and polar organics (see previous sections), a particular configuration has been developed for measuring the TWA concentration of metals. It consists of a Teflon watertight sampler body to hold in position a cellulose acetate diffusion-limiting membrane (0.45 μm pore size and thickness of 0.135 mm) that is placed over a 3M Empore™ chelating disk (47 mm diameter) as the receiving phase (Persson *et al.*, 2001; Blom *et al.*, 2002, 2003). Metal accumulation in the sampler is similar to that of the DGT and is based on the diffusion of metal species across the cellulose acetate membrane. The diffusion process is influenced by water temperature and by hydrodynamic conditions (water turbulence) at the surface of the sampler. As mentioned for the DGT, the WBL is important, but can this be compensated for by the use of laboratory-based calibration data (see below).

As highlighted above, one of the main factors affecting the reliability of the DGT and Chemcatcher is the availability of sound calibration data. DGT uses diffusion coefficients of metals in the hydrogel layer, whereas Chemcatcher uses metal-specific uptake rates, with both sets of values obtained under controlled laboratory conditions with constant aqueous metal concentrations. DGT does not depend on the concentration of other components in the solution; therefore, individual calibration in other media is unnecessary. However, understanding and calculating the diffusion coefficient for the type of hydrogel used is of critical importance. A number of studies have been undertaken to determine diffusion coefficients for the DGT for a range of analytes from metals to radionuclides and in environments from freshwaters to marine systems (Chang *et al.* 1998; Larner *et al.*, 2006; Scally *et al.*, 2006). The diffusion coefficient can be determined using a specially designed double-diffusion cell (Chang *et al*., 1998). A gel of known area A and thickness Δg provides the connection between the two cell compartments. The compartments are filled with the ion composition of interest dissolved in solutions of $NaNO_3$ or NaCl. One compartment contains a high concentration of the analyte of interest and the other side is at zero concentration (blank). The compartments are well mixed using stirrers and aliquots are withdrawn at set time intervals and concentrations of metals measured using appropriate analytical techniques (e.g. inductively coupled plasma mass spectrometry (ICP-MS)). The diffusion coefficient can be calculated from the slope of a linear plot of the measured mass M passing through the gel versus time t using

$$D = \frac{\Delta g}{C_b A} \frac{\Delta M}{\Delta t} \tag{13.8}$$

Garmo *et al*. (2003) assessed the diffusion coefficients of 55 different elements using the diffusion-cell method, and also calculated them from the limiting equivalent conductance property (Li and Gregory (1974) based on data tabulated by Robinson and Stokes (1959)). The limiting equivalent conductance-based diffusion coefficients express free diffusion in water without the restriction imposed by the gel. Larner and co-workers (Larner and Seen, 2005; Larner *et al*., 2006) determined diffusion coefficients by deploying DGT samplers exposed to 2 L of estuarine water for a range of times (6, 12, 24, 36, 48 and 72 h). The diffusion coefficients D were then calculated from a plot of the mass M of metal accumulated over deployment time t using

$$D = \frac{M \Delta g}{t A C_b} \tag{13.9}$$

where Δg is the thickness of the diffusion layer, with exposed area A, and C_b is the average concentration of metal in the bulk solution over the deployment time.

Uptake by the Chemcatcher is influenced by the water temperature and by hydrodynamic conditions (water turbulence) at the diffusional surface of the sampler. To overcome this, laboratory-based calibrations are undertaken in large (e.g. 200 L) plastic containers at a series of fixed temperatures and stirring speeds (simulated water turbulences) (Vrana *et al*., 2005a; Allan *et al*., 2007). The uptake rate R_s obtained can be expressed as a volume of water cleared of metal by the sampler per unit of time (L day^{-1}). These uptake rates then enable the TWA concentrations $C_{\mathrm{Chemcatcher}}$ to be calculated from the mass M accumulated in the Empore™ disk, the exposure time t and the uptake rate R_s:

$$C_{\mathrm{Chemcatcher}} = \frac{M}{R_s t} \tag{13.10}$$

Field trials in marine waters have been limited, with the emphasis being on research ad-hoc deployments with no systematic and integrated inter-calibration approach being undertaken. The first DGT deployments in a marine system were undertaken in the Menai Strait (UK) and the North Atlantic Ocean (Zhang and Davison, 1995). Six gel assemblies with a diffusion layer thickness of 0.5 mm were deployed for different time periods (1–6 h) and six gel assemblies with various types of diffusion layer were also deployed (5.33 h). A linear relationship was found for Fe, Mn and Zn between the mass accumulated and deployment time for both types of deployment. This was due to the degree of water movement and, hence, the gel thickness layer dominated over the WBL thickness. Zhang and Davison (1995) measured the concentration of Cu in the North Atlantic water at 40 m depth using DGT by deploying four assemblies with different gel thicknesses. The plots of measured mass against $1/\Delta g$ were linear, showing that there was adequate natural convection for the use of DGT. The mean value of 0.15 ± 0.01 μg L^{-1} of Cu is comparable to typical concentrations in ocean water and falls within the OSPAR background assessment criteria range. Other individual studies in marine waters have seen DGT devices deployed in the Antarctic (2002–2003) to assess concentrations of trace metals (Larner *et al*., 2006). They found that whilst DGT water sampling was restricted to quantification of Cd, Fe and Ni due to detection-limit problems, pre-concentration using Empore™ chelating disks provided results for an additional nine elements (Al, As, Co, Cu, Cr, Mn, Pb, Sn and Zn). The DGT results showed that heavy metal concentrations in the impacted Brown Bay were comparable to those in the non-impacted O'Brien Bay.

In Australia, numerous studies have been undertaken in freshwater, but very few in marine waters. Dunn *et al*. (2003, 2007) studied the potential for using DGT to measure Cu, Ni, Pb and Zn in a dynamic estuarine environment. The studies demonstrated that, with judicious selection of deployment times, DGT measurements enabled changes in heavy metal concentrations to be related to various cycles and events within the estuary. Dunn *et al*. (2003) found that DGT-labile measurements as a fraction of 0.45 μm filterable concentrations were 21 ± 2 % for Cu, 29 ± 11 % for Pb, 28 ± 5 % for Zn and 27 ± 12 % for Ni, demonstrating the capabilities of DGT for measuring speciation. In Australia, a comprehensive monitoring programme in Port Curtis, undertaken by Vision Environment (http://www.visionenvironment.com.au/), used DGT and oysters as biomonitors (Olivier *et al*., 2002). Spot water samples were also taken to assess water quality. Approximately 2500 oysters and 450 DGT units were deployed annually in a number of trials ranging from dredge and effluent monitoring to whole harbour assessment; this is arguably the largest concurrent use of these techniques in Australia. Results from these deployments have been used in geographic information system models by Vision Environment and show the need for long-term deployments using different techniques for metal monitoring. The two methods provided complementary data because DGT accumulated only the dissolved species, whereas the oysters, being filter feeders, accumulated both dissolved species and metals bound to particulate material.

Within the Mediterranean Sea there have been several studies using DGT. A large trial was undertaken by IFREMER on the south coast of France in the summer of 2008 (Gonzalez *et al*., 2009). Here, triplicate DGT units for the measurement of trace metals (Ag, Cd, Co, Cr, Cu, Ni, Pb and Zn) along with POCIS passive samplers for polar organics were deployed at over 15 sites for a period of up to 50 h. The study was arranged so that each group at the individual sites followed the same procedure and received identical sampling kits from the host laboratory. The study highlighted the variability of trace metals along the French coast and showed that this integrated approach has the potential for validating the use of passive samplers for monitoring the marine environment within the EU. This group has also pursued the same approach by deploying passive samplers (DGT, POCIS, stir-bar sorptive extraction devices) in areas of the Indian Ocean (Reunion, Mayotte, French Guiana).

Within the Mediterranean, Schintu *et al*. (2008) deployed mussels (*Mytilus galloprovincialis*) and DGT units in the coastal waters off Sardinia for three 1-month periods during 2004, 2005 and 2006. Concentrations of Cd, Cu, Ni and Pb were measured in areas differently impacted by anthropogenic activities. The results provided an estimate of the water quality with respect to dissolved metals and pointed out the usefulness of a combination of biomonitoring and DGT techniques for a better understanding of trace metal availability in coastal waters. Within the shallow coastal areas of the Aegean Sea (eastern Mediterranean), dissolved Cd and Cu partitioning and their interaction with organic matter were investigated by Scoullos *et al*. (2006) using DGT. They found that the percentage of DGT-labile Cu as a proportion of total dissolved Cu ranged from 13 to 34% during summer and from 23 to 36% during winter, whereas the corresponding percentage for DGT-labile Cd was higher in summer (38–68%) in comparison with winter (29–44%).

For Chemcatcher, the majority of work with trace metals has been undertaken in freshwater environments (e.g. River Meuse, Netherlands). Comparisons between the Chemcatcher and DGT were undertaken in a 28 day trial during which fluctuating concentrations of metals (Cd, Cu, Ni, Pb, Zn) were observed in frequent spot samples. This study assessed the potential of these two devices for the regulatory monitoring of trace metals in surface water (Allan *et al.*, 2008). It demonstrated that the two sampling devices provided similar information, and were able to integrate concentrations reliably even where there were fluctuations in concentration during the deployment period. The DGT- and Chemcatcher-based concentrations were compared with whole (unfiltered), filtered (0.45 μm) and ultrafiltered (5 kDa) concentrations and with the concentrations of the various species predicted using Visual MINTEQ software. The uptake by the samplers was consistent with the known and measured speciation behaviour and properties of the metals. Neither sampler gave reasonable predictions for the environmental concentration of lead, and this was attributed to the low concentrations available to the samplers. When compared with the use of infrequent spot sampling over a 28 day period, the passive samplers provided a more representative and reliable picture of water quality. Bank-side tank experiments using spiked river water to produce highly fluctuating concentrations of the same metals (Allan *et al.*, 2007) provided further evidence of the ability of the samplers to provide integrated information on concentrations of the metals over the deployment period.

There have been very few studies using passive samplers in natural waters for monitoring radionuclides. Currently, only Cs, Sr, Tc and U have been studied, with Cs, Sr and U concentrations assessed using the DGT in fresh waters (Chang *et al.*, 1998; Murdock *et al.*, 2001; Li *et al.*, 2006, 2007) and ^{99}Tc in seawater (French *et al.*, 2005). ^{99}Tc in seawater was measured using DGT with TEVA (Eichrom Technologies) resin as the binding agent. The accumulated ^{99}Tc activity per unit area of resin-gel was proportional to both the bulk solution activity and the exposure time for deployments of up to 2 weeks. The response of the DGT was found to be independent of solution chemistries over the pH range 3–8 and ionic strength range 0.01–1.3 M. Detection limits of 0.05 Bq L^{-1} and 0.025 Bq L^{-1}, for 2- and 4-week DGT deployments respectively, were calculated for ^{99}Tc (measured by liquid scintillation spectrometry). The authors commented that by using ICP-MS techniques to measure bound ^{99}Tc the detection limit could be reduced to 0.125 mBq L^{-1} for a 4-week deployment. These detection limits are sufficiently low for monitoring contaminated environments, including the Irish Sea. Uranium measurements have currently focused on using DGT for speciation measurements in freshwaters using different binding resins (Li *et al.*, 2006, 2007). There is, however, potential for the development of passive samplers to measure actinide concentrations in coastal waters in both clean and contaminated environments.

As has been demonstrated in a number of field studies, both the DGT and Chemcatcher, once calibrated, have the potential to measure a number of trace metals in various aquatic environments. However, careful consideration has to be given to the exact fraction of metal being taken up by these devices. Receiving phases such as Chelex 100 and Empore™ chelating disks when used with diffusive layers and membranes accumulate only a fraction of the total dissolved metal concentration (e.g. the free ion, and inorganically, and some weakly organically complexed metals) but not the strongly bound organically complexed metals (Warnken *et al.*, 2008, 2009). In order for passive samplers to be fully recognized and adopted for routine use in the marine environment there needs to be increased standardization and also fully integrated interlaboratory comparisons similar to the International Passive Sampling Interlaboratory Comparison (IPSIC 2010 – http://www.cslab.cz/) which is focusing on contaminated surface river water.

13.3.6 Performance of Passive Sampling Devices

A summary of the performance of the commonly used passive samplers for monitoring hydrophobic and hydrophilic organic and organo-metallic contaminants (including those on the OSPAR list of chemicals for priority action and the EU Priority Substances Directive) is provided in Table 13.1. This provides information on the suitability of each sampler and estimates of the limits of detection over 28-day deployments for nonpolar samplers, and 10 days for polar samplers. Estimates of the deployment time over which the samplers for nonpolar compounds remain integrative have been calculated. For extremely nonpolar compounds ($\log K_{ow} > 6$) the integrative phase of sampling is of the order of hundreds of days, and the limits of detection are in the low picograms per litre range. For the less nonpolar hydrophobic compounds ($4 < \log K_{ow} < 6$) the integrative period is in the range of several days to tens of days for the polymeric strip and SPMD samplers and the limits of detection are in the order of tens of picograms per litre. Limits of detection for the nonpolar Chemcatcher are an order of magnitude higher than those for the polymeric strip and SPMD samplers, but the integrative phase tends to be longer for the less nonpolar hydrophobic compounds. Information on the samplers for polar organic compounds is sparser. The estimated limits of detection for the POCIS and polar Chemcatcher are based on the assumption that average apparent sampling rates are 100 mL day^{-1} for the former and 50 mL day^{-1} for the latter, an analytical limit of detection of 1 ng and a 10-day exposure, over which the receiving phase will act as an infinite sink for most compounds, and hence the samplers will operate in integrative mode. The limits of detection for these samplers are generally in the low nanograms per litre range. It is noteworthy that the limits of detection of the SPMD and polymeric sheets for TBT are sufficiently low to make these samplers potentially useful in monitoring compliance for this pollutant, even though the new EU EQS is difficult to reach with standard analytical methods.

13.4 VALIDATION

The use of passive samplers in either equilibrium or integrative mode to measure concentrations of chemicals in water involves knowledge of either equilibrium and/or kinetic parameters for the particular type of sampler, and in some cases of the environmental conditions during the deployment period. The parameters are generally measured in laboratory calibration studies. Since the calculated concentrations of chemicals in the environment depend on the parameter values used, it is important that the latter are accurate and reliable. Currently, few laboratories are involved in the calibration of passive samplers, and because it is very labour intensive, there is a shortage of validated calibration data for many types of passive sampler.

Considerably more calibration data are available for passive samplers for nonpolar organic compounds than for those for polar organic compounds and inorganics. This reflects the relative maturity and greater experience with samplers for nonpolar organics, where accumulation in the receiving phase is by partitioning, and the availability of PRCs. The use of PRCs effectively enables an in-situ calibration, since the deployed samplers are used for both environmental sampling (through uptake of pollutants of interest) and calibration (through off-loading of PRCs). For this type of sampler, field validation is possible. One problem that needs to be addressed for very hydrophobic compounds is the modification of the freely dissolved concentration in the presence of dissolved organic carbon (DOC) that can bind a large proportion of some of these compounds. This can bias both calibration measurements and field evaluations. Since PRCs are not available for samplers for polar organics, this approach is not available for this class of passive samplers. For these samplers it is necessary to calibrate (using static, semi-static or through-flow laboratory-based systems) the samplers over a range of combinations of temperature and turbulence that covers the conditions that are likely to be met in field deployments. It is then possible to interpolate on the calibration surface to determine calibration parameters appropriate for a particular deployment. However, as flow and temperature can vary throughout a deployment, it may be necessary to use a range of parameters and to use these to estimate a range in which the true concentration would fall.

Table 13.1 Estimated performance of hydrophobic samplers (SPMDs, non-polar Chemcatcher, LDPE, and silicone strips) and hydrophilic samplers (POCIS, polar Chemcatcher) for the monitoring of contaminants listed by OSPAR (list of chemicals for priority action, update 2007) and the EU (Priority Substances Directive, 2008/105/EC).

	EU	OSPAR	log K_{ow}	Sampler suitability		SPMD 460 cm^2, 1 mL triolein, 85 μm LDPE			LDPE strip, 460 cm^2, 85 μm			Non-polar Chemcatcher 46 mm Ø, 450 μL octanol			Silicone strip 600 cm^2 × 0.5 mm			POCIS 47 mm Ø		Polar Chemcatcher 46 mm Ø C_{18}/SDB	
				Non-polar	Polar	Stage	LOD_{28} (ng L^{-1})	t_{int} (d)	Stage	LOD_{28} (ng L^{-1})	t_{int} (d)	Stage	LOD_{28} (ng L^{-1})	t_{int} (d)	Stage	LOD_{28} (ng L^{-1})	t_{int} (d)	Stage	LOD_{10} (ng L^{-1})	Stage	LOD_{10} (ng L^{-1})
Organo chlorines																					
Aldrin	EQS		6.75	+	–	p	0.003	>300	p	0.003	>300	p	0.08	>300	p	0.002	>300	–	–	–	–
4,4′-DDT	EQS		6.79	+	–	d^a	0.003	>300	d^b	0.003	>300	p	0.08	>300	p	0.002	>300	–	–	–	–
2,4′-DDT	EQS		6.79	+	–	d^a	0.003	>300	p	0.003	>300	p	0.08	>300	p	0.002	>300	–	–	–	–
4,4′-DDD	EQS		5.87	+	–	d^a	0.003	>300	d^b	0.003	100	p	0.09	>300	p	0.002	>300	–	–	–	–
4,4′-DDE	EQS		6.00	+	–	d^a	0.003	>300	$d^{b,c}$	0.003	200	p	0.08	>300	d^c	0.002	>300	–	–	–	–
Dieldrin	EQS		5.45	+	–	d^a	0.003	100	d^b	0.004	50	d^d	0.09	>300	p	0.002	>300	d^e	1	p	2
Endosulfan	PHS	A	3.83	+	+	d^f	0.03	3	p	0.07	1	d^d	0.3	10	p	0.008	8	p	1	d^g	2
Endrin	EQS		5.45	+	–	d^a	0.003	100	d^b	0.004	50	p	0.09	>300	p	0.002	>300	d^e	1	p	2
Hexachlorobenzene	PHS		5.50	+	–	d^a	0.003	100	d^a	0.004	50	d^d	0.09	>300	$d^{c,h}$	0.002	>300	–	–	–	–
Hexachlorocyclo-hexanes	PHS	A	3.80	+	+	d^a	0.03	3	p	0.08	1	d^d	0.3	9	p	0.009	8	p	1	p	2
PCBs dichloro	Rev	A	5.10	+	–	d^a	0.004	50	p	0.006	20	p	0.09	200	d^i	0.003	200	–	–	–	–
PCBs trichloro	Rev	A	5.70	+	–	d^a	0.003	200	$d^{b,c,j,k}$	0.004	90	p	0.09	>300	$d^{c,h,i}$	0.002	>300	–	–	–	–

(*continued overleaf*)

Table 13.1 (*continued*)

	EU	OS PAR	log K_{ow}	Sampler suitability		SPMD 460 cm^2, 1 mL triolein, 85 µm LDPE			LDPE strip, 460 cm^2, 85 µm			Non-polar Chemcatcher 46 mm Ø, 450 µL octanol			Silicone strip 600 cm^2 × 0.5 mm			POCIS 47 mm Ø		Polar Chemcatcher 46 mm Ø C_{18}/SDB	
				Non-polar	Polar	Stage	LOD_{28} (ng L^{-1})	t_{int} (d)	Stage	LOD_{28} (ng L^{-1})	t_{int} (d)	Stage	LOD_{28} (ng L^{-1})	t_{int} (d)	Stage	LOD_{28} (ng L^{-1})	t_{int} (d)	Stage	LOD_{10} (ng L^{-1})	Stage	LOD_{10} (ng L^{-1})
PCBs tetrachloro	Rev	A	5.95	+	–	d^{a}	0.003	>300	d^{b,c,j,k}	0.003	200	p	0.08	>300	d^{c,h,i}	0.002	>300	–	–	–	–
PCBs pentachloro	Rev	A	6.31	+	–	d^{a}	0.003	>300	d^{b,c,j,k}	0.003	>300	p	0.08	>300	d^{c,h,i}	0.002	>300	–	–	–	–
PCBs hexachloro	Rev	A	7.03	+	–	d^{a}	0.003	>300	d^{b,c,j,k}	0.003	>300	p	0.08	>300	d^{c,h,i}	0.002	>300	–	–	–	–
PCBs heptachloro	Rev	A	6.85	+	–	d^{a}	0.003	>300	d^{b,c,j}	0.003	>300	p	0.08	>300	d^{c,h,i}	0.002	>300	–	–	–	–
PCBs octachloro	Rev	A	7.25	+	–	d^{a}	0.003	>300	p	0.003	>300	p	0.08	>300	d^{i}	0.002	>300	–	–	–	–
PCBs nonachloro	Rev	A	7.63	+	–	d^{a}	0.003	>300	p	0.003	>300	p	0.08	>300	p	0.002	>300	–	–	–	–
PCBs decachloro	Rev	A	8.21	+	–	p	0.003	>300	p	0.003	>300	p	0.08	>300	p	0.002	>300	–	–	–	–
PCDDs monochloro	Rev	A	4.87	+	–	d^{a}	0.005	30	p	0.008	10	p	0.10	100	p	0.003	90	–	–	–	–
PCDDs dichloro	Rev	A	5.65	+	–	d^{a}	0.003	200	d^{k}	0.004	80	p	0.09	>300	p	0.002	>300	–	–	–	–
PCDDs trichloro	Rev	A	6.35	+	–	d^{a}	0.003	>300	p	0.003	>300	p	0.08	>300	p	0.002	>300	–	–	–	–
PCDDs tetrachloro	Rev	A	6.85	+	–	d^{a}	0.003	>300	p	0.003	>300	p	0.08	>300	p	0.002	>300	–	–	–	–

PCDDs pentachloro	Rev	A	7.40	+	–	d[a]	0.003	> 300	p	0.003	> 300	p	0.08	> 300	p	0.002	> 300	–	–	–	–
PCDDs hexachloro	Rev	A	7.80	+	–	d[a]	0.003	>300	p	0.003	>300	p	0.08	>300	p	0.002	>300	–	–	–	–
PCDDs heptachloro	Rev	A	8.00	+	–	d[a]	0.003	>300	p	0.003	>300	p	0.08	>300	p	0.002	>300	–	–	–	–
PCDDs octachloro	Rev	A	8.20	+	–	d[a]	0.003	>300	p	0.003	>300	p	0.08	>300	p	0.002	>300	–	–	–	–
PCDFs monochloro		A	4.82	+	–	d[a]	0.005	30	p	0.008	10	p	0.10	90	p	0.003	80	–	–	–	–
PCDFs dichloro		A	5.26	+	–	d[a]	0.004	80	p	0.005	30	p	0.09	300	p	0.003	200	–	–	–	–
PCDFs trichloro		A	5.70	+	–	d[a]	0.003	200	p	0.004	90	p	0.09	>300	p	0.002	>300	–	–	–	–
PCDFs tetrachloro		A	6.10	+	–	d[a]	0.003	>300	p	0.003	200	p	0.08	>300	p	0.002	>300	–	–	–	–
PCDFs pentachloro		A	6.50	+	–	d[a]	0.003	>300	p	0.003	>300	p	0.08	>300	p	0.002	>300	–	–	–	–
PCDFs hexachloro		A	7.00	+	–	d[a]	0.003	> 300	p	0.003	> 300	p	0.08	> 300	p	0.002	> 300	–	–	–	–
PCDFs heptachloro		A	7.40	+	–	d[a]	0.003	> 300	p	0.003	> 300	p	0.08	> 300	p	0.002	> 300	–	–	–	–
PCDFs octachloro		A	8.00	+	–	d[a]	0.003	> 300	p	0.003	> 300	p	0.08	> 300	p	0.002	> 300	–	–	–	–
PCNs monochloro		C	4.00	+	–	p	0.02	4	p	0.05	2	p	0.2	10	p	0.006	10	–	–	–	–
PCNs dichloro		C	4.53	+	–	p	0.007	10	p	0.01	6	p	0.1	50	p	0.003	40	–	–	–	–

(continued overleaf)

Table 13.1 (*continued*)

	EU	OS PAR	log K_{ow}	Sampler suitability		SPMD 460 cm², 1 mL triolein, 85 µm LDPE			LDPE strip, 460 cm², 85 µm			Non-polar Chemcatcher 46 mm Ø, 450 µL octanol			Silicone strip 600 cm² × 0.5 mm			POCIS 47 mm Ø		Polar Chemcatcher 46 mm Ø C_{18}/SDB	
				Non-polar	Polar	Stage	LOD_{28} (ng L^{-1})	t_{int} (d)	Stage	LOD_{28} (ng L^{-1})	t_{int} (d)	Stage	LOD_{28} (ng L^{-1})	t_{int} (d)	Stage	LOD_{28} (ng L^{-1})	t_{int} (d)	Stage	LOD_{10} (ng L^{-1})	Stage	LOD_{10} (ng L^{-1})
PCNs trichloro		C	5.35	+	–	p	0.004	100	p	0.004	40	p	0.09	>300	p	0.003	300	–	–	–	–
PCNs tetrachloro		C	5.90	+	–	p	0.003	>300	p	0.003	100	p	0.09	>300	p	0.002	>300	–	–	–	–
PCNs pentachloro		C	6.59	+	–	p	0.003	>300	p	0.003	>300	p	0.08	>300	p	0.002	>300	–	–	–	–
PCNs hexachloro		C	7.26	+	–	p	0.003	>300	p	0.003	>300	p	0.08	>300	p	0.002	>300	–	–	–	–
PCNs heptachloro		C	8.20	+	–	p	0.003	>300	p	0.003	>300	p	0.08	>300	p	0.002	>300	–	–	–	–
PCNs octachloro		C	8.50	+	–	p	0.003	>300	p	0.003	>300	p	0.08	>300	p	0.002	>300	–	–	–	–
Heptachloro-norbornene		C	5.67	+	–	p	0.003	200	p	0.004	80	p	0.09	>300	p	0.002	>300	–	–	–	–
Pesticides																					
Alachlor	PS		3.52	+	+	d^{f}	0.06	1	p	0.2	1	p	0.5	5	p	0.02	4	d^{e}	1	$d^{g,l}$	2
Atrazine	PS		2.61	–	+	–	–	–	–	–	–	–	–	–	–	–	–	d^{m}	1	$d^{g,n,o,p}$	2
Chlorfenvinphos	PS		3.81	+	+	p	0.03	3	p	0.08	1	p	0.3	9	p	0.009	8	p	1	d^{g}	2
Chlorpyrifos	PS		4.96	+	–	$d^{f,n}$	0.004	40	p	0.007	20	p	0.09	100	p	0.003	100	d^{e}	1	p	2
Dicofol	PS	A	5.81	+	–	p	0.003	300	p	0.004	100	p	0.09	>300	p	0.002	>300	–	–	–	–

Diuron	PS		2.68	–	+	–	–	–	–	–	–	–	–	–	–	–	–	d[e,m]	1	d[n,o,p]	2
Ethyl *O*-(*p*-nitrophenyl) phenyl phosphonothionate (EPN)		C	4.78	+	–	p	0.005	30	p	0.009	10	p	0.10	90	p	0.003	70	p	1	p	2
Flucythrinate		C	6.20	+	–	p	0.003	>300	p	0.003	300	p	0.08	>300	p	0.002	>300	–	–	–	–
Isodrin	EQS	C	6.75	+	–	p	0.003	>300	p	0.003	>300	p	0.08	>300	p	0.002	>300	d[e]	1	p	2
Isoproturon	PS		2.87	–	+	–	–	–	–	–	–	–	–	–	–	–	–	d[m]	1	d[p]	2
Methoxychlor		A	5.08	+	–	d[a]	0.004	50	p	0.006	20	p	0.09	200	p	0.003	100	–	–	–	–
Pentachlorophenol	PS	A	5.12	+	–	p	0.004	60	p	0.005	20	p	0.09	200	p	0.003	200	d[e,q]	1	–	–
Simazine	PS		2.18	–	+	–	–	–	–	–	–	–	–	–	–	–	–	d[e]	1	d[n,o]	2
Tetrasul		C	6.87	+	–	p	0.003	>300	p	0.003	>300	p	0.08	>300	p	0.002	>300	–	–	–	–
Trifluralin	PS	A	5.34	+	–	d[a]	0.004	100	p	0.004	40	p	0.09	>300	p	0.003	300	d[r]	1	–	–
Glyphosate	Rev		–4.47	–	+	–	–	–	–	–	–	–	–	–	–	–	–	p	1	p	2
Aminomethylphosphonic acid (AMPA)	Rev		–2.17	–	+	–	–	–	–	–	–	–	–	–	–	–	–	p	1	p	2
Bentazon	Rev		1.67	–	+	–	–	–	–	–	–	–	–	–	–	–	–	p	1	p	2
Mecoprop	Rev		2.94	–	+	–	–	–	–	–	–	–	–	–	–	–	–	p	1	d[p]	2
Quinoxyfen	Rev		5.69	+	–	p	0.003	200	p	0.004	90	p	0.09	>300	–	–	–	–	–	–	–

(continued overleaf)

Table 13.1 (*continued*)

	EU	OS PAR	log K_{ow}	Sampler suitability		SPMD 460 cm^2, 1 mL triolein, 85 µm LDPE			LDPE strip, 460 cm^2, 85 µm			Non-polar Chemcatcher 46 mm Ø, 450 µL octanol			Silicone strip 600 cm^2 × 0.5 mm			POCIS 47 mm Ø		Polar Chemcatcher 46 mm Ø C_{18}/SDB	
				Non-polar	Polar	Stage	LOD_{28} (ng L^{-1})	t_{int} (d)	Stage	LOD_{28} (ng L^{-1})	t_{int} (d)	Stage	LOD_{28} (ng L^{-1})	t_{int} (d)	Stage	LOD_{28} (ng L^{-1})	t_{int} (d)	Stage	LOD_{10} (ng L^{-1})	Stage	LOD_{10} (ng L^{-1})
Volatile organic compounds																					
1,2,3-Trichlorobenzene	PS	A	4.10	+	–	p	0.02	5	p	0.04	2	p	0.2	20	p	0.005	20	–	–	–	–
1,2,4-Trichlorobenzene	PS	A	4.10	+	–	p	0.02	5	p	0.04	2	p	0.2	20	p	0.005	20	–	–	–	–
1,2-Dichloroethane	PS		1.48	–	–	–	–	–	–	–	–	–	–	–	–	–	–	–	–	–	–
1,3,5-Trichlorobenzene	PS	A	4.10	+	–	p	0.02	5	p	0.04	2	p	0.2	20	p	0.005	20	–	–	–	–
Benzene	PS		2.13	–	–	–	–	–	–	–	–	–	–	–	–	–	–	–	–	–	–
Dichloromethane	PS		1.25	–	–	–	–	–	–	–	–	–	–	–	–	–	–	–	–	–	–
Trichloromethane	PS		1.97	–	–	–	–	–	–	–	–	–	–	–	–	–	–	–	–	–	–
Tetrachloromethane	EQS		2.83	–	–		–	–	–	–	–		–	–		–	–	–	–	–	–

Tetrachloroethene	EQS		2.97	–	–	–	–	–	–	–	–	–	–	–	–	–	–	–	–	–	–
Trichloroethene	EQS		2.47	–	–	–	–	–	–	–	–	–	–	–	–	–	–	–	–	–	–
Polyaromatic hydrocarbons																					
PAHs	PHS	A					–	–		–	–		–	–		–	–		1		2
Naphthalene	PS		3.40	+	+	d[a]	0.08	1	d[b]	0.2	0	p	0.7	4	p	0.02	3	p	1	p	2
Anthracene	PHS		4.50	+	–	d[a]	0.007	10	d[b,c,j]	0.02	5	d[c,d,s]	0.1	40	d[c,h]	0.003	40	–	–	–	–
Fluoranthene	PS		5.20	+	–	d[a]	0.004	70	d[b,c,j,k]	0.005	30	d[c,d,s]	0.09	200	d[c,h]	0.003	200	–	–	–	–
Benzo[*a*]pyrene	PHS		6.11	+	–	d[a]	0.003	>300	d[b,c,j]	0.003	200	d[d]	0.08	>300	d[c,h]	0.002	>300	–	–	–	–
Benzo[*b*]fluoranthene	PHS		6.11	+	–	d[a]	0.003	>300	d[b,c,j]	0.003	200	d[d,s]	0.08	>300	d[c,h]	0.002	>300	–	–	–	–
Benzo[*k*]fluoranthene	PHS		6.11	+	–	d[a]	0.003	>300	d[b,c,j]	0.003	200	d[d,s]	0.08	>300	d[c,h]	0.002	>300	–	–	–	–
Benzo[*ghi*]perylene	PHS		6.70	+	–	d[a]	0.003	>300	d[b,c]	0.003	>300	d[d]	0.08	>300	d[c,h]	0.002	>300	–	–	–	–
Indeno[1,2,3-*cd*] pyrene	PHS		6.70	+	–	d[a]	0.003	>300	d[b,c]	0.003	>300	d[d]	0.08	>300	d[c,h]	0.002	>300	–	–	–	–

(*continued overleaf*)

Table 13.1 (*continued*)

	EU	OS PAR	log K_{ow}	Sampler suitability		SPMD 460 cm^2, 1 mL triolein, 85 µm LDPE			LDPE strip, 460 cm^2, 85 µm			Non-polar Chemcatcher 46 mm Ø, 450 µL octanol			Silicone strip 600 cm^2 × 0.5 mm			POCIS 47 mm Ø		Polar Chemcatcher 46 mm Ø C_{18}/SDB	
				Non-polar	Polar	Stage	LOD_{28} ($ng\ L^{-1}$)	t_{int} (d)	Stage	LOD_{28} ($ng\ L^{-1}$)	t_{int} (d)	Stage	LOD_{28} ($ng\ L^{-1}$)	t_{int} (d)	Stage	LOD_{28} ($ng\ L^{-1}$)	t_{int} (d)	Stage	LOD_{10} ($ng\ L^{-1}$)	Stage	LOD_{10} ($ng\ L^{-1}$)
Other organohalogens																					
Pentabromodiphenylethers	PHS	A	6.81	+	–	d[t]	0.003	>300	p	0.003	>300	p	0.08	>300	d[i]	0.002	>300	–	–	–	–
Octabromodiphenylethers		A	8.63	+	–	d[t]	0.003	>300	p	0.003	>300	p	0.08	>300	p	0.002	>300	–	–	–	–
Decabromodiphenylether		A	9.97	+	–	d[t]	0.003	>300	p	0.003	>300	p	0.08	>300	p	0.002	>300	–	–	–	–
Decabromobiphenyl		A	8.58	+	–	p	0.003	>300	p	0.003	>300	p	0.08	>300	p	0.002	>300	–	–	–	–
Hexabromocyclododecane		A	5.60	+	–	p	0.003	200	p	0.004	70	p	0.09	>300	p	0.002	>300	–	–	–	–
1,3,5-Tribromo-2-(2,3-dibromo-2-methylpropoxy)-benzene		C	6.79	+	–	p	0.003	>300	p	0.003	>300	p	0.08	>300	p	0.002	>300	–	–	–	–
Hexachlorobutadiene	PHS		4.72	+	–	p	0.005	20	p	0.010	9	p	0.1	70	p	0.003	60	–	–	–	–
Pentabromoethyl benzene		C	7.48	+	–	p	0.003	>300	p	0.003	>300	p	0.08	>300	p	0.002	>300	–	–	–	–

Pentachloroanisole		C	5.45	+	–	d[u]	0.003	100	p	0.004	50	p	0.09	>300	p	0.002	>300	–	–	–	–
Pentachlorobenzene	PHS		5.17	+	–	d[a]	0.004	60	p	0.005	30	d[d]	0.09	200	p	0.003	200	–	–	–	–
2-Propenoic acid, (pentabromo)methyl ester		C	6.67	+	–	p	0.003	>300	p	0.003	>300	p	0.08	>300	p	0.002	>300	–	–	–	–
Short-chain chlorinated paraffins (SCCPs)	PHS	A	6–7	+	–	p	0.003	>300	p	0.003	>300	p	0.08	>300	p	0.002	>300	–	–	–	–
Tetrabromobisphenol A (TBBP-A)		A	5.90	+	–	p	0.003	>300	p	0.003	100	p	0.09	>300	p	0.002	>300	–	–	–	–
Endocrine disruptors																					
Nonylphenol	PHS	A	5.76	+	–	p	0.003	300	p	0.004	100	p	0.09	>300	p	0.002	>300	–	–	–	–
Nonylphenol ethoxylates (NPEOs)		A																			
$NPEO_2$			5.30	+	–	p	0.004	90	p	0.005	30	p	0.09	300	p	0.003	200	d[q]	1	p	2
$NPEO_5$			4.48	+	–	p	0.008	10	p	0.02	5	p	0.1	40	p	0.003	40	–	–	–	–
$NPEO_{11}$			2.83	–	+	–	–	–	–	–	–	–	–	–	–	–	–	p	1	–	–
Octylphenol	PS	A	5.50	+	–	p	0.003	100	p	0.004	50	p	0.09	>300	p	0.002	>300	d[e]	1	p	2

(*continued overleaf*)

Table 13.1 *(continued)*

	EU	OS PAR	log K_{ow}	Sampler suitability		SPMD 460 cm^2, 1 mL triolein, 85 µm LDPE			LDPE strip, 460 cm^2, 85 µm			Non-polar Chemcatcher 46 mm Ø, 450 µL octanol			Silicone strip 600 cm^2 × 0.5 mm			POCIS 47 mm Ø		Polar Chemcatcher 46 mm Ø C_{18}/SDB	
				Non-polar	Polar	Stage	LOD_{28} (ng L^{-1})	t_{int} (d)	Stage	LOD_{28} (ng L^{-1})	t_{int} (d)	Stage	LOD_{28} (ng L^{-1})	t_{int} (d)	Stage	LOD_{28} (ng L^{-1})	t_{int} (d)	Stage	LOD_{10} (ng L^{-1})	Stage	LOD_{10} (ng L^{-1})
4-(1,1, 3, 3-Tetramethylbutyl)-phenol	PS		5.28	+	–	p	0.004	80	p	0.005	30	p	0.09	300	p	0.003	200	–	–	–	–
Dibutylphthalate		A	4.50	+	–	p	0.007	10	p	0.02	5	p	0.1	40	p	0.003	40	–	–	–	–
Diethylhexylphthalate (DEHP)	PS	A	7.60	+	–	p	0.003	>300	p	0.003	>300	p	0.08	>300	p	0.002	>300	–	–	–	–
Other organic chemicals																					
1,5,9-Cyclododecatriene		B	5.45	+	–	p	0.003	100	p	0.004	50	p	0.09	>300	p	0.002	>300	–	–	–	–
2,4,6-Tri-*tert*-butylphenol		A	6.06	+	–	p	0.003	>300	p	0.003	200	p	0.08	>300	p	0.002	>300	–	–	–	–
3,3′-(Ureylenedimethylene) bis(3,5,5-trimethylcyclohexyl)-diisocyanate		C	7.31	+	–	p	0.003	>300	p	0.003	>300	p	0.08	>300	p	0.002	>300	–	–	–	–

4–(Dimethylbutylamino) diphenylamine (6PPD)		A	4.68	+	–	p	0.006	20	p	0.01	8	p	0.1	70	p	0.003	60	–	–	–	–
Clotrimazole		A	6.26	+	–	p	0.003	>300	p	0.003	>300	p	0.08	>300	p	0.002	>300	–	–	–	–
Cyclododecane		B	6.12	+	–	p	0.003	>300	p	0.003	200	p	0.08	>300	p	0.002	>300	–	–	–	–
Diosgenin		C	6.34	+	–	p	0.003	>300	p	0.003	>300	p	0.08	>300	p	0.002	>300	–	–	–	–
Xylene musk	Rev	A	4.45	+	–	p	0.008	10	p	0.02	5	p	0.1	40	p	0.004	30	–	–	–	–
Neodecanoic acid, ethenyl ester		A	4.54	+	–	p	0.007	20	p	0.01	6	p	0.1	50	p	0.003	40	–	–	–	–
Perfluorooctane sulfonic acid and its salts (PFOS)	Rev	A	NA	?	?	p	?	?	p	0.003	>300	p	?	?	p	?	?	p	?	p	?
Bisphenol-A	Rev		3.64	+	+	p	0.05	2	p	0.1	1	p	0.4	6	p	0.01	5	p	1	p	2
Metals and related compounds																					
Cadmium	PHS	A		–	–	–	–	–	–	–	–	–	–	–	–	–	–	–	–	–	–
Lead	PS	A		–	–	–	–	–	–	–	–	–	–	–	–	–	–	–	–	–	–

(*continued overleaf*)

Table 13.1 (*continued*)

	EU	OS PAR	log K_{ow}	Sampler suitability		SPMD 460 cm^2, 1 mL triolein, 85 µm LDPE			LDPE strip, 460 cm^2, 85 µm			Non-polar Chemcatcher 46 mm Ø, 450 µL octanol			Silicone strip 600 cm^2 × 0.5 mm			POCIS 47 mm Ø		Polar Chemcatcher 46 mm Ø C_{18}/SDB	
				Non-polar	Polar	Stage	LOD_{28} (ng L^{-1})	t_{int} (d)	Stage	LOD_{28} (ng L^{-1})	t_{int} (d)	Stage	LOD_{28} (ng L^{-1})	t_{int} (d)	Stage	LOD_{28} (ng L^{-1})	t_{int} (d)	Stage	LOD_{10} (ng L^{-1})	Stage	LOD_{10} (ng L^{-1})
Organic lead compounds	PS	A																			
tetramethyllead			2.97	–	+	–	–	–	–	–	–	–	–	–	–	–	–	–	–	–	–
tetraethyllead			4.15	+	–	p	0.01	6	p	0.04	2	p	0.2	20	p	0.005	20	–	–	–	–
Mercury	PHS	A		–	–		–	–		–	–		–	–		–	–	–	–	–	–
Organic mercury compounds	PHS	A																			
methylmercury			0.08	–	+	–	–	–	–	–	–	–	–	–	–	–	–	–	–	–	–
dimethylmercury			2.59	–	+	–	–	–	–	–	–	–	–	–	–	–	–	–	–	–	–
Nickel and its compounds	PS																				
Organic tin compounds		A																			
tributyltin	PHS		3.00	+	+	p	0.2	0	p	0.5	0	p	2	1	p	0.06	1	–	–	–	–
triphenyltin			3.00	+	+	p	0.2	0	p	0.5	0	p	2	1	p	0.06	1	–	–	–	–
Free cyanide	Rev		−0.69	–	–	–	–	–	–	–	–	–	–	–	–	–	–	–	–	–	–

Ethylenediaminetetraacetic acid (EDTA)	Rev	−3.86	–	–	–	–	–	–	–	–	–	–	–	–	–	–	–	–	–	–

EU: PS = priority substance, PHS = priority hazardous substance, EQS = non PS/non PHS for which an environmental quality standard (EQS) has been defined, Rev = compound under review for classification as a PS/PHS. **OSPAR:** A = produced and emitted in the OSPAR area, B produced but not emitted, C = not produced in the OSPAR area. **logK_{ow}:** best available *n*-octanol−water partition coefficient (mixed literature sources and estimation programmes). **Sampler suitability:** expected performance of nonpolar and polar samplers based on the log K_{ow} value of the compound (nonpolar: log $K_{ow} > 3$, polar: log $K_{ow} < 4$). **Development stage:** sampler suitability demonstrated (d + reference), likely (p), or doubtful (−). **LOD$_{28}$(non-polar samplers):** detection limit (ng L^{-1}) based on a method detection limit of 1 ng, an apparent water sampling rate of 25 mL day^{-1} cm^{-2} sampler surface area, an exposure time of 28 days, and a sampler−water partition coefficient that is equal to K_{ow}. **LOD$_{10}$(polar samplers):** detection limit (ng L^{-1}) based on a method detection limit of 1 ng, an apparent water sampling rate of 100 mL day^{-1} (POCIS) or 50 mL day^{-1} (nonpolar Chemcatcher), an exposure time of 10 days, and the assumption that the receiving phase acts as an infinite sink. t_{int}: timescale over which the sampler yields a time-integrated signal, estimated as the mean residence time of the compound in the sampler (1.4 × half-life time).

[a] Huckins *et al*. (2006).
[b] Anderson *et al*. (2008).
[c] Allan *et al*. (2009).
[d] Vrana *et al*. (2006).
[e] Alvarez *et al*. (2007).
[f] Alvarez *et al*. (2004b).
[g] Gunold, *et al*. (2007).
[h] Smedes (2007).
[i] Ter Laak *et al*. (2008).
[j] Fernandez *et al*. (2009).
[k] Adams *et al*. (2007).
[l] Schafer *et al*. (2008).
[m] Alvarez *et al*. (2004a).
[n] Shaw and Muller (2005).
[o] Stephens *et al*. (2009).
[p] Vermeirssen *et al*. (2009).
[q] Alvarez *et al*. (2005).
[r] Mazzella *et al*. (2007).
[s] Lobpreis *et al*. (2008).
[t] Booij *et al*. (2002).
[u] Herve *et al*. (1995).

In addition, the effects of biofouling of the diffusion surface and the influence of salinity need to be considered. This is potentially a problem in estuarine waters, where salinity can fluctuate between full seawater and freshwater twice a day. For samplers for hydrophilic analytes there is a need for interlaboratory trials to validate the calibration parameter surfaces, in addition to field trials to validate the field application of the samplers by end users.

Passive samplers for metals (the metals version of Chemcatcher and DGT) can be calibrated in different ways. Calculations of concentrations of metals using the DGT are based on tables of diffusion coefficients of the analytes in the hydrogel diffusion-limiting layer over a range of temperatures, and these are generally supplied by the manufacturer and have not been validated. The calculations assume that the WBL thickness makes a negligible contribution to the overall resistance to diffusion compared with that of the hydrogel, and hence the impact of variation in turbulence during deployment is minimal. Calibration of the Chemcatcher for metals is carried out in a similar way to that used for the hydrophilic Chemcatcher.

If passive sampling is to become accepted for use in a regulatory context, then there is a need for the development of validation methods, and an associated accreditation scheme for both those involved in calibration measurements and those using passive samplers in the field. This would need to be organized in a way similar to that in place for analytical chemistry techniques. However, there are some obstacles to be overcome. For laboratory calibrations, there is a need for large volumes of calibration water that could not feasibly be distributed in the same way as reference materials used in the validation of analytical methods. In the case of metals and organic contaminants with log $K_{ow} < 5$, this could possibly be based on an effervescent pill formulation (similar to that used for the dispersal of some pesticides in the spray mixer tank). For hydrophobic compounds (log $K_{ow} > 6$), passive dosing appears to be a promising technique (Rusina *et al.*, 2010). For field trials it may be possible to use reference sites that are well characterized and consistent in composition. When such sites have been established and characterized, it may be possible to use them for the calibration and validation of other monitoring technologies (e.g. sensors and other in-situ methods). Multiple comparisons of various monitoring methods were undertaken as part of a European Union funded project (http://www.swift-wfd.com), and indicated the potential utility of this approach.

Recently, there has been some progress in a number of areas. In 2010 there were two interlaboratory field trials for a range of passive sampling technologies at European riverine sites, with a third currently being planned for 2011. An interlaboratory field trial was undertaken in 2010, organized within the framework of AQUAREF (the organization coordinating French laboratories involved in water monitoring: http://www.aquaref.fr) and focused mainly on pesticides, PAHs and metals. This trial was widened to allow participation by laboratories across Europe. The sites used for this trial included two rivers and a marine lagoon. The main objective was to assess the potential role and efficiency of passive samplers for water pollutant measurements in surface and coastal water in the frame of the WFD.

The International Passive Sampling Inter-laboratory Comparison study (IPSIC 2010) was undertaken during the summer of 2010 and organized by the Public Health Institute, Ostrava, Czech Republic (http://en.zu.cz/) and CSlab, Czech Republic (http://www.cslab.cz/). Assessment of PCBs, PAHs, PBDEs, metals and OCPs in surface water was undertaken using SPMDs and DGT passive sampling devices.

These exercises will be followed by a complementary passive sampler intercalibration trial organized by the Network of Reference Laboratories for Monitoring of Emerging Environmental Pollutants (NORMAN project: http://www.norman-network.net). The focus of this exercise will be on the application of passive sampling for monitoring pollutants of emerging concern. In addition, a BSI publicly available specification (BSI, 2006) is available to provide guidance for end users on the field deployment of passive samplers. This has led to the consideration of the development of an ISO/CEN standard to provide guidance of the field application of passive samplers. This

ISO standard was approved and published in 2011: Water quality – Sampling Part 23: Guidance on passive sampling in surface waters (ISO 5667-23:2011).

13.5 FUTURE PERSPECTIVES

In recent times there have been significant steps towards extending the range of environmental chemicals for which passive sampling can be used. However, some challenges remain, and include monitoring dissociable polar compounds (e.g. acid herbicides and some pharmaceuticals) which are poorly sequestered by existing devices, and which are difficult to monitor using bottle sampling. It may be possible to use an ion-exchange resin receiving phase combined with a microporous diffusion-limiting membrane for this purpose. Preliminary data based on a Chemcatcher design with an anion-exchange Empore™ disk indicate that this is a feasible approach (Kingston, 2002). Another opportunity to extend the range is provided by the radionuclides that are difficult to monitor using conventional methods, and that could be sequestered using bespoke receiving phases that match the properties of individual nuclides (mainly actinides).

The PRC approach to calibration is well established for samplers for nonpolar compounds, and there is an urgent need to extend this to other classes of passive sampler to develop a unified approach to calibration and validation of field application. An alternative approach to the lack of PRCs for polar organics is the use of a continuous-flow integrative sampler that actively pumps water through a chamber containing a receiving phase. This ensures constant and consistent turbulence conditions and, hence, reduces the calibration problems normally encountered with samplers for metals and polar organics (Llorca *et al.*, 2009).

Recently, there has been an increasing interest in coupling chemical analysis with biological assays of extracts from passive samplers in order to provide representative information on average exposures to sequestered biologically active pollutants. This approach can inform those involved in risk assessment of the toxicological significance of environmental pollutant loads. In particular, passive sampling has been linked with assays for endocrine disruptors from a wide range of chemical classes. Extracts of passive-sampler receiving phases have been used in toxicity identification evaluation studies. The use of passive samplers in this approach can be advantageous, since over a deployment period far greater quantities of pollutants can be sequestered than can be extracted from bottle samples of water. This facilitates both the bioassays and the associated fractionation and instrumental analysis.

Passive sampling has been recognized as providing useful information on the presence of pollutants at sites of interest, by lowering the limits of detection for some compounds that are present at only trace levels. The method has been used in investigative mode to map relative concentrations of pollutants and to detect their sources. In the marine environment it has potential utility in identifying representative sites, even where conditions and concentrations fluctuate markedly over a tidal cycle, and in indicating where information based on infrequent spot samples may be unreliable. However, in order for this technology to be accepted for use within a framework of risk assessment in a regulatory context, it will be necessary to provide a network of accredited practitioners who can provide robust data within defined limits of uncertainty. It would also be necessary to redefine EQS concentrations in terms of the freely dissolved fractions available to drive diffusion. In order for significant progress to be made it will be necessary for regulatory organizations (such as OSPAR and HELCOM) and other bodies, such as ICES, to promote and support the incorporation of passive sampling into routine monitoring that underpins risk assessment and compliance monitoring. In the longer term it may be possible to establish global networks based on passive samplers to monitor trans-global movements and distribution of key pollutants. Such a network (GAPS programme) organized by the Stockholm Convention under the United Nations Environmental Programme is already established for monitoring atmospheric pollutants.

REFERENCES

Adams R. G., Lohmann R., Fernandez L. A. *et al.* (2007). *Environ. Sci. Technol.*, **41**, 1317–1323.

Aguilar-Martínez R., Greenwood R., Mills G. A. *et al.* (2008a). *Int. J. Environ. Anal. Chem.*, **88**, 75–90.

Aguilar-Martínez R., Palacios-Corvillo M. A., Greenwood R. *et al.* (2008b). *Anal. Chim. Acta.*, **618**, 157–167.

Allan I. J., Knutsson J., Guigues N. *et al.* (2007). *J. Environ. Monit.*, **9**, 672–681.

Allan I. J., Knutsson J., Guigues N. *et al.* (2008). *J. Environ Monit.*, **10**, 821–829.

Allan I. J., Booij K., Paschke A. *et al.* (2009). *Environ. Sci. Technol.*, **43**, 5383–5390.

Alvarez D. A., Petty J. D., Huckins J. N. *et al.* (2004a). *Environ. Toxicol. Chem.*, **23**, 1640–1648.

Alvarez, D. A., Cranor, W. L., Huckins, J. N. *et al.* (2004b). Assessment of organic contaminants in integrative samplers from Chesapeake Bay tributaries, USGS/Columbia Environmental Research Center, Columbia, Missouri, USA.

Alvarez D. A., Stackelberg P. E., Petty J. D. *et al.* (2005). *Chemosphere*, **61**, 610–622.

Alvarez D. A., Huckins J. N., Petty J. D. *et al.* (2007). Tool for monitoring hydrophilic contaminants in water: polar organic chemical integrative sampler, in *Passive Sampling Techniques in Environmental Monitoring* (eds R. Greenwood, G. A. Mills and B. Vrana), Elsevier, Amsterdam, pp. 171–197.

Anderson K. A., Sethajintanin D., Sower G. and Quarles L. (2008). *Environ. Sci. Technol.*, **42**, 4486–4493.

Arditsoglou A. and Voutsa D. (2008). *Environ. Pollut.*, **156**, 316–324.

Blom L. B., Morrison G. M., Kingston J. *et al.* (2002). *J. Environ. Monit.*, **4**, 258–262.

Blom L. B., Morrison G. M., Roux M. S. *et al.* (2003). *J. Environ. Monit.*, **5**, 404–409.

Boehm P. D., Page D. S., Brown J. S. *et al.* (2005). *Mar. Pollut. Bull.*, **50**, 740–750.

Booij K., Zegers B. N. and Boon J. P. (2002). *Chemosphere*, **46**, 683–688.

Booij K., Vrana B. and Huckins J. N. (2007). Theory, modelling and calibration of passive samplers used in water monitoring, in *Passive Sampling Techniques in Environmental Monitoring* (eds R. Greenwood, G. Mills and B. Vrana), Elsevier, Amsterdam, pp. 141–169.

Booij K., van Bommel R., Mets A. and Dekker R. (2006). *Chemosphere*, **65**, 2485–2492.

BSI (2006). BSI PAS 61. Publicly Available Specification: Determination of priority pollutants in surface water using passive sampling (PAS-61), May.

Bueno M. J. M., Hernando M. D., Agüera A. and Fernández-Alba A. R. (2009). *Talanta*, **77**, 1518–1527.

Chang L. Y., Davison W., Zhang H. and Kelly M. (1998). *Anal. Chim. Acta*, **368**, 243–253.

Davison W. and Zhang H. (1994). *Nature*, **367**, 546–548.

Davison W., Fones G. R., Harper M. P. *et al.* (2000). Dialysis, DET and DGT: in situ diffusional techniques for studying water, sediments and soils, in *In-situ Monitoring of Aquatic Systems: Chemical Analysis and Speciation* (eds J. Buffle and G. Horvai), IUPAC/John Wiley & Sons, Ltd, pp. 495–569.

Dunn R. J. K., Teasdale P. R., Warnken J. and Schleich R. R. (2003). *Environ. Sci. Technol.*, **37**, 2794–2800.

Dunn R. J. K., Teasdale P. R., Warnken J. and Arthur J. M. (2007). *Environ. Pollut.*, **148**, 213–220.

Esteve-Turrillas F. A., Yusà V., Pasor A. and de la Guardia M. (2008). *Talanta*, **74**, 443–457.

Fernandez L. A., Macfarlane J. K., Tcaciuc A. P. and Gschwend P. M. (2009). *Environ. Sci. Technol.*, **43**, 1430–1436.

Følsvik N., Brevik E. M. and Berge J. A. (2000). *J. Environ. Monit.*, **2**, 281–284.

Følsvik N., Brevik E. M. and Berge J. A. (2002). *J. Environ. Monit.*, **4**, 280–283.

French M. A., Zhang H., Pates J. M. *et al.* (2005). *Anal. Chem.*, **77**, 135–139.

Fux E., Marcaillou C., Mondeguer F. *et al.* (2008). *Harmful Algae*, **7**, 574–583.

Fux E., Bire R. and Hess P. (2009). *Harmful Algae*, **8**, 523–537.

Garmo O. A., Royset O., Steinnes E. and Flaten T. P. (2003). *Anal. Chem.*, **75**, 3573–3580.

Gonzalez J.-L., Tapie N., Munaron D. and Budzinski H. (2009). Rapport Convention Cadre Ifremer/AERM&C Utilisation échantillonneurs passifs – Méditerranée, Contract no. 08/3211080/F, June 2009, 90– pp.

Greenwood R., Mills G. and Vrana B. (eds) (2007a). *Passive Sampling Techniques in Environmental Monitoring*, Elsevier, Amsterdam.

Greenwood R., Mills G. A., Vrana B. *et al.* (2007b). Monitoring of priority pollutants in water using Chemcatcher passive sampling devices, in *Passive Sampling Techniques in Environmental Monitoring* (eds R. Greenwood, G. Mills and B. Vrana), Elsevier, Amsterdam, pp. 199–229.

Gunold R., Schäfer R. B., Paschke A. *et al.* (2008). *Environ. Pollut.*, **155**, 52–60.

Harman C., Tollefsen K.-E., Bøyum O. *et al.* (2008). *Chemosphere*, **72**, 1510–1516.

Harman C., Thomas K. V., Tollefsen K. E. *et al.* (2009). *Mar. Pollut. Bull.*, **58**, 1671–1679.

Herve S., Prest H. F., Heinonen P. *et al*. (1995). *Environ. Sci. Pollut. Res*., **2**, 24–29.
Huckins J. N., Tubergen M. W. and Manuweera G. K. (1990). *Chemosphere*, **20**, 533–552.
Huckins J. N., Petty J. D. and Booij K. (2006). *Monitors of Organic Chemicals in the Environment: Semipermeable Membrane Devices*, Springer, New York, pp. 1–223.
Huckins J. N., Petty J. D., Lebo J. A. *et al*. (2002). *Environ. Sci. Technol*., **36**, 85–91.
Hyne R. V. and Aistrope M. (2008). *Chemosphere*, **71**, 611–620.
Hyötyläinen T. and Riekkola M.-L. (2007). *Trends Anal. Chem*., **26**, 788–808.
ITRC (2005). Technology overview of passive sampler technologies, DSP-4: Interstate Technology & Regulatory Council, Authoring Team, Washington, DC.
Jonker M. T. O. and Koelmans A. A. (2001). *Environ. Sci. Technol*., **35**, 3742–3748.
Kingston J. (2002). PhD thesis, University of Portsmouth, Portsmouth, UK.
Kingston J. K., Greenwood R., Mills G. A. *et al*. (2000). *J. Environ. Monit*., **2**, 487–495.
Komarova T. V., Bartkow M. E., Rutishauser S. *et al*. (2009). *Environ. Pollut*. **157**, 731–736.
Kot-Wasik A., Zabiegala B., Urbanowicz M. *et al*. (2007). *Anal. Chim. Acta*, **602**, 141–163.
Larner B. L. and Seen A. J. (2005). *Anal. Chim. Acta*, **539**, 349–355.
Larner B. L., Seen A. J. and Snape I. (2006). *Chemosphere*, **65**, 811–820.
Lepom P., Brown B., Hanke G. *et al*. (2009). *J. Chromatogr. A*, **1216**, 302–315.
Leslie H. A., Ter Laak T. L., Busser F. J. M. *et al*. (2002). *Environ. Sci. Technol*., **36**, 5399–5404.
Li Y. and Gregory S. (1974). *Geochim. Cosmochim. Acta*, **38**, 703–714.
Li W. J., Zhao J. J., Li C. S. *et al*. (2006). *Anal. Chim. Acta*, **575**, 274–280.
Li W. J., Li C. S., Zhao J. J. and Cornett R. J. (2007). *Anal. Chim. Acta*, **592**, 106–113.
Llorca J., Gutiérrez G., Capilla E. *et al*. (2009). *J. Chromatogr. A*, **1216**, 5783–5792.
Lobpreis T., Vrana B., Dominiak E. *et al*. (2008). *Environ. Pollut*., **153**, 706–710.
Lohmann R. and Muir D. (2010). *Environ. Sci. Technol*., **44**, 860–864.
MacKenzie L. A. (2010). *Curr. Opinion Biotechnol*., **21**, 1–6.
MacKenzie L., Beuzenberg V., Holland P. *et al*. (2004). *Toxicon*, **44**, 901–918.
MacLeod S. L., McClure E. L. and Wong, C. S. (2007). *Environ. Toxicol. Chem*., **26**, 2517–2529.
Mazzella N., Dubernet J-F. and Delmas F. (2007). *J. Chromatogr. A*, **1154**, 42–51.
Murdock C., Kelly M., Chang L. Y. *et al*. (2001). *Environ. Sci. Technol*., **35**, 4530–4535.
Namiesnik J., Zabiegala B., Kot-Wasik A. *et al*. (2005). *Anal. Bioanal. Chem*., **381**, 279–301.
Ogata Y., Takeda H., Mizukawa K. *et al*. (2009). *Mar. Pollut. Bull*., **58**, 1437–1446.
Olivier F., Ridd M. and Klumpp D. (2002). *Mar. Pollut. Bull*., **44**, 1051–1062.
Panther J. G., Teasdale P. R., Bennett W. W. *et al*. (2010). *Environ. Sci. Technol*., **44**, 9419–9424.
Paschke A., Vrana B., Popp P. *et al*. (2007). Membrane-enclosed sorptive coating for the monitoring of organic compounds in water, in *Passive Sampling Techniques in Environmental Monitoring* (eds R. Greenwood, G. Mills and B. Vrana), Elsevier, Amsterdam, pp. 231–249.
Persson L. B., Morrison G. M., Friemann J. U. *et al*. (2001). *J. Environ. Monit*., **3**, 639–645.
Richardson B. J., Zheng G. J., Tse E. S. C. and Lam P. K. S. (2001). *Chemosphere*, **45**, 1201–1208.
Roach A. C., Muller R., Komarova T. *et al*. (2009). *Chemosphere*, **75**, 1243–1251.
Robinson R. A. and Stokes R. H. (1959). *Electrolytic Solutions*, Butterworths, London, UK.
Roose P. and Brinkman U. A. Th. (2005). *Trends Anal. Chem*., **24**, 897–926.
Rundberget T., Gustad E., Samdal I. A. *et al*. (2009). *Toxicon*, **53**, 543–550.
Rusina T. P., Smedes F., Koblizkova M. and Klanova J. (2010). *Environ. Sci. Technol*., **44**, 362–367.
Scally S., Davison W. and Zhang H. (2006). *Anal. Chim. Acta*, **558**, 222–229.
Schäfer R. B., Paschke A. and Liess M. (2008a). *J. Chromatogr. A*, **1203**, 1–6.
Schäfer R. B., Paschke A., Vrana B. *et al*. (2008b). *Water Res*., **42**, 2707–2717.
Schintu M., Durante L., Maccioni A. *et al*. (2008). *Mar. Pollut. Bull*., **57**, 832–837.
Scoullos M., Plavsic M., Karavoltsos S. and Sakellari A. 2006. *Estuarine Coastal Shelf Sci*., **67**, 484–490.
Shaw, M. and Müller, J. F. (2005). *Mar. Pollut. Bull*., **51**, 876–881.
Shaw M., Eaglesham G. and Mueller J. F. (2009a). *Chemosphere*, **75**, 1–9.
Shaw M., Negri A., Fabricius K. and Mueller J. F. (2009b). *Aquat. Toxicol*., **95**, 108–116.
Shaw M., Furnas M. J., Fabricius K. *et al*. (2010). *Mar. Pollut. Bull*., **60**, 113–122.
Smedes F. (2007). Monitoring of chlorinated biphenyls and polycyclic aromatic hydrocarbons by passive sampling in concert with deployed mussels, in *Passive Sampling Techniques in Environmental Monitoring* (eds R. Greenwood, G. Mills and B. Vrana), Elsevier, Amsterdam, pp. 407–448.

Söderström H., Lindberg R. H. and Fick J. (2009). *J. Chromatogr. A*, **1216**, 623–630.
Stephens B. S., Kapernick A. P., Eaglesham G. and Mueller J. F. (2009). *Mar. Pollut. Bull.*, **58**, 1116–1122.
Teasdale P. R., Hayward S. and Davison W. (1999). *Anal. Chem.*, **71**, 2186–2191.
Ter Laak, T. L., Busser, F. J. M. and Hermens, J. L. M. (2008). *Anal. Chem.*, **80**, 3859–3866.
Togola A. and Budzinski H. (2007). *Anal. Chem.*, **79**, 6734–6741.
Vermeirssen E. L. M., Asmin J., Escher B. I. *et al*. (2008). *J. Environ. Monit.*, **10**, 119–128.
Vermeirssen E. L. M., Bramaz N., Hollender J. *et al*. (2009). *Water Res.*, **43**, 903–914.
Vrana B., Mills G. A., Allan I. J. *et al*. (2005a). *Trends Anal. Chem.*, **24**, 845–868.
Vrana B., Mills G., Greenwood R. *et al*. (2005b). *J. Environ. Monit.*, **7**, 612–620.
Vrana B., Mills G. A., Dominiak E. and Greenwood R. (2006). *Environ. Pollut.*, **142**, 333–343.
Vrana B., Mills G. A., Kotterman M. *et al*. (2007). *Environ. Pollut.*, **145**, 895–904.
Warnken K. W., Zhang H. and Davison W. (2007). In situ monitoring and dynamic speciation measurements in solution using DGT, in *Passive Sampling Techniques in Environmental Monitoring* (eds R. Greenwood, G. Mills and B. Vrana), Elsevier, Amsterdam, pp. 251–278.
Warnken K. W., Davison W. and Zhang H. (2008). *Environ. Sci. Technol.*, **42**, 6903–6909.
Warnken K. W., Lawlor A. J., Lofts S. *et al*. (2009). *Environ. Sci. Technol.*, **43**, 7230–7236.
Wong C. S. and MacLeod S. L. (2009). *J. Environ. Monit.*, **11**, 923–936.
Wu R. S. S., Chan A. K. Y., Richardson B. J. *et al*. (2008). *Mar. Pollut. Bull.*, **57**, 236–244.
Zeng E. Y., Tsukada D. and Diehl D. W. (2004). *Environ. Sci. Technol.*, **38**, 5737–5743.
Zhang H. and Davison W. (1995). *Anal. Chem.*, **67**, 3391–3400.
Zhang H., Davison W., Gadi R. and Kobayashi T. (1998). *Anal. Chim. Acta*, **370**, 29–38.

Part 7
Conclusions

14
Conclusions and Outlook

Gert Verreet[a], Patrick Roose[b], and Philippe Quevauviller[c]

[a]*Flemish Government – Environment, Nature and Energy Department (LNE), International Policy Division, Brussels, Belgium*
[b]*Royal Belgian Institute of Natural Sciences, Management Unit of the North Sea Mathematical Models and the Scheldt estuary (MUMM), Oostende, Belgium*
[c]*Vrije Universiteit Brussel (VUB), IUPWARE, Department of Hydrology and Hydrological Engineering, Brussels, Belgium*

14.1 CONTEXTUAL DEVELOPMENTS

The marine environmental policy setting of chemical monitoring in the respective environment has undergone a considerable development over recent decades, as described in this book (Chapters 1–3). In Europe, the adoption of the Water Framework Directive (and related directives) and of the Marine Strategy Framework Directive is leading to a reframing of the traditional marine environmental monitoring and assessment activities of the coastal member states (e.g. Chapter 12). It is not yet clear to what extent Europe will have a harmonizing effect, but the European reframing may already have led, in certain countries, to a reduction in pre-existing monitoring effort, whereas in others it will be (for the first time) a compelling argument for setting up a properly operated monitoring programme. Where they already exist, the monitoring programmes form the core of the regular data creation.

Marine environmental policy itself evolved from being merely 'reactive' to acute pollution incidents and concerns to a more comprehensive and long-term undertaking that takes a broader view of forms of chemical pollution. The policy context has also changed in other aspects, *inter alia* by the fact that the new EU legislative framework is now geared towards achieving 'good status'; that is, taking an even more result-driven stance than the pre-existing strategies of the regional sea conventions. Likewise, the monitoring has been broadened. Monitoring and assessment procedures

Chemical Marine Monitoring: Policy Framework and Analytical Trends, First Edition.
Edited by Philippe Quevauviller, Patrick Roose, and Gert Verreet.

will need to underpin, in expectedly more frequent assessment cycles (in some cases with 'annual indicators'), reliable information creation on the questions whether environmental status is getting towards the target levels sufficiently quickly, or at least moving in the right direction, and if not, what impedes progress.

Temporal trend analysis is thus one of the main reasons why marine monitoring is being undertaken, and long-term time trends require stability and dedication. However, despite the increasing ambition of the marine environmental policy context, the day-to-day management of marine environmental monitoring may not always be conducive to such stability. Contaminants may 'go out of fashion', new concerns may override old ones, monitoring managers can decide to allocate resources to other priorities, thus threatening the maintenance of a particular (part of a) monitoring programme. Time series may risk termination, before delivering their 'optimum' information content. This factor and the so-called 'ecosystem based approach' (Chapter 7) are some of the key challenges for current chemical monitoring programmes from their policy context.

14.2 INTERNAL DEVELOPMENTS

The community of scientists involved in marine chemical research and monitoring has attributed significant efforts to the common methodological problems that are most affecting their operation: statistical tools for monitoring, programme design, more robust sampling methodologies and in-situ monitoring methods that harness technological advances (Chapters 8–10). Furthermore, given the complexity of marine environmental matrices and the often low concentrations that are found in the different compartments, the community has benefited from and made good use of the developments in the field of analytical chemistry (Chapters 4–7).

For a number of reasons, quality assurance (in all its manifestations) has been of very high concern to the chemical monitoring community. This includes the use of appropriate analytical standards, quality assurance and control procedures, use of appropriate (certified) reference materials and assuring quality in data management procedures (Chapters 1 and 11). In a sense, the marine chemistry community has played a leading role in the development and promotion of certain key aspects of quality assurance/quality control. Ultimately, highly competent and skilled personnel are a necessary condition for setting up and running a successful chemical monitoring programme.

These developments have resulted in a strengthened methodological base that can serve to underpin a multiplicity of monitoring programmes in different marine environmental policy contexts.

14.3 OUTLOOK

Without attempting to make a comprehensive 'SWOT' analysis (Strength, Weaknesses, Threats, Opportunities), of the types of monitoring described in this book, a 'SWOT' format may be a convenient way to discuss the current chemical monitoring outlook and associated challenges.

Strengths

- Policy backing by a range of more legally binding instruments, including important EU directives.
- Technological advances have opened up new possibilities.
- Strong scientific investment in the methodological basis (including pervasive 'quality assurance culture') for all aspects of chemical monitoring in past decades.
- Range of chemical monitoring techniques broadened significantly.

Weaknesses

- Long lead times from monitoring programme design to harvesting of results, due also in part to the inherent inertia of the system observed.
- 'Ecosystem-based approach' not yet fully connected to the chemical world: chemical assessment scope may be too narrowly focused on purely determining 'trends' rather than on systemic understanding of what happens to the contaminants in the marine (sub)system(s). See also, for example, different use of modelling in understanding of eutrophication and microcontaminant behaviour.
- Few routine monitoring programmes have established strong operational links between different types of chemical, biochemical and biological effects monitoring, although earlier expectations were that this would lead to an overall improvement of the level of understanding of the levels and effects of chemical contamination and pollution.

Threats

- 'Chemical pollution' often no longer (and luckily) a main issue for environmental managers; concerns crowded out by other pressures (e.g. overfishing, habitat degradation).
- Longer lists of possibly interesting contaminants, not compensated by more resources: need to do more with less.
- The question of cost effectiveness: resource constraints can reduce monitoring efforts and/or favour 'cheaper' methods.
- Discontinuities in data series may have long-term impacts.
- Disconnecting of data providers from data users due to upscaling of database systems: important data knowledge may be lost and results of data users may not benefit data providers.

Opportunities

- Ongoing intensification of assessments at international level provides opportunities to squeeze as much information as possible out of the available data, proving their value, and monitoring data providers can build their case on the assessment results.
- Viewing data series (in some ways) as 'wine maturing': making more explicit that data series accrue value with time, and which, unlike wine, can be drunk repeatedly without loss of either quality or quantity (and without drunkenness).

Index

References to figures are given in italic type. References to tables are given in bold type.

Chemical Marine Monitoring: Policy Framework and Analytical Trends, First Edition.
Edited by Philippe Quevauviller, Patrick Roose, and Gert Verreet.

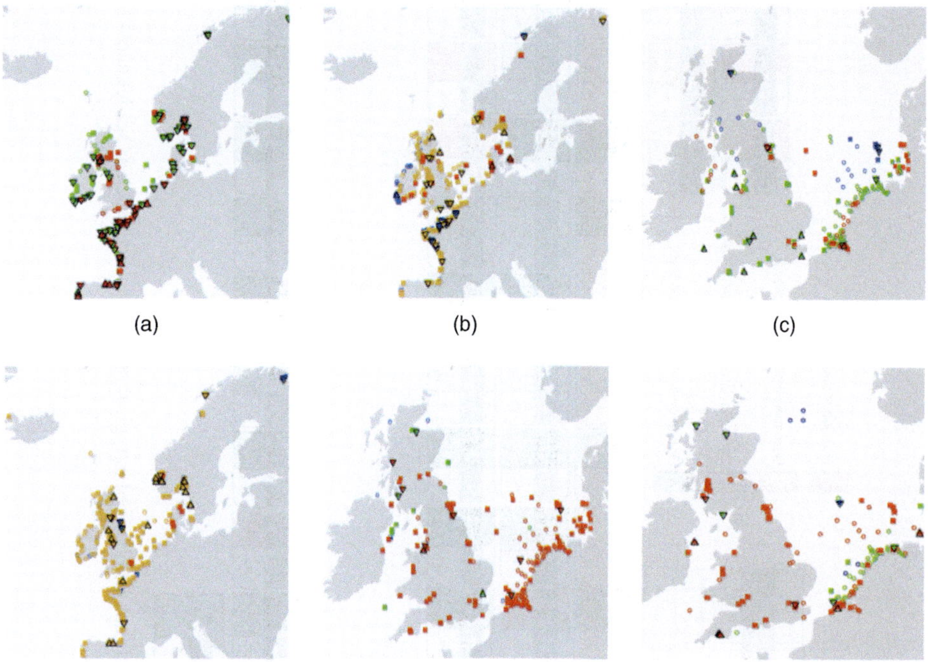

(a) (b) (c)

(d) (e) (f)

Figure 2.2 OSPAR temporal trend assessment for the period 1998–2007. Significant trends are indicated by the triangles ∇ (downward) and Δ (upward). Shading indicates status in last year of monitoring. Open circles indicate where insufficient data are available to assess trends (OSPAR, 2009). (a) PCBs in biota; (b) Pb in biota; (c) Cd in sediment; (d) Hg in biota; (e) Pb in sediment; (f) PAHs in sediment.

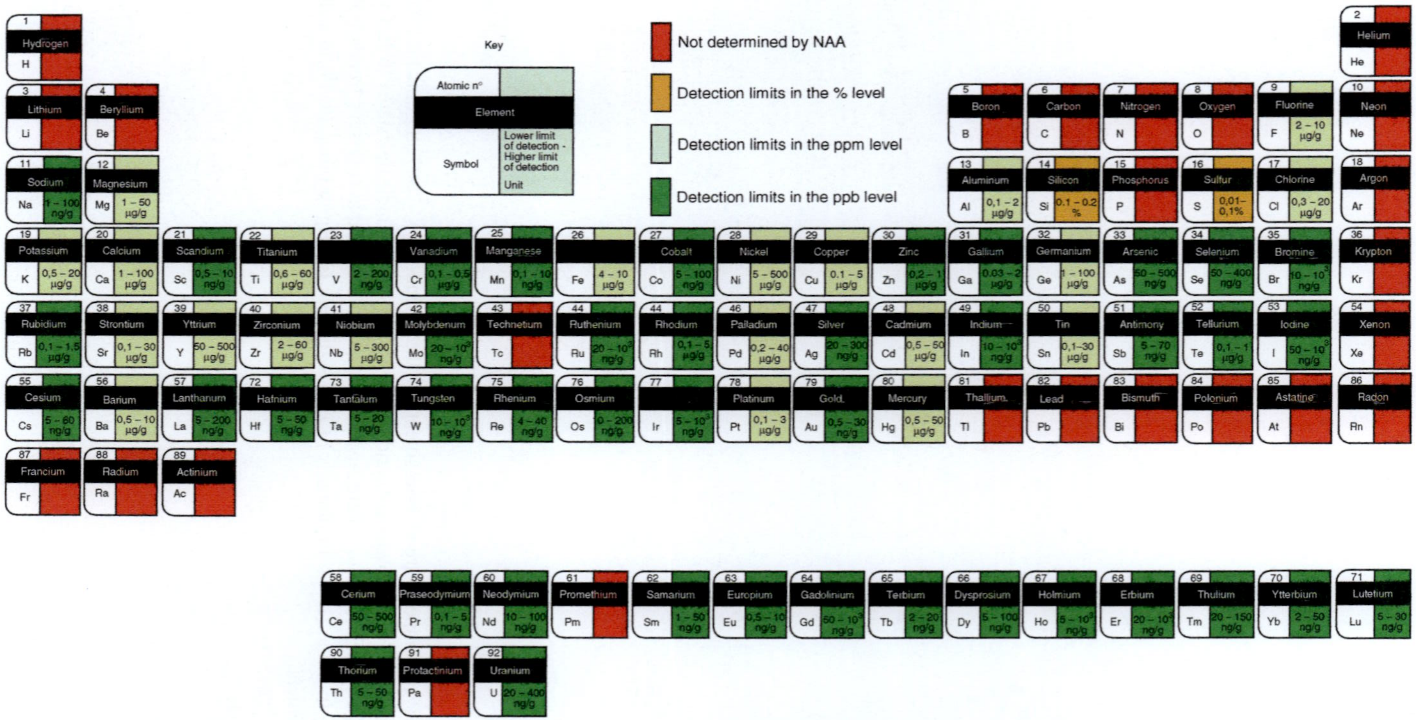

Figure 4.6 Detection limits in NAA (the lower value can be obtained in typical organic matrices such as sea lettuce – the higher value is representative for marine sediments).

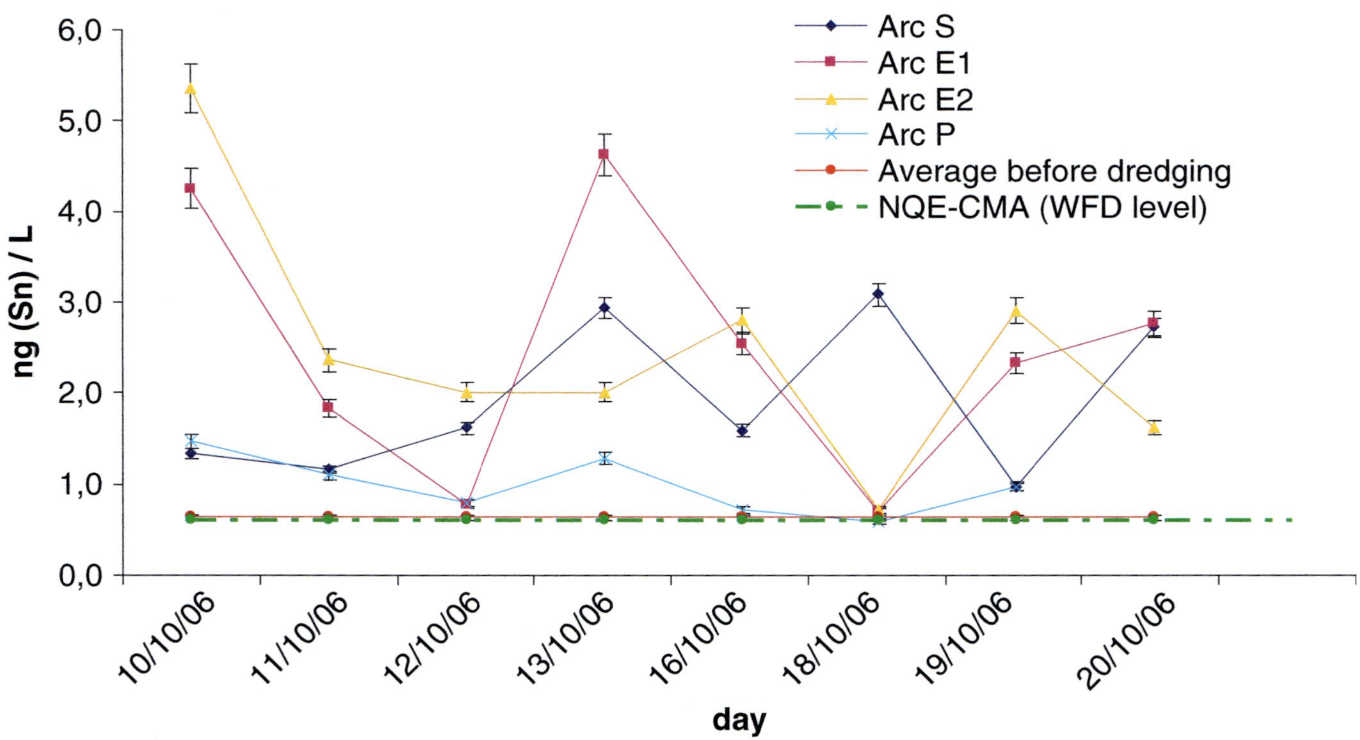

Figure 5.4 Environmental monitoring of TBT in a French harbour area during dredging operations. This graph presents the concentration of TBT in water samples (dissolved phase and SPM) which have been collected at different points of the port area (Arc S, Arc E1 and Arc E2). For each sampling station, TBT concentrations found are above the average concentration measured before dredging and also above the WFD level.

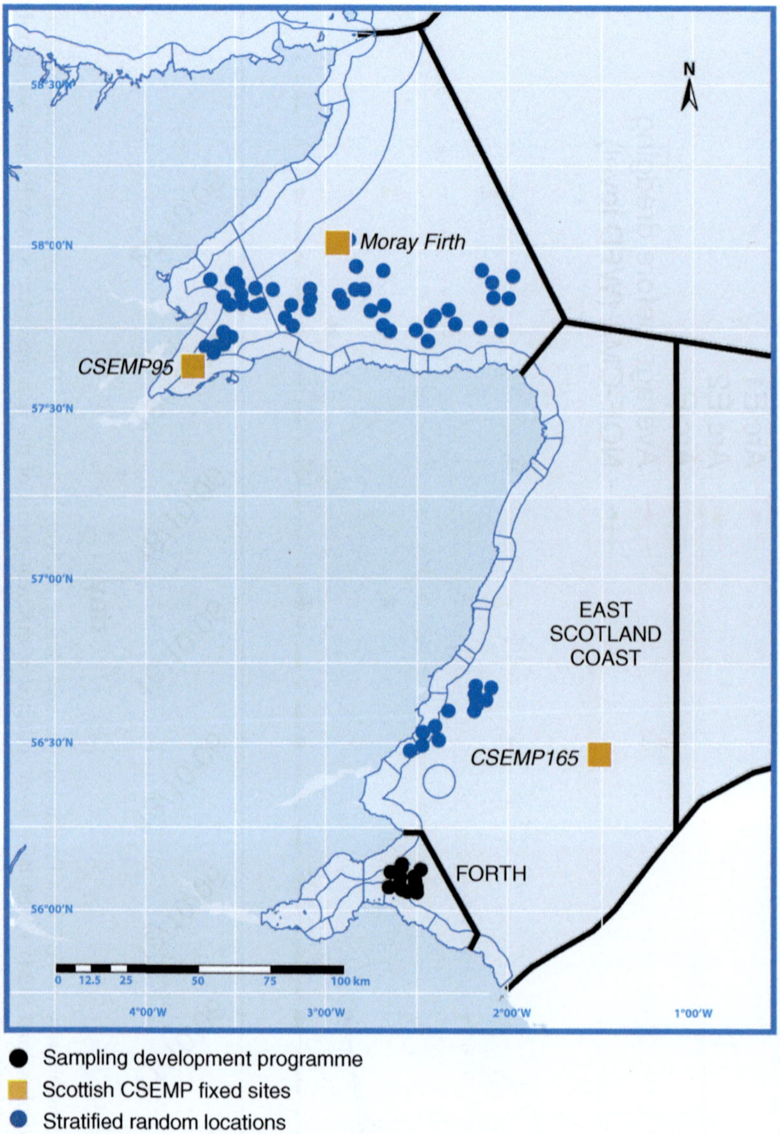

Figure 8.2 Assessing trends in the contaminant concentration of sediments along the east coast of Scotland has traditionally been done through repeated collection of five discrete samples at designated sites (squares) at a specified time of year. More recently, the sampling programme has been revised and now uses stratified random sampling of the muddy sediments (dots). Sites within a specific stratum are selected at random on an annual basis. This facilitates reporting on a regional basis. The figure covers the years 2006–2009.

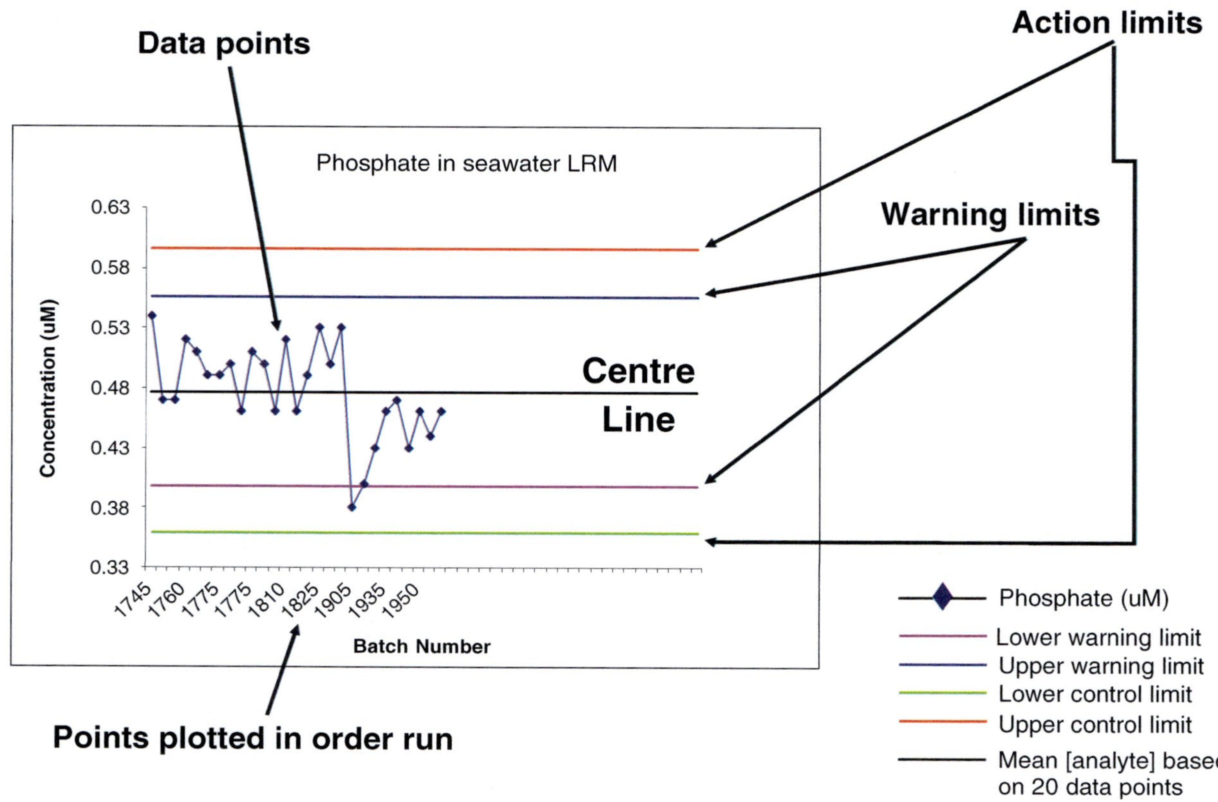

Figure 8.4 Example of a Shewart chart used as part of the internal quality assurance procedures when conducting classical chemical monitoring. A mean value (the centre line) is determined for the specific analyte using an LRM. Warning limits and action limits are usually defined at ±2 standard deviations and ±3 standard deviations respectively.

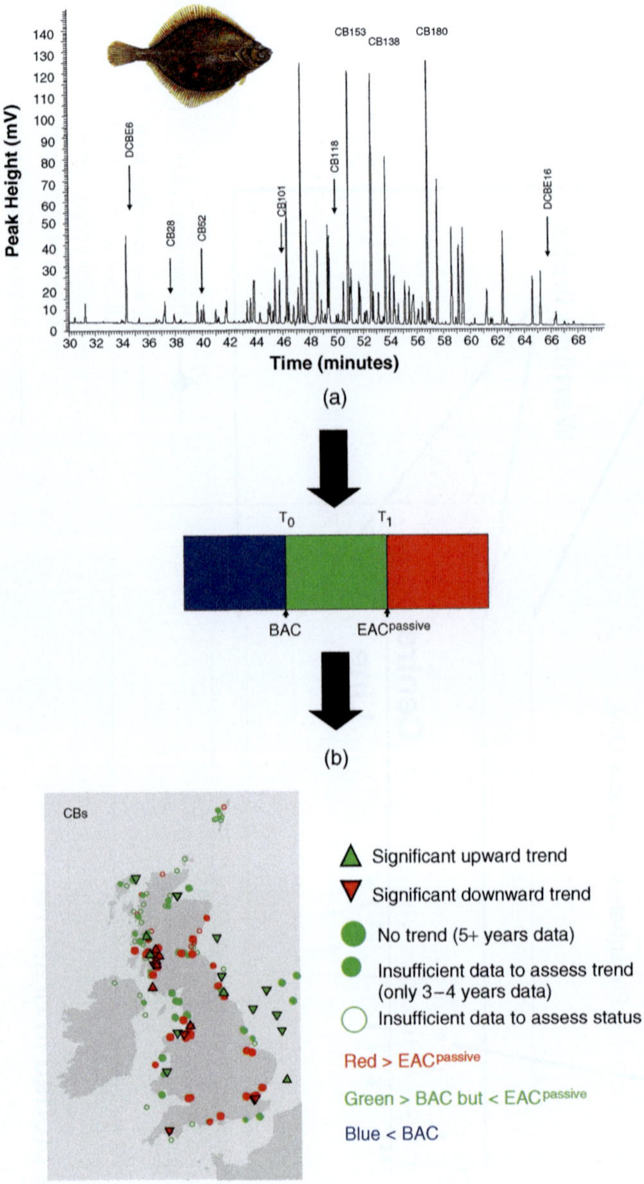

Figure 8.6 Determination of PCBs in plaice liver results in a chromatogram (a) from which a quantitative result can be determined. However, this must be compared with assessment criteria (b) in order that the data can be visualized in a clearly understandable manner (c). T_1 and T_2 are the transition points between the categories. BAC: background assessment concentration; $EAC^{passive}$: environmental assessment criteria calculated from the EAC for sediments based on partitioning theory.

Printed and bound by CPI Group (UK) Ltd, Croydon, CR0 4YY
16/09/2026
14963474-0002